Operator Theory: Advances and Applications
Vol. 118

Editor:
I. Gohberg

Operator Theory and Related Topics

Proceedings of the Mark Krein International Conference on Operator Theory and Applications, Odessa, Ukraine, August 18–22, 1997
Volume II

V. M. Adamyan
I. Gohberg
M. Gorbachuk
V. Gorbachuk
M. A. Kaashoek
H. Langer
G. Popov
Editors

Springer Basel AG

Editors:
V.M. Adamyan
Department of Theoretical Physics
University of Odessa
270026 Odessa
Ukraine

I. Gohberg
Department of Mathematical Sciences
Raymond and Beverly Sackler
Faculty of Exact Sciences
Tel Aviv University
69978 Ramat Aviv
Israel

M. Gorbachuk and V. Gorbachuk
Institute of Mathematics
National Academy of Sciences of Ukraine
Kyiv, Ukraine

M.A. Kaashoek
Department of Mathematics
Vrije Universiteit
De Boelelaan 1081a
1081 HV Amsterdam
The Netherlands

H. Langer
Department of Mathematics
Technical University of Vienna
Wiedner Hauptstrasse 8–10/1411
1040 Vienna
Austria

G. Popov
Institute of Mathematics, Economics and Mechanics
Odessa State University
270057 Dvoryanskaya str. 2
Odessa
Ukraine

1991 Mathematics Subject Classification 47-06

A CIP catalogue record for this book is available from the
Library of Congress, Washington D.C., USA

Deutsche Bibliothek Cataloging-in-Publication Data

Mark Krein International Conference on Operator Theory and Applications <1997, Odessa>: Proceedings of the Mark Krein International Conference on Operator Theory and Applications : Odessa, Ukraine, August 18 - 22, 1997 / V. M. Adamyan .... ed.. - Basel ; Boston ; Berlin : Birkhäuser
ISBN 978-3-0348-9557-6 ISBN 978-3-0348-8413-6 (eBook)
DOI 10.1007/978-3-0348-8413-6
Vol. 2. Operator theory and related topics. - 2000
(Operator theory ; Vol. 118)
ISBN 978-3-0348-9557-6

Originally published by Birkhäuser Verlag, Switzerland in 2000
Softcover reprint of the hardcover 1st edition 2000

Printed on acid-free paper produced from chlorine-free pulp. TCF ∞
Cover design: Heinz Hiltbrunner, Basel

ISBN 978-3-0348-9557-6

Preface

The present book is the second of the two volume Proceedings of the Mark Krein International Conference on Operator Theory and Applications. This conference, which was dedicated to the 90th Anniversary of the prominent mathematician Mark Krein, was held in Odessa, Ukraine from 18–22 August, 1997. The conference focused on the main ideas, methods, results, and achievements of M.G. Krein.

This second volume is devoted to operator theory and related topics. It opens with the bibliography of M.G. Krein and a number of survey papers about his work. The main part of the book consists of original research papers presenting the state of the art in operator theory and its applications.

The first volume of these proceedings, entitled Differential Operators and related Topics, concerns the other aspects of the conference. The two volumes will be of interest to a wide-range of readership in pure and applied mathematics, physics and engineering sciences.

Table of Contents

Operator Theory:
Advances and Applications, Vol. 118

Bibliography of Mark Grigor'evich Krein

[1] *Le systeme derive et les contours derives*, Zh. Nauchn. Issled. Kafedr. Odessa **2**, no. 3 (1926), 61–73.

[2] *On the Taylor series defining an analytic function which is regular in a domain bounded by several circles*, Izv. Fiz.-Mat. O.-va, Kazan' (3) **2** (1927), 50–57.

[3] *Uber den Satz von "Curvatura Integra"*, Izv. Fiz.-Mat. O.-va, Kazan' (3) **3** (1928), 36–47.

[4] *L'integrale de Stieltjes dans la theorie des contours convexes*, Izv. Fiz.-Mat. O.-va, Kazan' (3) **3** (1929), 81–93.

[5] *Sur l'aire mixte de deux ovales*, Zap. Prir.-Tekn. Otd. Akad. Nauk Ukr. SSR (3) **4** (1929), 128–130.

[6] *On normal operators in Hermitian spaces*, Izv. Fiz.-Mat. O.-va, Kazan' (3) **4** (1930), 71–84 (with F.R. Gantmakher).

[7] *Zur Strukturfrage von orthogonalen Matrizen*, Zap. Prir.-Tekn. Otd. Akad. Nauk Ukr. SSR (5) **4** (1930), 1–8 (with F.R. Gantmakher).

[8] *Erganzungen zu der Abhandlung "Zur Strukturfrage von orthogonalen Matrizen"*, Zap. Prir.-Tekn. Otd. Akad. Nauk Ukr. SSR **1** (1931), 103–108.

[9] *A contribution to the theory of symmetric polynomials*, Mat. Sb. (3) **40** (1933), 271–283.

[10] *On the theory of vibrations of multisupported beams*, Vestnik Inzh. i Tekhn. **4** (1933), 142–145.

[11] *On the spectrum of a Jacobi form in connection with the theory of torsional oscillations of shafts*, Mat. Sb. (4) **40** (1933), 455–466.

[12] *Uber eine neue Klasse von Hermiteschen Formen und uber eine Verallgemeinerung des trigonometrischen Momentem problems*, Izv. Akad. Nauk SSSR Otd. Mat. Estestven. Nauk **9** (1933), 1259–1275.

[13] *Uber eine Transformation der Bezoutiante, die Sturmschen Satze fuhrt*, Zap. Khar'kov. Mat. O.-va (4) **10** (1933), 33–40 (with M.A. Naimark).

[14] *On the nodes of harmonic oscillations of mechanical systems of a special form*, Mat. Sb. (2) **41** (1934), 339–348.

[15] *Uber Fouriersche Reihen beschrankter summierbarer Funktionen und ein neues Extremumproblem. 1*, Zap. Khar'kov. Mat. O.-va, (4) **9** (1934), 9–23 (with N.I. Akhiezer).

[16] *Uber Fouriersche Reihen beschrankter summierbarer Funktionen und ein neues Extremumproblem, 2*, Zap. Khar'kov. Mat. O.-va (4) **10** (1934), 3–32 (with N.I. Akhiezer).

[17] *A general method of forming the frequency equations of vibrating plane frames*, Trudy Inst. Inzh. Vodn. Transp., Odessa **1** (1935), 63–71 (with Ya.L. Nudel'man).

[18] *On integral kernels of Green function type*, Tr. Univ., Odessa **1** (1935), 39–50 (with F.R. Gantmakher).

[19] *On applications of the Bezoutian to problems of localization of roots of algebraic equations*, Tr. Univ., Odessa **1** (1935), 51–69 (with M.A. Naimark).

[20] *On a special class of differential operators*, Dokl. Akad. Nauk SSSR **2** (1935), 345–349.

[21] *Sur les derivees de noyaux de Mercer*, C.R. Acad. Sci., Paris **200** (1935), 797–799.

[22] *Sur une formule de quadrature de Tchebycheff*, C.R. Acad. Sci., Paris **200** (1935), 890–893 (with N.I. Akhiezer).

[23] *Sur les equations integrales chargees*, C.R. Acad. Sci., Paris **201** (1935), 24–26.

[24] *Sur quelques applications des noyaux de Kellog aux problemes d'oscillation*, Soobshch. Nauchn.-Issled. Inst. Mat. Mekh. Khar'kov. Gos. Univ. i Khar'kov. Mat. o-va (4) **11** (1935), 3–20.

[25] *Uber eine Transformation der reellen Toeplitzhen Formen und das Momentproblem in einem endlichen Intervalle*, Soobshch. Nauchn.-Issled. Inst. Mat. Mekh. Khar'kov. Gos. Univ. i Khar'kov. Mat. C.R. Acad. Sci., Paris **201** (1935), 24–26 (with N.I. Akhiezer).

[26] *On a special class of determinants in connection with Kellog's integral kernels*, Mat. Sb. **42** (1935), 501–508 (with F.R. Gantmakher).

[27] *Sur les matrices oscillatoires*, C.R. Acad. Sci. Paris **201** (1935), 577–579 (with F.R. Gantmakher).

[28] *The method of symmetric and Hermitian forms in the theory of separation of the roots of algebraic equations*, Khar'kov (1936), 1–41 (with M.A. Naimark); English transl.: Linear and Multilinear Algebra **10** (1981), 265–308.

[29] *On the Fourier series of bounded summable functions*, Proc. 2nd All-Union Math. Conf., vol. **2**, Leningrad, July 24–30, 1934, Moscow (1936), 151 (with N.I. Akhiezer).

[30] *On a generalization of the investigations of Academician A.A. Markov on limiting values of integrals*, Proc. 2nd All-Union Math. Conf., vol. **2**, Leningrad, July 24–30, 1934, Moscow (1936), 152–154.

[31] *The algebra of Hermitian forms in the theory of bounded functions*, Proc. 2nd All-Union Math. Conf., vol. **2**, Leningrad, July 24–30, 1934, Moscow (1936), 160.

[32] *On integral equations with oscillating fundamental functions*, Proc. 2nd All-Union Math. Conf., vol. **2**, Leningrad, July 24–30, 1934, Moscow (1936), 259–262.

[33] *On Green functions positive in the sense of Mercer*, Dokl. Akad. Nauk SSSR **1** (1935), 55–58.

[34] *On bilinear decompositions of symmetric kernels positive in the sense of Mercer*, Dokl. Akad. Nauk SSSR **1** (1936), 303–306 (with A.M. Danilevskii).

[35] *On two minimum problems connected with the moment problem*, Dokl. Akad. Nauk SSSR **1** (1936), 331–334 (with N.I. Akhiezer).

[36] *On oscillation differential operators*, Dokl. Akad. Nauk SSSR **4** (1936), 379–382.

[37] *Sur les vibrations propres des tiges dont l'une des extremites est encastree et l'autre libre*, Zap. Khar'kov. Mat. O.-va (4) **12** (1936), 3–11.

[38] *Das Momenteproblem bei der zusatzlichen Bedingung von A. Markoff*, Zap. Khar'kov. Mat. O.-va (4) **12** (1936), 13–36 (with N.I. Akhiezer).

[39] *Bemerkung zur Arbeit "Uber Fouriersche Reihen beschrankter summierbrer Funktionen und ein neues Extremumproblem"*, Zap. Khar'kov. Mat. O.-va (4) **12** (1936), 37–40 (with N.I. Akhiezer).

[40] *Sur quelques proprietes de noyaux de Kellog*, Zap. Khar'kov. Mat. O.-va (4) **13** (1936), 15–28 (with N.I. Akhiezer).

[41] *On P.F. Papkovich's paper "On a form of the basic differential equations of small oscillations of a system without gyroscopic terms"*, Prikl. Mat. Mekh. **1** (1936), 159–161 (with F.R. Gantmakher).

[42] *On the characteristic numbers of symmetric differentiable kernels*, Mat. Sb. **2** (44) (1937), 725–732.

[43] *Sur les developpements des fonctions arbitraires en series de fonctions fondamentales d'un probleme aux limites quelconque*, Mat. Sb. **2** (44) (1937), 923–934.

[44] *Sur les operateurs differentiels autoadjoints et leurs fonctions de Green simmetriques*, Mat. Sb. **2** (44) (1937), 1032–1072.

[45] *The moment problem on two intervals with the additional conditions of A.A. Markov*, Zap. Khar'kov. Mat. O.-va (4) **14** (1937), 47–60 (with N.I. Akhiezer).

[46] *On some properties of the Kellog's resolvent kernel*, Zap. Khar'kov. Mat. O.-va (4) **14** (1937), 61–74.

[47] *On positive additive functionals in linear normed spaces*, Zap. Khar'kov. Mat. O.-va (4) **14** (1937), 227–237.

[48] *On several questions concerning the geometry of convex sets belonging to a linear normed and complete space*, Dokl. Akad. Nauk SSSR **14** (1937), 5–8.

[49] *On the best approximation of differentiable periodic functions by trigonometric sums*, Dokl. Akad. Nauk SSSR **15** (1937), 107–112 (with N.I. Akhiezer).

[50] *Sur les matrices complement non-negatives et oscillatoires*, Comp. Math. **4** (1937), 445–476 (with F.R. Gantmakher).

[51] *Concerning the theory of vibrations of rod systems*, Trudy Inst. Mat., Odessa **12** (1938), 193–225 (with Ya.L. Nudel'man).

[52] *Some questions in the theory of moments*, Gos. Nauchn.-Tekhn. Izd-vo Ukr., Khar'kov (1938), (with N.I. Akhiezer); English Transl.: Transl. Math. Math. Monographs, vol. **2**, Amer. Math. Soc., Providence, R.I. (1962).

[53] *Some remarks on the coefficients of quadrature formulae of Gaussian type*, Trudy Odess'kogo Derzh. Univ., Mat. **2** (1938), 29–38 (with N.I. Akhiezer).

[54] *On the Nevanlinna-Pick problem*, Trudy Odess'kogo Derzh. Univ., Mat. **2** (1938), 63–77 (with P.G. Rekhtman).

[55] *On minimaximal properties of nodes of obertones of a vibrating rod*, Trudy Odess'kogo Derzh. Univ., Mat. **2** (1938), 103–112 (with Ya.L. Nudel'man).

[56] *On the theory of best approximation of periodic functions*, Dokl. Akad. Nauk SSSR **18** (1938), 245–250.

[57] *On the best approximation of continuous differentiable periodic functions on the full real axis*, Dokl. Akad. Nauk SSSR **18** (1938), 619–624.

[58] *On the linear operators that leave invariant a conic set*, Dokl. Akad. Nauk SSSR **23** (1939), 749–752.

[59] *On totally nonnegative Green functions of ordinary differential operators*, Dokl. Akad. Nauk SSSR **24** (1939), 220–223 (with G.M. Finkel'shtein).

[60] *On nonsymmetric oscillating Green functions of ordinary differential operators*, Dokl. Akad. Nauk SSSR **25** (1939), 643–646.

[61] *Oscillation theorems for ordinary linear differential operators of arbitrary order*, Dokl. Akad. Nauk SSSR **25** (1939), 717–720.

[62] *On the decomposition of a linear functional into positive components*, Dokl. Akad. Nauk SSSR **25** (1939), 721–724 (with Yu.I. Grosberg).

[63] *On some quadrature formulae of P.L. Chebyshev and A.A. Markov*, Memorial volume dedicated to D.A. Grave, Moscow (1940), 15–28 (with N.I. Akhiezer).

[64] *On weighted integral equations whose distribution functions are not monotonic*, Memorial volume dedicated to D.A. Grave, Moscow (1940), 88–103.

[65] *On a property of bases in Banach space*, Zap. Khar'kov. Mat. O.-va (4) **16** (1940), 106–110 (with D.P. Mil'man and M.A. Rutman).

[66] *Some remarks about three papers of M.S. Verblunsky*, Zap. Khar'kov. Mat. O.-va (4) **16** (1940), 47–60 (with N.I. Akhiezer).

[67] *On some minimum-problems in the class of Stepanoff almost periodic functions*, Zap. Khar'kov. Mat. O.-va (4) **17** (1940), 111–124 (with B.M. Levitan).

[68] *On extreme points of regularly convex sets*, Studia Math. **9** (1940), 133–138 (with D.P. Mil'man).

[69] *On regularly convex sets in the space conjugate to a Banach space*, Ann. Math. **41** (1940), 556–583 (with V.L. Shmul'an).

[70] *On the continuation problem for Hermitian-positive continuous functions*, Dokl. Akad. Nauk SSSR **26** (1940), 17–21.

[71] *On an intrinsic characteristic of the space of all continuous functions defined on a bicompact Hausdorff space*, Dokl. Akad. Nauk SSSR **27** (1940), 427–431 (with S.G. Krein).

[72] *Basic properties of normal conic sets in Banach space*, Dokl. Akad. Nauk SSSR **28** (1940), 13–17.

[73] *On the minimal decomposition of a linear functional into positive components*, Dokl. Akad. Nauk SSSR **28** (1940), 18–22.

[74] *On a ring of functions defined on a topological group*, Dokl. Akad. Nauk SSSR **29** (1940), 275–280.

[75] *On a special ring of functions*, Dokl. Akad. Nauk SSSR **29** (1940), 355–359.

[76] *On a new property of the Sturm-Lioville operator*, Trudy Univ., Odessa, Ser. Mat. **3** (1940), 21–32.

[77] *Oscillation Matrices and Small Oscillations of Mechanical Systems*, 2nd edition, Moscow, Leningrad (1941) (with F.R. Gantmakher).

[78] *On the theory of almost-periodic functions on a topological group*, Dokl. Akad. Nauk SSSR **30** (1941), 5–8.

[79] *On positive functionals on almost-periodic functions*, Dokl. Akad. Nauk SSSR **30** (1941), 9–12.

[80] *On a generalization of Plancherel's theorem to the case of Fourier integrals on a commutative topological group*, Dokl. Akad. Nauk SSSR **30** (1941), 482–486.

[81] *Sur l'espace des fonctions continues definies sur un bicompact de Hausdorff et ses sousespaces semi-ordonnes*, Mat. Sb. **13** (55) (1943), 1–38 (with S.G. Krein).

[82] *On the representation of functions by Fourier-Stieltjes integrals*, Uchen. Zap. Kuibyshev. Gos. Pedag. i Uchit. Inst. **7** (1943), 123–148.

[83] *On Hermitian operators with deficiency indices equal to one*, I, Dokl. Akad. Nauk SSSR **43** (1944), 339–342.

[84] *On Hermitian operators with deficiency indices equal to one*, II, Dokl. Akad. Nauk SSSR **44** (1944), 143–146.

[85] *On a remarkable class of Hermitian operators*, Dokl. Akad. Nauk SSSR **44** (1944), 191–195.

[86] *On a generalized moment problem*, Dokl. Akad. Nauk SSSR **44** (1944), 239–243.

[87] *On the logarithm of an infinitely decomposable Hermitian-positive function*, Dokl. Akad. Nauk SSSR **45** (1944), 99–102.

[88] *On the continuation problem for helical arcs in Hilbert space*, Dokl. Akad. Nauk SSSR **45** (1944), 147–150.

[89] *On a generalization of investigations of G.Szego, V.I.Smirnov and A.N. Kolmogorov*, Dokl. Akad. Nauk SSSR **46** (1945), 95–98.

[90] *On an extrapolation problem of A.N. Kolmogorov*, Dokl. Akad. Nauk SSSR **46** (1945), 339–342.

[91] *On self-adjoint extensions of bounded and semibounded Hermitian operators*, Dokl. Akad. Nauk SSSR **48** (1945), 323–326.

[92] *On the resolvents of a Hermitian operator with deficiency index* (m, m), Dokl. Akad. Nauk SSSR **52** (1946), 657–660.

[93] *On a general method of decomposing positive-definite kernels into elementary products*, Dokl. Akad. Nauk SSSR **53** (1946), 3–6.

[94] *On a theorem of M.Ya. Vygodstkii*, Mat. Sb. **18** (60) (1946), 447–450.

[95] *The basic theorems concerning the extension of Hermitian operators and some of their applications to the theory of orthogonal polynomials and the moment problem*, Usp. Mat. Nauk **2**, no. 3 (1947), 60–106 (with M.A. Krasnosel'skii).

[96] *On a general method of decomposing positive-definite kernels into elementary ones*, Usp. Mat. Nauk **2**, no. 3 (1947), p. 181.

[97] *The theory of self-adjoint extensions of semibounded Hermitian operators and its applications*. I, Mat. Sb. **20** (62) (1947), 431–498.

[98] *The theory of self-adjoint extensions of semibounded Hermitian operators and its applications*. II, Mat. Sb. **21** (63) (1947), 365–404.

[99] *On linear compact operators in functional spaces with two norms*, Zh. Inst. Mat. Akad. Nauk USSR **9** (1947), 104–129.

[100] *On the theory of entire functions of exponential type*, Izv. Akad. Nauk SSSR Ser. Mat. **11** (1947), 309–326.

[101] *Linear operators leaving invariant a cone in Banach space*, Usp. Mat. Nauk **3**, no. 1 (1948), 3–95 (with M.A. Rutman); English transl.: Amer. Math. Soc. Transl. (1) **10** (1962), 199–235.

[102] *Helical lines in an infinite-dimensional Lobachevsky space and Lorentz transformations*, Usp. Mat. Nauk **3**, no. 3 (1948), 158–160.

[103] *On some problems connected with Lyapunov's circle of ideas in stability theory*, Usp. Mat. Nauk **3**, no. 3 (1948), 166–169.

[104] *On Hermitian operators with directing functionals*, Sb. Trudov Inst. Mat. Akad. Nauk Ukr. SSR **10** (1948), 83–106.

[105] *On deficiency numbers of linear operators in Banach space and on certain geometric questions*, Sb. Trudov Inst. Mat. Akad. Nauk Ukr. SSR **11** (1948), 97–112 (with M.A. Krasnosel'skii and D.P. Mil'man).

[106] *Functional Analysis*, In: "Mathematics in the USSR over 30 years", Moscow, Leningrad (1948), 608–672 .

[107] *Soph'ya Kopvalevskaia*, Bolshevist. znam'a (1948), March 6: the portrait.

[108] *Fundamental aspects of the representation theory of Hermitian matrices with deficiency index* (m, m), Ukr. Mat. Zh. **1**, no. 2 (1949), 3–66; English transl.: Amer. Math. Soc. Tranl. (2) **97** (1970), 75–143.

[109] *Hermitian-positive kernels on homogeneous spaces, 1*, Ukr. Mat. Zh. **1**, no. 4 (1949), 64–98; English transl.: Amer. Math. Soc. Tranl. (2) **34** (1963), 69–108.

[110] *Infinite J-matrices and the matrix moment problem*, Dokl. Akad. Nauk SSSR **69** (1949), 125–128.

[111] *On entire almost-periodic functions of exponential type*, Dokl. Akad. Nauk SSSR **69** (1949), 285–287 (with B.Ya. Levin).

[112] *The duality principle for a bicompact group and square block-algebra*, Dokl. Akad. Nauk SSSR **69** (1949), 726–728.

[113] *Hermitian-positive kernels on homogeneous spaces, 2*, Ukr. Mat. Zh. **2**, no. 1 (1950), 10–59; English transl.: Amer. Math. Soc. Tranl. (2) **34** (1963), 109–164.

[114] *On an application of the fixed point principle in the theory of linear transformations of spaces with an indefinite metric*, Usp. Mat. Nauk **5**, no. 2 (1950), 180–190; English transl.: Amer. Math. Soc. Transl. (2) **1** (1955), 27–35.

[115] *Remark on a possible generalization of a theorem of A. Haar and A.N. Kolmogorov*, Usp. Mat. Nauk **5**, no. 1 (1950), 217–229 (with S.I. Zukhovitskii).

[116] *Generalization of a certain investigations of A.M. Lyapunov concerning differential equations with periodic coefficients*, Dokl. Akad. Nauk SSSR **73** (1950), 445–448.

[117] *On the Sturm-Liouville boundary value problem on the interval* $(0, \infty)$ *and on a class of integral equations*, Dokl. Akad. Nauk SSSR **73** (1950), 1125–1128.

[118] *On a one-dimensional singular boundary value problem of even order on the interval* $(0, \infty)$, Dokl. Akad. Nauk SSSR **74** (1950), 9–12.

[119] *On some investigations of A.M. Lyapunov concerning differential equations with periodic coefficients*, Dokl. Akad. Nauk SSSR **75** (1950), 495–498 (with K.P. Kovalenko).

[120] *Oscillation Matrices, Kernels and Small Oscillations of Mechanical Systems*, 2nd edition, Moscow, Leningrad (1950) (with F.R. Gantmakher).

[121] *Concerning the mathematical theory of the method of amplitude-phase modulation*, Sb. Trudov Inst. Elektrotekhn. Akad. Nauk SSSR **7** (1951), 96–104 (with B.I. Korenblyum and S.I. Tetel'baum).

[122] *On the theory of entire matrix-functions of exponential type*, Ukr. Mat. Zh. **3**, no. 2 (1951), 164–173.

[123] *Solution on the inverse Sturm-Liouville problem*, Dokl. Akad. Nauk SSSR **76** (1951), 21–24.

[124] *Determination of the density of a nonhomogeneous symmetric string from its frequency spectrum*, Dokl. Akad. Nauk SSSR **76** (1951), 345–348.

[125] *On the application of an algebraic proposition in the theory of monodromy matrices*, Usp. Mat. Nauk **6**, no. 1 (1951), 171–177.

[126] *On certain problems of the minimum and maximum of characteristic values and on the Lyapunov zones of stability*, Prikl. Mat. Mekh. **15** (1951), 323–348; English Transl.: Amer. Math. Soc. Transl. (2) **1** (1955), 163–187.

[127] *The ideas of P.L. Chebyshev and A.A. Markov in the theory of limit values of integrals and their further developments*, Usp. Mat. Nauk **6**, no. 4 (1951), 3–120 (with the redactional participation of P.G. Rekhtman); English Transl.: Amer. Math. Soc. Transl. (2) **12** (1959), 1–121.

[128] *On a generalization of the Schwarz and Lowener lemmas*, Zap. Mat. Otd. Fiz.-Mat. Fak. i Khar'kov. Mat. O.-va (4) **23** (1952), 95–101 (with N.I. Akhiezer).

[129] *Stability of the index of an unbounded operator*, Mat. Sb. **30** (72) (1952), 219–224 (with M.A. Krasnosel'skii).

[130] *On an undetermined case of the Sturm-Liouville boundary value problem on the interval* $(0, \infty)$, Izv. Akad. Nauk SSSR Ser. Mat. **16** (1952), 293–324.

[131] *On inverse problem for a nonhomogeneous string*, Dokl. Akad. Nauk SSSR **82** (1952), 669–672.

[132] *On a generalization of investigations of Stieltjes*, Dokl. Akad. Nauk SSSR **87** (1952), 881–884.

[133] *On some new problems of the oscillation theory for Sturmian systems*, Prikl. Mat. Mekh. **14** (1952), 555–568.

[134] *On a trace formula in perturbation theory*, Mat. Sb. **33** (75) (1953), 597–626.

[135] *On the transfer function of a one-dimensional second-order boundary value problem*, Dokl. Akad. Nauk SSSR **88** (1953), 405–408.

[136] *Analogue of the Chebyshev-Markov inequalities in a one-dimensional boundary value problem*, Dokl. Akad. Nauk SSSR **89** (1953), 5–8.

[137] *On some cases of effective determination of the density of a nonhomogeneous string from its spectral function*, Dokl. Akad. Nauk SSSR **93** (1953), 617–620.

[138] *On inverse problems of the theory of filters and λ-zones*, Dokl. Akad. Nauk SSSR **93** (1953), 767–770.

[139] *A fundamental approximation problem in the theory of extrapolation and filtration stochastic processes*, Dokl. Akad. Nauk SSSR **94** (1954), 13–16; English transl.: Selected Transl. in Math. Statistics and Probability, Inst. of Math. Statistics and Amer. Math. Soc. **4** (1963), 127–131.

[140] *On a method of effective resolution of an inverse boundary value problem*, Dokl. Akad. Nauk SSSR **94** (1954), 987–990.

[141] *On integral equations that generate second-order differential equations*, Dokl. Akad. Nauk SSSR **97** (1954), 21–24.

[142] *The theory of spectral and transfer functions of one-dimensional boundary value problems and its applications*, Usp. Mat. Nauk **9**, no. 3 (1954), p. 221.

[143] *Development in a new direction of the Chebyshev-Markov theory of limit values of integrals*, Usp. Mat. Nauk **10**, no. 1 (1955), 67–78 (with P.G. Rekhtman); English Transl.: Amer. Math. Soc. Transl. (2) **12** (1959), 123–135.

[144] *On a new method of solving linear integral equations of first and second kind*, Dokl. Akad. Nauk SSSR **100** (1955), 413–416.

[145] *On the determination of the potential of a particle from its S-function*, Dokl. Akad. Nauk SSSR **105** (1955), 433–436.

[146] *Continual analogues of propositions on polynomials orthogonal on the unit circle*, Dokl. Akad. Nauk SSSR **105** (1955), 637–640.

[147] *On tests for stable boundedness of solutions of periodic canonical systems*, Prikl. Mat. Mekh. **19** (1955), 641–680; English transl.: Amer. Math. Soc. Transl. (2) **120** (1983), 71–110.

[148] *The basic propositions of the theory of λ-zones of stability of a canonical system of linear differential equations with periodic coefficients*, Memorial volume dedicated to A.A. Andronov, Moscow (1955), 413–498; English transl.: Topics in Differential and Integral Equations and Operator Theory, Operator Theory: Advances and Applications, vol. **7**, Birkhauser, Basel (1983), 1–70.

[149] *On differential equations in Banach space*, Proc. 3rd All-Union Math. Conf., vol. **2** (1956), p. 11 (with M.A. Krasnosel'skii and S.G. Krein).

[150] *Ordinary differential equations with periodic coefficients*, Proc. 3rd All-Union Math. Conf., vol. **2** (1956), p. 11.

[151] *On some basic propositions of the theory of systems of integral equations on the half-line with kernels depending on the difference of the arguments*, Proc. 3rd All-Union Math. Conf., vol. **2** (1956), 37–38 (with I.C. Gohberg).

[152] *Evolution of the moment problem and the problem of oscillation of a string*, Proc. 3rd All-Union Math. Conf., vol. **2** (1956), p. 40.

[153] *Spectral theory of operators in spaces with an indefinite metric*, Usp. Mat. Nauk **11**, no. 4 (1956), 169–171 (with I.S. Iokhvidov).

[154] *Spectral theory of operators in spaces with an indefinite metric. 1*, Tr. Mosk. Mat. O.-va **5** (1956), 367–432 (with I.S. Iokhvidov); English transl.: Amer. Math. Soc. Transl. (2) **13** (1960), 105–175.

[155] *On the theory of accelerants and S-matrices of canonical differential systems*, Dokl. Akad. Nauk SSSR **111** (1956), 1167–1170.

[156] *Remark to the paper "Spectral theory of operators in spaces with an indefinite metric. 1"*, Tr. Mosk. Mat. O.-va **6** (1957), p. 486 (with I.S. Iokhvidov).

[157] *The basic propositions on defect numbers, root numbers and indices of linear operators*, Usp. Mat. Nauk **12**, no. 2 (1957), 43–118 (with I.C. Gohberg); English transl.: Amer. Math. Soc. Transl. (2) **13** (1960), 185–264.

[158] *On Bari bases in Hilbert spaces*, Usp. Mat. Nauk **12**, no. 3 (1957), 333–341.

[159] *On the characteristic function $A(\lambda)$ of a linear canonical system of second-order differential equations with periodic coefficients*, Prikl. Mat. Mekh. **21** (1957), 320–329.

[160] *On a continual analogue of a formula of Christoffel from the theory of orthogonal polynomials*, Dokl. Akad. Nauk SSSR **113** (1957), 970–973.

[161] *Integral equations on the half-line with kernel depending on the difference of the arguments*, Usp. Mat. Nauk **13**, no. 5 (1958), 3–120; English transl.: Amer. Math. Soc. Transl. (2) **22** (1962), 163–288.

[162] *Criterion for the discreteness of the spectrum of a singular string*, Izv. Vyssh. Uchebn. Zaved., Ser. Mat. **2** (1958), 136–153 (with I.S. Kats).

[163] *On a pair integral equation and its transpose. 1*, Teoret. i Prikl. Mat., L'vov **1** (1958), 58–81 (with I.C. Gohberg); English transl.: Topics in Differential and Integral Equations and Operator Theory, Operator Theory: Advances and Appl., vol. **7**, Birkhauser, Basel (1983), 243–274.

[164] *On the stability of the systems of partial indices of the Hilbert problem with several unknown functions*, Dokl. Akad. Nauk SSSR **119** (1958), 854–857 (with I.C. Gohberg).

[165] *Systems of integral equations on the half-line with kernel depending on the difference of arguments*, Usp. Mat. Nauk **13**, no. 2 (1958), 3–72 (with I.C. Gohberg); English transl.: Amer. Math. Soc. Transl. (2) **14** (1960), 217–287.

[166] *Correction to the paper "Systems of integral equations on the half-line with kernel depending on the difference of arguments"*, Usp. Mat. Nauk **13**, no. 4 (1958), p. 232 (with I.C. Gohberg).

[167] *On compact operators with spectrum concentrated at zero*, Dokl. Akad. Nauk SSSR **128** (1959), 227–230 (with I.C. Gohberg).

[168] *Criteria for completeness of the system of root vectors of a dissipative operator*, Usp. Mat. Nauk **14**, no. 3 (1959), 145–152 (with I.C. Gohberg); English transl.: Amer. Math. Soc. Transl. (2) **14** (1960), 217–287.

[169] *On an integral representation of a continuous Hermitian-indefinite function with a finite number of negative squares*, Dokl. Akad. Nauk SSSR **125** (1959), 31–34.

[170] *Spectral theory of operators in spaces with an indefinite metric.* 2, Tr. Mosk. Mat. O.-va **8** (1959), 413–496 (with I.S. Iokhvidov); English transl.: Amer. Math. Soc. Transl. (2) **34** (1963), 283–373.

[171] *On the theory of linear non-selfadjoint operators*, Dokl. Akad. Nauk SSSR **130** (1960), 253–256; English transl.: Soviet Math. Dokl. **1** (1960), 38–40.

[172] *Oszillationsmatrizen, Oszillationskerne und kleine Schwingungen Mechanischer Systeme*, Akademie Verlag, Berlin, (1960) (with F.R. Gantmakher).

[173] *Mark Aronovich Naimark: To 50-th birthday*, Usp. Mat. Nauk **14**, no. 2 (1960), 231–236 (with G.E. Shilov).

[174] *Hamiltonian systems of linear differential equations with periodic coefficients*, Proc. Internat. Symp. Nonlinear Vibrations, Kiev, vol. **1** (1961), 1–55 (with V.A. Yakubovich).

[175] *On the theory of transmission bands of periodic waveguides*, Prikl. Mat. Mekh. **25** (1961), 24–37 (with G.Ya. Lyubarski); English transl.: J. Appl. Math. Mech. **25** (1961), 29–48.

[176] *On the theory of triangular representations of non-selfadjoint operators*, Dokl. Akad. Nauk SSSR **137** (1961), 1034–1037 (with I.C. Gohberg); English transl.: Soviet Math. Dokl. **2** (1961), 392–395.

[177] *Transformations of kernels of integral equations and their influence on the spectra of these equations*, Ukr. Mat. Zh. **13**, no. 3 (1961), 12–38 (with I.C. Gohberg); English transl.: Amer. Math. Soc. Transl. (2) **35** (1964), 263–295.

[178] *On Volterra operators with imaginary component in one class or another*, Dokl. Akad. Nauk SSSR **137** (1961), 779–782 (with I.C. Gohberg); English transl.: Soviet Math. Dokl. **2** (1961), 983–986.

[179] *An extremum problem for polynomials*, Ann. Univ. Sci. Budapest Eotvos Sect. Math. 3–4 (1961), 9–14 (with N.I. Akhiezer).

[180] *Naum Il'ich Akhiezer: To 50-th birthday*, Usp. Mat. Nauk **16**, no. 4 (1961), 223–234 (with B.Ya. Levin).

[181] *On the theory of wave operators and scattering operators*, Dokl. Akad. Nauk SSSR **119** (1962), 475–478 (with M.Sh. Birman); English transl.: Soviet Math. Dokl. **3** (1962), 740–744.

[182] *On the problem of factoring operators in a Hilbert space*, Dokl. Akad. Nauk SSSR **147** (1962), 279–282 (with I.C. Gohberg); English transl.: Soviet Math. Dokl. **3** (1962), 1578–1582.

[183] *Analytic properties of multipliers of periodic canonical differential systems of positive type*, Izv. Akad. Nauk SSSR Ser. Mat. **26** (1962), 549–572 (with G.Ya. Lyubarskii). English transl.: Amer. Math. Soc. Transl. (2) **89** (1970), 1–20.

[184] *On perturbation determinants and a trace formula for unitary and self-adjoint operators*, Dokl. Akad. Nauk SSSR **144** (1962), 268–271; English transl.: Soviet Math. Dokl. **3** (1962), 707–710.

[185] *Hamiltonian systems of linear differential equations with periodic coefficients*, Proc. Internat. Symp. Nonlinear Vibrations, Kiev, vol. **1** (1963), 277–305 (with V.A. Yakubovich); English transl.: Amer. Math. Soc. Transl. (2) **120** (1983), 139–168.

[186] *Some topics of the theory of wave and scattering operators*, Joint Symposium Soviet-American Partial Differen. Equations, Novosibirsk (1963), 3–11 (with M.Sh. Birman).

[187] *On some new investigations on the perturbation theory of linear operators*, Proc. 4th All-Union Math. Conf., vol. **1** (1963), 133–134.

[188] *On the spectral function of a self-adjoint operator in a space with indefinite metric*, Dokl. Akad. Nauk SSSR **152** (1963), 39–42 (with H. Langer); English transl.: Soviet Math. Dokl. **4** (1963), 1236–1239.

[189] *On the theory of quadratic pencils of self-adjoint operators*, Dokl. Akad. Nauk SSSR **154** (1964), 1258–1261 (with H. Langer); English transl.: Soviet Math. Dokl. **5** (1964), 266–269.

[190] *Criteria for the completeness of the system of root vectors of a contraction*, Ukr. Mat. Zh. **16**, no. 1 (1964), 78–82 (with I.C. Gohberg). English transl.: Amer. Math. Soc. Transl. (2) **54** (1966), 119–124.

[191] *Lectures on the Theory of Stability of Solutions of Differential Equations in Banach Space*, Naukova Dumka, Kiev (1964).

[192] *On certain new studies in the perturbation theory for self-adjoint operators*, Proc. First Math. Summer School, Kanev, vol. **1**, Naukova Dumka, Kiev (1964), 103–187; English transl.: Topics in Differential and Integral Equations and Operator Theory, Operator Theory: Advances and Applications, vol. **7**, Birkhauser, Basel (1983), 107–172.

[193] *Some new investigations on the theory of nonselfadjoint operators*, Proc. 4th All-Union Math. Congr., vol. **2** (1964), 261–271 (with M.S. Brodskii, I.C. Gohberg, V.I. Matsaev); English transl.: Amer. Math. Soc. Transl. (2) **65** (1967), 237–251.

[194] *On the factorization of operators in Hilbert space*, Acta. Sci. Math. Szeged, 25, 1–2 (1964), 90–123 (with I.C. Gohberg); English transl.: Amer. Math. Soc. Transl. (2) **51** (1966), 155–188.

[195] *On a new application of the fixed point principle in the theory of operators in a space with indefinite metric*, Dokl. Akad. Nauk SSSR **154** (1964), 1023–1026; English transl.: Soviet Math. Dokl. **5** (1964), 224–228.

[196] *Introduction to the geometry of indefinite J-spaces and to the theory of operators in these spaces*, Proc. Second Math. Summer School, vol. **1**, Naukova Dumka, Kiev (1965), 15–92; English transl.: Amer. Math. Soc. Transl. (2) **93** (1970), 103–176.

[197] *Introduction to the Theory of Linear Non-Self-Adjoint Operators in Hilbert Space*, Nauka, Moscow (1965), (with I.C. Gohberg); English transl.: Transl. Math. Monographs, vol. **18**, Amer. Math. Soc., Providence, R.I. (1970).

[198] *On the theory of weighted integral equations*, Izv. Akad. Nauk Mold. SSSR **7** (1965), 40–46.

[199] *On a multiplicative representation of the characteristic functions of operators close to unitary ones*, Dokl. Akad. Nauk SSSR **161** (1965), 732–735 (with I.C. Gohberg); English transl.: Soviet Math. Dokl. **6** (1965), 1279–1283.

[200] *On some mathematical principles of the linear theory of damped oscillations of continua*, Proc. Internat. Sympos., Tbilisi, Appl. of Function Theory in Mech. of Continua, vol. **2** (1965), 283–322 (with H. Langer); English transl.: Integral Equations and Operator Theory **1**, no. 3 (1978), 364–399, and **1**, no. 4 (1978), 539–566.

[201] *Spectral theory of operators in spaces with an indefinite metric. 2*, Letter to the editor, Tr. Mosk. Mat. O.-va **15** (1965), 452–454 (with I.S. Iokhvidov).

[202] *Vitol'd L'vovich Shmulyan: To 50-th birthday and 20-th anniversary of the death*, Usp. Mat. Nauk **20**, no. 2 (1965), 131–133.

[203] *Feliks Ruvimovich Gantmakher*, Usp. Mat. Nauk **20**, no. 2 (1965), 149–157 (with M.A. Aizerman, L.A. Lyusternik, M.A. Naimark et al.).

[204] *Analytic problems and results of the theory of linear operators in Hilbert space*, Tez. Dokl. Mezhdunar. Congr. Mat., Moscow (1966), 22–26.

[205] *On some new Banach algebras and Wiener-Levy type theorems for Fourier series and integrals*, Mat. Issled. **1**, no. 1 (1966), 82–109; English transl.: Amer. Math. Soc. Transl. (2) **93** (1970), 177–199.

[206] *On some new results in the theory of resolvents of Hermitian operators*, Dokl. Akad. Nauk SSSR **169** (1966), 1269–1272 (with Sh.N. Saakyan); English transl.: Soviet Math. Dokl. **7** (1966), 1086–1089.

[207] *Plus-operators in a space with indefinite metric*, Mat. Issled. **1**, no. 1 (1966), 131–161 (with Yu.L. Shmulyan); English transl.: Amer. Math. Soc. Transl. (2) **85** (1969), 93–113.

[208] *On the distribution of roots of polynomials orthogonal on the unit circle with respect to a sign-alternating weight*, In: Function Theory, Function. Anal. and their Appl., no. 2 (1966), 131–137.

[209] *On a class of operators in a space with indefinite metric*, Dokl. Akad. Nauk SSSR **171** (1966), 34–37 (with Yu.L. Shmulyan); English transl.: Soviet Math. Dokl. **7** (1966), 1137–1141.

[210] *J-polar representation of plus operators*, Mat. Issled. **1**, no. 2 (1966), 172–210 (with Yu.L. Shmulyan); English transl.: Amer. Math. Soc. Transl. (2) **85** (1969), 115–143.

[211] *On triangular representations of linear operators and multiplicative representations of their characteristic functions*, Dokl. Akad. Nauk SSSR **175** (1967), 272–275 (with I.C. Gohberg); English transl.: Soviet Math. Dokl. **8** (1967), 831–834.

[212] *On a description of contraction operators similar to unitary ones*, Funkts. Anal. Prilozhen. **1**, no. 1 (1967), 38–60 (with I.C. Gohberg).

[213] *The description of all solutions of the truncated power moment problem and some problems of operator theory*, Mat. Issled. **2**, no. 2 (1967), 114–132; English transl.: Amer. Math. Soc. Transl. (2) **95** (1970), 219–234.

[214] *Theory and Applications of Volterra Operators in Hilbert Space*, Nauka, Moscow (1967) (with I.C. Gohberg); English transl.: Transl. Math. Monographs, vol. **24**, Amer. Math. Soc., Providence, R.I., 1970.

[215] *On linear-fractional transformations with operator coefficients*, Mat. Issled. **2**, no. 3 (1967), 64–96 (with Yu.L. Shmul'yan); English transl.: Amer. Math. Soc. Transl. (2) **103** (1974), 125–152.

[216] *Analytic problems and results in the theory of linear operators in Hilbert Space*, Proc. Internat. Congress Mathematicians, Moscow, 1966, Mir, Moscow (1968), 189–216; English transl.: Amer. Math. Soc. Transl. (2) **70** (1968), 68–72 and (2) **90** (1970), 181–209.

[217] *R-functions: analytic functions mapping the upper half-plane into itself*, Appendix 1 to the Russian transl. of F.V. Atkinson's book "Discrete and Continuous Boundary Problems", Mir, Moscow (1968), 629–647 (with I.S. Kats); English transl.: Amer. Math. Soc. Transl. (2) **103** (1974), 1–18.

[218] *On the spectral functions of the string*, Appendix 2 to the Russian transl. of F.V. Atkinson's book "Discrete and Continuous Boundary Problems", Mir, Moscow (1968), 648–737 (with I.S. Kats); English transl.: Amer. Math. Soc. Transl. (2) **103** (1974), 19–102.

[219] *Infinite Hankel matrices and the generalized problems of Caratheodory-Fejer and F. Riesz*, Funkts. Anal. Prilozhen. **2**, no. 1 (1968), 1–19 (with V.M. Adamyan and D.Z. Arov); English transl.: Function. Anal. Appl. **2**, no. 1 (1968), 1–18.

[220] *Infinite Hankel matrices and the generalized problems of Caratheodory-Fejer and Schur*, Funkts. Anal. Prilozhen. **2**, no. 4 (1968), 1–17 (with V.M. Adamyan and D.Z. Arov); English transl.: Function. Anal. Appl. **2**, no. 4 (1968), 269–281.

[221] *On the theory of S-matrices of canonical differential equations with summable potential*, Dokl. Akad. Nauk Arm. SSR **46**, no. 4 (1968), 150–155 (with F.E. Melik-Adamyan); English transl.: Topics in Differential and Integral Equations and Operator Theory, Operator Theory: Advances and Applications, vol. **7**, Birkhauser, Basel (1983), 295–302.

[222] *Boris Yakovlevich Levin: To 60-th birthday*, Usp. Mat. Nauk **23**, no. 5 (1968), 187–191 (with N.V. Efimov and I.V. Ostrovskii).

[223] *Bounded operators commuting with a contraction of class C_{00} with rank of unitarity one*, Funkts. Anal. Prilozhen. **3**, no. 3 (1968), 86–87 (with V.M. Adamyan and D.Z. Arov); English transl.: Function. Anal. Appl. **3**, no. 3 (1968), 242–243.

[224] *On angular localization of spectrum of a multiplicative integral in Hilbert space*, Funkts. Anal. Prilozhen. **3**, no. 1 (1969), 89–90.

[225] *General theorems on triangular representations of linear operators and multiplicative representations of their characteristic functions*, Funkts. Anal. Prilozhen. **3**, no. 4 (1969), 86–87 (with V.M. Brodskii and I.C. Gohberg); English transl.: Function. Anal. Appl. **3**, no. 4 (1969), 255–276.

[226] *Chebyshev-Markov inequalities in the theory of the spectral function of a string*, Mat. Issled. **5**, no. 1 (1970), 77–101; English transl.: Amer. Math. Soc. Transl. (2) **103** (1974), 103–124.

[227] *Definition and basic properties of the characteristic function of a G- node*, Funkts. Anal. Prilozhen. **4**, no. 1 (1970), 88–90 (with V.M. Brodskii and I.C. Gohberg); English transl.: Function. Anal. Appl. **4**, no. 1 (1970), 78–80.

[228] *The resolvent matrix of a Hermitian operator and the characteristic functions connected with it*, Funkts. Anal. Prilozhen. **4**, no. 3 (1970), 103–104 (with Sh.N. Saakyan); English transl.: Function. Anal. Appl. **4**, no. 3 (1970), 258–259.

[229] *Some applications of the theorem of the factorization of a unitary matrix*, Funkts. Anal. Prilozhen. **4**, no. 4 (1970), 73–75 (with F.E. Melik-Adamyan); English transl.: Function. Anal. Appl. **4**, no. 4 (1970), 327–329.

[230] *Some results and problems in the theory of stability of differential equations in a Banach space*, Proc. 5th Internat. Conf. Non-linear Oscillations, Kiev, 1969, vol. **1**, Akad. Nauk Ukr. SSR (1970), 332–347 (with Yu.L. Daletskii).

[231] *New inequalities for the characteristic numbers of integral equations with smooth kernels*, Mat. Issled. **5**, no. 1 (1970), 22–39 (with I.C. Gohberg); English transl.: Topics in Differential and Integral Equations and Operator Theory, Operator Theory: Advances and Appl. vol. **7**, Birkhauser, Basel (1983), 273–293.

[232] *Stability of Solutions of Differential Equations in Banach Space*, Nauka, Moscow (1970), (with Yu.L. Daletskii); English transl.: Transl. Math. Monographs, vol. **18**, Amer. Math. Soc., Providence, R.I. (1970).

[233] *Analytic properties of the Schmidt pairs of a Hankel operator and the generalized Schur-Takagi problem*, Mat. Sb. **85**, no. 1 (1971), 39–73 (with V.M. Adamyan and D.Z. Arov); English transl.: Math. USSR, Sb. **15** (1971), 31–73.

[234] *Infinite Hankel block-matrices and related extension problems*, Izv. Akad. Nauk Arm. SSR Ser. Mat. **6**, no. 2–3 (1971), 87–112 (with V.M. Adamyan and D.Z. Arov); English transl.: Amer. Math. Soc. Transl. (2) **111** (1978), 133–156.

[235] *Hankel operators, problems of extrapolation of matrix-functions and the theory of S-matrices*, All-Union Conf. Theory of Functions of a Complex Variable, Khar'kov (1971), 6–8 (with V.M. Adamyan and D.Z. Arov).

[236] *On the characteristic functions of an invertible operator*, Acta Sci. Math. Szeged **32**, no. 1–2 (1971), 141–164 (with V.M. Brodskii and I.C. Gohberg).

[237] *The defect subspaces and generalized resolvents of a Hermitian operator in a space* Π_k, *1*, Funkts. Anal. Prilozhen. **5**, no. 2 (1971), 59–71 (with H. Langer); English transl.: Function. Anal. Appl. **5**, no. 1 (1971), 136–146.

[238] *The defect subspaces and generalized resolvents of a Hermitian operator in a space* Π_k, 2, Funkts. Anal. Prilozhen. **5**, no. 3 (1971), 54–69 (with H. Langer); English transl.: Function. Anal. Appl. **5**, no. 3 (1971), 217–228.

[239] *Uber die verallgemeinerten Resolventen und die charackterische Funktion eines isometrischen Operators im Raume* Π_k, Coll. Math. Soc. Janos Bolyai, 5, Hilbert Space Operators and Operator Algebras (1971), 353–400 (with H. Langer).

[240] *The Markov Moment Problem and Extremal Problems*, Nauka, Moscow (1973), (with A.A. Nudel'man); English transl.: Transl. Math. Monographs, vol. **50**, Amer. Math. Soc., Providence, R.I. (1977).

[241] *On some new problems for functions of Hardy class and continual families of functions with double orthogonality*, Dokl. Akad. Nauk SSSR **209** (1967), 537–540 (with P.Ya. Nudel'man); English transl.: Soviet Math. Dokl. **14** (1973), 435–439.

[242] *On a conjecture of A.M. Lyapunov*, Funkts. Anal. Prilozhen. **7**, no. 3 (1973), 45–54; English transl.: Function. Anal. Appl. **7**, no. 3 (1973), 204–212.

[243] *Uber die Q-Funktionen eines* Π*-Hermiteschen Operators im Raume* Π_k, Acta Sci. Math. Szeged **34** (1973), 191–230 (with H. Langer).

[244] *On some spatial isoperimetric problems*, Kvant, no. 2 (1973), 22–26 (with A.A. Nudel'man).

[245] *A ballistic problem in cosmos*, Kvant, no. 5 (1973), 2–6 (with K.R. Kovalenko).

[246] *Approximation of functions in* $L_2(w_1, w_2)$ *by transfer functions of linear systems with minimal energy*, Problemy Pered. Inform. **11**, no. 2 (1975), 37–60 (with P.Ya. Nudel'man); English transl.: Problems of Inform. Transm. (1975), 124–142.

[247] *On the indefinite power moment problem*, Dokl. Akad. Nauk SSSR **226** (1976), 261–264 (with H. Langer); English transl.: Soviet Math. Dokl. **17** (1976), 90–93.

[248] *On generalized resolvents and resolvent matrices of positive Hermitian operators*, Dokl. Akad. Nauk SSSR **231** (1976), 1063–1066 (with I.E. Ovcharenko); English transl.: Soviet Math. Dokl. **17** (1976), 1705–1709.

[249] *On the theory of generalized resolvents of non-densely defined Hermitian contractions*, Dokl. Akad. Nauk Ukr. SSR, ser. A, no. 10 (1976), 881–884 (with I.E. Ovcharenko).

[250] *On nonlinear integral equations which play a role in the theory of Wiener-Hopf equations. 1*, Mat. Issled., no. 42 (1976), 47–90 and 2, Mat. Issled., no. 45 (1977), 67–92 ; English transl.: Topics in Differential and Integral Equations and Operator Theory, Operator Theory: Advances and Appl., vol. **7**, Birkhauser, Basel (1983), 173–242.

[251] *On Q-functions and SC-resolvents of non-densely defined Hermitian contractions*, Sib. Mat. Zh. **18**, no. 5 (1977), 1032–1056 (with I.E. Ovcharenko); English transl.: Sib. Mat. J. **18**, no. 5 (1977), 728–746.

[252] *Remark on a theorem of V.A. Yakubovic: "Frequency theorem for the case when...",* Sib. Mat. Zh. **18**, no. 6 (1977), 1411–1412; English transl.: Sib. Mat. J. **18**, no. 6 (1977), 1001–1002.

[253] *On some new results in the theory of factorization of matrix-functions on the unit circle,* Dokl. Akad. Nauk SSSR **234** (1977), 287–290 (with I.M Spitkovskii); English transl.: Soviet Math. Dokl. **18** (1977), 641–645

[254] *On an interpolation problem connected with the Stieltjes moment problem,* Dokl. Akad. Nauk Ukr. SSR, Ser. A., no. 12 (1977), 1068–1072 (with A.A. Nudel'man).

[255] *Uber einige Fortsetzungsprobleme, Die eng mit Hermiteschen Operatoren in Raume Π_k zusammenhangen. I, Einige Funktionneklassen und ihre Darstellungen,* Math. Nachricht. **77** (1977), 187–236 (with H. Langer).

[256] *Boris Yakovlevich Levin: To 70-th birthday,* Usp. Mat. Nauk **32**, no. 5 (1977), 210–213 (with N.I. Akhiezer, N.V. Efimov, M.A. Lavrent'ev et al.).

[257] *Inverse problems for Q-functions and resolvent matrices of positive Hermitian operators,* Dokl. Akad. Nauk SSSR **242** (1978), 521–524 (with I.E. Ovcharenko).

[258] *Approximation of functions of the class L^∞ by functions of the class $H^\infty + C$,* Research on linear operators and function theory: 99 unsolved problems on linear and complex analysis, Proc. Sci. Sem. LOMI, Leningrad **81** (1978), 190–192 (with V.M. Adamyan and D.Z. Arov).

[259] *On measurable Hermitian-positive functions,* Mat. Zametki **23**, no. 1 (1978), 79–89; English transl.: Math. Notes **23**, no. 1 (1978), 45–50.

[260] *On the factorization of α-sectorial matrix-functions on the unit circle,* Mat. Issled., no. 47 (1978), 41–63 (with I.M. Spitkovskii).

[261] *Uber einige Fortsetzungsproblems, die engt mit Hermiteschen Operatoren in Raume $\Pi_k$zusammenhagen. II, Verallgemeinerte Resolventen, u-Resolventen und ganze Operatoren,* J. Functional Anal. **30** (1978), 390–447 (with H. Langer).

[262] *Introduction to the Theory of Linear Non-Self-Adjoint Operators in Hilbert Space,* 2nd edition, Amer. Math. Soc., Providence, R.I. (1978) (with I.C. Gohberg).

[263] *On some extension problems which are closely related to the theory of Hermitian operators in a space Π_k. III. Indefinite analogues of the Hamburger and Stieltjes moment problems,* Part 1, Beitrage Anal., no. 14 (1979), 25–40 (with H. langer).

[264] *Some propositions on the Fourier-Plancherel and Paley-Winener type obtained by methods in the theory spectral of functions,* Funkts. Anal. Prilozhen. **13**, no. 4 (1979), 79–80 (with A.A. Nudel'man); English transl.: Function. Anal. Appl. **13**, no. 4 (1979), 301–303.

[265] *On direct and inverse problem for the boundary-dissipation frequencies of a non-uniform string,* Dokl. Akad. Nauk SSSR **247** (1979), 1046–1049 (with A.A. Nudel'man); English transl.: Soviet Math. Dokl. **20** (1979), 838–841.

[266] *On some extension problems which are closely related to the theory of Hermitian operators in a space Π_k. III. Indefinite analogues of the Hamburger and Stieltjes moment problems,* Part 2, Beitrage Anal. no. 15 (1980), 27–45 (with H. Langer).

[267] *Stable plus-operators in 3-spaces, Linear Operators,* Mat. Issled., no. 54 (1980), 67–83 (with Yu.L. Shmul'yan).

[268] *Mark Aronovich Naimark: the obituary,* Usp. Mat. Nauk **35**, no. 4 (1980), 135–140 (with I.M. Gel'fand, M.I. Graev, D.P. Zhelobenko et al.).

[269] *Some propositions on analytic matrix-functions related to the theory of operators in a space Π_k,* Acta. Sci. Math., Szeged **43** (1981), 181–205 (with H. Langer).

[270] *Continual analogues of orthogonal polynomials with respect to an indefinite weight on the unit circle and extension problems related to them*, Dokl. Akad. Nauk, SSSR **258** (1981), 537–542 (with H. Langer); English transl.: Soviet Math. Dokl. **23** (1981), 553–557.

[271] *On representation of entire functions positive on the real axis, or on a semi-axis, or outside a finite interval*, Mat. Issled., no. 61 (1981), 40–59 (with A.A Nudel'man);. English transl.: Amer. Math. Soc. Transl. (2) **127** (1986), 17–32.

[272] *The problem of search of minimum of entropy in undetermined extension problems*, Funkts. Anal. Prilozhen. **15**, no. 2 (1981), 61–64 (with D.Z. Arov): English transl.: Function. Anal. **15**, no. 2 (1981), 123–126.

[273] *Spectral shift functions that arise in perturbations of a positive operator*, J. Operator Theory **6**, no. 1 (1981), 155–191 (with V.A. Yavryan).

[274] *On nonstandard variational problems of determination of optimal form of a vessel*, Proc. Anniversary Scientific Session, Bulgarian Institute of Ship Hydrodynamics, Sofia (1981), pp. 77–1 77–5 (with V.G. Sizov).

[275] *Remarkable limits generated by classical means*, Kvant, no. 9 (1981), 13–15 (with A.A. Nudel'man).

[276] *Naum Il'ich Akhiezer: the obituary*, Usp. Mat. Nauk **36**, no. 4 (1981), 183–184 (with A.N. Kolmogorov, B.Ya. Levin, B.M. Levitan et al.).

[277] *Feliks Aleksandrovich Berezin /1931–1980/: the obituary*, Usp. Mat. Nauk **36**, no. 4 (1981), 183–184 (with N.N. Bogolyubov, I.M. Gel'fand, B.R. Dobrushin et al.).

[278] *Introduction to the Spectral Theory of Operators in Spaces with Indefinite Metric*, Math. Research, vol. **9**, Akademie -Verlag, Berlin (1982) (with I.S. Iokhvidov and H. Langer).

[279] *On the theory of inverse problems for canonical differential equations*, Dokl. Akad. Naur Ukr. SSR, Ser. A, no. 2 (1982), 14–18 (with I.E. Ovcharenko).

[280] *Wiener-Hopf equations whose kernels admit an integral representation in terms of exponentials*, Izv. Akad. Nauk Arm. SSR **17**, no. 4 (1982), 307–327 and **17**, no. 5 (1982), 335–375 (with Yu. L. Shmul'yan); English transl.: Soviet J. Contemp. Math. Anal. **17**, no. 4 (1982), 71–93 and **17**, no. 5 (1982), 1–42.

[281] *On the calculation of entropy functionals and of their minima in undetermined continuation problems*, Acta Sci. Math. Szeged **45** (1983), 33–50 (with D.Z. Arov).

[282] *Some generalizations of Szego's first limit theorem*, Anal. Mat. **9**, no. 1 (1983), 23–41 (with I.M. Spitkovskii).

[283] *The Borsuk-Ulam theorem*, Kvant, no. 8 (1983), 20–25 (with A.A. Nudel'man).

[284] *Introduction to the Theory of Linear Non-Self-Adjoint Operators in Hilbert Space*, 3d edition, Amer. Math. Soc., Providence, R.I. (1983) (with I.C. Gohberg).

[285] *Topics in Differential and Integral Equations and Operator Theory*, Operator Theory: Adv. Appl., vol. **7**, Birkhauser Verlag, Basel (1983).

[286] *Integral Hankel operators and related continuation problems*, Izv. Akad. Nauk Arm. SSR, Mat. **19**, no. 4 (1984), 311–332 and **19**, no. 5 (1984), 339–360 (with F.E. Melik-Adamyan): English transl.: Sovet J.Contemp. Math. Anal. **19**, no. 5 (1984), 1–22.

[287] *On the special representation of a polynomial positive on a system of closed intervals*, Preprint 28–84 Fiz. Tekn. Inst. Nizkikh Temper. Akad. Nauk Ukr. SSR, Khar'kov (1984) (with B.Ya. Levin and A.A. Nudelman).

[288] *Some function theoretic problems connected with the theory of spectral measures of isometric operators*, Lecture Notes., Math., vol. **1043** (1984), 160–163 (with V.M. Adamyan and D.Z. Arov).

[289] *Approximation of bounded functions by elements of* $H^{\infty} + C$, Lecture Notes Math., vol. **1043** (1984), 254–256 (with V.M. Adamyan and D.Z. Arov).

[290] *Some problems connected with the Szego limit theorems*, Lecture Notes Math., vol. **1043** (1984), 285–288 (with I.M. Spitkovskii).

[291] *Banach algebras of functions generated by the set of all almost periodic polynomials whose exponents belong to a given interval*, Lecture Notes Math., vol. **1043** (1984), 632–635.

[292] *On some continuation problems which are closely related to the theory of operators in spaces* Π_k. *Part 4. Continuous analogues of orthogonal polynomials on the unit circle with respect to an indefinite weight and related continuation problems for some classes of functions*, J. Operator Theory **13** (1985), 299–417 (with H. Langer).

[293] *A diophantine equation of academician A.A. Markov*, Kvant, no. 4 (1985), 13–16 and no. 5 (1985), 59.

[294] *Iosif Semenovich Iokhvidov /1919–1984/: the obituary*, Usp. Mat. Nauk **40**, no. 6 (1985), 131–132 (with Yu.M. Berezanskii, S.L. Sobolev and E.M. Semenov).

[295] *Matrix-continual analogues of the Schur and Caratheodory-Toeplitz problems*, Izv. Akad. Nauk Arm. SSR, Ser. mat. **21**, no. 2 (1986), 107–141 (with F.E. Melik-Adamyan); English Transl.: Soviet J. Contemp. Math. Anal. **21**, no. 2 (1986).

[296] *A supplement to the paper of M.G. Krein and Yu. L. Shmul'yan "Wiener-Hopf equations kernels admit an integral representation in terms of exponentials,"* Izv. Akad. Nauk. Arm. SSR, Mat. **21** (1986) no. 3 (with Yu.L. Shmul'yan).

[297] *On pertutbation determinants and a trace formula for certain classes of operator pairs*, J. Operator Theory **17**, no. 1 (1987), 129–188.

[298] *Boris Yakovlevich Levin: To 80-th birthday*, Usp. Mat. Nauk **42**, no. 4 (1987), 207–210 (with I.M. Gel'fand, V.A. Marchenko, N.K. Nikol'skii et al.).

[299] *Yurii L'vovich Daletskii: To 60-th birthday*, Usp. Mat. Nauk **42**, no. 4 (1987), 213–214 (with Yu.M. Berezanskii, I.M. Gel'fand S.G. Krein et al.).

[300] *Introduction to the Theory of Linear Non-Self-Adjoint Operators in Hilbert Space*, 4th edition, Amer. Math. Soc., Providence, R.I. (1988) (with I.C. Gohberg).

[301] *On some spectral properties of nonhomogeneous string with dissipative boundary conditions*, J. Operator Theory **22** (1989), 369–395 (with A.A. Nudel'man).

[302] *Reminiscences, Operator Theory: Advances and Applications* **40** (1989), 60–61 (The Gohberg Anniversary Collection).

[303] *On special representations of polynomials that are positive on a system of closed intervals and some applications*, Funct. Anal. Optimization and Math. Economics, N.-Y., Oxford: Oxford Univ. Press (1990) (with B.Ya. Levin and A.A. Nudel'man).

Operator Theory:
Advances and Applications, Vol. 118

M.G. Krein's Contributions to Prediction Theory

Harry Dym

Prediction is difficult, especially the future.
Niels Bohr

The paper presents a brief survey of M.G. Krein's contributions to prediction theory and some related literature.

1 Introduction

This paper is an expanded version of a lecture which was delivered in Odessa in August, 1997, at a conference in honor of M.G. Krein, on the ninetieth anniversary year of his birth. It is my pleasure once again to thank the organizers of this conference for inviting me to present that talk. I never had the privilege of meeting M.G. Krein, so I cannot present the inside story of his work on prediction theory. I shall instead try to (1) give some indication of the nature of the problems under consideration and (2) convey some idea of how the accidental discovery of one short Doklady note by M.G. Krein influenced the book [DMcK3] on prediction that was written with Henry McKean.

Let $-\infty < x_t < \infty$, be a real, stationary, Gaussian process with zero mean and finite variance. Assume further that x_t is mean square continuous and let

$$R(t) = E\{x_{t+s}\bar{x}_s\} = E\{x_t\bar{x}_0\},$$

where E denotes the expectation of the indicated term. It is then readily checked that:

(1) $R(t)$ is continuous on $(-\infty, \infty)$.

(2) $R(t) = R(-t) = \overline{R(t)}$.

(3) $R(t)$ defines a positive semidefinite kernel in the sense that

$$\begin{aligned}\sum_{i,j=1}^{n} c_i R(t_i - t_j)\bar{c}_j &= \sum_{i,j=1}^{n} E\{c_i x_{t_i}\bar{x}_{t_j}\bar{c}_j\} \\ &= E\left\{\left|\sum_{i=1}^{n} c_i x_{t_i}\right|^2\right\} \geq 0\end{aligned}$$

for every choice of complex coefficients $c_1, \ldots, c_n$ and real numbers $t_1, \ldots, t_n$.

(4) $|R(t)| \leq R(0)$.

Therefore, by the Bochner-Khintchine theorem,

$$R(t) = \int_{-\infty}^{\infty} e^{i\lambda t} d\Delta(\lambda)$$

for some odd nondecreasing function $\Delta(\lambda)$ with

$$\Delta(\infty) = R(0) < \infty.$$

Hence,

$$\begin{aligned} E\{x_t \bar{x}_s\} &= R(t-s) = \int_{-\infty}^{\infty} e^{i\lambda(t-s)} d\Delta(\lambda) \\ &= \langle e_t, e_s \rangle_{L^2(d\Delta)}, \end{aligned}$$

where

$$e_t(\lambda) = e^{i\lambda t}.$$

Thus problems of mean square approximation in the stochastic sense translate to problems of trigonometric approximation in $L^2(d\Delta)$. In particular,

$$E\left\{\left|x_t - \sum_{j=1}^{n} c_j x_{t_j}\right|^2\right\} = \int_{-\infty}^{\infty} \left|e_t - \sum_{j=1}^{n} c_j e_{t_j}\right|^2 d\Delta,$$

and hence

$$\inf_{t_j \in I} E\left\{\left|x_t - \sum c_j x_{t_j}\right|^2\right\} = \inf_{t_j \in I} \int_{-\infty}^{\infty} \left|e_t - \sum c_j e_{t_j}\right|^2 d\Delta,$$

where the infimum is taken over all finite linear combinations.

There are three basic cases of interest:

(1) $t > 0$ and $I = (-\infty, 0]$.

(2) $t > T > 0$ and $I = [-T, T]$.

(3) $t \in (-T, T)$ and $I = (-\infty, -T] \cup [T, \infty)$.

In this paper we shall focus mainly on the second case: prediction of the future from a finite segment of the past. To warm up, however, we shall begin with a brief discussion of the first case in the next section. The third case, the case of interpolation, is treated in the book [DMcK3], which also contains much more information than will be presented here on the other two cases.

In later sections it will prove convenient to relax the assumptions that $\Delta(\lambda)$ is bounded and to work with odd nondecreasing functions $\Delta(\lambda)$ on $\mathbb{R}$ which are subject to the restriction

$$\int_{-\infty}^{\infty} \frac{d\Delta(\lambda)}{1+\lambda^2} < \infty \tag{1.1}$$

and the following subspaces of

$$\begin{aligned} Z(\Delta) &= L^2(\mathbb{R}, d\Delta): \\ Z^{ab}(\Delta) &= \operatorname{cls}\left\{\int_{t_1}^{t_2} e_s ds : a < t_1 < t_2 < b\right\} \text{ in } Z(\Delta), \\ Z^T(\Delta) &= Z^{-T,T}(\Delta) \text{ and } Z^{T+}(\Delta) = \bigcap_{\varepsilon>0} Z^{T+\varepsilon}(\Delta), \end{aligned}$$

in which cls = closed linear span.

It is readily checked that

$$\int_{-\infty}^{\infty} \left| \int_{t_1}^{t_2} e_s ds \right|^2 d\Delta \longrightarrow 0 \text{ as } t_2 - t_1 \longrightarrow 0$$

and hence that the space $Z^{ab}(\Delta)$ is not changed by allowing $t_1 = a$ if a is finite and/or $t_2 = b$ if b is finite. Moreover, if $\Delta(\lambda)$ is bounded, then

$$Z^{ab}(\Delta) = \operatorname{cls}\{e_t : a < t < b\} \text{ in } Z(\Delta),$$

and once again the space is not changed by allowing $t = a$ if a is finite and/or $t = b$ if b is finite.

2 Prediction Using the Full Infinite Past

Let $\Delta(\lambda)$ be absolutely continuous with density $\Delta'(\lambda)$, which is even, nonnegative and summable. Let us suppose further that the Szegö condition

$$\int_{-\infty}^{\infty} \frac{\log \Delta'(\lambda)}{1+\lambda^2} d\lambda > -\infty \tag{2.1}$$

is fulfilled. Then $\Delta'(\lambda)$ admits a factorization

$$\Delta'(\lambda) = |h(\lambda)|^2,$$

where $h(\lambda)$ is an outer function in the Hardy space H^2 with respect to the upper half plane. Consequently, if $t > 0 \geq t_j$, then

$$\begin{aligned} &\int_{-\infty}^{\infty} \left| e_t(\lambda) - \sum c_j e_{t_j}(\lambda) \right|^2 \Delta'(\lambda) d\lambda \\ &= \int_{-\infty}^{\infty} \left| e_{-t}(\lambda) - \sum \bar{c}_j e_{-t_j}(\lambda) \right|^2 |h(\lambda)|^2 d\lambda \\ &= \int_{-\infty}^{\infty} \left| e_{-t}h - \sum \bar{c}_j e_{-t_j} h \right|^2 d\lambda. \end{aligned}$$

Thus, as h is outer, the infimum over all such finite sums with $t_j \leq 0$ is equal to

$$\begin{aligned}
&\inf_{f\in H^2}\int_{-\infty}^{\infty}|e_{-t}h-f|^2d\lambda\\
&=\inf_{f\in H^2}\left\{2\pi\int_{-\infty}^{0}|(e_{-t}h)^\vee(s)|^2ds+2\pi\int_{0}^{\infty}|(e_{-t}h)^\vee(s)-f^\vee(s)|^2ds\right\}\\
&=2\pi\int_0^t|h^\vee(s)|^2ds,
\end{aligned}$$

where

$$f^\vee(s)=\frac{1}{2\pi}\int_{-\infty}^{\infty}e^{-i\lambda s}f(\lambda)d\lambda$$

designates the inverse Fourier transform of $f(\lambda)$. This last formula gives a neat and simple expression for the prediction error (based on the full infinite past) when the Szegö condition is met. The error is zero, when the Szegö condition fails. The formula is given by Krein in [Kr2] without proof "for want of space".

Discrete Analogues

We remark that much the same argument shows that if $\Delta(\theta)$ is an absolutely continuous monotone nondecreasing function on the interval $[0, 2\pi]$ with density

$$\Delta'(\theta)=\left|\sum_{j=0}^{\infty}h_je^{ij\theta}\right|^2,$$

where $h(\zeta)=\sum_{j=0}^{\infty}h_j\zeta^j$ is an outer function in the Hardy space H^2 with respect to the disc, then for every integer $\nu \geq 0$,

$$\inf_{c_0,c_1,\dots}\int_0^{2\pi}\left|e^{-i\nu\theta}-\sum c_je^{ij\theta}\right|^2|h(e^{i\theta})|^2d\theta=\sum_{j=0}^{\nu-1}|h_j|^2,$$

In the special case that $\nu = 1$, the "error" can be expressed directly in terms of Δ' with the help of the supplementary formula

$$|h_0|^2=\exp\left\{\frac{1}{2\pi}\int_0^{2\pi}\log|h(e^{i\theta})|^2d\theta\right\},$$

which is characteristic of outer functions. The subsequent formula

$$\begin{aligned}
&\inf_{c_0,c_1,\dots}\left\{\int_0^{2\pi}\left|e^{-i\theta}-\sum c_je^{ij\theta}\right|^2\Delta'(\theta)d\theta\right\}\\
&=\exp\left\{\frac{1}{2\pi}\int_0^{2\pi}\log\Delta'(\theta)d\theta\right\}
\end{aligned}\tag{2.2}$$

was obtained by Szegö [Sz] in 1920 (by different methods) as a byproduct of his work on polynomial approximation. Kolmogorov [Ko1]–[Ko3] generalized Szegö's theorem to the case of arbitrary monotone Δ in his work on prediction theory for stationary stochastic sequences. Formula (2.2) displays the fact that

$$\text{span}\{e^{ij\theta} : j = 0, 1, \ldots\} \quad \text{in} \quad L^p(d\Delta)$$

is equal to the full space if and only if

$$\int_0^{2\pi} \log \Delta'(\theta) d\theta = -\infty,$$

for the case $p = 2$. Krein [Kr1] showed that this conclusion is valid for all finite $p \geq 1$. Norbert Wiener [Wi] obtained Kolmogorov's results independently for the case of absolutely continuous Δ and also solved the corresponding continuous parameter prediction problem which was discussed earlier. Wiener's work first appeared as a classified report in the Radiation Laboratories in 1942. This report was affectionately known as the "Yellow Peril", partially because the cover was yellow and perhaps even more so because the material was considered very difficult. [The Hardy space proof presented above was not known at that time, at least in the West; this was before Beurling's famous 1949 Acta paper [Be], though after Smirnov [Sm].]

Synchronicity

The fact that Wiener got Norman Levinson interested in this class of problems will be seen to be a significant event in the development of at least part of this story. Indeed, some twenty years later Levinson combined forces with Henry P. McKean, a former student of William Feller, to write the paper [LMcK]. That paper was the platform on which the initial research which was to ultimately lead to the book [DMcK3] rested.

3 Prediction Using a Finite Segment of the Past

Let us suppose first that $\Delta(\lambda)$ is bounded on $\mathbb{R}$. Then the prediction problem based on a finite segment of the past is equivalent to finding the orthogonal projection of $e_t(\lambda)$ for $t > 0$ onto the span$\{e_t : -2T \leq t \leq 0\}$ in $Z(\Delta)$. Since

$$\int_{-\infty}^{\infty} \left| e_t(\lambda) - \sum c_j e_{t_j}(\lambda) \right|^2 d\Delta(\lambda)$$
$$= \int_{-\infty}^{\infty} \left| e_{t+T}(\lambda) - \sum c_j e_{t_j+T}(\lambda) \right|^2 d\Delta(\lambda),$$

this is equivalent to finding the orthogonal projection of $e_t(\lambda)$ for $t > T$ onto $Z^T(\Delta)$. It turns out that the auxiliary subspace $Z^{T+}(\Delta)$ plays a significant role in the analysis of this problem. In particular, the inclusion

$$Z^T(\Delta) \subseteq Z^{T+}(\Delta)$$

may be proper and a major difficulty is to describe the gap in some reasonable way. This problem does not occur in the prediction based on the infinite past because

$$\bigcap_{\varepsilon>0} Z^{-\infty\varepsilon}(\Delta) = Z^{-\infty 0}(\Delta).$$

Subspaces of $Z(\Delta)$

There are four descriptions of subspaces of $Z(\Delta)$ which come into play:

(1) $Z^{ab}(\Delta)$, i.e., spans of exponentials (or integrals of exponentials).

(2) $I^T(\Delta)$ = the set of entire functions of exponential type $\leq T$ whose restriction to $\mathbb{R}$ belongs to $Z(\Delta)$.

$I^{T-}(\Delta)$ = the set of entire functions of exponential type $< T$ whose restriction to $\mathbb{R}$ belongs to $Z(\Delta)$.

(3) de Branges spaces of entire functions.

(4) Krein spaces, i.e., images of generalized cosine, sine transforms, adapted to $Z(\Delta)$.

The existence of a connection between spans of exponentials and entire functions of exponential type is not so surprising if one keeps the Paley-Wiener theorem in mind. In the present more general setting, the following alternative, which is due to Krein [Kr6], holds:

Theorem 3.1 *Either $Z^{T+}(\Delta) = Z(\Delta)$, or $Z^{T+}(\Delta) = I^T(\Delta)$ (i.e., the last equality prevails if $Z^{T+}(\Delta)$ is a proper closed subspace of $Z(\Delta)$).*

The remaining two descriptions of subspaces of $Z(\Delta)$ require additional discussion, which will be sketched below.

De Branges Spaces

Let $E(\lambda)$ be an entire function such that

(3.1) $$|E(\lambda)| > |E(\overline{\lambda})| \quad \text{for} \quad \lambda \in \mathbb{C}_+.$$

Then $E^{-1}E^{\#}$ is an inner function (with respect to $\mathbb{C}_+$). Let $\mathcal{B}(E)$ denote the set of entire functions f such that

(3.2) $$E^{-1}f \in H_2 \ominus (E^{-1}E^{\#})H_2$$

endowed with the inner product

$$(3.3) \qquad \langle f, g\rangle_{\mathcal{B}(E)} = \int_{-\infty}^{\infty} \overline{(E^{-1}g)(\lambda)}(E^{-1}f)(\lambda)d\lambda.$$

Then $\mathcal{B}(E)$ is a reproducing kernel Hilbert space with reproducing kernel

$$(3.4) \qquad \Lambda_\omega(\lambda) = \frac{E(\lambda)E(\omega)^* - E^\#(\lambda)E^\#(\omega)^*}{-2\pi i(\lambda - \bar{\omega})}.$$

The last statement means that

$$(3.5) \qquad \begin{aligned} &(1)\ \ \Lambda_\omega \in \mathcal{B}(E) \quad \text{and} \\ &(2)\ \ \langle f, \Lambda_\omega\rangle_{\mathcal{B}(E)} = f(\omega) \end{aligned}$$

for every choice of $\omega \in \mathbb{C}$ and $f \in \mathcal{B}(E)$.

The proof in this direction is basically an exercise in the use of Cauchy's theorem for H_2 functions.

There is a converse statement which lies deeper. But first, let us note that $\mathcal{B}(E)$ is closed under the isometries

$$(3.6) \qquad f \longrightarrow \frac{\lambda - \bar{\omega}}{\lambda - \omega} f(\lambda) \quad \text{when} \quad f(\omega) = 0$$

and

$$(3.7) \qquad f \longrightarrow f^\#(\lambda) = f(\bar{\lambda})^*.$$

Theorem 3.2 *If $\mathcal{H}$ is a reproducing kernel Hilbert space of entire functions with reproducing kernel $\Lambda_\omega(\lambda)$ such that $\mathcal{H}$ is closed under the two isometries* (3.6) *and* (3.7)*, then $\mathcal{H}$ is equal to a de Branges space $\mathcal{B}(E)$, where $E(\lambda)$ may be chosen equal to*

$$(3.8) \qquad E(\lambda) = \sqrt{\pi}(1 - i\lambda)\Lambda_i(\lambda)\{\Lambda_i(i)\}^{-\frac{1}{2}}.$$

Theorem 3.2 is due to de Branges; for proofs see pp. 57–59 of [dB6] and/or pp. 224–226 of [DMcK3].

4 Strings

A string is either a pair $\{\ell, m(x)\}$ or a triple $\{\ell, m(x), k\}$, where:

$0 < \ell \le \infty$ designates the length of the string,

$m(x)$, the mass distribution function of the string, is a nondecreasing right continuous function on $[0, \ell)$ and

$0 \le k \le \infty$ is a constant (that determines right hand boundary conditions) which only comes into play if $\ell + m(\ell-) < \infty$.

It is assumed throughout that:

$m(0-) = 0,$

$0 < m(x) < \infty$ for $0 < x < \ell,$

ℓ is a growth point of m (as is 0 by the preceding line) and

$m[\ell] = 0$, if $\ell = \infty$ (i.e., there is no point mass at ℓ if $\ell = \infty$).

The string is said to be short if $\ell + m(\ell-) < \infty$; it is said to be long if $\ell + m(\ell-) = \infty$.

With each string we associate a pair of functions

$$\begin{aligned} A(x,\lambda) &= 1 - \lambda^2 \int_0^x \int_{0-}^u A(s,\lambda)dm(s)du \\ B(x,\lambda) &= \lambda \int_{0-}^x A(s,\lambda)dm(s). \end{aligned}$$

The function $A(x,\lambda)$ is the analogue of $\cos\lambda x$ and $B(x,\lambda)$ is the analogue of $\sin\lambda x$. Indeed, if $m(x) = x$ for $x \in [0,\ell)$, then $A(x,\lambda) = \cos\lambda x$ and $B(x,\lambda) = \sin\lambda x$ for all such points.

For any string $\{\ell, m(x), k\}$ and any point $x \in [0,\ell)$, $A(x,\lambda)$ and $B(x,\lambda)$ are entire functions of exponential type

$$\int_0^x \sqrt{m'(s)}ds$$

in the variable λ. Moreover,

$$\begin{aligned} &A(x,0) = 1, && B(x,0) = 0, \\ &A(x,-\lambda) = A(x,\lambda), && B(x,-\lambda) = -B(x,\lambda), \\ &A^{\#}(x,\lambda) = A(x,\lambda), && B^{\#}(x,\lambda) = B(x,\lambda) \end{aligned}$$

and the pair of equations can be reexpressed as the system

$$\begin{aligned} A(x,\lambda) &= 1 - \lambda \int_0^x B(u,\lambda)du \\ B(x,\lambda) &= \lambda \int_{0-}^x A(s,\lambda)dm(s). \end{aligned}$$

Let

$$E(x,\lambda) = A(x,\lambda) - iB(x,\lambda).$$

Then

$$E^{\#}(x,\lambda) = A(x,\lambda) + iB(x,\lambda) = E(x,-\lambda)$$

for every point $\lambda \in \mathbb{C}$ and, for real λ,

$$|E(x,\lambda)|^2 = A(x,\lambda)^2 + B(x,\lambda)^2 = |E^{\#}(x,\lambda)|^2 \quad (\lambda \in \mathbb{R}).$$

Moreover,

$$\begin{aligned}\frac{1}{\pi}\int_{0-}^{x} A(s,\lambda)A(s,\omega)^{*}dm(s) + \frac{1}{\pi}\int_{0}^{x} B(s,\lambda)B(s,\omega)^{*}ds \\ = \frac{A(x,\lambda)B(x,\omega)^{*} - B(x,\lambda)A(x,\omega)^{*}}{\pi(\lambda-\bar{\omega})} \\ = \frac{E(x,\lambda)E(x,\omega)^{*} - E^{\#}(x,\lambda)E^{\#}(x,\omega)^{*}}{-2\pi i(\lambda-\omega^{*})}.\end{aligned}$$

Therefore,

$$|E(x,\omega)| > |E(x,\bar{\omega})|$$

for every point $\omega \in \mathbb{C}_{+}$. Thus there is a de Branges space $\mathcal{B}(E_x)$ associated with the string for each $x \in [0,\ell)$, with ℓ included too if $\ell + m(\ell-) < \infty$. Moreover, the de Branges spaces corresponding to growth points are nested by isometric inclusion.

5 Spectral Functions

Let $\Delta(\lambda)$ be an odd nondecreasing function on $\mathbb{R}$ which is subject to the growth constraint (1.1). Then $\Delta(\lambda)$ is said to be a spectral function for the string $\{\ell, m(x)\}$ if the A-transform

$$\hat{f}(\lambda) = \int_{0-}^{\ell} f(s)A(s,\lambda)dm(s) \tag{5.1}$$

is an isometry of $L^2([0,\ell]; dm)$ into $L^2(\mathbb{R}, d\Delta) = Z(\Delta)$ (apart from a factor of $\sqrt{\pi}$):

$$\frac{1}{\pi}\int_{-\infty}^{\infty} |\hat{f}(\lambda)|^2 d\Delta(\lambda) = \int_{0-}^{\ell} |f(s)|^2 dm(s). \tag{5.2}$$

The spectral function $\Delta(\lambda)$ is said to be a principal spectral function if the mapping is onto the space

$$Z(\Delta)_{\text{even}} = \{\text{the even functions in } Z(\Delta)\}. \tag{5.3}$$

For example, the function $\Delta(\lambda) = \lambda$ is a spectral function for the string with mass distribution $m(x) = x$ on the interval $0 \le x \le \ell$ for every choice of $\ell > 0$. However, it is a principal spectral function only if $\ell = \infty$.

Every string $\{\ell, m(x)\}$ has at least one spectral function.

If $\ell + m(\ell-) = \infty$, then the string has exactly one spectral function and it is a principal spectral function.

If $\ell + m(\ell-) < \infty$, there are uncountably many spectral functions and a one-parameter family $\Delta_k(\lambda)$, $0 \le k \le \infty$ of principal spectral functions corresponding to the right hand boundary condition

$$\begin{aligned} kA^{+}(\ell,\lambda) + A(\ell,\lambda) &= 0 \quad \text{for} \quad 0 \le k < \infty \\ A^{+}(\ell,\lambda) &= 0 \quad \text{for} \quad k = \infty. \end{aligned}$$

There is also a B-transform

$$f^{\Delta}(\lambda)=\int_0^{\ell} f(s)B(s,\lambda)ds \tag{5.4}$$

which is defined for the class of functions $f \in L_2([0,\ell],dx)$ which are constant on those intervals across which m (or equivalently B) is constant. Moreover,

$$\frac{1}{\pi}\int_{-\infty}^{\infty}|f^{\Delta}(\lambda)|^2 d\Delta(\lambda)=\int_0^{\ell}|f(s)|^2 ds, \tag{5.5}$$

i.e., this mapping is an isometry (apart from a factor of $\sqrt{\pi}$) into the space

$$Z(\Delta)_{\text{odd}}=\{\text{the odd functions in } Z(\Delta)\}. \tag{5.6}$$

The main facts are that every odd nondecreasing function $\Delta(\lambda)$ on $\mathbb{R}$ that is subject to (1.1) is the principal spectral function of exactly one string $\{\ell, m(x), k\}$, where $0\le k\le\infty$ and is superfluous if $\ell+m(\ell-)=\infty$. Furthermore, if $Z^{T+}(\Delta)$ is a proper closed subspace of $Z(\Delta)$, then

(1) $T<\int_0^{\ell}\sqrt{m'(s)}ds,$

$$\text{(2) } Z^T(\Delta)_{\text{even}}=\left\{\int_{0-}^{x_1(T)-} f(s)A(s,\lambda)dm(s) : f\in L_2(dm)\right\}, \tag{5.7}$$

$$\text{(3) } Z^{T+}(\Delta)_{\text{even}}=\left\{\int_{0-}^{x_2(T)+} f(s)A(s,\lambda)dm(s) : f\in L_2(dm)\right\}, \tag{5.8}$$

where

$$x_1(T) = \min\left\{x : \int_0^x \sqrt{m'(s)}ds=T\right\} \tag{5.9}$$

and

$$x_2(T) = \max\left\{x : \int_0^x \sqrt{m'(s)}ds=T\right\}. \tag{5.10}$$

Similar conclusions hold for the odd functions:

$$\text{(4) } Z^T(\Delta)_{\text{odd}}=\left\{\int_0^{x_1(T)} g(s)B(s,\lambda)ds : g\in L_2(ds)\right\} \tag{5.11}$$

and

$$\text{(5) } Z^{T+}(\Delta)_{\text{odd}}=\left\{\int_0^{x_2(T)} g(s)B(s,\lambda)ds : g\in L_2(ds)\right\}, \tag{5.12}$$

with the understanding that in (5.11) and (5.12), g is constant on mass free intervals. Moreover, all these transforms can be inverted: there are inverse transforms. You can see what they ought to be by substituting the formula

$$\Lambda_\omega^T(\lambda) = \frac{1}{\pi}\int_0^{x_1(T)-} A(s,\lambda)A(s,\omega)^* dm(s) + \frac{1}{\pi}\int_0^{x_1(T)} B(s,\lambda)B(s,\omega)^* ds$$

into

$$\begin{aligned} f(\omega) &= \langle f, \Lambda_\omega^T\rangle_\Delta \\ &= \int_{-\infty}^{\infty} f(\lambda)\Lambda_\omega^T(\lambda)^* d\Delta(\lambda) \end{aligned}$$

and interchanging the orders of integration.

Notice that formulas (5.7), (5.8), (5.11) and (5.12) yield a description of the gap between $Z^{T+}(\Delta)$ and $Z^T(\Delta)$ in terms of the indicated transforms. The images of these transforms are the Krein spaces referred to in Section 3.

6 Strings and Finite Past Prediction

Let $\Delta(\lambda)$ be an odd nondecreasing function on $\mathbb{R}$ which is subject to the growth constraint (1.1). Then the finite prediction problem is equivalent to taking $f \in Z(\Delta)$ and computing

(1) the orthogonal projection $P_T f$ of f onto $Z^T(\Delta)$ and

(2) the prediction error $\|f - P_T\|_\Delta$.

The basic strategy is to identify $\Delta(\lambda)$ as the principal spectral function of a string $\{\ell, m(x), k\}$. Then

$$\begin{aligned} f(\lambda) &= \int_{0-}^{\ell} \varphi(s)A(s,\lambda)dm(s) + \int_0^{\ell+k} \psi(s)B(s,\lambda)ds, \\ (P_T f)(\lambda) &= \int_{0-}^{x_1(T)-} \varphi(s)A(s,\lambda)dm(s) + \int_0^{x_1(T)} \psi(s)B(s,\lambda)ds \end{aligned}$$

and

$$\|f - P_T f\|_\Delta^2 = \pi\int_{x_1(T)-}^{\ell} |\varphi(s)|^2 dm(s) + \pi\int_{x_1(T)}^{\ell+k} |\psi(s)|^2 ds,$$

where the first two sets of integrals converge in the mean square sense, and $\psi(s)$ is constant on mass-free intervals.

This is essentially the whole story, but effective computations are far from simple. Some examples are considered in Section 6.10 of [DMcK3]. They are based in part on a set of rules that help to solve (some cases of) the inverse spectral problem which are mostly adapted from Krein [Kr5] and are presented in Section 6.9. Another fairly general approach to solving the inverse spectral problem which has potential application to prediction theory that has not yet been exploited is developed in [DKr1] and [DKr2].

7 Some Personal Recollections

My role in the investigations which ultimately led to the writing of the book [DMcK3] with Henry P. McKean began somewhat sporadically at MIT in the academic year 1965–1966 (following my Ph.D.). At Henry's suggestion, I started to work my way through the Acta paper by Levinson-McKean [LMcK]. The objective was to simplify and extend that paper with the help of the machinery introduced by L. de Branges [dB1]–[dB5]; [dB6] appeared later. The analysis focused on the case in which $\Delta(\lambda)$ was bounded and absolutely continuous on $\mathbb{R}$ with a derivative $\Delta'(\lambda)$ which satisfied the Szegö condition (2.1). This served to insure that the spaces $Z^{T+}(\Delta)$ were proper closed subspaces of $Z(\Delta)$ for every $T > 0$ and was the case which was of most interest to probability theory.

In this instance it was shown in Levinson-McKean [LMcK] that $Z^{T+}(\Delta) = I^T(\Delta)$, and that evaluation at the point i was a bounded linear functional. This suggested that $Z^{T+}(\Delta)$ might be a reproducing kernel Hilbert space of entire functions of the kind studied by L. de Branges, as indeed turned out to be the case. It was further conjectured that, at least in the Szegö case under consideration, equality prevailed in the selfevident inclusion $Z^T(\Delta) \subseteq Z^{T+}(\Delta)$. Fortunately, we were unable to prove this. At the same time it was completely unclear what could sit between two such spaces.

In [DMcK1], which was based largely on work carried out at Rockefeller University in the academic year 1966–1967, it was shown that if $Z^T(\Delta) = Z^{T+}(\Delta)$ for all $T > 0$, then $Z^{T+}(\Delta)$ could also be described in terms of a pair of transforms analogous to the classical cosine and sine transforms. The kernels of these transforms were obtained from the solution of a second order differential equation, as in the classical case. A number of examples were worked out using a modified version of the Gelfand-Levitan method to solve the inverse spectral problem for a form of the string equation with continuous mass distribution.

All of this work was carried out in complete ignorance of M.G. Krein's fundamental paper [Kr6]. This paper was discovered by Loren Pitt who was a Postdoctoral Fellow at Rockefeller University somewhere between the writing of [DMcK1] and [DMcK2]. Unfortunately, while this paper was clearly relevant, there were no proofs either in the paper itself or in the references, to a number of earlier Doklady notes. I remember that Henry McKean asked Peter Lax whether or not M.G. Krein was reliable. The answer was of course yes, which is surely no surprise to the attendees of this conference. But, we came to this subject from Stochastic Processes, not from Operator Theory. Moreover, there was little contact with the former Soviet Union at that time.

The task of merging M.G. Krein's results with our own work was far from trivial. Fortunately, the fates were kind. It turned out that operators corresponding to strings had been studied independently by W. Feller because of their role in birth and death processes and diffusions [Fe1]–[Fe4]. Henry was a student of Feller.

The basic problem was to show that the correspondence between strings and principal spectral functions was one to one and onto. The most difficult part of this program was to show that if two strings had the same principal spectral function, then they were identical. The main tool was a theorem of de Branges which can be formulated as follows: Let $\mathcal{B}(E_1)$ and $\mathcal{B}(E_2)$ be two de Branges spaces such that

(1) $\mathcal{B}(E_j)$ sits isometrically inside $Z(\Delta)$ and

(2) $\mathcal{B}(E_j)$ is closed under the map

$$f \longrightarrow (R_i f)(\lambda) = \frac{f(\lambda) - f(i)}{\lambda - i}$$

for $j = 1, 2$. Then either $\mathcal{B}(E_1) \subseteq \mathcal{B}(E_2)$ or $\mathcal{B}(E_2) \subseteq \mathcal{B}(E_1)$.

The close to final version of the manuscript of [DMcK3] was written when Henry visited Israel for several months in the first half of 1973. It was submitted to the publisher in September 1973, some two weeks or so before the Yom-Kippur war broke out. There were some later refinements, but not many. In particular, it was basically finished before the translation of the Kac-Krein [KaKr1], [KaKr2] papers became available. As best as I can remember, they had little or no influence on the text.

8 "Predicting" the Past

M.G. Krein's work on prediction theory is but a small chapter in his overall list of accomplishments. Nevertheless it is a remarkable achievement. The formulation in [Kr6] is the essence of the whole story. It is impossible for an outsider to know precisely how Krein arrived at these results. It would seem, at least partially, to be an outgrowth of his work with Gantmacher [GKr] and a general interest in boundary value problems and inverse problems; see [Kr3]–[Kr5] and [Kr7] for hints. Probably the best written sources of information are the sections on Historical and Bibliographical Remarks on pages 96–99 of [KaKr2] and Appendix 3 of the recent book by the Gorbachuks [GoGo].

It would also seem likely that Krein's work on extension problems (particularly the extension of positive functions and helical functions) and his work on entire operators played a significant role in his thinking; see [Kr2] for connections with extensions.

References

[Ak] N.I. Akhiezer, *Theory of Approximation*, Ungar, New York, 1956.

[Be] A. Beurling, *On two problems concerning linear transformations in Hilbert space*, Acta Math. **81** (1949), 239–250.

[dB1] L. de Branges, *Some Hilbert space of entire functions*, Proc. Amer. Math. Soc. **10** (1959), 840–846.

[dB2] L. de Branges, *Some Hilbert spaces of entire functions. I*, Trans. Amer. Math. Soc. **96** (1960), 259–295.

[dB3] L. de Branges, *Some Hilbert spaces of entire functions. II*, Trans. Amer. Math. Soc. **99** (1961), 118–152.

[dB4] L. de Branges, *Some Hilbert spaces of entire functions. III*, Trans. Amer. Math. Soc. **100** (1961), 73–115.

[dB5] L. de Branges, *Some Hilbert spaces of entire functions. IV*, Trans. Amer. Math. Soc. **105** (1962), 43–83.

[dB6] L. de Branges, *Hilbert spaces of entire functions*, Englewood Cliffs, NJ, 1968.

[Do] J.L. Doob, *Stochastic Processes*, Wiley, New York, 1952.

[D] H. Dym, *An introduction to de Branges spaces of entire functions with applications to differential equations of the Sturm-Liouville type*, Adv. Math. **5** (1970), 395–471.

[DKr1] H. Dym and N. Kravitsky, *On the inverse spectral problem for the string equation*, Integral Equations Operator Theory **1** (1978), 270–277.

[DKr2] H. Dym and N. Kravitsky, *On recovering the mass distribution of a string from its spectral function*, "Topics in Functional Analysis", Adv. Math. Supp. **3** (1978), 45–90.

[DMcK1] H. Dym and H.P. McKean, *Applications of de Branges spaces of integral functions to the prediction of stationary Gaussian processes*, Illinois J. Math. **14** (1970), 299–343.

[DMcK2] H. Dym and H.P. McKean, *Extrapolation and interpolation of stationary Gaussian processes*, Ann. Math. Stat. **41** (1970), 1817–1844.

[DMcK3] H. Dym and H.P. McKean, *Gaussian Processes*, Function Theory and the Inverse Spectral Problem, Academic Press, New York, 1976.

[Fe1] W. Feller, *On second order differential operators*, Ann. of Math. (2) **61** (1955), 90–105.

[Fe2] W. Feller, Generalized second order differential operators and their lateral conditions, *Illinois J. Math.* **1** (1957), 459–504.

[Fe3] W. Feller, *On the equation of the vibrating string*, J. Math. Mech. **8** (1959), 339–348.

[Fe4] W. Feller, *The birth and death processes as diffusion processes*, J. Math. Pures Appl. (9) **38** (1959), 301–345.

[GKr] F.R. Gantmacher and M.G. Krein, *Oscillation Matrices and Kernels and Small Oscillations of Mechanical Systems*, 2nd ed., GITTL, Moscow 1950; German transl., Akademie-Verlag, Berlin, 1960.

[GoGo] M.L. Gorbachuk and V.L. Gorbachuk, *M.G. Krein's Lectures on Entire Operators*, Birkhäuser, Basel, 1997.

[KaKr1] I.S. Kac and M.G. Krein, R-functions–analytic functions mapping the upper half plane into itself, Supplement I to the Russian translation of F.V. Atkinson, *Discrete and Continuous Boundary Problems*, Mir, Moscow, 1968, pp. 629–647; English transl. Amer. Math. Soc. Transl. (2) **103** (1974), 1–18.

[KaKr2] I.S. Kac and M.G. Krein, On the spectral functions of the string, Supplement II to the Russian translation of F.V. Atkinson, *Discrete and Continuous Boundary Problems*, Mir, Moscow, 1968, pp. 648–737; English transl. Amer. Math. Soc. Transl. (2) **103** (1974), 19–102.

[Ko1] A.N. Kolmogorov, *Sur l'interpolation et extrapolation des suites stationnaires*, C. R. Acad. Sci. Paris **208** (1939), 2043–2045.

[Ko2] A.N. Kolmogorov, *Stationary sequences in Hilbert space*, Bull. Math. Univ. Moscow **2** (1941), 1–40.

[Ko3] A.N. Kolmogorov, *Interpolation and Extrapolation von stationären zuffälligen Folgen.*, Izv. Akad. Nauk SSSR **5** (1941), 3–14.

[Kr1] M.G. Krein, *On a generalization of the investigations of G.* Szegö, W.M. Smirnov and A.N. Kolmogorov, Dokl. Akad. Nauk SSSR **46** (1945), 95–98.

[Kr2] M.G. Krein, *On a problem of extrapolation of A. N. Kolmogorov*, Dokl. Akad. Nauk SSSR **46** (1945), 306–309.

[Kr3] M.G. Krein, *On inverse problems for a non-homogeneous cord*, Dokl. Akad. Nauk SSSR **82** (1952), 669–672.

[Kr4] M.G. Krein, *On a generalization of an investigation of Stieltjes*, Dokl. Akad. Nauk SSSR **87** (1952), 881–884.

[Kr5] M.G. Krein, *On some cases of the effective determination of the density of a non-homogeneous string from its spectral function*, Dokl. Akad. Nauk SSSR **93** (1953), 8617–620.

[Kr6] M.G. Krein, *On a fundamental approximation problem in the theory of extrapolation and filtration of stationary random processes*, Dokl. Akad. Nauk SSSR **94** (1954), 13–16; [English transl: Amer. Math. Soc. Selected Transl. Math. Statist. Prob. **4** (1964), 127–131.

[Kr7] M.G. Krein, *On a method for the effective solution of the inverse boundary-value problem*, Dokl. Akad. Nauk SSSR **94** (1954), 987–990.

[LMcK] N. Levinson and H.P. McKean, *Weighted trigonometrical approximation on the line with application to the germ field of a stationary Gaussian noise*, Acta. Math. **112** (1964), 99–143.

[Sz] G. Szegö, *Beiträge zur Theorie der Toeplitzchen Formen*, Math. Z. **6** (1920), 167–202.

[Wi] N. Wiener, *Extrapolation, Interpolation and Smoothing of Stationary Time Series*, Wiley, New York, 1949.

Department of Theoretical Mathematics
The Weizmann Institute of Science
Rehovot 76100
Israel

AMS Classification Numbers: 34A55, 34B24, 34L05, 60G10, 60G25.

Operator Theory:
Advances and Applications, Vol. 118

M.G. Krein's Contribution to the Moment Problem

Adolf A. Nudelman

In the paper a survey of M.G. Krein's works on the moment problem is presented. Connections of this works with his investigations in other fields and their interrelations are considered.

The moment problem played a special role in the creative activity of Mark Grigorievich Krein. Being an eminent expert in the classical works of Chebyshev, Markov, Korkin, Zolotarev, Stieltjes, and others he grasped the meaning of these works from the contemporary standpoint, developed and generalized them, and found new, sometimes unexpected treatments, and new fields of applications.

M.G. Krein took an interest in the moment problem in the beginning of thirties, later he permanently returned to it, up to his last works. His results and methods roused wide resonance and took wide acknowledgment. But the special role of the moment problem in the creative activity of M.G. Krein consists not only in this. Perhaps more important that its results and methods served him as the tools for orientation while building new theories such as the extension theory for Hermitian operators, the theory of entire operators, the theory of prolongation of Hermitian positive functions and helicoidal arcs, the spectral theory of non-homogeneous strings, and so on, in which M.G. Krein has obtained fundamental results. In turn, these new theories exerted the converse influence on the moment problem, with the aid of them he has obtained new results in the moment problem even in the classical case.

1 Some Definitions

Let a sequence $\{u_k(t)\}_0^n \subset \mathbf{C}[a,b](n \leq \infty)$ be given. For a nonnegative measure $d\sigma$ numbers

$$\int_a^b u_k(t)d\sigma(t)$$

are called *the moments* of the functions u_k.

The moment problem is stated as follows. Let the number sequence $\{c_k\}_0^n$ be given. The questions are:

1. Whether this sequence is a moment sequence? In other words, whether there exists a measure $d\sigma \geq 0$ such that

$$c_k = \int_a^b u_k(t)d\sigma(t), k = 0, 1, \dots n?$$

If so, the moment problem is *solvable* and the measure $d\sigma$ is *a solution.*

2. If the moment problem is solvable, whether it has a unique solution or not, in other words is the problem *determinate* or *indeterminate*?

3. If the problem is indeterminate, describe the set of all solutions or at least the simplest ones.

In the case

$$u_k(t) = t^k$$

the moment problem is called *the power moment problem.* Moreover, the power moment problem is called

the Hamburger moment problem if $a = -\infty, b = \infty$;

the Stieltjes moment problem if $a = 0, b = \infty$;

the Hausdorff moment problem if $-\infty < a < b < \infty$.

The moment problem is called *trigonometric* if $u_k(t) = e^{ikt}, a = 0, b = 2\pi$.

One can immediately demonstrate one of the applications of the general moment problem proposed by Krein.

Let us consider a class $\mathcal{G}$ of functions F which admit the representation

$$F(z) = \int_a^b G(z,t)\, d\sigma(t)$$

where $G(z,t)$ is a given kernel and $d\sigma$ is some nonnegative measure. The interpolation problem

$$F(z_k) = c_k, \quad F \in \mathcal{G}, \quad k = 0, 1, \dots, n$$

is reduced to the moment problem if we put $u_k(t) = G(z_k, t)$. In particular, the simplest solutions of the interpolation problem correspond to the simplest solutions of the moment problem. Functions of some important classes admit such integral representation. For instance,

1. The class $\mathcal{R}$ of the Nevanlinna functions, i.e. functions which are holomorphic in the open half-plane $\mathbb{C}_+$ and map it in the closed upper halfplane. It is known (see e.g. [KN1]) that a function F belongs to $\mathcal{R}$ if and only if it admits a representation

$$F(z) = \alpha + \beta z + \int_{-\infty}^{\infty} \left(\frac{1}{t-z} - \frac{t}{1+t^2} \right) d\sigma(t) \tag{1}$$

where $\alpha \in \mathbb{R}$, $\beta \geq 0$, $\int_{-\infty}^{\infty} (1+t^2)^{-1} d\sigma(t) < \infty$.

The next two classes were introduced by M.G. Krein [K8], [K10] (see also [KN1]).

2. The class $\mathcal{S}$ consists of those $F \in \mathcal{R}$ which are holomorphic on $(-\infty, 0)$ and take there nonnegative values. A function F belongs to $\mathcal{S}$ if and only if it admits the representation

$$F(z) = \gamma + \int_0^{\infty} \frac{d\sigma(t)}{t-z} \tag{2}$$

where $\gamma = F(-\infty) \geq 0$ and $d\sigma(t) \geq 0$. Functions of such form had arisen in the famous Stieltjes researches on continued fractions [S].

3. The class $\mathcal{R}\,[a,b]$ consists of those $F \in \mathcal{R}$ which are holomorphic on $(-\infty, a) \cup (b, \infty)$, nonnegative in $(-\infty, a)$, and non-positive in (b, ∞). A function F belongs to $\mathcal{R}\,[a,b]$ if and only if it admits the representation

$$F(z) = \int_a^b \frac{d\sigma(t)}{t-z}.$$

2 Classical Moment Problem and L-problem

Before the time when N.I. Akhiezer and M.G. Krein began the investigations on the moment problem the following scheme was traditional.

The initial object was infinite continued fractions of a special form, say of the following form:

$$-\cfrac{b_1}{z-a_1-\cfrac{b_2}{z-a_2-\cfrac{b_3}{z-a_3-\dots}}}$$

where a_j are real and b_j are positive numbers. To the n-th truncated fraction corresponds an approximant $P_n(z)/Q_n(z)$ where $P_n(z)$, $Q_n(z)$ are polynomials of degrees $n-1$ and n respectively. All zeros of these polynomials are real, simple and alternate. The expansions of the approximants $P_n(z)/Q_n(z)$ into series in powers of z^{-1} generates the series

(3) $$-\sum_0^\infty \frac{s_k}{z^{k+1}}$$

and on the other hand the decomposition of $P_n(z)/Q_n(z)$ into partial fractions generates an integral

(4) $$\int_{-\infty}^{\infty} \frac{d\sigma(t)}{t-z}.$$

It occurs that

(5) $$s_k = \int_{-\infty}^{\infty} t^k \, d\sigma(t), \quad k = 0, 1, \dots$$

The series (3) is a formal expansion of the integral (4). In reality the series (3) defines the asymptotic behavior of the integral as $z \to \infty$ in $\mathbb{C}_+$ on non-tangential pathes. The denominators $Q_n(z)$ are orthogonal relatively to $d\sigma$:

$$\int_{-\infty}^{\infty} Q_m(t) Q_n(t) \, d\sigma(t) = 0 \quad (m \neq n).$$

In the series of works which was summarized in the book [AK] N.I. Akhiezer and M.G. Krein proposed an approach which does not use continued fractions. It has based on equivalency of the power moment problem to the problem of finding a function F of the appropriate class with prescribed asymptotic behavior at ∞:

$$F(z) \sim -\sum_{k=0}^{\infty} \frac{s_k}{z^{k+1}}.$$

These classes are:

$\mathcal{R}$ for the Hamburger problem,

$\mathcal{S}$ for the Stieltjes problem,

$\mathcal{R}\,[a,b]$ for the Hausdorf problem.

N.I. Akhieser and M.G. Krein studied the so called *canonical solutions* of the truncated moment problem (for $\{s_k\}_0^n, n < \infty$). These solutions form an one-parametric family of discrete measures the supports of which coincide with the sets of zeros of quasiorthogonal polynomials $\alpha Q_{\nu+1}(z) + \beta Q_\nu(z)$.

The L -problem is the problem of finding a function $f \in L_\infty(a,b)$ which satisfy the relations

$$c_k = \int_a^b u_k(t) f(t) dt, \quad k = 1, \ldots, n, \quad \|f\|_\infty \le L$$

for the given numbers $\{c_k\}_0^n$ and $L > 0$. A.A. Markov [M] considered the power L-problem under slightly different condition ($0 \le f(t) \le L$) by very cumbersome tools by using an expansion of functions of the form

$$\exp \frac{1}{L} \int_a^b \frac{f(t)\,dt}{t-z}, \quad \frac{z-a}{t-z} \exp \frac{1}{L} \int_a^b \frac{f(t)\,dt}{t-z}$$

in continued fractions.

N.I. Akhiezer and M.G. Krein discovered the analytical background of Markov's results and owing to it they found a simple and transparent way, which gives the opportunity not only to reobtain Markov's results but to develop them further. Moreover, on this way one can investigate the trigonometric L-problem. This approach is based on the following elementary fact. For $F \in \mathcal{R}$ and $z \in \mathbb{C}_+$ we have $0 \le \arg F(z) \le \pi$ so $\log F \in \mathcal{R}$. From this one can easily deduce that besides the representation (1) there exists the representation

$$F(z) = C \exp \int_{-\infty}^{\infty} \left(\frac{1}{t-z} - \frac{t}{1+t^2} \right) \varphi(t)\,dt$$

where $C \ge 0,\ 0 \le \varphi(t) \le 1$.

Slightly modified this formula allows to establish a connection between the L-problem

$$c_k = \int_{-\infty}^{\infty} t^k f(t)\,dt, \quad k = 0, \ldots, n, \quad \|f\|_\infty \le L$$

and the Hamburger moment problem

$$s_k = \int_{-\infty}^{\infty} t^k \, d\sigma(t), \quad k = 0, 1, \dots, n+1, \quad d\sigma \geq 0.$$

and to construct L-canonical solutions starting from the canonical solutions of the corresponding Hamburger moment problem.

The power L-problem was used by N.I. Akhiezer and M.G. Krein to consider the Korkin-Zolotarev extremal problem and to find an answer to the very difficult question concerning the existence of the Chebyshev and the Markov types quadrature formulas. We take in mind the formulas of the form

$$\int_{-1}^{1} f(t)\, q(t)\, dt = L \sum_{j=1}^{m} (f(\eta_j) - f(\xi_j))$$

where $\int_{-1}^{1} q(t)\, dt = 0$, $\{\xi_j\}_1^m$, $\{\eta_j\}_1^m$ are the sequences of alternating numbers laying in $[-1, 1]$. These formulas must be precise for polynomials of degree $\leq 2m$ (the Markov type formula) or of degree $\leq 2m + 1$ (the Chebyshev type formula).

3 Abstract *L*-problem

As it is well known a linear bounded functional F acting in the space $L_1(a, b)$ has a form

$$F(u) = \int_a^b u(t) f(t)\, dt$$

where $f \in L_\infty(a, b)$ and $\|F\| = \|f\|_\infty$. Thus, the L-problem

$$c_k = \int_a^b u_k(t) f(t)\, dt, \quad k = 0, 1, \dots, n, \quad \|f\|_\infty \leq L$$

can be restated as follows.

Let a sequence $\{c_k\}_0^n$ of numbers, a sequence $\{u_k\}_0^n \subset L_1(a, b)$, and a positive number L be given. Find linear functional F such that

$$F(u_k) = c_k, \quad k = 0, 1, \dots, n, \quad \|F\| \leq L$$

We obtain an abstract L-problem by changing $L_1(a, b)$ on an abstract normed space $\mathbf{B}$.

M.G. Krein considered two following dual problems. Let two sequences $\{c_k\}_0^n \subset \mathbb{R}$ and $\{u_k\}_0^n \subset \mathbf{B}$ be given.

I. Find

$$\Lambda = \min\{\|F\| : F(u_k) = c_k, \quad k = 0, 1, \dots, n\}.$$

II. Find

$$\frac{1}{\mathrm{M}} = \min \left\{ \left\| \sum_0^n \alpha_k u_k \right\| : \sum_0^n \alpha_k c_k = 1 \right\}.$$

It occurs that $\Lambda = \mathrm{M}$.

This allows to investigate many extremal problems by the same method, using joint consideration of some general extremal problem together with the dual problem. In our time such approach is usual but it must be taken into account that these results were obtained in 1936, that is this work was one of the first in which this approach was proposed.

A particular case of problem II is a problem of the best approximation of the given function f by generalized polynomials $P = \sum_0^{n-1} a_k u_k$.

To this end it is sufficient to put

$$f(t) = u_n(t), \quad c_0 = \ldots = c_{n-1} = 0, \quad c_n = 1.$$

The last conditions imply the equality $\alpha_n = 1$, and there are no restrictions on the coefficients $\alpha_k = -a_k$. So in this case the problem II reduces to minimization of $\|f - P\|$.

The abstract L-problem obtained a wide resonance in the control theory. Several monographs and large number of papers are devoted to its applications in the control theory (see e.g. [Kra], [VIP]).

4 Markov-Krein's and Chebyshev-Markov's Inequalities

Simultaneously M.G. Krein investigated the moment problem in some other direction. Starting from methods of Riesz and Carathéodory he perfected and developed Markov's investigations on the canonical solutions theory, made them simple and transparent owing to using the modern geometric and function-theoretic methods, and considerably extended the region of their applications [K1], [K8]. The paper [K8] was developed in the books [KS] and [KN1]. In these books various modifications of moment problem and associated extremal problems are considered with applications to extremal problems arising in various fields of analysis, the constructive theory of functions, the theory of analytic functions, mathematical programming, functional analysis, problems in the geometry of $\mathbb{R}^n$ and so on. Below we give a sketch of Krein's approach.

Let a sequence $\{u_k(t)\}_0^n \subset C(a, b)$ be fixed. Let K denotes the set of all sequences $\mathbf{c} = \{c_k\}_0^n \subset \mathbb{R}^{n+1}$ which admit the representations

$$c_k = \int_a^b u_k(t) d\sigma(t), k = 0, 1, \ldots, n$$

and $\mathbf{U}$ denotes the curve $\{\mathbf{u}(t) = \{u_k(t)\}_0^n : a \le t \le b\}$. It occurs that K is the conic hull of $\mathbf{U}$. From the analytical point of view this means that $\mathbf{c} \in K$ if and only if for a generalized polynomial $P(t) = \sum_0^n \alpha_k u_k(t)$ its non-negativity on $[a, b]$ implies the inequality $\sum_0^n \alpha_k c_k \ge 0$.

Especially refined results can be obtained in the case where the functions $\{u_k(t)\}_0^n$ form a T-system, that is where every generalized polynomial $P(t) = \sum_0^n \alpha_k u_k(t)$, like an algebraic polynomial, has on $[a, b]$ at most n zeros. In this case the point $\mathbf{c}$ belongs to the boundary of K if and only if there exists a nonzero polynomial $P_0(t) = \sum_0^n \alpha_k^0 u_k(t) \geq 0 \quad (a \leq t \leq b)$, such that $\sum_0^n \alpha_0^k c_k = 0$. In this and only in this case the solution $d\sigma$ is unique and its support consists of in some sense minimal number of points.

Let $\mathbf{c} = \{c_k\}_0^n$ be an interior point of K and let $V(\mathbf{c})$ be the set of all solutions of the corresponding moment problem:

$$V(\mathbf{c}) = \left\{ d\sigma \geq 0 : \quad c_k = \int_a^b u_k(t)\, d\sigma(t), \quad k = 0, 1, \ldots, n \right\}.$$

It is required to find maximum and minimum of the integral

$$\int_a^b u_{n+1}(t)\, d\sigma(t)$$

where $d\sigma$ runs through the set $V(\mathbf{c})$.

In the case where a function $u_{n+1}(t)$ is a T-extension of the T-system $\{u_k(t)\}_0^n$ M.G. Krein proposed a very elegant solution of this problem.

Together with the cone $K \subset \mathbb{R}^{n+1}$ associated with $\{u_k(t)\}_0^n$, let us consider the cone $\tilde{K} \subset \mathbb{R}^{n+2}$ associated with $\{u_k(t)\}_0^{n+1} \quad (a \leq t \leq b)$ as well as the set I of points $(c_0, c_1, \ldots, c_n, \gamma) \in \tilde{K}$. The set I is a segment, the endpoints $(c_0, c_1, \ldots, c_n, \underline{c}_{n+1})$ and $(c_0, c_1, \ldots, c_n, \bar{c}_{n+1})$ of which belong to the boundary of $\tilde{K}$ and so uniquely determine measures $d\underline{\sigma}$ and $d\bar{\sigma}$ for which

$$\int_a^b u_{n+1}(t)\, d\underline{\sigma}(t) \leq \int_a^b u_{n+1}(t)\, d\sigma(t) \leq \int_a^b u_{n+1}(t)\, d\bar{\sigma}(t)$$

for all $d\sigma \in V(\mathbf{c})$.

The structures of the extremal measures $d\underline{\sigma}$ and $d\bar{\sigma}$ are well studied, and in the case of the power moment problem for these measures one can calculate their supports and masses. In particular, for any solution $d\sigma$ of the Stieltjes moment problem we have

$$\int_0^\infty \frac{d\underline{\sigma}(t)}{t+x} \leq \int_0^\infty \frac{d\sigma(t)}{t+x} \leq \int_0^\infty \frac{d\bar{\sigma}(t)}{t+x} \quad (x > 0) \tag{6}$$

since the function $1/(t+x)$ is a T-extension of the T-system $\{t^k\}_0^n$ on $[0, \infty)$. These are *the Stieltjes inequalities.*

The measures $d\underline{\sigma}(t)$ and $d\bar{\sigma}(t)$ are contained in the one-parametric family of the simplest (canonical) solutions (as it was noted yet, in the case of the power moment problem the support of a canonical solution coincides with the set of zeros of some quasiorthogonal polynomial). For the given $\xi \in (a, b)$ there exists one and only one canonical solution $d\sigma_\xi$ such that $\xi \in \operatorname{supp}(\sigma_\xi)$. The measure $d\sigma_\xi$ gives a solution of the following extremal problem.

Find

$$\max(\min) \left(\int_a^\xi u_{n+1}(t)\, d\sigma(t) : \quad \sigma \in V(\mathbf{c}) \right).$$

Under some supplementary assumptions concerning the system $\{u_k(t)\}_0^{n+1}$ the following inequalities are true

$$\left.\begin{aligned} \int_a^{\xi-0} u_{n+1}(t)\,d\sigma(t) &\geq \int_a^{\xi-0} u_{n+1}(t)\,d\sigma_\xi(t) \\ \int_a^{\xi+0} u_{n+1}(t)\,d\sigma(t) &\leq \int_a^{\xi+0} u_{n+1}(t)\,d\sigma_\xi(t) \end{aligned}\right. \qquad (d\sigma \in V(\mathbf{c})).$$

In the particular case where $u_{n+1} \equiv 1$ this inequalities give estimation of $\sigma(\xi \pm 0)$ (the famous *Chebyshev-Markov inequalities*).

Later M.G. Krein extended the Chebyshev-Markov inequalities on spectral functions of the Sturm-Liouville problem [K10] and on spectral functions of non-homogeneous string [K12].

5 Moment Problem and Symmetric Operators

M. Stone proposed the following operator approach to the Hamburger moment problem

$$s_k = \int_{-\infty}^{\infty} t^k\,d\sigma(t), \quad k = 0, 1, \ldots.$$

Let $\mathcal{D}$ denotes a linear manifold of algebraic polynomials. Let us introduce on $\mathcal{D}$ the inner product by the formula

$$(t^k, t^l) = s_{k+l}$$

by supposing positive definiteness of all quadratic forms

$$\sum_0^n s_{k+l}\xi_k\bar{\xi}_l, \quad n = 0, 1, \ldots.$$

The last condition is necessary and sufficient for the solvability of the Hamburger moment problem. The completion of $\mathcal{D}$ is a Hilbert space $\mathcal{H}$. Let us define on $\mathcal{D}$ the operator A which sends the polynomial $P(t)$ to the polynomial $tP(t)$. This operator is symmetric, its defect indices are either $(0, 0)$ or $(1, 1)$, so A has self-adjoint extensions. Let $\tilde{E}(t)$ is a spectral decomposition of unity of some self-adjoint extension $\tilde{A}$ of A. Then we can write

$$s_k = (t^k, 1) = (A^k 1, 1) = (\tilde{A}^k 1, 1) = \int_{-\infty}^{\infty} t^k\,d(\tilde{E}(t)1, 1).$$

So $d\sigma(t) = d(\tilde{E}(t)1, 1)$ is a solution of the Hamburger moment problem. It occurs, conversely, that every solution of this problem can be represented in such a form, that is there is a one-to-one correspondence between solutions of the Hamburger moment problem and self-adjoint extensions of A, and the moment problem is indeterminate if and only if the defect indices are $(1, 1)$.

M.G. Krein built a full extension theory of symmetric operators with defect indices (1, 1) [K3] and the moment problem was an evident prompter. The main relations similar to ones in the moment problem were obtained, in particular, the formula describing generalized resolvents of all self-adjoint extensions in the case of moment problem turns into the Nevanlinna formula describing the set of all solutions of the indeterminate Hamburger moment problem.

Soon after M.G. Krein elaborated the extension theory of symmetric operators with defect indices (m, m) [K6] owing of which the results on the operator power moment problem was obtained [K7], [KKr], see also [B].

The Hamburger moment problem was prompter when M.G. Krein created an elegant theory of entire operators (see [GG]) which contains as particular cases the Hamburger moment problem, the prolongation problem of Hermitian positive functions and helicoidal arcs, and many other problems.

At the end of forties M.G. Krein built the theory of self-adjoint extensions of positive symmetric operators with preserving of positivity taking in the mind different applications in problems of mathematical physics. In particular, obtained by him formulas allow to give new effective methods of construction the Green functions for different boundary problems. A point of departure was the fact that the problem of self-adjoint extension of positive symmetric operator S with preserving positivity can be reduced to the problem of self-adjoint extension with preserving of norm of a nondensely defined bounded operator A. This time methods were others but undoubtedly the moment problem was behind the scenes. To confirm this it is enough to formulate the following assertion.

A positive operator S has two extremal positive extensions S_μ and S_{M} such that an operator $\tilde{S}$ is positive extension of S if and only if for any $f \in \mathcal{H}$ and $x > 0$ the following inequalities are true

$$((S_\mu + xI)^{-1} f, f) \le ((\tilde{S} + xI)^{-1} f, f) \le ((S_{\mathrm{M}} + xI)^{-1} f, f).$$

They can be rewritten as follows

$$\int_0^\infty \frac{d(E_\mu(t) f, f)}{t + x} \le \int_0^\infty \frac{d(\tilde{E}(t) f, f)}{t + x} \le \int_0^\infty \frac{d(E_M(t) f, f)}{t + x}$$

For the moment problem these are exactly the Stieltjes inequalities (6).

The formula describing all generalized resolvents of positive extensions contains as a particular case the formula describing the set of all solutions of the indeterminate Stieltjes moment problem which was unknown at that time.

Analogously the extension theory of symmetric operators with preserving the norm contains as a particular case the truncated moment problem on a finite interval $[a, b]$

$$s_k = \int_a^b t^k \, d\sigma(t), \quad k = 0, \dots n, \quad n = 2\nu$$

for even $n = 2\nu$. As a consequence the formula describing the set of all solutions of this moment problem was obtained. This formula was also unknown at that time.

Later M.G. Krein found an elementary proof of this formula [K11]. By another way, rather elementary too, it was obtained in our joint monograph [KN1]. This way one can obtain it also in the case of odd $n = 2\nu + 1$.

By developing extension theory of symmetric operators in the Pontryagin space Π_κ M.G. Krein and H. Langer considered the indefinite power moment problem [KL]. It can be again formulated in the form

$$F(z) \sim -\sum_0^\infty \frac{s_k}{z^{k+1}}$$

but now the analytic function F defines the kernel $\frac{F(z)-\overline{F(\zeta)}}{z-\overline{\zeta}}$ having κ negative squares in $\mathbb{C}_+$. This means that for any $\{z_k\}_1^m \subset \mathbb{C}_+, m = 1, 2, \ldots$ the matrix $(\frac{F(z_k)-\overline{F(z_l)}}{z_k-\overline{z}_l})$ has at most κ negative eigenvalues and there exists $\{\zeta_k\}_1^m \subset \mathbb{C}_+$ for which this matrix has exactly κ negative eigenvalues.

6 The Stieltjes Moment Problem and the Spectral Theory of Non-homogeneous String

M.G. Krein had a wonderful ability to discover connections between regions which were at the first sight very far one from another. One of the very impressive examples of such ability is a mechanical interpretation of the Stieltjes investigations on continued fractions and moment problem, which later became the base in creation of the spectral theory of non-homogeneous string.

Stieltjes investigated fractions of the following type:

$$\cfrac{1}{-a_1 z + \cfrac{1}{a_2 + \cfrac{1}{-a_3 z + \cfrac{1}{a_4 + \cdots}}}}.$$

Taking into account the structure of the fraction one must distinguish approximants of odd and even orders

$$\frac{P_{2n-1}(z)}{Q_{2n-1}(z)}, \quad \frac{P_{2n}(z)}{Q_{2n}(z)}.$$

The nominators and the denominators of these fractions satisfy the following relations:

$$\text{(7)} \qquad \frac{u_{2n} - u_{2n-2}}{a_{2n}} = u_{2n-1}, \qquad \frac{u_{2n-1} - u_{2n-3}}{a_{2n-1}} = -zu_{2n-2}.$$

The fractions $P_m(z)/Q_m(z)$ generate the series (3) coefficients of which $\{s_k\}_0^\infty$ form the Stieltjes moment sequence. Let us note that $P_m(z)$ and $Q_m(z)$ as well as the entries a_m can be easily expressed in terms of $\{s_k\}_0^\infty$.

M.G. Krein proposed to consider a thread wits beads stretched by unit tensile stress, a_{2n-1} are masses of beads, a_{2n} are distances between neighboring beads (the first bead is in the left end of the thread). Such system is called *the Stieltjes string* if the beads have an unique point of condensation, namely the right end $(x = L = \sum a_{2n})$ of thread.

It occurs that values y_k of oscillation amplitudes of k-th bead satisfy the same relations (7).

Now the most part of Stieltjes results has a mechanical interpretation. For instance, the Stieltjes fraction diverges and so the corresponding Stieltjes moment problem is indeterminate if and only if the series $\sum a_m$ converges. This means that in this case the Stieltjes string has finite length $L = \sum a_{2n}$ and finite total mass $\sum a_{2n-1}$.

Starting from Stieltjes strings M.G. Krein created the spectral theory of the string having an arbitrary mass distribution $M(x)$ $(0 \le x \le L)$ (see [KK] or [DMcK]).

Amplitude functions of such a string are solutions of the equation

$$\frac{d^2y}{dx\,dM(x)} + zy = 0.$$

Let $\varphi(x, z)$ and $\psi(x, z)$ be solutions corresponding to the free left end or to the fixed left end $(x = 0)$ respectively. A measure $d\sigma$ is called *the spectral measure* of the string if the generalized Fourier transform $f \to F$:

$$F(\lambda) = \int_0^L \varphi(x, \lambda) f(x)\,dM(x)$$

maps isometrically the space $L_2(0, L; dM)$ into the space $L_2(0, \infty; d\sigma)$.

The spectral measure is unique if and only if the length L or/and total mass $M(L)$ of string are infinite: $L+M(L) = \infty$. In opposite case all spectral measures can be found using the formula establishing one-to-one correspondence between spectral measures $d\sigma$ and functions $\omega \in S$ (including $\omega \equiv \infty$):

$$\gamma + \int_0^\infty \frac{d\sigma(t)}{t - z} = \frac{\psi'(L + 0, z)\omega(z) + \psi(L, z)}{\varphi'(L + 0, z)\omega(z) + \varphi(L, z)} \tag{8}$$

were γ is the length of the string segment which is free of masses and has the left endpoint in $x = 0$. The measures $d\underline{\sigma}$ and $d\overline{\sigma}$ corresponding to $\omega \equiv 0$ and $\omega \equiv \infty$ respectively have the following extremal property: for any spectral measure of the string $d\sigma$ the Stieltjes inequalities (6) hold. Moreover, Krein's results on inverse spectral problem for strings imply that for any spectral measure $d\sigma$ of the string S there exists some prolongation of S to the right such that the extremal measure $d\underline{\sigma}^*$ of the prolonged string S^* coincides with $d\sigma$.

Let us return to the Stieltjes string. Let $x_n = \sum_{k=1}^{n-1} a_{2k}$, $(a_0 = 0)$ be a coordinate of n-th bead then

$$\begin{aligned} \psi(x_n, z) &= P_{2n}(z), & \psi'(x_n + 0) &= P_{2n+1}(z), \\ \varphi(x_n, z) &= Q_{2n}(z), & \varphi'(x_n + 0) &= Q_{2n+1}(z). \end{aligned}$$

It occurs that the set of solutions of the Stieltjes moment problem coincide with the set of spectral measures of the corresponding Stieltjes string, so the set of all solutions of the indeterminate Stieltjes moment problem can be obtained by (8) were now $\gamma = 0$ and

$$\begin{aligned} \psi(L,z) &= \lim_{n\to\infty} P_{2n}(z), & \psi'(L+0,z) &= s \lim_{n\to\infty} P_{2n+1}(z) \\ \varphi(L,z) &= \lim_{n\to\infty} Q_{2n}(z), & \varphi'(L+0,z) &= \lim_{n\to\infty} Q_{2n+1}(z) \end{aligned}.$$

7 Some Applications

The last series of works using results and methods of moment problem began by short communication [KN2] (the detailed exposition see [KN4]) where the following interpolation problem in the class $\mathcal{S}$ with nodes in negative semiaxis of real axis was investigated:

$$(9) \qquad F(-x_k) = c_k, \quad x_k > 0, \quad k = 1, 2, \ldots, \quad F \in \mathcal{S}, \quad \sum_{k=1}^{\infty} x_k^{-1} < \infty.$$

Without loss of generality one can suppose that $c_k \to 0$ $(k \to \infty)$, in this case the problem (9) is the moment problem

$$(10) \qquad c_k = \int_0^{\infty} \frac{d\sigma(t)}{t + x_k}, \quad k = 1, 2, \ldots$$

since in (2) $\gamma = 0$. A sequence $\{s_k\}_{k=0}^{\infty}$ is constructed in such a way that the problem (9) and the Stieltjes moment problem

$$(11) \qquad s_k = \int_0^{\infty} t^k \, d\tau(t), \quad k = 0, 1, \ldots$$

are solvable simultaneously. Moreover, let $P(t) = \prod_{k=1}^{\infty} (1 + t/x_k)$, then to each solution $d\sigma(t)$ of (10) the solution $d\,\tau(t) = d\,\sigma(t)/P(t)$ of (11) corresponds and conversely, if $d\tau(t)$ is a solution of (11) then $d\sigma(t) = P(t)\,d\tau(t)$ is the solution of (10). In particular, extremal measures $d\underline{\sigma}$ and $d\overline{\sigma}$ correspond to extremal measures $d\underline{\tau}$ and $d\overline{\tau}$ respectively. The case of non-real nodes was considered too. Results of [KN2] were used in the molecular interaction theory [NT], [Ka].

The problem (9) was a prompter when the following question was considered. It is known that every algebraic polynomial $P(z)$ which is positive on $[0, \infty)$ admits the representation

$$(12) \qquad P(z) = A^2(z) + zB^2(z)$$

where $A(z)$ and $B(z)$ are real polynomials, their roots are real, simple and alternate. These conditions are equivalent to the condition $B/A \in \mathcal{S}$ (up to sign). But an entire functions $P(z)$ positive on $[0, \infty)$ not always admits the representation

(12) where $A(z)$ and $B(z)$ are real entire functions such that $B/A \in \mathcal{S}$. Indeed, let $Z = \{z_k\}$ be the set of zeros of $P(z)$ located in $\mathbb{C}_+ \cup (-\infty, 0)$. The representation (12) implies $B(z_k)/A(z_k) = i/\sqrt{z_k}$, $z_k \in Z$. Thus, the interpolation problem

$$(13) \qquad F(z_k) = \frac{i}{\sqrt{z_k}}, \quad z_k \in Z, \quad F \in \mathcal{S}$$

must be indefinite (since both functions $B(z)/A(z)$ and $i/\sqrt{z}$ are solutions of (13)) so the necessary condition of existence (12) is the Blaschke condition $\sum_{z_k \in Z} \Im(-1/\sqrt{z_k}) < \infty$ which occurs the sufficient condition too. In contrast to the case were $P(z)$ is an algebraic polynomial, the representation (12) of the entire function which satisfies the Blaschke condition, is not unique. All pairs (A, B) $(B/A \in \mathcal{S})$ of entire functions that realize (12) can be expressed in terms of the pair (A_0, B_0) such that B_0/A_0 is the extremal solution of the problem (13). These results were essentially used in solving an inverse spectral problem for the string [KN5], [KN6].

In [KLN1] and [KLN2] some generalizations of the representation (12) for polynomials that are positive on a system of closed intervals were investigated. Let us put

$$E = [0, \infty) \Big\backslash \bigcup_{j=1}^{g} (a_j, b_j) (0 < a_1 < b_1 < \ldots < a_g < b_g < \infty),$$

$$a(z) = \prod_{j=1}^{g} (z - a_j), \quad b(z) = \prod_{j=1}^{g} (z - b_j).$$

An analogue of representation (12) for polynomial of degree N that is positive on E is

$$(14) \qquad P(z) = A^2(z) + z a(z) b(z) B^2(z)$$

in which the zeros of real polynomials $A(z)$ and $B(z)$ of degree $[N/2]$ and $([N - 2g - 1)/2]$, respectively, are simple, belong to $[0, \infty)\backslash E$, and are alternating in such a way that

$$(15) \qquad \frac{A(z)}{z b(z) B(z)} \in \mathcal{S}_+(E)$$

where $\mathcal{S}_+(E)$ is the set of those functions in $\mathcal{S}$ that are holomorphic and nonnegative in $[0, \infty)\backslash E$. The criterion of existence the representation (14) satisfying (15) was obtained using the extremal solution of the following interpolation problem

$$F(z_k) = v(z_k), \quad z_k \in Z$$

where $v(z) = \sqrt{-a(z)/zb(z)}$, Z is the set of roots of $P(z)$ situated in $\mathbb{C}_+ \cup ([0, \infty)\backslash E)$ and of the associated moment problem $\{s_k\}_0^{N-1}$ where

$$(16) \qquad s_k = \frac{1}{\pi} \int_E t^k \frac{\sqrt{a(t)/tb(t)}}{P(t)} \, dt.$$

The representation exists if and only if the numbers b_j, $j = 1, 2, \ldots, g$, are roots of the polynomial $Q_{2\mu+1}(z)$, ($\mu = [(N-1)/2]$) (see Section 6). Moreover, under these conditions $A(z) = Q_{2\nu}(z)$ ($\nu = [N/2]$), $B(z) = Q_{2\mu+1}(z)$.

The same conditions arose in the following problem. Let S_E be a string with extremal spectral measure

$$d\underline{\sigma}_E(t) = \begin{cases} \frac{1}{\pi}\sqrt{a(t)/tb(t)} & \text{for } t \in E \\ 0 & \text{for } t \in [0,\infty)\backslash E, \end{cases}$$

$S_{E,P}$ be a string with extremal spectral measure $d\underline{\tau}_{E,P}(t) = d\underline{\sigma}_E(t)/P(t)$ where $P(t)$ is a polynomial of degree N positive on E. The measure $d\underline{\tau}_{E,P}$ has moments (16), therefore the initial segment $[0, l)$ of S_E is a thread with beads. Let $\tilde{S}$ be the remaining part of S_E. The following question is posed: under which conditions does string S_E coincide (after a shift by l) with S_E? The answer is: $\tilde{S} = S_E$ if and only if the polynomial $P(t)$ admits the representation (14).

In [KLN1] and [KLN2] the main facts were obtained by two methods: one uses the theory of moments, the other uses the theory of Abel integrals. In [KLN2] other representations of $P(t)$ and some new application was considered.

It is impossible to describe in a short article all Krein's results connected with applications and/or generalizations of results and methods of the moment problem, so some fundamental fields such as the continual analogues of the trigonometric moment problem, the theory of entire operators, the method of directed functionals, inverse problems of spectral theory, and so on were omitted. But I hope that the content of this paper shows that the moment problem is an important and essential part of the rich and wide scientific legacy of Mark Grigorievich Krein.

References

[AK] N.I. Akhiezer and M.G. Krein, *Some questions in the theory of moments*, GONTI, Khar'kov, 1938 (Russian).

[B] Yu.M. Berezanskii, *The expansion on eigenfunctions of selfadjoint operators,* Naukova Dumka, Kiev, 1965 (Russian).

[DMcK] H. Dym and H.P. McKean, *Gaussian processes, function theory, and inverse spectral problem,* Academic Press, 1976.

[GG] V.I. Gorbachuk and M.L. Gorbachuk, *M.G. Krein's lectures on entire operators,* Birkhäuser Verlag, 1997.

[K1] M.G. Krein, *On a generalization of the studies of Academician A.A. Markov on limiting values of integrals,* Proc. Second All-Union Math. Congr. (Leningrad, 1934), Izdat. Akad. Nauk SSSR, Moscow, vol. **2** (1935), 152–154 (Russian).

[K2] M.G. Krein, *On a prolongation problem of Hermitian positive continuous functions*, Dokl. Akad. Nauk SSSR vol. **26** (1940), no. 1, 17–21 (Russian).

[K3] M.G. Krein, *On Hermitian operators defect indices of which equal to one*, I, Dokl. Akad. Nauk SSSR vol. **43** (1944), no. 8, 339–342; II, ibid., vol. **44**, no. 4, 143–146 (Russian).

[K4] M.G. Krein, *The theory of self-adjoint extensions of semi-bounded Hermitian operators and its applications*, I, Mat. Sb. vol. **20** (1947), no. 3, 431–495; II, ibid., vol. **21**, no. 3, 365–404 (Russian).

[K5] M.G. Krein, *On Hermitian functionals with directed functionals*, Zbirn. prats' Inst. Matem. AN Ukr. SSR, 1948, vyp. 10, 83–106 (Russian).

[K6] M.G. Krein, *Fundamental assertions of the representation theory of Hermitian operators with defect indices (m,m)*, Ukr. Mat. J., vol. **1** (1949), no. 2, 3–66 (Russian).

[K7] M.G. Krein, *Infinite J-matrices and matrix moment problem*, Dokl. Akad. Nauk SSSR, vol. **69** (1949), no. 2, 125–128 (Russian).

[K8] M.G. Krein, *Ideas of P.L.Chebyshev and A.A. Markov in the theory of limiting values of integrals and their further development*, Uspekhi Mat. Nauk, vol. **6** (1951), no. 4, 3–120 (Russian).

[K9] M.G. Krein, *On a generalization of investigations of Stieltjes*, Dokl. Akad. Nauk SSSR, vol. **87** (1952), no. 6, 881–884 (Russian).

[K10] M.G. Krein, *An analogue of the Chebyshev-Markov inequalities in the one-dimensional boundary problem*, Dokl. Akad. Nauk SSSR, vol. **88** (1953), no. 3, 405–408 (Russian).

[K11] M.G. Krein, *The description of all solutions of the truncated power moment problem and some problems of operator theory*, Mat. Issled., vol. **2** (1967), no. 2, 114–132 (Russian).

[K12] M.G. Krein, *The Chebyshev-Markov inequalities in the spectral theory of the string*, Mat. Issled. vol. **5** (1970), no. 1, 77–101 (Russian).

[Ka] I.G. Kaplan, *Introduction to the theory of molecular interactions*, "Nauka", Moscow, 1982 (Russian).

[KKr] M.G. Krein and M.A. Krasnoselskii, *Fundamental theorems on the extension of Hermitian operators and certain of their applications to the theory of orthogonal polynomials and the problem of moments*, Uspekhi Mat. Nauk, vol. **2** (1947), no. 3, 60–106 (Russian).

[KK] I.S. Kac and M.G. Krein, *On spectral functions of the string*, Appendix 2 to the Russian translation of the book by F.V. Atkinson "Discrete and continuous boundary problems", "Mir", Moscow (1968), 648–737 (Russian).

[KL] M.G. Krein and H. Langer, *On indefinite power moment problem*, Dokl. Akad. Nauk SSSR, vol. **226** (1976), no. 2, 261–264 (Russian).

[KLN1] M.G. Krein, B.Ya Levin and A.A. Nudelman, *On a special representation of a polynomial positive on a system of closed intervals*, Phys.-Tech. Inst. of Low Temperatures, Khar'kov, Preprint (1984), 28–84.

[KLN2] M.G. Krein, B.Ya. Levin and A.A. Nudelman, *On special representations of polynomials that are positive on a system of closed intervals, and some applications*, Functional Analysis, Optimization and Mathematical Economics (Ed. L. Leifman), Oxford Univ. Press, 1990, 56–114.

[KN1] M.G. Krein and A.A. Nudelman, *The Markov moment problem and extremal problems*, "Nauka", Moscow, 1973 (Russian). English translation: Translation of Mathematical Monographs AMS, vol. **50** 1977.

[KN2] M.G. Krein and A.A. Nudelman, *On an interpolation problem related to the Stieltjes moment problem,* Dokl. Akad. Nauk Ukr. SSR, Ser. A no. 12 (1977), 1063–1072 (Russian).

[KN3] M.G. Krein and A.A. Nudelman, *On representation of entire functions positive on the real axis, or on semiaxis, or outside of finite interval,* Mat. Issled. (1981), no. 61, 40–59 (Russian).

[KN4] M.G. Krein and A.A. Nudelman, *An interpolation problem in the class of Stieltjes functions and its connection with other problems,* Integral Equations and Operator Theory, vol. **30**, 251–278.

[KN5] M.G. Krein and A.A. Nudelman, *On direct and inverse problems for the boundary frequencies of dissipation of the nonhomogeneous string,* Dokl. Acad. Nauk SSSR, vol. **247** (1979), no. 5, 1046–1049 (Russian).

[KN6] M.G. Krein and A.A. Nudelman, *Some spectral properties of nonhomogeneous string with a dissipative boundary condition,* Journal of Operator Theory, vol. **22** (1989), 369–395 (Russian).

[KS] S.J. Karlin and W.J. Studden, *Tchebysheff systems: with application in analysis and statistics*, Interscience Publishers, 1966.

[Kra] N.N. Krasovskii, *The movement control theory,* Moscow, "Nauka", 1968 (Russian).

[M] A.A. Markov, *Nouvelles applications des fractions continues,* Math. Ann., vol. **47** (1896), 579–597.

[NT] A.A. Nudelman and A.V. Tulub, *Some extremal problems in the theory of the Van-der-Vaals forces,* Teoreticheskaya i Matematicheskaya Fizika, vol. **39** (1979), no. 3, 359–367 (Russian).

[S] T. Stieltjes, *Recherches sur les fraction continues,* Annales de la Faculté de Sci. de Toulouse, vol. **VIII** (1894), 1–122; vol. **IX** (1895), 1–47.

[VIP] F.P. Vasil'ev, A.Z. Ishmuhamedov and M.M. Potapov, *Generalized method of moments in optimal control problems,* Moscow, Noscow University, 1989.

A.A. Nudelman
ul. Balkovskaya, 141/187
270005, Odessa
Ukraine

MSC Primary 30E05, 44A60, Secondary 30E20, 47A57, 73K03

Operator Theory:
Advances and Applications, Vol. 118

Solution of the Truncated Matrix Hamburger Moment Problem According to M.G. Krein

V.M. Adamyan and I.M. Tkachenko*

The matrix version of the truncated Hamburger moment problem is considered. The criterium of its solvability is specified and a matrix version of the scalar Nevanlinna formula describing all possible solutions is established. As a main tool methods of the theory of extensions of symmetric operators and, in particular, the M.G. Krein formula for generalized resolvents are used.

1 Introduction

Let

$$\sigma(t) = (\sigma_{ik}(t))_{i,k=1}^{m} \ (m < \infty; -\infty < t < \infty)$$

be a Hermitian matrix functions satisfying conditions:

For $t'' \geq t'$ and any complex numbers $(\xi_k)_1^m$

(1) $$\sum_{i,k=1}^{m} (\sigma_{ik}(t'') - \sigma_{ik}(t'))\overline{\xi}_i \xi_k \geq 0;$$

(2) $$\sigma(-\infty) = 0, \ \sigma(t-0) = \sigma(t).$$

We will use the symbol S_m to denote the class of all matrix functions σ satisfying (1) and (2). We call elements of S_m *(matrix) distributions*. Let $\mathbf{C}_m$ denote the linear space of column vectors $\boldsymbol{\xi} = (\boldsymbol{\xi}_1, \ldots, \xi_m)^T$ with complex components and the standard scalar product

$$(\boldsymbol{\xi}, \boldsymbol{\eta}) = \sum_{j=1}^{m} \overline{\eta}_j \xi_j.$$

For each $\boldsymbol{\xi} \in \mathbf{C}_m$ the scalar function

$$\sigma_{\xi}(t) = (\sigma(t)\boldsymbol{\xi}, \boldsymbol{\xi}) = \sum_{i,k=1}^{m} \sigma_{ik}(t)\overline{\xi}_i \xi_k$$

is a non-negative measure on the real axis $\mathbb{R}$.

*This author was supported by the Government of Ukraine and the USA Civillian Research and Development Foundation, Award UM1 - 298.

Let us consider an ordered set $\widehat{C}_n := (C_0, \ldots, C_{2n})$ of $m \times m$ Hermitian matrices with the property:

$$\sum_{j,k=0}^{s} (C_{j+k}\xi_k, \xi_j) \geq 0,\ 0 \leq s \leq n, \tag{3}$$

whatever the vectors $\xi_0, \ldots, \xi_s$ from $\mathbf{C}_m$ are. The purpose of this work is to describe the subset $S_m(\widehat{C}_n) \subset S_m$ of all matrix distributions σ, such that

$$\int_{-\infty}^{\infty} t^k \, d\sigma(t) = C_k, k = 0, 1, 2, \ldots, 2n. \tag{4}$$

We call $C_0, \ldots, C_{2n}$ *(matrix) moments* of the distribution σ. The problem of description of the subset $S_m(\widehat{C}_n)$ by the given *finite* set of matrix moments $\widehat{C}_n$, we call it the truncated Hamburger moment problem. Remember that by virtue of the M.G. Krein theorem for $n = \infty$ Eq. (3) is equivalent to the claim that $\widehat{C}_\infty$ is the sequence of power moments of some matrix distribution [6].

The exhaustive solution of the scalar version of formulated problem has been found by R. Nevanlinna still in early twenties [9]. Let us state for further use the following extract from known results for the scalar case, see [7], [11], [1], [8].

Let a set $\hat{c}_n$ of complex number $c_0, \ldots, c_{2n}$ have the property: for any non-zero set of complex numbers $\xi_0, \ldots, \xi_n$ $(\max_{0 \leq j \leq n} |\xi_j| > 0)$

$$\sum_{j,k=0}^{n} (c_{j+k}\xi_k, \xi_j) > 0.$$

Then there exists an infinite set $s(\hat{c}_n)$ of non-negative measures σ on the real axis such that

$$\int_{-\infty}^{\infty} t^k \, d\sigma(t) = c_k \ ,\ k = 0, 1, 2, \ldots, 2n. \tag{5}$$

Let $(D_k(t))_{k=0}^n$ be the finite set of polynomials constructed according to the formulas

$$D_0 = \frac{1}{\sqrt{c_0}}, \quad D_k(t) = \frac{1}{\sqrt{\Delta_k \Delta_{k-1}}} \det \begin{Vmatrix} c_0 & \cdots & c_k & 1 \\ c_1 & \cdots & c_{k+1} & t \\ \vdots & \vdots & \vdots & \vdots \\ c_{k+1} & \cdots & c_{2k} & t^k \end{Vmatrix},$$

$$\Delta_k = \det \begin{Vmatrix} c_0 & \cdots & c_k \\ \vdots & \vdots & \vdots \\ c_{k+1} & \cdots & c_{2k} \end{Vmatrix}, \ k = 1, 2, \ldots, n. \tag{6}$$

Polynomials D_k form an orthonormal system with respect to each σ-measure from $s(\hat{c}_n)$. Let

$$D_0^* \equiv 0, \ D_k^*(t) = \int_{-\infty}^{\infty} \frac{D_k(t) - D_k(s)}{t - s} \, d\sigma(s), \ k = 1, \ldots, n,$$

be the corresponding set of conjugate polynomials. Then the formula

$$\text{(7)} \qquad \int_{-\infty}^{\infty} \frac{d\sigma(t)}{z-t} = \frac{D_{n+1}^*(z) + \omega(z) D_n^*(z)}{D_{n+1}(z) + \omega(z) D_n(z)}, \quad \text{Im } z > 0,$$

establishes a one-to-one correspondence between the elements of $s(\hat{c}_n)$ and the elements $\omega(z)$ of the class of all Nevanlinna functions in the upper half-plane, i.e. functions, which are holomorphic in the upper half-plane, map it into itself, and are such that $\omega(z)/z \to 0$ as $z \to \infty$ inside any angel $\varepsilon < \arg z < \pi - \varepsilon$, $0 < \varepsilon < \pi$.[1]

Description of all matrix measures having a given set $\widehat{C}_n$ as their first $2n+1$ moments was obtained in [5]. The approach to this problem in [5] was based on the method of matrix inequalities developed by V. Potapov. Despite of visible efficiancy of algebraic methods in some applications[2], the results obtained in [5] for matrix versions of the moment problem are not so transparent and elegant as in the scalar case.

This paper is intended as an attempt to find such matrix analogue of the Nevanlinna formula (7), which would not differ much in its form from that for the scalar case. We do not and can not pretend here a great originality. Note that the Nevanlinna-type description formula for all solution of the matrix Hamburger moment problem in the form of linear fractional transformations with matrix orthogonal polynomials as coefficients for the cases where the quadrqtic form (3) is strictly positive was earlier established by H. Dym [4] using the original methods of reproducing kernel space. Actually M.G. Krein in [6] was not far from a solution of this question. It is worth to remind that, in general, for M.G. Krein moment problems were always a source of creative search and a quest of inspirations. The aim of this work was to demonstrate again how some of M.G. Krein's ideas apparently from the abstract operator theory when focused on a concrete question of analysis, immediately bring about concrete and complete results.

In Section 2 we specify the solvability criterium for the truncated moment problem. Though the power moment problems have been posed more then hundred years ago it was pointed out much later that the classical "full" moment problems does not include the truncated problems as a special case [2], [3]. The refinement of the solvability criterium obtained in [3] for the truncated scalar moment problems is reproduced for the matrix Hamburger moment problem in Section 2 using the usual approach of M. Stone.

In Section 3 we consider only the non-degenerate case, where the block Hankel matrix $(C_{j+k})_{j,k=0}^n$ have the maximal rank $m(n+1)$. Here using the matrix analogue of orthogonal polynomials with respect to a matrix measure on the real axis, we describe the subset of so called canonical solutions $\subset S_m(\widehat{C}_n)$.

---

[1] Moments c_{2n+1} and c_{2n+2} are immaterial part of (7). Their variation leads only to the variation of $\omega(z)$.

[2] See: [10].

Let us denote by $\mathbb{N}_0$ the Nevanlinna set of all holomorphic in the upper half-plane $l \times l$, $1 \leq l \leq m$, matrix functions $\Omega(z)$ with the following properties:

$$\frac{1}{2i}(\Omega(z) - \Omega(z)^*) \geq 0, \ \text{Im } z > 0;$$

$$\lim_{z \to \infty} \frac{1}{z}\Omega(z) = 0, \varepsilon < \arg z < \pi - \varepsilon, \ 0 < \varepsilon < \pi.$$

In Section 4 for the non-degenerate case and for the degenerate case where

$$mn \leq \ rank \ (C_{j+k})^n_{j,k=0} < m(n+1)$$

we give a matrix version of the Nevanlinna formula, which like (7) establishes a one-to-one correspondence between elements of $S_m(\widehat{C}_n)$ and elements of $\mathbb{N}_0$.

2 Existence of Solutions of the Truncated Hamburger Problem for Matrix Moments

The *truncated Hamburger problem* for matrix moments is formulated in the following way:

Given a set of Hermitian $m \times m$ matrices

$$\{C_0, C_1, C_2, \ldots, C_{2n}\}, \quad n = 0, 1, 2, \ldots. \tag{8}$$

To find all matrix measures $\sigma(t)$ such that

$$\int_{-\infty}^{\infty} t^k \, d\sigma(t) = C_k \ , \ k = 0, 1, 2, \ldots, 2n. \tag{9}$$

The next statement for $n = \infty$ was actually proven in [6], where the standard Hamburger matrix moment problem for matrix distributions was considered as a simple application of the developed theory of extensions of symmetric operators. This is the specified statement with a quite analogous proof for the truncated Hamburger problem.

Theorem 1 *A system of Hermitian matrices* (8) *admits representation* (9) *if and only if the block-Hankel matrix $(C_{k+j})^n_{k,j=0}$ is non-negative and for any set*

$$\xi_0, \ldots, \xi_s \in \mathbf{C}_m, \ 0 \leq s \leq n-1,$$

the condition

$$\sum_{j,k=0}^{s} (C_{j+k}\xi_k, \xi_j) = 0 \tag{10}$$

implies

$$\sum_{j,k=0}^{s} (C_{j+k+2}\xi_k, \xi_j) = 0. \tag{11}$$

Proof: Suppose that the relations (9) hold. For an arbitrary set of m-dimensional complex vectors $\xi_0, \xi_1, \xi_2, \dots, \xi_n$ we define

$$\mathbf{P}(t) = \xi_0 + \xi_1 t + \xi_2 t^2 + \cdots + \xi_s t^s. \tag{12}$$

By (9)

$$\sum_{k,j=0}^{s} (C_{j+k}\,\xi_k, \xi_j) = \int_{-\infty}^{+\infty} (d\sigma(t)\mathbf{P}(t), \mathbf{P}(t)) \geq 0. \tag{13}$$

Hence the block matrix $(C_{k+j})_{k,j=0}^{n}$ is non-negative. If for some set $\xi_0, \dots, \xi_s \in \mathbf{C}_m$, $0 \leq s \leq n-1$, Eq. (10) holds, then for the vector polynomial $\mathbf{P}(t)$ defined by (12) we have:

$$\int_{-\infty}^{\infty} (d\sigma(t)\mathbf{P}(t), \mathbf{P}(t)) = 0$$

and hence,

$$\sum_{k,j=0}^{s} (C_{j+k+2}\,\xi_k, \xi_j) = \int_{-\infty}^{+\infty} t^2\,(d\sigma(t)\mathbf{P}(t), \mathbf{P}(t)) = 0.$$

Assume now that the matrix $(C_{k+j})_{k,j=0}^{n}$ satisfies the theorem conditions. Construct a finite-dimensional unitary space $\mathcal{C}$ of column vectors $\widehat{\xi} = (\xi_0, \xi_1, \xi_2, \dots, \xi_n)^T$, $\xi_0, \dots, \xi_n \in \mathbf{C}_m$ taking the bilinear functional

$$\langle \widehat{\xi}, \widehat{\eta} \rangle = \sum_{k,j=0}^{n} (C_{k+j}\,\xi_k, \eta_j) \tag{14}$$

as a scalar product. Let $\mathcal{C}_0$ be the subspace of $\mathcal{C}$ consisting of all vectors $\widehat{\xi}$ such that $\|\widehat{\xi}\| := \sqrt{\langle \widehat{\xi}, \widehat{\xi} \rangle} = 0$. If $\widehat{\eta} = \widehat{\xi} + \widehat{\xi}_0$, where $\widehat{\xi} \in \mathcal{C}$, $\widehat{\xi}_0 \in \mathcal{C}_0$, than, due to the Schwartz inequality $\langle \widehat{\xi}, \widehat{\xi}_0 \rangle = 0$ and hence $\|\widehat{\eta}\| = \|\widehat{\xi}\|$. Let us denote by $\widetilde{\mathcal{C}}$ the factor -space $\mathcal{C}/\mathcal{C}_0$. For $\widehat{\eta} = \widehat{\xi} + \mathcal{C}_0$ we set $\|\widehat{\eta}\|_{\widetilde{\mathcal{C}}} = \|\widehat{\xi}\|$. With this definition of the norm $\widetilde{\mathcal{C}}$ is a finite-dimensional Hilbert space. Let $\widetilde{\mathcal{D}}_0$ be subspace of $\widetilde{\mathcal{C}}$, each element of which is a class containing vectors of the form $(\xi_0, \xi_1, \dots, \xi_{n-1}, 0)^T$. For a class of $\widetilde{\mathcal{D}}_0$ containing some vector $\widehat{\xi} = (\xi_0, \xi_1, \dots, \xi_{n-1}, 0)^T$ we denote by $A_0\widehat{\xi}$ the class containing $(0, \xi_0, \xi_1, \dots, \xi_{n-1})^T$. By virtue of our assumptions $\widehat{\xi} \in \widetilde{\mathcal{D}}_0 \cap \widetilde{\mathcal{C}}_0$ makes $A_0\widehat{\xi} \in \widetilde{\mathcal{C}}_0$. Therefore A_0 is a linear operator A_0 from $\widetilde{\mathcal{D}}_0$ into $\widetilde{\mathcal{C}}$. If $\widehat{\xi} = (\xi_0, \xi_1, \dots, \xi_{n-1}, 0)^T$ and $\widehat{\eta} = (\eta_0, \eta_1, \dots, \eta_{n-1}, 0)^T$ are representatives of some classes from $\widetilde{\mathcal{D}}_0$, then

$$\langle A_0\widehat{\xi}, \widehat{\eta} \rangle = \sum_{j,k=0}^{n-1} (C_{j+k+1}\xi_k, \eta_j) = \langle \widehat{\xi}, A_0\widehat{\eta} \rangle,$$

i.e. A_0 is a symmetric operator. Hence there exists a self-adjoint operator A acting in $\widetilde{\mathcal{C}}$ or in some Hilbert space $\mathcal{H}$, which is an extension of $\widetilde{\mathcal{C}}$ such that $\widetilde{\mathcal{D}}_0$ belongs to the domain $\widetilde{\mathcal{D}}_A$ of A and

(15) $$A_{|\widetilde{\mathcal{D}}_0} = A_0.$$

Let $\mathbf{E}_t$, $-\infty < t < \infty$, be the unity decomposition of A. Taking the set of classes $\{\widehat{\mathbf{e}}_{1,0} + \widetilde{\mathcal{C}}_0, \ldots, \widehat{\mathbf{e}}_{m,n} + \widetilde{\mathcal{C}}_0\} \subset \widetilde{\mathcal{C}} \subseteq \mathcal{H}$, where

$$\widehat{\mathbf{e}}_{1,0} := (\mathbf{e}_1, 0, \ldots, 0)^T, \ldots, \widehat{\mathbf{e}}_{m,n} := (0, \ldots, 0, \mathbf{e}_m)^T$$

and

$$\mathbf{e}_1 := (1, 0, \ldots, 0)^T, \ldots, \mathbf{e}_m = (0, \ldots, 0, 1)^T$$

is a canonical basis of $\mathbf{C}_m$ let us introduce the non-decreasing $m \times m$ matrix function $\sigma(t)$, $-\infty < t < \infty$, of bounded variation with elements

(16) $$\sigma_{r,s}(t) := (\mathbf{E}_t \widehat{\mathbf{e}}_{s,0}, \widehat{\mathbf{e}}_{r,0}),\ 1 \leq r, s \leq m.$$

Due to (15),

$$\widehat{\mathbf{e}}_{s,l} = A_0^l \widehat{\mathbf{e}}_{s,0} = A^l \widehat{\mathbf{e}}_{s,0},\ 1 \leq s \leq m;\ 0 \leq l \leq n.$$

Hence

$$\begin{aligned}(C_{l+k})_{r,s} &= (C_{l+k}\mathbf{e}_s, \mathbf{e}_r) = \langle \widehat{\mathbf{e}}_{s,l}, \widehat{\mathbf{e}}_{r,k} \rangle \\ &= \langle A^l \widehat{\mathbf{e}}_{s,0}, A^k \widehat{\mathbf{e}}_{r,0} \rangle = \int_{-\infty} t^{l+k} d\sigma_{r,s}(t), \quad 0 \leq l, k \leq n.\end{aligned}$$

□

3 Matrix Orthogonal Polynomials and Canonical Solutions

Let A be a self-adjoint operator acting in $\widetilde{\mathcal{C}}$ and satisfying the condition (15). It is evident that $A = A_0$ if $\widetilde{\mathcal{D}}_0$ coincides with $\widetilde{\mathcal{C}}$. Otherwise A is a symmetric extension of A_0 without going out of $\widetilde{\mathcal{C}}$. Let $\mathbf{E}_{A,t}$ be the unity decomposition of such A. The matrix function $\sigma_A(t)$, elements of which are expressed through $\mathbf{E}_{A,t}$ by formulas (16), is some solution of the truncated Hamburger problem. All solutions of this problem obtained using such self-adjoint extensions of A_0 acting in $\widetilde{\mathcal{C}}$ will be called *canonical*. Observe that the number of points of growth for any canonical solution $\sigma_A(t)$ does not exceed $\dim \widetilde{\mathcal{C}} \leq (n+1)m$. From now on we will assume that the quadratic form (3) is non-degenerate. Then the subspace $\mathcal{C}_0$ of $\mathcal{C}$ contains only the zero-vector, $\mathcal{C} = \widetilde{\mathcal{C}}$ and $\dim \mathcal{C} = (n+1)m$. If this is the case, the truncated Hamburger problem is called *non-degenerate*.

Let $\mathbf{E}_t$, $-\infty < t < \infty$, be the unity decomposition of some fixed canonical self-adjoint extension A. Since the vectors

(17) $$\widehat{\mathbf{e}}_{s,j} = A^j \widehat{\mathbf{e}}_{s,0} = \int_{-\infty}^{\infty} t^j \, d\mathbf{E}_t \widehat{\mathbf{e}}_{s,0},\ j = 0, 1, \ldots, n;\ s = 1, \ldots, m$$

form a basis of the Hilbert space , any vector $\widehat{f} \in C$ is representable in the form

$$\widehat{\mathbf{f}} = \sum_{r=1}^{m} \int_{-\infty}^{\infty} f_r(t)\, d\mathbf{E}_t \widehat{e}_{r,0}, \tag{18}$$

where $f_r(t)$ are polynomials of degrees not exceeding n. Thus, as usual, C can be realized as a Hilbert space $\mathcal{L}_n$ of vector functions

$$\mathbf{f}(t) = (f_1(t), \dots, f_m(t))^T,$$

components of which are polynomials of degrees not higher than n and the scalar product $\langle .,. \rangle$ is defined as :

$$\langle \mathbf{f}, \mathbf{g} \rangle = \int_{-\infty}^{\infty} \mathbf{g}(t)^* \, d\sigma_A(t)\, \mathbf{f}(t) = \sum_{s,r=1}^{m} \int_{-\infty}^{\infty} \overline{g_s(t)}\, f_r(t) d\sigma_{A,sr}(t). \tag{19}$$

Set

$$\Gamma_s = \begin{pmatrix} C_0 & C_1 & \cdots & C_s \\ C_1 & C_2 & \cdots & C_{s+1} \\ \vdots & \vdots & \vdots & \vdots \\ C_s & C_{s+1} & \cdots & C_{2s} \end{pmatrix},$$

$$\Xi_s = \Big(\underbrace{0, \dots, 0}_{s}, I\Big)^T, \Theta_s(t) = \left(I, tI, \dots, t^s I\right), s = 0, 1, \dots n,$$

where I is the $m \times m$ - unit matrix. According to our assumptions the matrices Γ_s are invertible. Let us introduce $m \times m$ - matrix polynomials

$$D_s(t) = \Theta_s(t) \Gamma_s^{-1} \Xi_s, \; s = 0, \dots, n.$$

By definition of $D_s(t)$ for each matrix measure $\sigma(t)$ satisfying conditions (9)

$$\int_{-\infty}^{\infty} t^r \, d\sigma(t) D_s(t) = \sum_{l=0}^{s} C_{r+l} (\Gamma_s^{-1})_{ls} = \delta_{rs} I, \; r \leq s, 0 \leq s \leq n, \tag{20}$$

where δ_{rs} is the Kronecker symbol. (20) and their adjoints

$$\int_{-\infty}^{\infty} t^r \, D_s(t)^* d\sigma(t) = \delta_{rs} I, \; r \leq s, 0 \leq s \leq n, \tag{21}$$

yield the orthogonality relations

$$\int_{-\infty}^{\infty} D_{s'}(t)^* \, d\sigma(t) D_s(t) = \delta_{s's} (\Gamma_s^{-1})_{ss}, \; 0 \leq s', s \leq n. \tag{22}$$

Observe that due to Eq. (21) the condition

$$\int_{-\infty}^{\infty} D_s(t)^* d\sigma(t)\mathbf{x}(t) = 0, \quad 0 \leq s \leq n,$$

for some vector polynomial $\mathbf{x} \in \mathcal{L}_n$ of a degree not exceeding n, is equivalent to the identity $\mathbf{x}(t) \equiv 0$. Thus, each vector polynomial $\mathbf{h} \in \mathcal{L}_n$ admits the unique representation of the form

$$\mathbf{h}(t) = \sum_0^n D_s(t)\mathbf{h}_s, \tag{23}$$

where

$$\mathbf{h}_s = (\Gamma_s^{-1})_{ss}^{-1} \int_{-\infty}^{\infty} D_s(t)^* d\sigma(t)\mathbf{h}(t). \tag{24}$$

By construction A_0 and its self-adjoint extension A act in $\mathcal{L}_n$ on vector polynomials of degree not exceeding $n-1$ as the multiplication operator on the variable t. In particular, for vector polynomials $(D_s\mathbf{h})(t) \equiv D_s(t)\mathbf{h}$, $1 \leq s \leq n-1$, $\mathbf{h} \in \mathbf{C}_m$, using Eqs. (20–24) we obtain

$$(AD_s\mathbf{h})(t) = (A_0D_s\mathbf{h})(t) = tD_s(t)\mathbf{h} = D_{s-1}(t)\mathbf{h} + D_s(t)R_s\mathbf{h} + D_{s+1}(t)L_s\mathbf{h},$$

where

$$R_s = (\Gamma_s^{-1})_{ss}^{-1} \int_{-\infty}^{\infty} tD_s(t)^* d\sigma(t)D_s(t), \; L_s = (\Gamma_{s+1}^{-1})_{s+1,s+1}^{-1}(\Gamma_s^{-1})_{ss}. \tag{25}$$

Setting $D_{-1}(t) \equiv 0$, we obtain that matrix polynomials $D_s(t)$ satisfy the system of recursion relations

$$tD_s(t) = D_{s-1}(t) + D_s(t)R_s + D_{s+1}(t)L_s, \; 0 \leq s \leq n-1. \tag{26}$$

¿From the property of symmetry of A and (25), we see that

$$\begin{aligned} \langle AD_nh, D_sq\rangle &= \langle D_nh, A_0D_sq\rangle = \langle D_nh, D_{s-1}q\rangle \\ &+ \langle D_nh, D_sQ_sq\rangle + \langle D_n\mathbf{h}, D_{s+1}R_s\mathbf{q}\rangle = 0, \; 0 \leq s \leq n-2; \end{aligned}$$

$$\begin{aligned} \langle AD_n\mathbf{h}, D_{n-1}\mathbf{q}\rangle &= \langle D_n\mathbf{h}, D_nR_{n-1}\mathbf{q}\rangle \\ &= \langle (\Gamma_{n-1}^{-1})_{n-1,n-1}\mathbf{h}, \mathbf{q}\rangle, \; \mathbf{h}, \mathbf{q} \in \mathbf{C}_m. \end{aligned} \tag{27}$$

Writing representation (23) for a vector polynomial $(AD_n\mathbf{h})(t)$, $\mathbf{h} \in \mathbf{C}_m$ and applying (27), we obtain the representation

$$(AD_n\mathbf{h})(t) = D_{n-1}(t)\mathbf{h} + D_n(t)R_A\mathbf{h}. \tag{28}$$

By (24) for the quadratic form of matrix R_A we get

$$((\Gamma_n^{-1})_{nn}R_A\mathbf{h},\mathbf{q})_{\mathbf{C}_m} = \left((\Gamma_n^{-1})_{nn}(\Gamma_n^{-1})_{nn}^{-1} \int_{-\infty}^{\infty} D_n(t)^* d\sigma(t)(AD_n\mathbf{h})(t),\mathbf{q}\right)_{\mathbf{C}_m} \tag{29}$$
$$= \langle AD_n\mathbf{h}, D_n\mathbf{q}\rangle = \overline{((\Gamma_n^{-1})_{nn}R_A\mathbf{q},\mathbf{h})_{\mathbf{C}_m}}$$
$$= (R_A^*(\Gamma_n^{-1})_{nn}\mathbf{h},\mathbf{q})_{\mathbf{C}_m},\quad \mathbf{h},\mathbf{q}\in\mathbf{C}_m.$$

Hence R_A in (28) admits representation

$$R_A = (\Gamma_n^{-1})_{nn}G_A,\ G_A = G_A^*. \tag{30}$$

We are now in a position to deduce an explicit representation for the resolvent of A in $\mathcal{C}$. To this end, let us consider the equation

$$(A - z)\widehat{\mathbf{f}} = \widehat{\mathbf{g}}, \tag{31}$$

where $\widehat{\mathbf{g}}$ is arbitrary vector of $\mathcal{C}$ and Im $z \neq 0$. According to the above $\mathcal{L}_n$-realization (17–19) for $\widehat{\mathbf{g}}$ and $\widehat{\mathbf{f}}$, there exist unique vector polynomials $\mathbf{g}(t)$ and $\mathbf{f}(t;z)$ of degree not exceeding n, which are isomorphic images of $\widehat{\mathbf{g}}$ and $\widehat{\mathbf{f}}$ in $\mathcal{L}_n$. In turn, $\mathbf{f}(t;z)$ can be uniquely represented in the form

$$\mathbf{f}(t;z) = \mathbf{f}_0(t;z) + D_n(t)\mathbf{h}(z), \tag{32}$$

where $\mathbf{f}_0(t;z)$ is the vector polynomial of degree not exceeding $n-1$, which is the image of the orthogonal projection of $\widehat{\mathbf{f}}$ onto the above subspace $\mathcal{D}_0$, and $D_n(t)\mathbf{h}(z)$ is the image of the orthogonal projection of $\widehat{\mathbf{f}}$ onto the defect subspace of A_0. Since the image of A_0 in $\mathcal{L}_n$ is simply the multiplication operator on "t" Eq. (31) in $\mathcal{L}_n$ with account of (28) has the form:

$$(t - z)\mathbf{f}_0(t;z) + [D_n(t)(R_A - z) + D_{n-1}(t)]\mathbf{h}(z) = \mathbf{g}(t). \tag{33}$$

We see that

$$[D_n(z)(R_A - z) + D_{n-1}(z)]\mathbf{h}(z) = \mathbf{g}(z). \tag{34}$$

Proposition 2 *For* Im $z \neq 0$ *the matrix function*

$$Y_n(z) := D_n(z)(R_A - zI) + D_{n-1}(z) \tag{35}$$

is invertible and, and thus,

$$\mathbf{h}(z) = Y_n(z)^{-1}\mathbf{g}(z). \tag{36}$$

Proof: Suppose that the matrix $Y_n(z_0)$ is non-invertible for some non-real z_0. Then there exists a non-trivial solution $\mathbf{x}_0 \in \mathbf{C}_m$ of the equation $Y_n(z_0)\mathbf{x}_0 = 0$. Taking such $\mathbf{x}_0$, let us consider the vector polynomial

$$\mathbf{x}_0(t; z_0) := (D_n(t)(R_A - z_0 I) + D_{n-1}(t))\,\mathbf{x}_0.$$

By construction, $\mathbf{x}_0(z_0; z_0) = 0$. However $\mathbf{x}_0(t; z_0)$ is not identically equal to zero for $\mathbf{x}_0 \neq 0$. Indeed, by (22), the vector polynomials $D_n(t)(R_A - z_0 I)\mathbf{x}_0$ and $D_{n-1}(t)\,\mathbf{x}_0$ are orthogonal in $\mathcal{L}_n$. Therefore, the identity

$$D_n(t)(R_A - z_0 I)\mathbf{x}_0 \equiv -D_{n-1}(t)\,\mathbf{x}_0$$

can be valid only if $\mathbf{x}_0 = 0$. Hence, $\mathbf{x}_0(t; z_0)$ is a non-zero vector polynomial such that $\mathbf{x}_0(z_0; z_0) = 0$. Thus,

$$\frac{1}{t - z}\mathbf{x}_0(t; z_0)$$

is a non-zero vector polynomial of a degree not exceeding $(n - 1)$. Consider now the vector polynomial

$$\mathbf{y}(t) := -\frac{1}{t - z}\mathbf{x}_0(t; z_0) + D_n(t)\,\mathbf{x}_0, \tag{37}$$

which is also non-zero since the summands in (37) are orthogonal in $\mathcal{L}_n$. Let $\widehat{\mathbf{y}} \in \mathcal{C}$ be the vector, the $\mathcal{L}_n$-realization of which coincides with $\mathbf{y}(t)$. Then for the $\mathcal{L}_n$-realization $\mathbf{y}_1(t)$ of the vector $\widehat{\mathbf{y}}_1 = (A - z_0)\widehat{\mathbf{y}}$, we obtain

$$\mathbf{y}_1(t) = -\mathbf{x}_0(t; z_0) + (D_n(t)(R_A - z_0 I) + D_{n-1}(t))\,\mathbf{x}_0 \equiv 0.$$

Hence z_0 is a non-real eigenvalue of A and $\widehat{\mathbf{y}}$ is a non-zero eigenvector related to this eigenvalue. This is a contradiction since A is a self-adjoint operator. □

Equations (32, 33) and Proposition 3.1 make it obvious that

$$((A - z)^{-1}\widehat{\mathbf{g}})(t)(= \mathbf{f}(t; z)) = \frac{1}{t - z}\mathbf{g}(t) - \frac{1}{t - z}Y_n(t)Y_n(z)^{-1}\mathbf{g}(z) \tag{38}$$

For the canonical matrix distribution $\boldsymbol{\sigma}_A(t) = (\sigma_{A,sr}(t))_{s,r=1}^m$ generated by a canonical extension A, let us consider the $m \times m$ matrix function

$$\mathbf{F}_A(z) = \int_{-\infty}^{\infty} \frac{1}{t - z}\,d\sigma_A(t), \text{ Im } z \neq 0. \tag{39}$$

By definition of the unity decomposition for elements $F_{sr}(z)$ of $\mathbf{F}(z)$, we have

$$\begin{aligned} F_{sr}(z) &= \int_{-\infty}^{\infty} \frac{1}{t - z}\,d\sigma_{A,sr}(t) = \int_{-\infty}^{\infty} \frac{1}{t - z}\,d(\mathbf{E}_t\widehat{\mathbf{e}}_{r,0}, \widehat{e}_{s,0}) \\ &= \langle (A - zI)^{-1}\widehat{\mathbf{e}}_{r,0}, \widehat{\mathbf{e}}_{s,0}\rangle, \ 1 \leq r, s \leq m, \text{ Im } z \neq 0 \end{aligned} \tag{40}$$

Observe that for the above $\mathcal{L}_n$-images of vectors, $\widehat{\mathbf{e}}_{r,0}, \widehat{\mathbf{e}}_{s,0}$ are zero-order vector polynomials $\mathbf{e}_{r,0}(t) \equiv \mathbf{e}_r, \mathbf{e}_{s,0}(t) \equiv \mathbf{e}_s$. Substituting $\widehat{\mathbf{e}}_{r,0}$ into (38) as $\widehat{\mathbf{g}}$ and calculating of the $\mathcal{L}_n$- scalar product of the obtained expression on $\widehat{\mathbf{e}}_{s,0}$, we get

$$(41) \qquad F_{A,sr}(z) = \int_{-\infty}^{\infty} \frac{1}{t-z} d(\sigma_A(t)\,(Y_n(t) - Y_n(z))\,Y_n(z)^{-1}\mathbf{e}_r, \mathbf{e}_s).$$

Let us denote by $E_s(z), 0 \le s \le n$, the adjoint matrix polynomial defined by

$$(42) \qquad E_s(z) := \int_{-\infty}^{\infty} \frac{1}{t-z} d\sigma(t)\,(D_s(t) - D_s(z)),$$

where any matrix measure satisfying (9) can stand as $\sigma(t)$. Observe that matrix polynomials $E_s(z)$ satisfy the same relations (26, 27) as $D_s(t)$ do. Taking into account the equality

$$\int_{-\infty}^{\infty} d\sigma(t) D_n(t) = 0$$

we conclude from (41) and (30) that

$$(43) \qquad \begin{aligned} \int_{-\infty}^{\infty} \frac{1}{t-z} d\sigma_A(t) &= (\mathbf{F}_A(z)) \\ &= (E_n(z)(R_A - zI) + E_{n-1}(z)) \\ &\quad \times (D_n(z)\,(R_A - zI) + D_{n-1}(z))^{-1}, \\ R_A &= (\Gamma_n^{-1})_{nn} G_A, \; G_A = G_A^*, \; \text{Im } z \ne 0. \end{aligned}$$

Hermitian matrix G_A is the only parameter in the right hand side of (43), which is responsible for the choice of a certain canonical self-adjoint extension of A_0. Let us denote by $\mathcal{N}$ the defect subspace $\mathcal{N} = \mathcal{C} \ominus \mathcal{D}$ of A_0. By (29) $P_{\mathcal{N}} A_{|\mathcal{N}}$ is isomorphic to the operator, which is represented in the natural basis of $\mathbf{C}_m$ by the matrix $(\Gamma_n^{-1})_{nn}^{-1/2} G_A (\Gamma_n^{-1})_{nn}^{-1/2}$. Since there are no restrictions on $P_{\mathcal{N}} A_{|\mathcal{N}}$ except of the property of self-adjointness, any Hermitian matrix can play the part of G_A. Applying the Stieltjes inversion formula, one can reconstruct the measure σ_A corresponding to a given parameter G_A. Besides, *matrix functions* $\mathbf{F}_{A_1}(z)$ *and* $\mathbf{F}_{A_2}(z)$ *obtained as the linear fractional transformations in the right hand side of* (43) *over different Hermitian parameters* $G_{A_1} \ne G_{A_2}$ *and respectively related canonical solutions* $\sigma_{A_1}(t)$ *and* $\sigma_{A_2}(t)$, *cannot coincide.*

To see that the last claim is true, let us first indicate that the matrix polynomials appearing in (33) obey the identities

$$(44) \qquad D_{n-1}(\overline{z})^* E_n(z) - E_{n-1}(\overline{z})^* D_n(z) \equiv (\Gamma_n^{-1})_{nn};$$

$$(45) \qquad D_n(\overline{z})^* D_n(z) - E_n(\overline{z})^* E_n(z) \equiv 0;$$

$$(46) \qquad D_{n-1}(\overline{z})^* D_{n-1}(z) - E_{n-1}(\overline{z})^* E_{n-1}(z) \equiv 0.$$

Actually these relations follow from the fact that

$$\mathbf{F}_A(\overline{z})^* = \mathbf{F}_A(z)$$

independently of the choice of the Hermitian parameter G_A. But also they follow immediately from the definition and already noticed properties of these matrix polynomials. We give here only the proof of (44); other identities can be verified analogously.

By definition of $E_s(z)$, for a non-real z, we can write

$$\begin{aligned}\Delta_n(z) &:= D_{n-1}(\overline{z})^* E_n(z) - E_{n-1}(\overline{z})^* D_n(z)\\ &= D_{n-1}(\overline{z})^* \int_{-\infty}^{\infty} \frac{1}{t-z} d\sigma(t) D_n(t)\\ &\quad - \int_{-\infty}^{\infty} \frac{1}{t-z} D_{n-1}(t)^* d\sigma(t) D_n(z).\end{aligned}$$

Writing

$$\frac{1}{t-z} = -\frac{1}{z} - \cdots - \frac{t^{n-1}}{z^n} + \frac{t^n}{z^n(t-z)}$$

and applying (20, 21), we get for $z \to \infty$

$$\Delta_n(z) = \frac{1}{z^n} D_n(z) + O\left(\frac{1}{z}\right) = (\Gamma_n^{-1})_{nn} + O\left(\frac{1}{z}\right).$$

But $\Delta_n(z)$ is a polynomial of a degree not exceeding $2n-1$. Hence $\Delta_n(z) \equiv (\Gamma_n^{-1})_{nn}$.

Using (44–46), we see that

$$\begin{aligned}&\mathbf{F}_{A_2}(z) - \mathbf{F}_{A_1}(z) = \mathbf{F}_{A_2}(\overline{z})^* - \mathbf{F}_{A_1}(z)\\ &= ((R_{A_2}^* - zI) D_n(\overline{z})^* + D_{n-1}(\overline{z})^*)^{-1} ((R_{A_2}^* - zI) E_n(\overline{z})^* + E_{n-1}(\overline{z})^*)\\ &\quad - (E_n(z)(R_{A_1} - zI) + E_{n-1}(z)) \times (D_n(z)(R_{A_1} - zI) + D_{n-1}(z))^{-1}\\ &= ((R_{A_2}^* - zI) D_n(\overline{z})^* + D_{n-1}(\overline{z})^*)^{-1} (G_{A_2} - G_{A_1})\\ &\quad \times (D_n(z)(R_{A_1} - zI) + D_{n-1}(z))^{-1}.\end{aligned} \tag{47}$$

Therefore $\mathbf{F}_{A_2}(z) = \mathbf{F}_{A_1}(z)$ if and only if $G_{A_2} = G_{A_1}$. Thus we proved

Theorem 3 *The Nevanlinna-type formula* (43) *establishes in the non-degenerate case the one-to-one correspondence between the set of all Hermitian $m \times m$ matrix and the set of all canonical solutions of the truncated Hamburger $m \times m$ matrix moment problem.*

Further, if in fact all the moments up to the $2(n+1)$-th are known and the quadratic form of the block Hankel matrix $(C_{j+k})_{j,k=0}^{n+1}$ remains positive definite, we can introduce, as above, matrix polynomials $D_{n+1}(t)$, $E_{n+1}(t)$. Using the expression

$$tD_n(t)=D_{n-1}(t)+D_n(t)R_n+D_{n+1}(t)L_n$$

and the same expression for $tE_n(t)$, where R_n, L_n are defined like in (25), we can write (43) for the canonical solutions in the usual form of the Nevanlinna formula:

$$\begin{aligned} \mathbf{F}_A(z) &= (E_{n+1}(z)+E_n(z)\widehat{R}_A)(D_{n+1}(z)+D_n(z)\widehat{R}_{\tilde{A}})^{-1}, \\ \widehat{R}_A &= \widehat{G}_A(\Gamma_{n+1}^{-1})_{n+1,n+1}, \\ \widehat{G}_A &= \widehat{G}_A^* = R_n(\Gamma_n^{-1})_{nn}^{-1} - (\Gamma_n^{-1})_{nn}^{-1}G_A(\Gamma_n^{-1})_{nn}^{-1} \end{aligned} \tag{48}$$

Note that $\widehat{G}_A$ in (48), as before, runs the set of all Hermitian $m \times m$ matrices.

4 Description of all Solutions

We have pointed out already that each canonical or non-canonical self-adjoint extension of A_0 generates some solution of the truncated Hamburger problem. On the other hand, it is clear due to the above development, that each solution $\sigma(t)$ of this problem generates some self-adjoint extension A of A_0, which is unitary equivalent to the multiplication operator by the independent variable "t" in the space $\mathcal{L}_\sigma$ of all square integrable m-vector functions with respect to the matrix measure σ. The one-to-one connection between the simple part of A [3] and σ is established in the general case by (39, 40), where as A there can appear any self-adjoint extension of A_0. What we need further to obtain the complete description of all solutions of the truncated Hamburger problem, starting from (39, 40), is an explicit formula for generalized resolvents

$$W_z(A) := P_{\mathcal{C}}(A-z)^{-1}_{|\mathcal{C}},$$

where $P_{\mathcal{C}}$ is the orthogonal projection on the introduced above subspace $\mathcal{C}$ in the extended Hilbert space $\mathcal{H}$, where the self-adjoint extension A acts. But this is only a special case of the renown formula of M.G. Krein [6]. Let A_1 be some canonical extension of A_0 and let as before $P_{\mathcal{N}}$ be the orthoprojector onto the defect subspace $\mathcal{N}$. Set

$$Q_1(z) = P_{\mathcal{N}}(A_1-z)^{-1}_{|\mathcal{N}}$$

Then *the formula of M.G. Krein*

$$\begin{aligned} W_z(A) &= (A_1-z)^{-1} - (A_1-z)^{-1} \\ &\qquad P_{\mathcal{N}}(Q_1(z)+R(z))^{-1}P_{\mathcal{N}}(A_1-z)^{-1}, \operatorname{Im}\ z \geq 0 \end{aligned} \tag{49}$$

---

[3]The simple part of A is the restriction of A to its least invariant subspace containing all vectors $e_{s,0}$, $0 \leq s \leq m$.

establishes the one-to-one correspondence between the set of all generalized resolvents of A_0 and the set of all operator Nevanlinna holomorphic in the upper half-plane operator functions, the values of which are operators acting in $\mathcal{N}$.

To describe the whole set of matrix functions $\mathbf{F}_A(z)$, it remains now to substitute the expression (49) into (39, 40), and make rather simple calculations using explicit representations of $\mathcal{N}$, $(A_1 - z)^{-1}$and $Q_1(z)$ in the space of vector polynomials $\mathcal{L}_n$. We note that

$$\mathcal{N} \subset \mathcal{C} \longleftrightarrow \{\mathbf{f}(t) : \mathbf{f}(t) = D_n(t)\mathbf{x}, \mathbf{x} \in \mathbf{C}_m\} \subset \mathcal{L}_n,$$

and according to (36)

$$P_{\mathcal{N}}(A_1 - z)^{-1}\widehat{\mathbf{g}} \longleftrightarrow (D_n(z)(R_{A_1} - z) + D_{n-1}(z))^{-1}\mathbf{g}(z).$$

Hence

$$P_{\mathcal{N}}(A_1 - z)^{-1}\widehat{\mathbf{e}}_{s,0} \longleftrightarrow (D_n(z)(R_{A_1} - z) + D_{n-1}(z))^{-1}\mathbf{e}_s$$

and

$$Q_1(z) \longleftrightarrow (D_n(z)(R_{A_1} - z) + D_{n-1}(z))^{-1}D_n(z).$$

¿From these calculations with account of the last relations and Theorem 3 the following theorem results.

Theorem 4 *The Nevanlinna-type formula*

$$\begin{aligned} \int_{-\infty}^{\infty} \frac{1}{z-t} d\sigma_A(t) &= (E_n(z)(R_A(z) - zI) + E_{n-1}(z)) \\ &\quad \times (D_n(z)(R_A(z) - zI) + D_{n-1}(z))^{-1}, \\ R_A(z) &= (\Gamma_n^{-1})_{nn}\Omega(z), \ \text{Im}\ z \neq 0. \end{aligned} \tag{50}$$

establishes in the non-degenerate case the one-to-one correspondence between the set Nevanlinna $m \times m$ matrix functions $\Omega(z) \in \mathrm{N}_0$ and the set of all solutions of the truncated Hamburger $m \times m$ matrix moment problem.

In conclusion we discuss shortly the degenerate case, where the block Hankel matrix Γ_n is not invertible and satisfies conditions of Theorem 1. For simplicity we will assume that nevertheless Γ_{n-1} is invertible. The general case would demand only of piling up of in reality immaterial details. By Theorem 1 and our assumptions there exist at least one canonical extension A of A_0 acting in the introduced above factor-space $\widetilde{\mathcal{C}}$ and hence at least one matrix measure σ_A satisfying conditions (9). Since Γ_n is non-invertible there exist non-zero vector polynomials $\mathbf{g}(t)$ of degree n such that

$$\int_{-\infty}^{\infty} \mathbf{g}(t)^* \, d\sigma_A(t)\, \mathbf{g}(t) = 0. \tag{51}$$

Let us denote by $\mathcal{L}_n^0$ the linear set of such polynomials and by $\mathcal{L}_n$ the pre-Hilbert space of all vector polynomials of degree not exceeding n. Observe that by invertibility of Γ_{n-1} each non-zero polynomial satisfying the condition (51) has degree not less then n. Let us define further

$$\mathbf{N} := \left\{ \mathbf{y} : \mathbf{y} = \lim_{t\to\infty} \frac{1}{t^n}\mathbf{g}(t),\ \mathbf{g}(t) \in \mathcal{L}_n^0 \right\} \subseteq \mathbf{C}_m,$$

and $\mathbf{M} = \mathbf{N}^\perp$, where symbol $\perp$ means the orthogonal complement in $\mathbf{C}_m$ to the corresponding subspace with respect to the standardly defined euclidean scalar product. In the sequel $P_\mathbf{N}$ and $P_\mathbf{M}$ stand for the orthoprojectors in $\mathbf{C}_m$ on the subspaces $\mathbf{N}$ and $\mathbf{M}$, respectively. Observe that the introduced subspaces do not depend of the choice of the extension A.

Let us consider the $m(n+1)\times m$ matrix Ω_0 organized like the one column block matrix

$$\Omega_0 = \begin{pmatrix} X_0 \\ \dots \\ X_{n-1} \\ P_\mathbf{N} \end{pmatrix},$$

which satisfies the equation

$$\Gamma_n \Omega_0 = 0.$$

By this matrix we form the matrix polynomial

$$D_{n0}(t) = X_0 + tX_1 + \dots + t^{n-1}X_{n-1} + t^n P_\mathbf{N}.$$

Note that for each $\mathbf{h} \in \mathbf{M}$ we have $D_{n0}(t)\mathbf{h} = 0$. Indeed, for arbitrary $\mathbf{h} \in \mathbf{C}_m$

$$\left(\int_{-\infty}^{\infty} D_{n0}(t)^* d\sigma(t)(AD_{n0}\mathbf{h})(t), \mathbf{h} \right)_{\mathbf{C}_m} = \sum_{j,k=0}^{n} (C_{j+k}X_k\mathbf{h}, X_j\mathbf{h})_{\mathbf{C}_m} = 0.$$

But for $\mathbf{h} \in \mathbf{M}$ the last equality has form

$$\text{(52)} \qquad \sum_{j,k=0}^{n-1} (C_{j+k}\xi_k, \xi_j)_{\mathbf{C}_m} = 0,\ \xi_k = X_k\mathbf{h},\ k = 0, \dots, n-1.$$

Since Γ_{n-1} is invertible matrix it follows from (52) that $P_\mathbf{N}\mathbf{h} = 0$ entails $X_k\mathbf{h} = 0$, $k = 0, \dots, n-1$.

Let us consider now the block column matrix

$$\Omega_1 = \begin{pmatrix} Z_0 \\ \dots \\ Z_{n-1} \\ Z_n \end{pmatrix},$$

with $m \times m$ matrix entries, which satisfies the equation

$$\Gamma_n \Omega_1 = \begin{pmatrix} 0 \\ \dots \\ 0 \\ P_{\mathbf{M}} \end{pmatrix}, \tag{53}$$

where 0 stands for $m \times m$ null matrix and the condition

$$Z_n P_{\mathbf{N}} = 0. \tag{54}$$

Observe that (53) is solvable since by construction of $\mathbf{M}$ for each $\mathbf{h} \in \mathbf{M}$ the $(n+1)m$ component vector

$$\begin{pmatrix} 0 \\ \dots \\ 0 \\ \mathbf{h} \end{pmatrix}$$

is orthogonal in the euclidean metric of $\mathbf{C}_{(n+1)m}$ to the null-space of Γ_n. It follows from (53) and (54) that

$$\Delta_n Z_n = P_{\mathbf{M}}, \ \Delta_n = C_{2n} - \sum_{j,k=0}^{n-1} C_{j+n} (\Gamma_{n-1}^{-1})_{j,k} C_{k+n}. \tag{55}$$

It is worth mentioning that

$$\Delta_n P_{\mathbf{N}} = 0. \tag{56}$$

We leave it to the reader to verify that the matrix polynomial

$$D_{n1}(t) = Z_0 + t Z_1 + \dots + t^n Z_n$$

has properties

$$\int_{-\infty}^{\infty} t^s \, d\sigma_A(t) \, D_{n1}(t) = \begin{cases} 0, \ 0 \le s \le n-1, \\ P_{\mathbf{M}}, \ s = n. \end{cases},$$

$$\int_{-\infty}^{\infty} D_{n1}(t)^* d\sigma_A(t) \, D_{n1}(t) = Z_n, \ D_{n1}(t) P_{\mathbf{N}} \equiv 0.$$

Images of all canonical extensions of A_0 in $\mathcal{L}_n$ or in $\mathcal{L}_n / \mathcal{L}_n^0$ may differ only by their action on vector polynomials of the form $D_{n1}(t)\mathbf{h}$, $\mathbf{h} \in \mathbf{M}$. Proceeding analogously to the derivation of expression (28) we get for the canonical extension A that

$$(A D_{n1} \mathbf{h})(t) = D_{n1}(t) Z_n G_A \mathbf{h} + D_{n-1}(t) P_{\mathbf{M}} \mathbf{h}, \mathbf{h} \in \mathbf{M}, \tag{57}$$

where G_A some Hermitian operator acting in $\mathbf{M}$.

For getting of the expression for the resolvent of the canonical extension A let us consider the matrix polynomial

$$Y_n(z) = D_{n1}(z)(Z_n G_A - z) + D_{n-1}(z) P_{\mathbf{M}} + D_{n0}(z). \tag{58}$$

Proposition 5 *For* Im $z \neq 0$ *the matrix function* $Y(z)$ *is invertible.*

Proof: Suppose that for some $\mathbf{h} = \mathbf{h}_0 + \mathbf{h}_1$, $\mathbf{h}_0 \in \mathbf{N}, \mathbf{h}_1 \in \mathbf{M}$ we have $Y_n(z_0)\mathbf{h} = 0$, Im $z_0 \neq 0$. Let us form the vector polynomial

$$\begin{aligned} \mathbf{y}(t) &= -D_{n1}(t)\mathbf{h}_1 + \frac{1}{t - z_0}(D_{n1}(t)(Z_n G_A - z_0)\mathbf{h}_1 + D_{n-1}(t)\mathbf{h}_1 \\ &\quad + D_{n0}(t)\mathbf{h}_0). \end{aligned}$$

By (57) and the fact that the image of A_0 is the multiplication operator on the variable t we obtain

$$(A\mathbf{y})(t) - z_0 \mathbf{y}(t) = D_{n0}(t)\mathbf{h}_0.$$

Since the class, which contains $D_{n0}(t)\mathbf{h}_0$, is the zero element of $\mathcal{L}_n/\mathcal{L}_n^0$ we infer that $\mathbf{y}(t)$ is the image of eigenvector of A related to eigenvalue z_0. But A is a selfadjoint operator. Hence $\mathbf{y}(t)$ belongs to the class of zero elements i.e., there exists $\mathbf{q} \in \mathbf{N}$ such that

$$\mathbf{y}(t) + D_{n0}(t)\mathbf{q} \equiv 0. \tag{59}$$

But

$$\mathbf{y}(t) + D_{n0}(t)\mathbf{q} \underset{t\to\infty}{=} t^n(-Z_n\mathbf{h}_1 + \mathbf{q}) + O(t^n).$$

Hence $\mathbf{h}_1 = 0$ and $\mathbf{q} = 0$. We see that the polynomial

$$\mathbf{y}(t) = \frac{1}{t - z_0} D_{n0}(t)\mathbf{h}_0$$

of degree not exceeding $n - 1$ belongs the zero class. But by invertibility of Γ_{n-1} this is possible only if $\mathbf{h}_0 = 0$. □

Now we can state that for the resolvent of arbitrary canonical extension A the expression (38) remains true but with Y_n defined by (58). Using this expression we immediately obtain the description formula for canonical solutions of the truncated Hamburger problem and then establish the matrix analog of the Nevanlinna formula (7) for the degenerate case.

Theorem 6 *Let* Γ_n *is not invertible, but* $\Gamma_{n-1} > 0$.

If $0 < \dim \mathbf{M} = l < m$ *then the Nevanlinna-type formula*

$$\begin{aligned}\int_{-\infty}^{\infty} \frac{1}{z-t} d\sigma_A(t) &= (E_{n1}(z)(R_A(z)-z) + E_{n-1}(z)P_{\mathbf{M}} + E_{n0}(z)) \\ &\quad \times (D_{n1}(z)(R_A(z)-z) + D_{n-1}(z)P_{\mathbf{M}} + D_{n0}(z))^{-1}, \\ E_{ni}(t) &= \int_{-\infty}^{\infty} \frac{1}{z-t} d\sigma_A(t)(D_{ni}(z) - D_{ni}(t)), \ i = 0, 1 \\ R_A(z) &= Z_n S^* \Omega(z) S, \ \operatorname{Im}\ z \neq 0,\end{aligned}$$

where S *is any unitary mapping of* $\mathbf{M}$ *onto* $\mathbf{C}_l$ *establishes the one-to-one correspondence between the set Nevanlinna* $l \times l$ *matrix functions* $\Omega(z) \in \mathbb{N}_0$ *and the set of all solutions of the truncated Hamburger* $m \times m$ *matrix moment problem.*

If $l = 0$ *then the unique solution of the truncated Hamburger problem is given by the formula*

$$\int_{-\infty}^{\infty} \frac{1}{z-t} d\sigma(t) = E_{n0}(z) \cdot D_{n0}(z)^{-1}.$$

References

[1] N.I. Akhiezer, *The classical moment problem and some related questions in Analysis*, Hafner Publishing Company, N. Y. (1965).

[2] T. Ando, *Truncated moment problems for operators*, Acta Sci. Math. **31** (1970), 319–334.

[3] R.E. Curto and L.A. Fialkow, *Recursiveness, positivity, and truncated moment problems*. Houston J. Math. **17**, 603 (1991); see also:
R.E. Curto and L.A. Fialkow, *Solutions of the truncated moment problem*. Mem. Amer. Math. Soc. **119** (1996).

[4] H. Dym, *On Hermitian Block Hankel Matrices, Matrix Polynomials, the Hamburger Moment Problem, Interpolation and Maximum Entropy*, Integral Equations and Operator Theory **12** (1989), 757–812.

[5] I.V. Kovalishina, *New aspects of the classical problem of moments*. Izv. Akad. Nauk. Ser. Mat. **47**, 455 (1983) [Math. USSR Izv. **22**, 419 (1984)].

[6] M.G. Krein, *The theory of extensions of semibounded Hermitian operators and its applications*, Mat. Sbornik (Russian) **20** (1947), 431–495, see also [7].

[7] M.G. Krein and A.A. Nudel'man, *The Markov moment problem and extremal problems*, Trans. of Math. Monographs **50**, Amer. Math. Soc., Providence, R.I. (1977).

[8] H.J. Landau (Ed.), *Moments in Mathematics*, Proc. Sympos. Applied Math. **37**, Amer. Math. Soc., Providence, R.I. (1987).

[9] R. Nevanlinna, *Asymptotische Entwickelungen beschränkter Funktionen und das Stieltjessche Momentenproblem*. Ann. Acad. Sci. Fenn. A **18**, 5 (1922).

[10] J. Ortner, V.M. Rylyuk and I.M. Tkachenko, *Reflectivity of cold dense plasmas*. Phys. Rev. E **50**, 4937 (1994); see also:

V.M. Adamyan and I.M. Tkachenko, *High-frequency electrical conductivity of a collisional plasma.* High Temp. (USA) **21**, 420 (1983).

[11] J.A. Shohat and J.D. Tamarkin, *The problem of moments*, Mathematical Surveys **1**, Amer. Math. Soc., Providence, R.I. (1943) (4th Ed. 1970).

Department of Theoretical Physics
University of Odessa
270026 Odessa
Ukraine

Department of Applied Mathematics
Polytechnic University of Valencia
E-46071 Valencia
Spain

MSC Primary 30E05, 44A60, Secondary 30E20, 47A57, 73K03

Operator Theory:
Advances and Applications, Vol. 118

Extreme Points of a Positive Operator Ball

T. Ando

The celebrated Krein-Milman theorem guarantees that a compact convex set in a topological linear space has so many extreme points as their closed convex hull coincides with the set itself. It is, however, usually difficult to determine and parametrize all extreme points. In this paper we will determine extreme points of a positive operator ball $\{X \geq 0\,;\ X^2 \leq A^2\}$ with $A \geq 0$ in the space of bounded selfadjoint operators on a Hilbert space.

1 Introduction

Let $\mathcal{H}$ be a (complex) Hilbert space with inner product $\langle\cdot,\cdot\rangle$, and $\mathcal{B}(\mathcal{H})$ the space of (bounded linear) operators on $\mathcal{H}$. For $T \in \mathcal{B}(\mathcal{H})$ its usual (operator) norm is denoted by $\|T\|$. An operator T is called a *contraction* if $\|T\| \leq 1$. When $\dim(\mathcal{H}) = n < \infty$, with respect of a suitably chosen orthonormal base, $\mathcal{B}(\mathcal{H})$ is identified with the space of $n \times n$ matrices with (spectral) norm.

For selfadjoint operators A, B the order relation $A \geq B$ means that $A - B$ is *positive* (semi-definite). Therefore $A \geq 0$ means positivity of A. We write $A > B$ to mean that $A \geq B$ and $A - B$ is invertible, which is equivalent to $A - B \geq \epsilon I$ for some $\epsilon > 0$ where I is the identity operator.

In various problems of operator theory the set of all solutions appears in the form of a so-called operator ball (see [AAK], [Smu]). The *operator ball* $\mathcal{K}(C; A, B)$ with center $C \in \mathcal{B}(\mathcal{H})$, left radius $A \geq 0$ and right radius $B \geq 0$ is defined as

$$\mathcal{K}(C; A, B) \overset{def}{=} \{C + AZB\,;\ \|Z\| \leq 1\}.$$

To treat an operator of the form AZB with a contraction Z the following well known observation (see [FoF], XVI. 1) is quite useful.

Suppose that an operator X is written as a block operator-matrix

$$X = \begin{bmatrix} X_{11} & X_{12} \\ X_{21} & X_{22} \end{bmatrix}$$

with respect to an *orthogonal decomposition* $\mathcal{H} = \mathcal{H}_1 \oplus \mathcal{H}_2$. Then $X \geq 0$ if and only if $X_{11} \geq 0$, $X_{22} \geq 0$ and there is a contraction Z from $\mathcal{H}_2$ to $\mathcal{H}_1$ such that

$$X_{12} = X_{11}^{1/2} Z X_{22}^{1/2} \quad \text{and} \quad X_{21} = X_{22}^{1/2} Z^* X_{11}^{1/2}.$$

This result is formulated in the following form, convenient for later use.

Lemma 1.1

$$X = \begin{bmatrix} X_{11} & X_{12} \\ X_{21} & X_{22} \end{bmatrix}$$

is positive if and only if $X_{11} \geq 0$, $X_{22} \geq 0$, $X_{21} = X_{12}^*$ *and*

$$\lim_{\epsilon \downarrow 0} X_{12}(X_{22} + \epsilon I_2)^{-1} X_{21} \leq X_{11}$$

where I_2 *is the identity operator on* $\mathcal{H}_2$.

It causes no confusion to write $X_{12}X_{22}^{-1}X_{21}$ for $\lim_{\epsilon\downarrow 0} X_{12}(X_{22} + \epsilon I_2)^{-1}X_{21}$. And $X_{11} - X_{12}X_{22}^{-1}X_{21}$ is sometimes called the *Schur complement* of X for its (2,2)-component X_{22}. The Schur complement for the $(1, 1)$-component is defined correspondingly.

Now when center $C = 0$, the operator ball $\mathcal{K}(0; A, B)$ can be simply defined as

$$\mathcal{K}(0; A, B) = \left\{ X ; \begin{bmatrix} A^2 & X \\ X^* & B^2 \end{bmatrix} \geq 0 \right\}.$$

Further when $B > 0$ the correspondence $X \mapsto B^{-1}XB^{-1}$ establishes an affine bijection form $\mathcal{K}(0; A, B)$ to $\mathcal{K}(0; (B^{-1}A^2B^{-1})^{1/2}, I)$. From this point of view we will study in this paper only the operator ball $\mathcal{K}(0; A, I)$, and more specifically its subset of positive operators $\{X \geq 0 ; X \in \mathcal{K}(0; A, I)\}$, called *positive operator ball.* We shall denote this positive operator ball by $\Delta_2(A)$ in accordance with a notation in a previous paper [An3]

(1.1) $$\Delta_1(A) \overset{def}{=} \{X ; A \geq X \geq 0\}.$$

By Lemma 1.1 we can define $\Delta_2(A)$ as

(1.2) $$\Delta_2(A) \overset{def}{=} \left\{ X \geq 0 ; \begin{bmatrix} A^2 & X \\ X & I \end{bmatrix} \geq 0 \right\} = \{X \geq 0 ; A^2 \geq X^2\}.$$

It is obvious that both $\Delta_1(A)$ and $\Delta_2(A)$ are weakly compact convex sets.

When $A = I$, obviously $\Delta_2(I)$ coincides with $\Delta_1(A)$. For general $A \geq 0$ we have only inclusion

(1.3) $$\Delta_2(A) \subset \Delta_1(A).$$

This is a consequence of the basic fact (see [Löw]) that for $X, Y \geq 0$

(1.4) $$Y^2 \geq X^2 \implies Y \geq X.$$

As the converse implication in (1.4) does not hold in general, $\Delta_2(A)$ is a proper subset of $\Delta_1(A)$.

Denote by $\mathrm{ex}\Delta_1(A)$ and $\mathrm{ex}\Delta_2(A)$ the set of *extreme points* of $\Delta_1(A)$ and $\Delta_2(A)$ respectively. According to the classical theorem of Krein-Milman (see [KrM])

$\mathrm{ex}\Delta_1(A)$ (resp. $\mathrm{ex}\Delta_2(A)$) is so large as its weakly closed convex hull coincides with $\Delta_1(A)$ (resp. $\Delta_2(A)$).

Our main interest in this paper is to study the structure of $\mathrm{ex}\Delta_2(A)$. In Section 2 we establish an intrinsic characterization for $X \in \Delta_2(A)$ to be an extreme point. In Section 3 we treat finite dimensional cases and give a complete parametrization of $\mathrm{ex}\Delta_2(A)$ for the case $\dim(\mathcal{H}) \leq 3$.

In the rest of this section, let us recall several useful operations in the cone of positive operators; *parallel addition* and *short*, which will become powerful tools in our study (cf. [AnD] and [PeS]).

Parallel addition is a binary operation $X, Y \mapsto X : Y$ in the cone of positive operators ($X : Y$ is called *parallel sum*), defined as

$$\langle u, (X:Y)u\rangle = \inf\{\langle v, Xv\rangle + \langle w, Yw\rangle\ ;\ v + w = u\} \quad (u \in \mathcal{H}). \tag{1.5}$$

For $X, Y > 0$ this definition reduces to

$$X : Y = \{X^{-1} + Y^{-1}\}^{-1}. \tag{1.6}$$

Also it is clear from definition (1.5) that parallel addition is *commutative* and *associative*. It is obvious that

$$X \geq X : Y \geq 0 \quad \text{and} \quad Y \geq X : Y \geq 0, \tag{1.7}$$

and, the map $(X, Y) \mapsto X : Y$ is *monotone*;

$$X_1 \geq X_2 \geq 0,\ Y_1 \geq Y_2 \geq 0 \implies X_1 : Y_1 \geq X_2 : Y_2, \tag{1.8}$$

and further it is *continuous from above* in the sense

$$X_k \downarrow X \geq 0,\ Y_k \downarrow Y \geq 0 \implies X_k : Y_k \downarrow X : Y, \tag{1.9}$$

where $X_k \downarrow X$ means that $X_1 \geq X_2 \geq \dots$ and X_k converges to X strongly as $k \to \infty$. Finally, remark that for $\alpha, \beta > 0$

$$(\alpha X) : (\beta X) = (\alpha : \beta)X = \frac{\alpha\beta}{\alpha + \beta} X. \tag{1.10}$$

Given an operator T, denote by $\mathrm{ran}(T)$ its *range*, and by $\ker(T)$ its *kernel*. There is close relationship between range inclusion and operator inequality (see [Dou], [FiW]).

Lemma 1.2 *For operators S, T, inclusion* $\mathrm{ran}(S) \subset \mathrm{ran}(T)$ *occurs if and only there is a constant $\gamma > 0$ such that $SS^* \leq \gamma TT^*$.*

In this connection an exact characterization of $\mathrm{ran}(X : Y)^{1/2}$ is known (see [AnD], [FiW]);

$$\mathrm{ran}(X : Y)^{1/2} = \mathrm{ran}(X^{1/2}) \cap \mathrm{ran}(Y^{1/2}). \tag{1.11}$$

As a consequence, we have the following useful criterion.

Lemma 1.3 *For $X, Y \geq 0$*

$$X : Y = 0 \iff \operatorname{ran}(X^{1/2}) \cap \operatorname{ran}(Y^{1/2}) = \{0\}.$$

Remark that if $X \geq 0$ is of finite rank then $\operatorname{ran}(X) = \operatorname{ran}(X^{1/2})$.

Another important operation is *short*. It is a theorem of M. Krein (see [Kre] and [AT1]) that for any closed subspace $\mathcal{M}$ and any operator $X \geq 0$ there always exists the maximum among all $Y \geq 0$ such that $Y \leq X$ and $\operatorname{ran}(Y) \subset \mathcal{M}$. This maximum operator will be denote by $[P]X$, with orthoprojection P to $\mathcal{M}$, and is called the *short* of X to $\mathcal{M} = \operatorname{ran}(P)$. Since by Lemma 1.2 inclusion $\operatorname{ran}(Y) \subset \mathcal{M}$ is equivalent to $Y \leq \gamma P$ for some $\gamma = \gamma(Y) > 0$, we can write

$$(1.12) \qquad \begin{aligned} [P]X &= \max\{Y\,;\ 0 \leq Y \leq X \text{ and} \\ &\qquad Y \leq \gamma P \text{ for some } \gamma = \gamma(Y) > 0\}. \end{aligned}$$

The following is seen from the construction of $[P]X$ (see [Kre]).

Lemma 1.4 *For an orthoprojection P and $X \geq 0$*

$$[P]X = X^{1/2} Q X^{1/2}$$

where Q is the orthoprojection to $\{u \in \mathcal{H}\,;\ X^{1/2}u \in \operatorname{ran}(P)\}$.

We have more explicit form for the short in term of Schur complement (see [An2], [AT1]).

Lemma 1.5 *Let P be an orthoprojection, and let $X \geq 0$ be written in a block operator-matrix*

$$\begin{bmatrix} X_{11} & X_{12} \\ X_{21} & X_{22} \end{bmatrix}$$

with respect to the orthogonal decomposition $\mathcal{H} = \operatorname{ran}(P) \oplus \ker(P)$. Then the short $[P]X$ is identified with the Schur complement of X for the (2,2)-component in the sense

$$[P]X = \begin{bmatrix} X_{11} - X_{12}X_{22}^{-1}X_{21} & 0 \\ 0 & 0 \end{bmatrix}.$$

2 Characterization

First let us remark that, by Lemma 1.1, for $A, X \geq 0$ the condition

$$(2.1) \qquad \begin{bmatrix} A^2 & X \\ X & I \end{bmatrix} \geq 0$$

is equivalent to $A^2 \geq X^2$ as well as to $I - XA^{-2}X \geq 0$. Here, if A is not invertible, $XA^{-2}X$ is understood as the increasing limit $\lim_{\epsilon \downarrow 0} X(A^2 + \epsilon I)^{-1}X$.

In this connection the following characterization of extremity is quite natural.

Theorem 2.1 *Let $A \geq 0$ and $X \in \Delta_2(A)$. Then X is an extreme point of $\Delta_2(A)$ if and only if*

$$X : (A^2 - X^2) : (I - XA^{-2}X) = 0. \tag{2.2}$$

Proof: The following is a proof for the case $A > 0$. But by continuity from above of parallel addition (1.9), similar arguments are valid for general $A \geq 0$.

Suppose first that X satisfies (2.2). To prove extremity of X, we have to show that

$$X \pm B \in \Delta_2(A) \implies B = 0.$$

Since the minimum 0 and the maximum A are obviously extreme points, let $X \neq 0$ and $X \neq A$. Since $(X \pm B)^2 \leq A^2$ implies

$$X^2 + B^2 = \frac{1}{2}\{(X + B)^2 + (X - B)^2\} \leq A^2,$$

we have

$$B^2 \leq A^2 - X^2. \tag{2.3}$$

Similarly since $(X \pm B)A^{-2}(X \pm B) \leq I$ implies

$$\|A\|^{-2}B^2 \leq BA^{-2}B \leq I - XA^{-2}X,$$

with $\alpha \stackrel{def}{=} \|A\|^{-2}$, we have

$$\alpha B^2 \leq I - XA^{-2}X. \tag{2.4}$$

Finally, from $X \pm B \geq 0$ it follows that there is a selfadjoint contraction C such that $B = X^{1/2}CX^{1/2}$. Then we have

$$B^2 = X^{1/2}CXCX^{1/2} \leq \|X\|X,$$

so that with $\beta \stackrel{def}{=} \|X\|^{-1}$

$$\beta B^2 \leq X. \tag{2.5}$$

By monotony of parallel addition (1.8), it follows from (2.3)–(2.5) and (1.10) that

$$(1 : \alpha : \beta)B^2 \leq X : (A^2 - X^2) : (I - XA^{-2}X) = 0,$$

which leads to $B = 0$. Therefore X is an extreme point of $\Delta_2(A)$.

Suppose conversely that X is an extreme point of $\Delta_2(A)$ and let

$$D \stackrel{def}{=} X : (A^2 - X^2) : (I - XA^{-2}X). \tag{2.6}$$

We have to prove that $D = 0$. Suppose, by contradiction, that $D \neq 0$. Then there is a rank-one operator F such that $0 \leq F \leq D$. Then (2.6) implies via (1.7)

$$0 \leq F \leq I - XA^{-2}X,$$

which, in its turn, implies by Lemma 1.1

$$\begin{bmatrix} A^2 & X \\ X & I-F \end{bmatrix} \geq 0,$$

so that

$$\begin{bmatrix} A^2 & X \\ X & I-\epsilon F \end{bmatrix} \geq 0 \qquad (0 < \epsilon < 1)$$

with invertible $I - \epsilon F$. Then again by Lemma 1.1 we have

$$0 \leq A^2 - X(I - \epsilon F)^{-1} X \qquad (0 < \epsilon < 1). \tag{2.7}$$

Further since $I + \epsilon F \leq (I - \epsilon F)^{-1}$ it follows from (2.7) that

$$0 \leq A^2 - X(I + \epsilon F)X$$

or equivalently

$$\epsilon X F X \leq A^2 - X^2. \tag{2.8}$$

Then by Lemma 1.2 it follows from (2.8) that

$$\operatorname{ran}(X F^{1/2}) \subset \operatorname{ran}(A^2 - X^2)^{1/2}.$$

On the other hand, $0 \leq F \leq A^2 - X^2$ implies, again by Lemma 1.2,

$$\operatorname{ran}(F^{1/2} X) \subset \operatorname{ran}(F^{1/2}) \subset \operatorname{ran}(A^2 - X^2)^{1/2}.$$

Therefore we have

$$\operatorname{ran}(X F^{1/2} + F^{1/2} X) \subset \operatorname{ran}(A^2 - X^2)^{1/2}.$$

Since $X F^{1/2} + F^{1/2} X$ is a selfajoint operator of finite rank,

$$\operatorname{ran}|X F^{1/2} + F^{1/2} X|^{1/2} = \operatorname{ran}|X F^{1/2} + F^{1/2} X| = \operatorname{ran}(X F^{1/2} + F^{1/2} X)$$

where $|Y|$ is the positive square root of Y^*Y for general Y. Therefore we can conclude

$$\operatorname{ran}|X F^{1/2} + F^{1/2} X|^{1/2} \subset \operatorname{ran}(A^2 - X^2)^{1/2},$$

which implies again by Lemma 1.2

$$\epsilon |X F^{1/2} + F^{1/2} X| \leq A^2 - X^2 \quad \text{for small } \epsilon > 0. \tag{2.9}$$

Now since $0 \leq F \leq D$, it follows from (2.6) and (2.9) that

$$\begin{aligned} (X \pm \epsilon^2 F^{1/2})^2 &= X^2 \pm \epsilon^2 (X F^{1/2} + F^{1/2} X) + \varepsilon^4 F \\ &\leq X^2 + (\epsilon + \epsilon^4)(A^2 - X^2) \\ &= A^2 - (1 - \epsilon - \epsilon^4)(A^2 - X^2), \end{aligned}$$

which implies

$$(X \pm \epsilon^2 F^{1/2})^2 \leq A^2 \quad \text{for small } \epsilon > 0. \tag{2.10}$$

Finally, since $0 \leq F \leq X$ and F is of finite rank,

$$\epsilon^2 F^{1/2} \leq X \qquad \text{for small } \epsilon > 0.$$

Therefore $X \pm \epsilon^2 F^{1/2} \geq 0$, so that by (2.10) both $X \pm \epsilon^2 F^{1/2}$ belong to $\Delta_2(A)$ and

$$X = \frac{(X + \epsilon^2 F^{1/2}) + (X - \epsilon^2 F^{1/2})}{2},$$

which contradicts extremity of X. This contradiction proves $D = 0$. □

The second part of the above proof shows the following.

Corollary 2.2 *If $X \in \Delta_2(A)$ is not an extreme point, then there are $Y, Z \in \Delta_2(A)$ such that $Y \leq X \leq Z$ and both $X - Y$ and $Z - X$ are of finite rank.*

Let us present typical examples of extreme points of $\Delta_2(A)$. First, remark that A and 0 are the maximum and the minimum extreme points.

Suppose $\ker(A) = \{0\}$. The map $X \mapsto X^{1/2}$ is a concave, monotone bijection from $\Delta_1(A^2)$ onto $\Delta_2(A)$ (see [Löw], [An1]). Further the map $X \mapsto AXA$ gives an affine, monotone bijection from $\Delta_1(I)$ onto $\Delta_1(A^2)$. Therefore the map

$$\Phi(X) \overset{def}{=} (AXA)^{1/2} \tag{2.11}$$

is a concave, monotone bijection from $\Delta_1(I)$ onto $\Delta_2(A)$.

Theorem 2.3 *If* $\ker(A) = \{0\}$, *then the bijection* Φ *transforms* $\mathrm{ex}\Delta_1(I)$ *into* $\mathrm{ex}\Delta_2(A)$.

Proof: It is well-known that $\mathrm{ex}\Delta_1(I)$ consists of all orthoprojections. Let us show that $(AQA)^{1/2}$ is in $\mathrm{ex}\Delta_2(A)$ for any orthoprojection Q. For this, we have to prove that if, for some $Y, Z \in \Delta_2(A)$,

$$(AQA)^{1/2} = \frac{Y + Z}{2} \tag{2.12}$$

then $Y = Z = X$. Since the square formation is a convex map (cf. [An1]), (2.12) implies

$$AQA \leq \frac{Y^2 + Z^2}{2}. \tag{2.13}$$

Since by assumption $Y^2, Z^2 \leq A^2$, with $C \overset{def}{=} \Phi^{-1}(Y)$, $D \overset{def}{=} \Phi^{-1}(Z)$ we can conclude from (2.13) via $\ker(A) = \{0\}$ that

$$Q \leq \frac{C + D}{2} \quad \text{with } 0 \leq C, D \leq I.$$

It is easy to see that for an orthoprojection Q such an inequality is possible only when $C, D \geq Q$. If $C \neq Q$ or $D \neq Q$, since Φ is a monotone bijection, we have

$$\frac{Y+Z}{2} = \frac{\Phi(C)+\Phi(D)}{2} \neq \frac{\Phi(Q)+\Phi(Q)}{2} = (AQA)^{1/2},$$

contradicting (2.12). Therefore $C = D = Q$, and $Y = Z = X$. □

Let P be an orthoprojection. Since $[P]A^2 \leq A^2$ by (1.12),

$$\Delta_2(([P]A^2)^{1/2}) \subset \Delta_2(A)$$

so that we have obviously

$$\mathrm{ex}\Delta_2(A) \cap \Delta_2(([P]A^2)^{1/2}) \subset \mathrm{ex}\Delta_2(([P]A^2)^{1/2})$$

However, the converse inclusion holds too.

Theorem 2.4 *For any orthoprojection P the following identity holds:*

$$\mathrm{ex}\Delta_2(A) \cap \Delta_2(([P]A^2)^{1/2}) = \mathrm{ex}\Delta_2(([P]A^2)^{1/2}).$$

Proof: Take $X \in \mathrm{ex}\Delta_2(([P]A^2)^{1/2})$, and suppose that

$$X = \frac{Y+Z}{2} \quad \text{for some } Y, Z \in \Delta_2(A).$$

We have to prove $Y = Z = X$.

Since by (1.12) $X^2 \leq [P]A^2 \leq \gamma^2 P$ for some $\gamma > 0$, we have $X \leq \gamma P$ by (1.4), which implies $Y, Z \leq 2\gamma P$, so that $Y^2, Z^2 \leq 4\gamma^2 P$ because Y, Z commute with P in this case. Then again by (1.12) we can see that $Y^2, Z^2 \leq [P]A^2$ so that $Y, Z \in \Delta_2(([P]A^2)^{1/2})$. Since $X \in \mathrm{ex}\Delta_2(([P]A^2)^{1/2})$, we can conclude that $Y = Z = X$. □

In (1.3) we pointed out inclusion $\Delta_2(A) \subset \Delta_1(A)$. In this connection the following result is of some interest.

Theorem 2.5 *We have*

$$\mathrm{ex}\Delta_1(A) \cap \Delta_2(A) = \{AP\ ;\ P \text{ orthoprojection commuting with } A\}.$$

Proof: An extreme point of $\Delta_1(A)$ is of the form $A^{1/2}PA^{1/2}$ for an orthoprojection P (see [An3], [AT2], [ErL], [GrM]) with $\mathrm{ran}(P) \subset \ker(A)^{\perp}$. If $A^{1/2}PA^{1/2}$ is in $\Delta_2(A)$, then

$$A^{1/2}PAPA^{1/2} \leq A^2$$

or equivalently $PAP \leq A$. Then with $P^{\perp} \overset{def}{=} I - P$ we have

$$0 \leq P^{\perp}AP + PAP^{\perp} + P^{\perp}AP^{\perp},$$

which, by $A \geq 0$, is possible only when

$$P^{\perp}AP = PAP^{\perp} = 0$$

or equivalently

$$AP = PA = A^{1/2}PA^{1/2}.$$

Thus we have proved that the left side of the relation in question is contained in the right side. The reversed inclusion is obvious. □

3 Finite Dimensional Case

It is difficult to visualize or parametrize all extreme points of $\Delta_2(A)$ in contrast to the case of $\Delta_1(A)$. In this section we assume that $\mathcal{H}$ is of *finite dimension* and $A > 0$.

Lemma 3.1 *Let X be an invertible element of $\Delta_2(A)$ with $A > 0$. Then X is an extreme point of $\Delta_2(A)$ if and only if*

$$\text{rank}(Q^{\perp}XQ) = \text{rank}(Q^{\perp}), \tag{3.1}$$

where Q is the orthoprojection to $\ker(A^2 - X^2)$, *and* $Q^{\perp} = I - Q$.

Remark that since

$$\text{ran}(Q^{\perp}XQ) \subset \text{ran}(Q^{\perp}),$$

by finite dimensionality (3.1) is equivalent to

$$\text{ran}(Q^{\perp}XQ) = \text{ran}(Q^{\perp}). \tag{3.2}$$

Proof: Since X is invertible by assumption, $\text{ran}(X^{1/2}) = \mathcal{H}$ so that by Lemma 1.3 extremity condition (2.2) in Theorem 2.1 is equivalent to the following requirements:

$$\text{ran}(A^2 - X^2)^{1/2} \cap \text{ran}(I - XA^{-2}X)^{1/2} = \{0\},$$

or by finite dimensionality of $\mathcal{H}$

$$\text{ran}(A^2 - X^2) \cap \text{ran}(I - XA^{-2}X) = \{0\},$$

which, by duality in finite dimensional space, is converted to the following relation:

$$\ker(A^2 - X^2) + \ker(I - XA^{-2}X) = \mathcal{H}. \tag{3.3}$$

We claim the following general identity:

$$\ker(I - XA^{-2}X) = \text{ran}(XQ). \tag{3.4}$$

First, if $u \in \ker(I - XA^{-2}X)$, that is, $u = XA^{-2}Xu$, with $v \overset{def}{=} A^{-2}Xu$ we have

$$X^2v = X \cdot XA^{-2}Xu = Xu = A^2v,$$

hence $v \in \ker(A^2 - X^2)$ and $u = Xv$, which shows that the left side of (3.4) is contained in the right side. Conversely, if $u = Xv$ with $v \in \ker(A^2 - X^2)$, that is, $A^2v = X^2v$, we have $Xu = X^2v = A^2v$, so that $XA^{-2}Xu = Xv = u$, which shows that the right side of (3.4) is contained in the left side, proving the claim.

Now it follows from (3.4) that condition (3.3) is equivalent to

$$\mathrm{ran}(Q) + \mathrm{ran}(XQ) = \mathcal{H},$$

which reduces to (3.2). □

When X is not invertible, we can consider X, and $([P]A^2)^{1/2}$ instead of A on $\mathrm{ran}(X)$, where P is the orthoprojection to the (closed) subspace $\mathrm{ran}(X)$. By Theorem 2.4, $X \in \Delta_2(([P]A^2)^{1/2})$ is an extreme point of $\Delta_2(A)$ if and only if it is an extreme point of $\Delta_2(([P]A^2)^{1/2})$. Since both X and $([P]A^2)^{1/2}$ are invertible on $\mathrm{ran}(X)$, we can deduce the following assertion from Lemma 3.1.

Theorem 3.2 *Let $X \in \Delta_2(A)$. Then X is an extreme point of $\Delta_2(A)$ if and only if*

(3.5) $$\mathrm{rank}(Q^{\perp}XQ) = \mathrm{rank}(Q^{\perp}).$$

where Q is the orthoprojection to $\ker([P]A^2 - X^2)$ *with P orthoprojection to* $\mathrm{ran}(X)$.

Theorem 3.2 is useful to construct an extreme point of $\Delta_2(A)$, not of the form $(APA)^{1/2}$ with orthoprojection P.

Theorem 3.3 *Let Q_1 and Q_2 be mutually annihilating non-zero orthoprojections. If Q_2 is of rank one and $Q_1A^2Q_2 \neq 0$, then for every $0 < \lambda < 1$*

(3.6) $$X_\lambda \overset{def}{=} \{A(Q_1 + \lambda Q_2)A\}^{1/2}$$

is an extreme point of $\Delta_2(A)$.

Proof: Let P_λ be the orthoprojection to $\mathrm{ran}(X_\lambda)$. Since by Lemma 1.2

$$\begin{aligned}\mathrm{ran}(X_\lambda) &= \mathrm{ran}(A(Q_1 + \lambda Q_2)^{1/2}) = A\mathrm{ran}(Q_1 + \lambda Q_2)^{1/2} \\ &= A\mathrm{ran}(Q_1 + \lambda Q_2) = A\mathrm{ran}(Q_1 + Q_2),\end{aligned}$$

$Au \in \mathrm{ran}(X_\lambda)$ is equivalent to $u \in \mathrm{ran}(Q_1 + Q_2)$. Therefore by Lemma 1.4

(3.7) $$[P_\lambda]A^2 = A(Q_1 + Q_2)A.$$

Now since by (3.6) and (3.7)

$$[P_\lambda]A^2 - X_\lambda^2 = (1-\lambda)AQ_2A,$$

we have

$$[P_\lambda]A^2 \geq X_\lambda^2, \quad \text{rank}([P_\lambda]A^2 - X_\lambda^2) = 1.$$

Then with Q the orthoprojection to $\ker([P_\lambda]A^2 - X_\lambda^2)$, we can see

$$\text{rank}(Q^\perp) = \text{rank}(Q_2) = 1.$$

Therefore to check extremity condition (3.5) it suffices to prove $Q^\perp X_\lambda Q \neq 0$.

First, since $Q_1 A^2 Q_2 \neq 0$ by assumption, it is immediate from (3.6) and (3.7) to see that X_λ^2 does not commute with $[P_\lambda]A^2$.

Next, we claim $Q^\perp X_\lambda^2 Q \neq 0$. In fact, if $Q^\perp X_\lambda^2 Q = 0$, that is, Q commutes with X_λ, it follows from

$$([P_\lambda]A^2 - X_\lambda^2)Q = 0$$

that $Q^\perp \cdot [P_\lambda]A^2 \cdot Q = 0$, that is, Q commutes with $[P_\lambda]A^2$. On the other hand, since

$$\text{rank}(Q^\perp X_\lambda^2 Q^\perp) = \text{rank}(Q^\perp \cdot [P_\lambda]A^2 \cdot Q^\perp) = 1,$$

there is a constant $\alpha \geq 0$ such that

$$Q^\perp X_\lambda^2 Q^\perp = \alpha Q^\perp \cdot [P_\lambda]A^2 \cdot Q^\perp.$$

These show that

$$X_\lambda^2 = (Q + \alpha Q^\perp)[P_\lambda]A^2,$$

which leads to a contradictory conclusion that X_λ^2 commutes with $[P_\lambda]A^2$. This proves the claim. Finally, $Q^\perp X_\lambda^2 Q \neq 0$ is possible only if $Q^\perp X_\lambda Q \neq 0$. This completes the proof. □

When $\dim(\mathcal{H}) \leq 3$, Theorem 3.3 yields a complete parametrization of $\text{ex}\Delta_2(A)$.

Theorem 3.4 *Suppose that* $1 \leq \dim(\mathcal{H}) \leq 3$ *and* $A > 0$. *Then every* $X \in \text{ex}\Delta_2(A)$ *is one of the following forms:*

(3.8) $$X = (APA)^{1/2} \quad \text{for some orthoprojection } P,$$

and

(3.9) $$X = \{A(P + \lambda Q)A\}^{1/2}$$

for some orthoprojections P, Q *and* $0 < \lambda < 1$ *such that*

$$PQ = 0, \; PA^2Q \neq 0 \quad \textit{and} \quad \text{rank}(Q) = 1.$$

Proof: When $X = 0$ or $\dim(\mathcal{H}) = 1$ the assertion is obvious. Therefore we assume that $2 \leq \dim(\mathcal{H}) \leq 3$ and $X \neq 0$.

Suppose first that X is invertible. If $X = A$, then X is of the form (3.8) with $P = I$. Now suppose that $X \neq A$. Then $A^2 - X^2 \neq 0$ and $A^2 - X^2 \not> 0$. We claim $\ker(A^2 - X^2) \neq \{0\}$. In fact, if otherwise, $A^2 > X^2$ implies also $I > XA^{-2}X$, so that

$$X : (A^2 - X^2) : (I - XA^{-2}X) > 0,$$

which contradicts extremity of X by Theorem 2.1.

Let Q_1 be the orthoprojection to $\ker(A^2 - X^2)$. Since $2 \leq \dim(\mathcal{H}) \leq 3$, from the above we have $1 \leq \operatorname{rank}(Q_1) \leq 2$. Further by Lemma 3.1 we have

$$\operatorname{rank}(Q_1^\perp X Q_1) = \operatorname{rank}(Q_1^\perp).$$

Since $\operatorname{rank}(Q_1^\perp) = 1$ or 2 according as $\operatorname{rank}(Q_1) = 2$ or 1, we are lead to the conclusion that $\operatorname{rank}(Q_1^\perp) = 1$. Then since $A^2 - X^2 \geq 0$ and $(A^2 - X^2)Q_1 = 0$,

$$A^2 - X^2 = \rho Q_1^\perp \quad \text{for some } \rho > 0.$$

Therefore, with respect to the orthogonal decomposition

$$\mathcal{H} = \operatorname{ran}(Q_1) \oplus \operatorname{ran}(Q_1^\perp)$$

the operators A^2 and X^2 have the following block representations:

$$A^2 = \begin{bmatrix} B_{11} & B_{12} \\ B_{21} & \alpha \end{bmatrix} \quad \text{and} \quad X^2 = \begin{bmatrix} B_{11} & B_{12} \\ B_{21} & \beta \end{bmatrix} \tag{3.10}$$

where $\alpha > \beta > 0$. By Lemma 1.5 it follows from (3.10) that

$$A^2 - [Q_1^\perp]A^2 = \begin{bmatrix} B_{11} & B_{12} \\ B_{21} & B_{21}B_{11}^{-1}B_{12} \end{bmatrix} = X^2 - [Q_1^\perp]X^2,$$

so that

$$X^2 = \{A^2 - [Q_1^\perp]A^2\} + [Q_1^\perp]X^2. \tag{3.11}$$

Since $\operatorname{rank}(Q_1^\perp) = 1$, both $[Q_1^\perp]A^2$ and $[Q_1^\perp]X^2$ are positive scalar multiples of $Q_1^\perp$, so that

$$[Q_1^\perp]X^2 = \lambda[Q_1^\perp]A^2 \quad \text{for some } 0 < \lambda < 1. \tag{3.12}$$

On the other hand, with orthoprojection Q to the one-dimensional subspace $\operatorname{ran}(A^{-1}\, Q_1^\perp)$, by Lemma 1.4

$$[Q_1^\perp]A^2 = AQA, \tag{3.13}$$

hence with $P \stackrel{def}{=} Q^\perp$

$$A^2 - [Q_1^\perp]A^2 = APA, \tag{3.14}$$

Combining relations (3.10) to (3.14), we can conclude

$$X = \{A(P + \lambda Q)A\}^{1/2} \quad \text{for some } 0 < \lambda < 1.$$

Here P annihilates Q by definition, and $PA^2Q \neq 0$, that is, Q does not commute with A^2, hence nor with A. In fact, if otherwise, $X = AP + \sqrt{\lambda}AQ$, which is obviously not extreme in $\Delta_2(A)$, a contradiction. Thus we have proved that X is of the form (3.9).

When X is not invertible, applying the above arguments to $([R]A^2)^{1/2}$ instead of A, where R is the orthoprojection to $\mathrm{ran}(X)$, we can conclude that either

$$X = ([R]A^2)^{1/2} \tag{3.15}$$

or

$$X = \{([R]A^2)^{1/2}(\tilde{P} + \lambda\tilde{Q})([R]A^2)^{1/2}\}^{1/2}, \tag{3.16}$$

where $\tilde{P}$, $\tilde{Q}$ are orthoprojections of rank one such that

$$\tilde{P}\tilde{Q} = 0, \; \tilde{P} \cdot [R]A^2 \cdot \tilde{Q} \neq 0, \; \text{and } [R]A^2 = ([R]A^2)^{1/2}(\tilde{P} + \tilde{Q})([R]A^2)^{1/2}.$$

Since, with S orthoprojection to $\mathrm{ran}(A^{-1}X)$, by Lemma 1.4

$$[R]A^2 = ASA$$

there is unitary U such that

$$([R]A^2)^{1/2} = USA.$$

Then, with

$$P \overset{def}{=} U^*\tilde{P}U, \quad \text{and} \quad Q \overset{def}{=} U^*\tilde{Q}U$$

we have $S = P + Q$. Therefore, according as (3.15) or (3.16), X takes one of the following forms

$$X = (ASA)^{1/2}$$

or

$$X = \{A(P + \lambda Q)A\}^{1/2}$$

where $PQ = 0$, $PA^2Q \neq 0$, and $\mathrm{rank}(Q) = 1$. This completes the proof. □

References

[AAK] V.M. Adamjan, D.Z. Arov and M.G. Krein, *Infinite block-Hankel matrices and some related problems of extension,* Izv. Akad. Nauk Armenian SSR, Math. (1971), 87–112 = English Translation in Trans. Series, Amer. Math. Soc. **111** (1978), 133–156.

[AnD] W.N. Anderson Jr. and R.J. Duffin, *Series and parallel addition of matrices,* J. Math. Anal. Appl. **3** (1969), 576–594.

[AT1] W.N. Anderson and G.E. Trapp, *Shorted operators II,* SIAM J. Appl. Math. **28** (1988), 60–71.

[AT2] W.N. Anderson and G.E. Trapp, *The extreme points of a set of positive semidefinite operators,* Linear Alg. Appl. **106** (1988), 209–217.

[An1] T. Ando, *Concavity of certain maps on positive definite matices and applications to Hadamard prodcuts,* Linear Alg. Appl. **26** (1979), 203–241.

[An2] T. Ando, *Generalized Schur complements,* Linear Alg. Appl. **27** (1979), 173–180.

[An3] T. Ando, *Extreme points of an intersection of operator intervals,* Proc. Inter. Math. Conf. '94, Kaohsing, Taiwan, pp. 1–18, World Scientific, 1996.

[Dou] R.G. Douglas, *On majorization, factorization and range inclusion of operators on Hilbert spaces,* Proc. Amer. Math. Soc. **17** (1966), 413–415.

[ErL] S.L. Eriksson and H. Leutwiller, *A potential theoretic approach to parallel addition,* Math. Ann. **274** (1986), 301–317.

[FiW] P.A. Fillmore and J.P. Williams, *Operator ranges,* Adv. in Math. **7** (1971), 251–281.

[FoF] C. Foiaş and A.E. Frazho, *The commutant lifting approach to interpolation problems,* Operator Theory: Advances and Applications vol. 44, Birkhäuser Verlag, Basel, 1990.

[GM] W.L. Green and T.D. Morley, *The extreme points of order intervals of positive operators,* Adv. in Appl. Math. **15** (1994), 360–370.

[Kre] M.G. Krein, *Theory of selfadjoint extensions of semi-bounded Hermitian operators and its applications I,* Mat. Sbornik **20** (62) (1947), 431–495.

[KrM] M.G. Krein and D.P. Milman, *On extreme points of regular convex sets,* Studia Math. **9** (1940), 133–138.

[Löw] K. Löwner, *Über monotone Matrixfunktionen,* Math. Z. **38** (1934), 177–216.

[PeS] E.L. Pekarev and Ju.L. Smul'jan, *Parallel addition and parallel subtraction of operators,* Izv. Aka. Nauk SSSR, Ser. Mat. **40** (1976), 366–387 (Russian), English translation in Math. USSR Izv. **10** (1976), 280–337.

[Smu] Ju.L. Smul'jan, *Operator balls,* Function theory, Funcional Anal. Appl. (Kharkov) **6** (1968), 68–81 (Russian) = English Translation in Integral Equ. Operator Theory **13** (1990), 865–885.

This work is supported in part by Grand-in-Aid for Scientific Research 10640183.

Hokusei Gakuen University
Faculty of Economics
Sapporo 004-8631
Japan
ando@hokusei.ac.jp

Primary 47D20, Secondary 46A55

Operator Theory:
Advances and Applications, Vol. 118

M-Accretive Extensions of Sectorial Operators and Krein Spaces

Y.M. Arlinskii

For a closed densely defined sectorial and coercive operator a description of all maximal accretive extensions is obtained. Applications to a second order partial elliptic differential operator are given.

1

As is well known [14] a linear operator S in a complex Hilbert space H is said to be accretive if $Re\,(Sf, f) \geq 0$ for all vectors f in the domain $\mathcal{D}(S)$, and maximal accretive (m-accretive) if it is accretive and has no accretive extensions in H. Maximal accretive operators have dense domains, their resolvent sets contain the left half-plane, and such operators generate C_0-contractive semigroups $\exp(-tS)$ for $t \geq 0$.

Every accretive operator has an m-accretive extension. The problem of the description of all m-accretive extensions was examined by R.S. Phillips in [21–22] who suggested to use for this purpose the geometry of Krein spaces. The Phillips's approach was applied in [11] for a positive definite minimal differential operator generated by an even order regular differential expression on a segment of the real axis in a weighted $\mathcal{L}^2$ space, and in [19–20] for an abstract positive definite symmetric operator with a finite defect number.

In this paper we use basic concepts of the Phillips's theory and elaborate methods of [11], [19–20] for a closed sectorial operator with the vertex at the origin [14]. For the case of a densely defined symmetric and nonnegative operator, its nonnegative selfadjoint extensions were studied by M.G. Krein in his fundamental papers [16–17] and later by M.I. Vishik [27] and M.S. Birman [9]. Descriptions of all proper m-accretive and proper m-sectorial extensions were obtained in [15], [18], [24–26], [1], [10] by using the method of abstract boundary conditions [12–13]. A characterization of all proper m-accretive extensions of a nonnegative linear relation was given in [2].

We will use the following notations: $\mathcal{L}(H_1, H_2)$ is the Banach space of all continuous linear operators acting from the Hilbert space H_1 into the Hilbert space H_2 and $\mathcal{L}(H) = \mathcal{L}(H, H)$. The domain, range and null-space of a linear operator T is denoted by $\mathcal{D}(T)$, $\mathcal{R}(T)$ and $Ker(T)$, respectively, $T[u, v]$ is the closure of the sesquilinear form $(T\cdot, \cdot)$ determined by a sectorial operator T, $\mathcal{D}[T]$ is the domain of $T[u, v]$ and finally $T[u] = T[u, u]$ is the corresponding quadratic form.

If T is an m-sectorial operator then [14]

$$T = T_R^{1/2}(I + iG)T_R^{1/2}, \; T[u, v] = ((I + iG)T_R^{1/2}u, T_R^{1/2}v), \quad u, v \in \mathcal{D}[T],$$

where T_R is the "real part" of T, i.e. the nonnegative selfadjoint operator associated with the closed form $T_R[u, v] = (T[u, v] + \overline{T[v, u]})/2$ and $G = G^* \in \mathcal{L}(\overline{R(T)})$. According to the First and Second Representation Theorems [14] we have equalities

$$\mathcal{D}[T] = \mathcal{D}[T_R] = \mathcal{D}(T_R^{1/2}).$$

2

Consider the set $\mathbf{H} = H \times H$ as a Hilbert space with the inner product

$$(\vec{u}, \vec{v}) = (u_1, v_1) + (u_2, v_2),$$

and as a Krein space with the indefinite inner product

$$Q(\vec{u}, \vec{v}) = (u_2, v_1) + (u_1, v_2)$$

for $\vec{u} = \langle u_1, u_2 \rangle$, $\vec{v} = \langle v_1, v_2 \rangle$.

Recall that a closed subspace $\mathbf{S}$ in $\mathbf{H}$ is called a linear relation (l.r.) in H (see [23]). The Q-orthogonal complement of $\mathbf{S}$ coincides with $-\mathbf{S}^*$ where $\mathbf{S}^*$ is the adjoint l.r. to $\mathbf{S}$. It is clear that a l.r. $\mathbf{S}$ is accretive if and only if $Q(\vec{u}, \vec{u}) \geq 0$ for all $\vec{u} \in \mathbf{S}$, i.e. the subspace $\mathbf{S}$ is nonnegative [7] with respect to the indefinite inner product, and $\mathbf{S}$ is m-accretive if and only if it is a maximal nonnegative subspace.

Let $\mathbf{S}$ be an accretive l.r., let $\mathfrak{N} = Ker(\mathbf{S}^* + I)$ be the defect subspace of $\mathbf{S}$ and let

$$\mathbf{N}_+ = \{\langle \phi, \phi \rangle, \; \phi \in \mathfrak{N}\}.$$

Consider an extension of $\mathbf{S}$ defined as follows

$$\mathbf{T} = \mathbf{S} \dot{+} \mathbf{N}_+. \tag{1}$$

One can easily check that $\mathbf{T}$ is an m-accretive extension of $\mathbf{S}$ and

$$-\mathbf{S}^* = -\mathbf{T}^* \oplus \mathbf{N}_+,$$

where the decomposition is both orthogonal and Q-orthogonal. The subspace $\mathbf{N}_+$ is positive and intrinsically complete with respect to Q-norm, but the maximal nonpositive lineal $-\mathbf{T}^*$ need not be intrinsically complete. Let $\mathbf{M}_-$ be the intrinsic completion of $-\mathbf{T}^*$. Following to [21–22], [11] we define an abstract boundary space $\hat{\mathbf{H}}$ by the equality

$$\hat{\mathbf{H}} = \mathbf{M}_- \oplus \mathbf{N}_+.$$

According to R.S. Phillips [21–22] the formula

$$-\mathbf{L} = \hat{\mathbf{L}} \cap \{-\mathbf{S}^*\}$$

gives a one-to-one correspondence between m-accretive restrictions **L** of $\mathbf{S}^*$ and maximal nonpositive subspaces $\hat{\mathbf{L}}$ of $\hat{\mathbf{H}}$. More precisely, the formula

$$\mathbf{L} = (\mathbf{I} + \mathbf{K})\mathbf{T}^*$$

gives a one-to-one correspondence between m-accretive restrictions of $\mathbf{S}^*$ and linear operators $\mathbf{K} : \mathbf{T}^* \to -\mathbf{N}_+$ which satisfy the condition

$$-Q(\mathbf{K}\vec{u}, \mathbf{K}\vec{u}) \leq Q(\vec{u}, \vec{u}), \ \vec{u} \in \mathbf{T}^*.$$

In fact, a vector $\vec{v} = \langle v, v' \rangle \in \mathbf{L}$ can be represented as $\vec{v} = \vec{u} + \vec{\phi}$, where $\vec{u} = \langle u, u' \rangle \in \mathbf{T}^*$ and $\vec{\phi} = \langle \phi, -\phi \rangle, \phi \in \mathfrak{N}$. The equalities $\mathcal{R}(\mathbf{L}+I) = \mathcal{R}(\mathbf{T}+I) = H$ imply that there exists an operator $\mathbf{K} : \mathbf{T}^* \to -\mathbf{N}_+$ such that $\vec{\phi} = \mathbf{K}\vec{u}$ and $-Q(\mathbf{K}\vec{u}, \mathbf{K}\vec{u}) \leq Q(\vec{u}, \vec{u}), \ \vec{u} \in \mathbf{T}^*$.

Let S be a closed accretive operator and let **S** be the graph of S. Then the m-accretive extension T whose graph **T** is defined by (1), takes the form

$$\mathcal{D}(T) = \mathcal{D}(S)\dot{+}\mathfrak{N}, \ T(f + \varphi) = Sf + \varphi, \ f \in \mathcal{D}(S), \varphi \in \mathfrak{N}. \tag{2}$$

Therefore the following relation

$$\mathbf{L} = \{\langle u + Ku, T^*u - Ku \rangle, u \in \mathcal{D}(T^*)\}$$

gives a one-to-one correspondence between all m-accretive restrictions **L** of $\mathbf{S}^*$ and all linear operators $K : \mathcal{D}(T^*) \to \mathfrak{N}$ satisfying the condition $\|Ku\|^2 \leq Re\,(T^*u, u), \quad u \in \mathcal{D}(T^*)$.

3

Next we give a realization of $\hat{\mathbf{H}}$ when **S** is the graph of a densely defined closed sectorial operator S with the vertex at the origin [14]. Let S_F and S_N be the Friedrichs [14] and the von Neumann-Krein [3–4] m-sectorial extensions of S. The operators S_F and S_N have representations

$$S_F = S_{FR}^{1/2}(I + iG_F)S_{FR}^{1/2}, \ S_N = S_{NR}^{1/2}(I + iG_N)S_{NR}^{1/2},$$

where S_{FR} and S_{NR} are real parts of S_F and S_N, G_F and G_N are bounded self-adjoint operators.

Proposition 1 [6]. *The extension T of S defined by* (2), *is m-sectorial if and only if the following inclusion holds:*

$$\mathcal{D}(S^*) \subseteq \mathcal{D}[S_N]. \tag{3}$$

Consider $\mathcal{D}[S_N]$ as a Hilbert space with the norm $(\|u\|^2 + Re\,S_N[u])^{1/2}$. Then $\mathcal{D}[S_F]$ is a subspace of $\mathcal{D}[S_N]$.

Definition 1 [3]. A pair $\{\mathcal{H}, \Gamma\}$ is called a boundary pair for the sectorial operator S if $\mathcal{H}$ is a Hilbert space and $\Gamma \in \mathcal{L}(\mathcal{D}[S_N], \mathcal{H})$ such that $Ker\, \Gamma = \mathcal{D}[S_F]$, $\mathcal{R}(\Gamma) = \mathcal{H}$.

Proposition 2 [6]. *Let S be a closed densely defined sectorial operator and let $\{\mathcal{H}, \Gamma\}$ be a boundary pair for S. Then there exists a linear operator $G : \mathcal{D}(S^*) \cap \mathcal{D}[S_N] \to \mathcal{H}$ such that*

(4) $$S_N^*[u, v] = (S^*u, v) - (Gu, \Gamma v)_{\mathcal{H}}, \quad u \in \mathcal{D}(S^*) \cap \mathcal{D}[S_N],\ v \in \mathcal{D}[S_N].$$

Definition 2 [6]. The triplet $\{\mathcal{H}, G, \Gamma\}$ in Proposition 2 is called a boundary value space (b.v.s.) of S^*.

Observe that from (4) it follows $Ker\, G = \mathcal{D}(S_N)$. If $\mathcal{H}_+ \subset \mathcal{H} \subset \mathcal{H}_-$ is a rigged Hilbert space [8] and $\{\mathcal{H}_-, \Gamma\}$ is a boundary pair for S, then the linear operator G in (4) can be chosen as taking values in $\mathcal{H}_+$. In such a case the triplet $\{\mathcal{H}_+ \subset \mathcal{H} \subset \mathcal{H}_-, G, \Gamma\}$ is also called a b.v.s. of S^*.

Let the condition (3) be satisfied. Then [3] $\mathcal{D}[S_N] = \mathcal{D}[S_F] \dot{+} \mathfrak{N}$. Define operators connected with a b.v.s. $\{\mathcal{H}, G, \Gamma\}$ of S^*:

(5) $$\gamma = (\Gamma|\mathfrak{N})^{-1}, \ A = -G\gamma, \ B = S_{NR}^{1/2}\gamma.$$

Then $\gamma \in \mathcal{L}(\mathcal{H}, H) \cap \mathcal{L}(\mathcal{H}, \mathcal{D}[S_N]), \ A \in \mathcal{L}(\mathcal{H}), \ B \in \mathcal{L}(\mathcal{H}, H)$.

Proposition 3 *Let* (3) *be satisfied and let $\{\mathcal{H}, G, \Gamma\}$ be a b.v.s for S^*. Then the domain of T^* can be described as follows:*

(6) $$\mathcal{D}(T^*) = \{u \in \mathcal{D}(S^*) : Gu = A^*\Gamma u - 2B^* S_{NR}^{1/2} u\}.$$

Proof: A vector $u \in \mathcal{D}(S^*)$ belongs to $\mathcal{D}(T^*)$ if and only if the equality

$$(S^*u, \varphi) + (u, S^*\varphi) = 0$$

holds for all $\varphi \in \mathfrak{N}$. Representing φ as $\varphi = \gamma e, \ e \in \mathcal{H}$ and using (4), (5), we get

$$\begin{aligned} 0 &= S_N^*[u, \gamma e] + (Gu, e)_{\mathcal{H}} + S_N[u, \gamma e] + (\Gamma u, G\gamma e)_{\mathcal{H}} \\ &= (2B^* S_{NR}^{1/2} u - A^*\Gamma u + Gu, e)_{\mathcal{H}}. \end{aligned}$$

This implies (6). □

Denote $H_0 = \overline{\mathcal{R}(S)}$ and define a Krein space

$$\mathfrak{H} = H_0 \times \mathcal{H} \times \mathcal{H}$$

with the inner product

$$Q_1(\vec{f}, \vec{g}) = -2(f, g) - (e_1, h_2)_{\mathcal{H}} - (e_2, h_1)_{\mathcal{H}}$$

for $\vec{f} = \langle f, e_1, e_2\rangle$, $\vec{g} = \langle g, h_1, h_2\rangle$, $f, g \in H_0$, $e_1, e_2, h_1, h_2 \in \mathcal{H}$ and define a linear operator $\Phi : -\mathbf{S}^* \to \mathfrak{H}$ by the relation

$$\Phi\langle u, -S^*u\rangle = \langle S_{NR}^{1/2}u, \Gamma u, Gu\rangle, \quad u \in \mathcal{D}(S^*).$$

It follows from (4) that Φ preserves indefinite inner products. Let $\overline{\Phi}$ is the continuation of Φ on $\hat{\mathbf{H}}$.

Theorem 1 *Let S be a closed densely defined sectorial operator satisfying the condition* (3). *Then $\mathcal{R}(\overline{\Phi}) = \mathfrak{H}$.*

Proof: Denote

$$\mathfrak{H}_- = \{\langle h, e, A^*e - 2B^*h\rangle, \ h \in H_0, e \in \mathcal{H}\}, \ \mathfrak{H}_+ = \{\langle Be, e, -Ae\rangle, e \in \mathcal{H}\}.$$

The subspaces $\mathfrak{H}_-$ and $\mathfrak{H}_+$ are Q_1-orthogonal and $\Phi\mathbf{N}_+ = \mathfrak{H}_+$. From Proposition 3 it follows that $\Phi(-\mathbf{T}^*) \subseteq \mathfrak{H}_-$. Let us show that $\overline{\Phi}\mathbf{M}_- = \mathfrak{H}_-$. Relations (4) and (5) imply for all $e \in \mathcal{H}$

$$Re\,(Ae, e)_{\mathcal{H}} - \|Be\|^2 = \|\gamma e\|^2. \tag{7}$$

Hence for a vector $\vec{f} = \langle h, e, A^*e - 2B^*h\rangle$ we have

$$-Q_1(\vec{f}, \vec{f}) = 2\|h - Be\|^2 + 2\|\gamma e\|^2. \tag{8}$$

Let $u \in \mathcal{D}(T^*)$. Then for $\vec{g} = \Phi\langle u, -T^*u\rangle$ from (6) we get

$$-Q_1(\vec{g}, \vec{g}) = 2\|S_{NR}^{1/2}(u - \gamma\Gamma u)\|^2 + 2\|\gamma\Gamma u\|^2. \tag{9}$$

By Proposition 1 the operator T is m-sectorial. Since $\mathfrak{N} \subset \mathcal{D}[S_N]$, we have $\mathcal{D}[T] = \mathcal{D}[S_N]$ (see [3]). Therefore $\mathcal{D}(T^*)$ is a core of $S_{NR}^{1/2}$ and if a sequence $\{u_n\} \subset \mathcal{D}(T^*)$ converges in the Hilbert space $\mathcal{D}[S_N]$ to a vector u, from (9) the sequence $\{\Phi\langle u_n, -T^*u_n\rangle\}$ Q_1-converges in $\mathfrak{H}_-$ to the vector $\langle S_{NR}^{1/2}u, \Gamma u, A^*\Gamma u - 2B^*S_{NR}^{1/2}u\rangle$. Thus,

$$\overline{\Phi}\mathbf{M}_- \supseteq \{\langle S_{NR}^{1/2}u, \Gamma u, A^*\Gamma u - 2B^*S_{NR}^{1/2}u\rangle, \ u \in \mathcal{D}[S_N]\}.$$

Let $u \in \mathcal{D}[S_N]$, $\vec{g} = \langle S_{NR}^{1/2}u, \Gamma u, A^*\Gamma u - 2B^*S_{NR}^{1/2}u\rangle$. Then

$$-Q_1(\vec{f} - \vec{g}, \vec{f} - \vec{g}) = 2\|h - S_{NR}^{1/2}u - B(e - \Gamma u)\|^2 + \|\gamma(e - \Gamma u\|^2.$$

Choosing $u = \varphi + \gamma e$, where $\varphi \in \mathcal{D}[S_F]$, $e \in \mathcal{H}$, we have

$$-Q_1(\vec{f} - \vec{g}, \vec{f} - \vec{g}) = 2\|h - S_{NR}^{1/2}\varphi - Be\|^2.$$

Taking into account that $H_0 = \overline{\mathcal{R}(S_{NR}^{1/2})}$, $\mathcal{R}(B) \subset H_0$ and by using the relation [3]:

$$\inf\{Re\, S_N[u-\varphi],\ \varphi \in \mathcal{D}[S_F]\}, \quad u \in \mathcal{D}[S_N],$$

we get that $\overline{\Phi \mathbf{M}_-} = \mathfrak{H}_-$.

Next we show that $\mathfrak{H} = \mathfrak{H}_- \dot{+} \mathfrak{H}_+$. Let $\vec{f} = \langle f, h_1, h_2 \rangle$ be an arbitrary vector from $\mathfrak{H}$. It follows from (7) that $(A + A^* - 2B^*B)^{-1} \in \mathcal{L}(\mathcal{H})$. Therefore the system of linear equations

$$x + Bz = f,\ y + z = h_1,\ A^*y - 2B^*x - Az = h_2$$

has unique solutions $x \in H_0$, $y, z \in \mathcal{H}$. This means that

$$\vec{f} = \langle x, y, A^*y - 2B^*x \rangle + \langle Bz, z, -Az \rangle.$$

So, $\mathcal{R}(\overline{\Phi}) = \mathfrak{H}$. □

4

From Theorem 1 and R. Phillips's results [21–22] it follows that the formula

$$-\mathbf{L} = \{\vec{u} \in -\mathbf{S}^* : \Phi\vec{u} \in \mathfrak{L}^-\}$$

gives a one-to-one correspondence between all maximal nonpositive subspaces $\mathfrak{L}^-$ in $\mathfrak{H}$ and graphs $\mathbf{L}$ of all m-accretive restrictions L of S^*.

Observe that subspaces

$$\mathfrak{H}^+ = \{\langle 0, -h, h \rangle, h \in \mathcal{H}\},\ \mathfrak{H}^- = \{\langle f, h, h \rangle, f \in H_0, h \in \mathcal{H}\}$$

are maximal positive and maximal negative with respect to Q_1-inner product. From [7] it follows that an arbitrary maximal nonnegative subspace in $\mathfrak{H}$ takes the form

$$\mathfrak{L}^+ = \{\langle Xh, -(I+Y)h, (I-Y)h \rangle, h \in \mathcal{H}\}, \tag{10}$$

where operators $X \in \mathcal{L}(\mathcal{H}, H_0)$ and $Y \in \mathcal{L}(\mathcal{H})$ satisfy the condition

$$\|Xh\|^2 + \|Yh\|^2_{\mathcal{H}} \le \|h\|^2_{\mathcal{H}}. \tag{11}$$

As is well known [7] a subspace $\mathfrak{L}^-$ is maximal nonpositive in a Krein space if and only if it is orthogonal to a maximal nonnegative subspace $\mathfrak{L}^+$ with respect to the indefinite inner product.

Theorem 2 *Let S be a closed densely defined sectorial operator satisfying the condition* (3) *and let $\{\mathcal{H}, G, \Gamma\}$ be a b.v.s. for S^*. Then the formula*

$$\mathcal{D}(L) = \{u \in \mathcal{D}(S^*) : (I+V)Gu - (I-V)\Gamma u = 2\Psi u\} \tag{12}$$

gives a one-to-one correspondence between all m-accretive restrictions of the operator S^* and all pairs $\langle V, \Psi \rangle$ where V is a contraction in $\mathcal{H}$ and Ψ is a linear operator from $\mathcal{D}(S^*)$ into $\mathcal{H}$ such that for all $u \in \mathcal{D}(S^*)$, $h \in \mathcal{H}$

$$|(\Psi u, h)_{\mathcal{H}}|^2 \leq Re\, S_N[u](\|h\|^2_{\mathcal{H}} - \|V^* h\|^2_{\mathcal{H}}). \tag{13}$$

Proof: Let **L** be the graph of an m-accretive restriction L of S^*, then the image $\Phi(-\mathbf{L})$ is Q_1-orthogonal to some maximal nonnegative subspace $\mathfrak{L}^+$ in $\mathfrak{H}$, which has a representation (10) with the condition (11). Therefore, for all $u \in \mathcal{D}(L)$ and all $h \in \mathcal{H}$

$$-2(S^{1/2}_{NR} u, Xh) - (\Gamma u, (I - Y)h)_{\mathcal{H}} + (Gu, (I + Y)h)_{\mathcal{H}} = 0.$$

Putting $V = Y^*$, $\Psi = X^* S^{1/2}_{NR}$, we get (12) and (13). □

5

Next, we give a description in terms of abstract boundary conditions of all m-accretive extensions of a closed densely defined coercive sectorial operator.

An accretive operator S is called coercive if there exists $m > 0$ such that

$$Re\,(Sf, f) \geq m\|f\|^2,\ f \in \mathcal{D}(S).$$

Let S be a closed densely defined sectorial and coercive operator S. Then the Friedrichs extension S_F of S is also a coercive and $Re\,(S_F f, f) \geq m\|f\|^2$, $f \in \mathcal{D}(S_F)$. Hence

$$\mathcal{D}(S^*) = \mathcal{D}(S^*_F) \dot{+} Ker\, S^*.$$

From the decomposition (see [3])

$$\mathcal{D}[S_N] = \mathcal{D}[S_F] \dot{+} Ker\, S^* \tag{14}$$

it follows that the condition (3) is satisfied.

Proposition 4 *Let S be a closed densely defined sectorial and coercive operator. Then the domain of any m-accretive extension of S is contained in $\mathcal{D}[S_N]$.*

Proof: We will use the equality [3]:

$$\mathcal{D}[S_N] = \left\{ u \in H : \sup_{f \in \mathcal{D}(S)} \frac{|(Sf, u)|^2}{Re\,(Sf, f)} < \infty \right\}.$$

Let $\widetilde{S}$ be an m-accretive extension of S. Represent $\mathcal{D}(\widetilde{S})$ as $\mathcal{D}(\widetilde{S}) = \mathcal{D}(S) \dot{+} M$, where M is a linear manifold in $\mathcal{D}(\widetilde{S})$. Then $\widetilde{S}(f + \varphi) = Sf + \widetilde{S}\varphi$, $f \in \mathcal{D}(S)$, $\varphi \in M$ and for all real t:

$$\begin{aligned} Re\,(\widetilde{S}(tf + \varphi), tf + \varphi)) &= t^2 Re\,(Sf, f) + t\{[Re\,(Sf, \varphi) \\ &\quad + Re\,(\widetilde{S}\varphi, f)\} + Re\,(\widetilde{S}\varphi, \varphi) \geq 0. \end{aligned}$$

Hence

$$|Re\,(Sf,\varphi)+Re\,(\widetilde{S}\varphi,f)|\le 2\sqrt{Re\,(Sf,f)}\sqrt{Re\,(\widetilde{S}\varphi,\varphi)}.$$

It follows that

$$|Re\,(Sf,\varphi)|\le 2\sqrt{Re\,(Sf,f)}\sqrt{Re\,(\widetilde{S}\varphi,\varphi)}+|Re\,(S_{FR}^{-1/2}\widetilde{S}\varphi,S_{FR}^{1/2}f)|.$$

Thus, for all $f\in\mathcal{D}(S),\varphi\in M$ we have $|(Sf,\varphi)|^2\le a\,Re\,(Sf,f)$. Hence $\varphi\in\mathcal{D}[S_N]$. □

In [3] it was proved that if S is a closed densely defined sectorial and coercive operator then the operator defined by the relation

$$S_*=S^*|(\mathcal{D}(S_F^*)\cap\mathcal{D}(S_N^*))$$

is also a closed and densely defined sectorial operator. Moreover, the Friedrichs and the von Neumann-Krein extensions of S_* coincide with S_F^* and S_N^*, respectively. Let S_*^* be the adjoint operator. Then

$$\mathcal{D}(S_*^*)=\mathcal{D}(S_F)\dot{+}Ker\,S^*.$$

Let $\{\mathcal{H},\Gamma\}$ be a boundary pair for S. Put $Z_0=(\Gamma|Ker\,S^*)^{-1}$. The operator $Z_0\Gamma$ is the projection onto $Ker\,S^*$ with respect to the decomposition (14) and (see [3]) the relation

$$S_N[u,v]=S_F[u-Z_0\Gamma u,v-Z_0\Gamma v],\ u,v\in\mathcal{D}[S_N] \tag{15}$$

holds. It is clear that

$$S_*^*u=S_F(u-Z_0\Gamma u),\ u\in\mathcal{D}(S_*^*). \tag{16}$$

Proposition 5 *Let S be a closed densely defined sectorial and coercive operator and let $\{\mathcal{H},G,\Gamma\}$ be a b.v.s.for S^*. If $\widetilde{S}$ is an extension of S such that $\mathcal{D}(\widetilde{S})\subseteq\mathcal{D}[S_N]$, then for all $u\in\mathcal{D}(\widetilde{S})$ and all $v\in\mathcal{D}(\widetilde{S}^*)$ the following relation holds:*

$$S_N^*[v,u-S_F^{-1}\widetilde{S}u]+(Gv,\Gamma u)_{\mathcal{H}}-(\Gamma v,Z_0^*\widetilde{S}u)_{\mathcal{H}}=0. \tag{17}$$

Proof: From (4) and (15) we have

$$\begin{aligned}&S_N^*[v,u-S_F^{-1}\widetilde{S}u]+(Gv,\Gamma u)_{\mathcal{H}}-(\Gamma v,Z_0^*\widetilde{S}u)_{\mathcal{H}}\\&=(\widetilde{S}^*v,u)-S_F^*[v-Z_0\Gamma v,S_F^{-1}\widetilde{S}u]-(Z_0\Gamma v,\widetilde{S}u)\\&=(\widetilde{S}^*v,u)-(v-Z_0\Gamma v,\widetilde{S}u)-(Z_0\Gamma v,\widetilde{S}u)=0\end{aligned}$$

for $u\in\mathcal{D}(\widetilde{S})$ and $v\in\mathcal{D}(\widetilde{S}^*)$. □

Let $\{\mathcal{H}, \Gamma\}$ be a boundary pair for S. Put $G_* = Z_0^* S_*^*$. Then from (15) and (16) it follows that

$$S_N[u, v] = (S_*^* u, v) - (G_* u, \Gamma v)_{\mathcal{H}}, \quad u \in \mathcal{D}(S_*^*), v \in \mathcal{D}[S_N].$$

Thus, the triplet $\{\mathcal{H}, G_*, \Gamma\}$ forms a b.v.s. for the operator S_*^*. Let us define a quadratic functional

$$\mu[\varphi] = \sup\{Re\, S_F[2\varphi - f, f], \ f \in \mathcal{D}[S_F]\}. \tag{18}$$

Since

$$Re\, S_F[2\varphi - f, f] = \|(I + iG_F)S_{FR}^{1/2}\varphi\|^2 - \|S_{FR}^{1/2} f - (I + iG_F)S_{FR}^{1/2}\varphi\|^2,$$

we get

$$\mu[\varphi] = \|(I + iG_F)S_{FR}^{1/2}\varphi\|^2. \tag{19}$$

For a coercive S the supremum in (19) is attained with a unique vector $f_0 \in \mathcal{D}[S_F]$ such that $S_{FR}^{1/2} f_0 = (I + iG_F)S_{FR}^{1/2}\varphi$ and $\mu[\varphi] = \|S_{FR}^{1/2} f_0\|^2$. If S is a positive definite operator then $\mu[\varphi] = S_F[\varphi]$.

Theorem 3 *Let S be a densely defined, closed, sectorial and coercive operator and let $\{\mathcal{H}, G_*, \Gamma\}$ be a b.v.s. of S_*^*. Then the relations*

$$\begin{aligned} \mathcal{D}(\widetilde{S}) &= \{u \in \mathcal{D}[S_N] : u - Z_0\Gamma u + 2C\Gamma u \in \mathcal{D}(S_F); \\ &\qquad G_*(u + 2C\Gamma u) \in \mathbf{W}(\Gamma u)\}, \end{aligned} \tag{20}$$

$$\widetilde{S}u = S_F(u - Z_0\Gamma u + 2C\Gamma u) \tag{21}$$

establish a one-to-one correspondence between all m-accretive extensions $\widetilde{S}$ of S and all pairs $\langle \mathbf{W}, C\rangle$ where $\mathbf{W}$ is an m-accretive l.r. in $\mathcal{H}$ and C is a linear operator from $\mathcal{D}(\mathbf{W})$ into $\mathcal{D}[S_F]$, such that

$$\mu[Ch] \le Re\,(\mathbf{W}(h), h)_{\mathcal{H}}, \ h \in \mathcal{H}. \tag{22}$$

The extension $\widetilde{S}$ has a bounded inverse if and only if the corresponding l.r. $\mathbf{W}$ has a bounded inverse W^{-1}; in this case

$$\widetilde{S}^{-1} = S_F^{-1} + (Z_0 - 2C)W^{-1}Z_0^*. \tag{23}$$

Proof: Let $\widetilde{S}$ be an m-accretive extension of S. Then $\widetilde{S}^*$ is an m-accretive restriction of S^*. Consider a b.v.s. $\{\mathcal{H}, G, \Gamma\}$ of S^*. The image Φ of the graph of the operator $-\widetilde{S}^*$ is Q_1-orthogonal to some maximal nonnegative subspace $\mathfrak{L}^+$ of $\mathfrak{H}$ which takes the form (11) with the condition (12). Let $\mathbf{W} = \{\langle (I+Y)h, (I-Y)h\rangle, \ h \in \mathcal{H}\}$. Then $\mathbf{W}$ is an m-accretive l.r. From (17), Propositions 4, 5 and the definition of Φ it follows that for $u \in \mathcal{D}(\widetilde{S})$ the vector

$$\left\langle \frac{1}{2}(I + iG_N)S_{NR}^{1/2}(u - S_F^{-1}\widetilde{S}u), \Gamma u, -Z_0^*\widetilde{S}u \right\rangle$$

belongs to $\mathfrak{L}^{+}$. Therefore for some $h \in \mathcal{H}$

$$\begin{aligned} &\frac{1}{2}(I+iG_N)S_{NR}^{1/2}(u-S_F^{-1}\widetilde{S}u)=Xh,\\ &\Gamma u=-(I+Y)h,\ -Z_0^*\widetilde{S}u=(I-Y)h. \end{aligned} \tag{24}$$

It follows that $\langle \Gamma u, Z_0^*\widetilde{S}u\rangle \in \mathbf{W}$. Let $\varphi = u - S_F^{-1}\widetilde{S}u - Z_0\Gamma u$. Then $u - Z_0\Gamma u - \varphi \in \mathcal{D}(S_F)$ and

$$\begin{aligned} \widetilde{S}u &= S_F(u-Z_0\Gamma u-\varphi)=S_*^*(u-\varphi),\\ Z_0^*\widetilde{S}u &= G_*(u-\varphi),\ S_{NR}^{1/2}(u-S_F^{-1}\widetilde{S}u)=S_{NR}^{1/2}(\varphi+Z_0\Gamma u)=S_{NR}^{1/2}\varphi. \end{aligned}$$

Since for all $f \in \mathcal{D}[S_F]$ it holds $S_{NR}^{1/2}f = U_F S_{FR}^{1/2}f$ with some isometry U_F from H onto H_0 and $U_F G_F = G_N U_F$ (see [3]), we have $\varphi = 2S_{FR}^{-1/2}(I + iG_F)^{-1}U_F^{-1}Xh$. Let us define a linear operator K by the relation:

$$K(I+Y)e = Xe,\ e \in \mathcal{H}.$$

It follows from the condition (12) that the operator K is correctly defined, acts from $\mathcal{D}(\mathbf{W})$ into H_0 and $\|Kg\| \le Re\,(\mathbf{W}(g), g)_{\mathcal{H}}$ for all $g \in \mathcal{D}(\mathbf{W})$. From (24)

$$\varphi = -2S_{FR}^{-1/2}(I+iG_F)^{-1}U_F^{-1}K\Gamma u.$$

Let $Cg = S_{FR}^{-1/2}(I+iG_F)^{-1}U_F^{-1}Kg,\ g \in \mathcal{D}(\mathbf{W})$. Then the operator C acts from $\mathcal{D}(\mathbf{W})$ into $\mathcal{D}[S_F]$ and $\varphi = -2C\Gamma u$. From (19) it follows that C satisfies the condition (22) and relations (20), (21) hold.

Conversely, let $\widetilde{S}$ is defined by (20), (21) and the condition (22) is satisfied. Then for $u, v \in \mathcal{D}(\widetilde{S})$ we have

$$\begin{aligned} (\widetilde{S}u, v) &= (S_*^*(u+2C\Gamma u, v) = S_N[u+2C\Gamma u, v]\\ &\quad + (G_*(u+2C\Gamma u), \Gamma v)_{\mathcal{H}}\\ &= S_N[u,v]+2S_N[C\Gamma u, v]+(\mathbf{W}(\Gamma u), \Gamma v)_{\mathcal{H}}, \end{aligned}$$

$$\begin{aligned} Re\,(\widetilde{S}u, u) &= \|S_{NR}^{1/2}u+2(I+iG_N)S_{NR}^{1/2}C\Gamma u\|^2\\ &\quad + Re\,(\mathbf{W}\Gamma u, \Gamma u)_{\mathcal{H}} - \|(I+iG_N)S_{NR}^{1/2}C\Gamma u\|^2 \ge 0. \end{aligned}$$

Thus, $\widetilde{S}$ is an accretive extension of S. Represent $\mathbf{W}$ as

$$\mathbf{W} = \{\langle (I+Y)h, (I-Y)h >, h \in \mathcal{H}\},$$

where $Y \in \mathcal{L}(\mathcal{H})$ is a contraction and let

$$Xh = U_F(I+iG_F)S_{FR}^{1/2}C(I+Y)h,\ h \in \mathcal{H}.$$

Then from (22) and (19) the subspace

$$\mathfrak{L}^+ = \{\langle Xh, -(I+Y)h, (I-Y)h\rangle, h \in \mathcal{H}\}$$

is a maximal nonnegative in the Krein space $\mathfrak{H}$. Let $\mathfrak{L}^-$ be the Q_1-orthogonal complement to $\mathfrak{L}^+$ and let L be an m-accretive restriction of S^* defined by the relation

$$-\mathbf{L} = \{\vec{v} \in -\mathbf{S}^* : \Phi\vec{v} \in \mathfrak{L}^-\}.$$

As was shown the right-hand sides of (20) and (21) determine the adjoint operator L^* which is an m-accretive extension of S. Thus, $\widetilde{S}$ coincides with L^*.

Since $Ker\, Z_0^* = H_0$, $\mathcal{R}(Z_0^*) = \mathcal{H}$, then an m-accretive extension $\widetilde{S}$ has a bounded inverse if and only if $\mathcal{R}(Z_0^*\widetilde{S}) = \mathcal{H}$. From (21) it follows

$$Z_0^*\widetilde{S}u = Z_0^*S_F(u - Z_0\Gamma u + 2C\Gamma u) = G_*(u + 2C\Gamma u).$$

Therefore by (20), $\mathcal{R}(Z_0^*\widetilde{S}) = \mathcal{H}$ if and only if $\mathcal{R}(\mathbf{W}) = \mathcal{H}$. Since $\mathbf{W}$ is an m-accretive l.r., we have that $\mathcal{R}(\mathbf{W}) = \mathcal{H}$ iff $\mathbf{W}$ has a bounded inverse operator W^{-1}. In this case from (20) for all $h \in H$:

$$\widetilde{S}^{-1}h = u,\ S_F^{-1} = u - (Z_0 - 2C)\Gamma u,\ \Gamma u = W^{-1}G_*S_F^{-1}h = W^{-1}Z_0^*h.$$

Hence we get (23). □

Remark 1 In [5] it was proved that $\widetilde{S}$ is an m-sectorial extension of S if and only if it takes the form (20) and (21) where $\mathbf{W}$ is an m-sectorial l.r. in $\mathcal{H}$ and for some $\delta \in [0, 1)$

$$\mu[Ch] \le \delta\, Re\, (\mathbf{W}(h), h)_{\mathcal{H}},\ h \in \mathcal{H}$$

For this purpose in [5] we used a parametrization of all closed sectorial forms associated with m-sectorial extensions of a sectorial operator obtained in [3].

Remark 2 If S is a positive symmetric operator then condition (22) takes the form

$$S_F[Ch] \le Re\, (\mathbf{W}(h), h),\ h \in \mathcal{H}.$$

Letting $C = 0$ we get all proper m-accretive (m-sectorial) extensions $\widetilde{S}$ ($S \subset \widetilde{S} \subset S^*$) which were described in a similar form in [15], [18], [24–26], [1], [10]. When S is a positive definite symmetric operator with finite defect number a decription of its all m-accretive extensions was obtained in [19–20] in a form similar to (20)–(21).

Denote by P the orthogonal projection in H onto $Ker\, S^*$ and by Π_F and Π_0 the projections in $\mathcal{D}[S_N]$ onto $\mathcal{D}[S_F]$ and $Ker\, S^*$ with respect to the decomposition (14). Put

$$\mathcal{H} = Ker\, S^*,\ \Gamma = \Pi_0,\ G_*u = PS_F\Pi_F u,\quad u \in \mathcal{D}(S_*^*).$$

Then $\{\mathcal{H}, \Gamma\}$ and $\{\mathcal{H}, G_*^*, \Gamma\}$ form a boundary pair and a b.v.s for S and S_*^*, respectively. Let M is the operator part of an m-accretive l.r $\mathbf{W}^{-1}$ in $Ker\, S^*$ and let

$$\mathfrak{M} = Ker\mathbf{W} = \mathbf{W}^{-1}(0) = Ker\, S^* \ominus \overline{\mathcal{D}(M)}.$$

Then relations (20)–(22) imply the following result:

Theorem 4 *The relations*

$$\mathcal{D}(\widetilde{S}) = \mathcal{D}(S)\dot{+}(S_F^{-1} + (I - 2C)M)\mathcal{D}(M)\dot{+}\mathfrak{M},$$

$$\widetilde{S}(f + S_F^{-1}h + (I - 2C)Mh + e) = Sf + h, \;\; f \in \mathcal{D}(S), h \in \mathcal{D}(M), e \in \mathfrak{M}$$

give a one-to-one correspondence between all m-accretive extensions of a closed densely defined sectorial and coercive operator S and all triplets $\langle \mathfrak{M}, M, C\rangle$ where $\mathfrak{M}$ is a subspace of $Ker\, S^*$, M is an m-accretive operator in the subspace $Ker\, S^* \ominus \mathfrak{M}$ with the domain $\mathcal{D}(M)$ and C is a linear operator from $\mathcal{R}(M)$ into $\mathcal{D}[S_F]$ which meets the condition

$$\mu[CMh] \leq Re\,(Mh, h), h \in \mathcal{D}(M).$$

The extension $\widetilde{S}$ is an m-sectorial operator if and only if the corresponding M is an m-sectorial operator in $Ker\, S^* \ominus \mathfrak{M}$ and for some $\delta \in [0, 1)$

$$\mu[CMh] \leq \delta\, Re\,(Mh, h), h \in \mathcal{D}(M).$$

The extension $\widetilde{S}$ has a bounded inverse if and only if $\mathcal{D}(M) = Ker\, S^*$; in this case

$$\widetilde{S}^{-1} = S_F^{-1} + (I - 2C)MP.$$

6

Next we give an application of previous results to a description of all m-accretive extensions of the minimal operator generated by second order elliptic differential expression.

Let Ω be a bounded domain in $\mathbf{R}^n$ $(n \geq 2)$ with C^∞ boundary $\partial\Omega$. Consider in Ω the differential expression given by

$$\begin{aligned} l(u) \;=\; & -\sum_{j=1}^{n} \frac{\partial}{\partial x_j}\left(\sum_{k=1}^{n} a_{jk}(x)\frac{\partial u}{\partial x_k} + \alpha_j(x)u\right) \\ & + \sum_{k=1}^{n} a_k(x)\frac{\partial u}{\partial x_k} + a(x)u. \end{aligned} \tag{25}$$

The coefficients in (25) are supposed to be in $C^\infty(\overline{\Omega})$. Let $Q(x) = \|q_{jk}(x)\|$ be an $(n+1)\times(n+1)$ matrix-function with entries

$$\begin{aligned} q_{j0}(x) &= \alpha_j(x),\ q_{0k}(x) = a_k(x),\ q_{00}(x) = a(x), \\ q_{jk}(x) &= a_{jk}(x),\ j,k = 1,\dots,n \end{aligned}$$

Assume that the condition of uniform ellipticity is fulfilled:

$$Re\left(\sum_{j,k=0}^{n} q_{jk}(x)\xi_k\overline{\xi_j}\right) \geq c\sum_{k=0}^{n}|\xi_k|^2 \tag{26}$$

for all $x \in \Omega$ and $\xi = (\xi_0, \dots, \xi_n)^T \in \mathbf{C}^{n+1}$, where $c > 0$.

Denote by $H^1(\Omega)$, $H^2(\Omega)$, $\overset{0}{H^1}(\Omega)$, $\overset{0}{H^2}(\Omega)$ the Sobolev spaces [8]. Let S be the minimal [8] operator generated by the expression (25) in $\mathcal{L}^2(\Omega)$. Under the condition (25) the operator S is sectorial and coercive, $\mathcal{D}(S) = \overset{0}{H^2}(\Omega)$. The Friedrichs extension S_F is the Dirichlet operator and $\mathcal{D}[S_F] = \overset{0}{H^1}(\Omega)$, $\mathcal{D}(S_F) = \overset{0}{H^1}(\Omega) \cap H^2(\Omega)$ [8]. The the adjoint operator S^* is the maximal operator generated in $\mathcal{L}^2(\Omega)$ by the adjoint differential expression

$$l^+(u) = -\sum_{j=1}^{n}\frac{\partial}{\partial x_j}\left(\sum_{k=1}^{n}\overline{a_{kj}(x)}\frac{\partial u}{\partial x_k} + \overline{a_j(x)}u\right) + \sum_{k=1}^{n}\overline{\alpha_k(x)}\frac{\partial u}{\partial x_k} + \overline{a(x)}u.$$

For the von Neumann-Krein extension S_N we have

$$\mathcal{D}(S_N) = \overset{0}{H^2}(\Omega)\dot{+}Ker\, S^*,\ S_N|Ker\, S^* = 0,\ \mathcal{D}[S_N] = \overset{0}{H^1}(\Omega)\dot{+}Ker\, S^*.$$

The linear manifold $H^1(\Omega) \cap Ker\, S^*$ is dense in $Ker\, S^*$ [5] and the trace operator

$$H^1(\Omega) \ni u \longrightarrow u|\partial\Omega \in H^{1/2}(\partial\Omega)$$

has a continuation to the operator $\Gamma : \mathcal{D}[S_N] \to \mathcal{H}_-$, where $\mathcal{H}_-$ is a Hilbert space which contains $\mathcal{L}^2(\partial\Omega)$ and $\mathcal{R}(\Gamma) = \mathcal{H}_-$ [5]. The pair $\{\mathcal{H}_-, \Gamma\}$ forms a boundary pair for S. Denote by Z_0h the unique solution of the boundary value problem

$$S^*u = l^+(u) = 0,\ \Gamma u = h,\ h \in \mathcal{H}_-.$$

Then $Z_0 = (\Gamma|Ker S^*)^{-1}$. Let $\mathcal{H}_+ \subset \mathcal{L}^2(\partial\Omega) \subset \mathcal{H}_-$ be a rigged Hilbert space.

Denote by $\vec{n}(x)$ the field of normal unit vectors of $\partial\Omega$ directed into the interior of Ω and let

$$\frac{\partial u}{\partial \nu}\Bigg|\partial\Omega = \sum_{j=1}^{n}\left(\sum_{k=1}^{n} a_{jk}(x)\frac{\partial u}{\partial x_k} + \alpha_j(x)u\right)\cos(\vec{n}(x), x_j)\Bigg|\partial\Omega.$$

Then [5] the operator

$$G_* u = \frac{\partial(u - Z_0 \Gamma u)}{\partial\nu}\bigg|\partial\Omega$$

is defined for $u \in \mathcal{D}(S_*^*) := \mathcal{D}(S_F)\dot{+}Ker\, S^*$ and takes values in $\mathcal{H}_+$. As was shown in [5] the triplet

$$\{\mathcal{H}_+ \subset \mathcal{L}^2(\partial\Omega) \subset \mathcal{H}_-, G_*, \Gamma\}$$

forms a b.v.s. for S_*^*. The sesquilinear form

$$\begin{aligned} J[u, v] &= \int_\Omega \left\{ \sum_{j,k=1}^n a_{jk}(x) \frac{\partial u}{\partial x_k} \overline{\frac{\partial v}{\partial x_j}} \right. \\ &\quad \left. + \sum_{k=1}^n \left(a_k(x) \frac{\partial u}{\partial x_k} \overline{v} + \alpha_k(x) u \overline{\frac{\partial v}{\partial x_k}} \right) + a(x) u \overline{v} \right\} dx \end{aligned}$$

is closed and sectorial on $\overset{0}{H^1}(\Omega)$ and on $H^1(\Omega)$. The quadratic functional $\mu[\varphi]$ defined by (18) in this case takes the form

$$\mu[\varphi] = \max\{Re\, J[2\varphi - u, u],\ u \in \overset{0}{H^1}(\Omega)\}.$$

Applying Theorem 3 we get

Theorem 5 *All m-accretive extensions of the minimal operator S generated by the differential expression* (25) *under the condition* (26) *take the form:*

$$\begin{aligned} \mathcal{D}(\widetilde{S}) &= \left\{ u : u - (Z_0 - 2C)\Gamma u \in \mathcal{D}(S_F), \right. \\ &\qquad \left. \frac{\partial(u - (Z_0 - 2C)\Gamma u)}{\partial\nu}\bigg|\partial\Omega \in \mathbf{W}(\Gamma u) \right\}, \\ \widetilde{S} u &= l(u - (Z_0 - 2C)\Gamma u) \end{aligned}$$

where $\mathbf{W}$ *is m-accretive l.r.,* $\mathcal{R}(\mathbf{W}) \subseteq \mathcal{H}_+$, $C : \mathcal{D}(\mathbf{W}) \to \overset{0}{H^1}(\Omega)$ *and*

$$\mu[Ch] \le Re\,(\mathbf{W}(h), h)_{\mathcal{L}^2(\partial\Omega)}, \quad h \in \mathcal{D}(\mathbf{W}).$$

The extension $\widetilde{S}$ *is m-sectorial if* $\mathbf{W}$ *is an m-sectorial l.r. and*

$$\mu[Ch] \le \delta\, Re\,(\mathbf{W}(h), h)_{\mathcal{L}^2(\partial\Omega)}, \quad h \in \mathcal{D}(\mathbf{W})$$

for some $\delta \in [0, 1)$.

References

[1] Y.M. Arlinskii, *Positive boundary value spaces and sectorial extensions of non-negative symmetric operator*, Ukrainian Math. J. **40** (1988), no. 6, 22–27.

[2] Y.M. Arlinskii, *On proper accretive extensions of positive linear relations*, Ukrainian Math. J. **47** (1995), no. 6, 723–730.

[3] Y.M. Arlinskii, *Maximal sectorial extensions and associated with them closed forms*, Ukrainian Math. J. **48** (1996), no. 6, 723–739.

[4] Y.M. Arlinskii, *Extremal extensions of sectorial linear relations*, Matematichnii Studii **7** (1997), no. 1, 81–96.

[5] Y.M. Arlinskii, *Abstract boundary conditions for maximal sectorial extensions of sectorial operators*, Math. Nachr., to appear.

[6] Y.M. Arlinskii, *On functions connected with sectorial operators and their extensions*, Int. Equat. Oper. Theory **33** (1999), no. 2, 125–152.

[7] T.Y. Azizov and I.S. Iokhvidov, *Foundations of the Theory of Linear Operator in Spaces with Indefinite Metric*, Moscow, 1986.

[8] Y.M. Berezanskii, *Expansions in Eigenfunction of Selfadjoint Operators*, Amer. Math. Soc. Providence, 1968.

[9] M.S. Birman, On the selfadjoint extensions of positive definite operators, Mat. Sbornik **38** (1956), 431–450.

[10] V.A. Derkach, M.M. Malamud and E.R. Tsekanovskii, *Sectorial extensions of positive operator and characteristic function*, Ukrainian Math. J. **41** (1989), no. 2, 151–158.

[11] W.D. Evans and I. Knowles, *On the extensions problem for accretive differential operators*, J. Funct. Anal. **63** (1985), no. 3, 276–298.

[12] V.I. Gorbachuk and M.L.Gorbachuk, *Boundary Value Problems for Differential-Operator Equations*, Naukova Dumka, Kiev, 1984.

[13] V.I. Gorbachuk, M.L. Gorbachuk and A.N. Kochubei, *Extension theory of symmetric operators and boundary value problems*, Ukrainian Mat. J. **41** (1989), no. 10, 1298–1313.

[14] T. Kato, Perturbation Theory for Linear Operators, Springer-Verlag, 1966.

[15] A.N. Kochubei, *On extensions of positive definite symmetric operator*, Dokl. Acad. Science Ukraine, Ser. A, no. 3 (1979), 168–171.

[16] M.G. Krein, *The theory of selfadjoint extensions of semibounded hermitian transformations and its applications. I*, Mat. Sbornik **20** (1947), no. 3, 431–495.

[17] M.G. Krein, *The theory of selfadjoint extensions of semibounded hermitian transformations and its applications. II*, Mat. Sbornik **21** (1947), no. 3, 365–404.

[18] V.A. Mihailets, *On general boundary problems for differential equations in Hilbert space*, Spectral Theory of Differential-Operator Equations, Inst. Math. Acad. Science Ukraine, Kiev, 1986, 15–24.

[19] O.Ya. Milyo and O.G. Storozh, *On general form of maximal accretive extension of positive definite operator*, Dokl. Acad. Science Ukraine, no. 6 (1991), 19–22.

[20] O.Ya. Milyo and O.G. Storozh, *Maximal accretive extensions of positive definite operator with finite defect number*, Lviv University, 1993, 31 pages. Deposited in GNTB of Ukraine 28.10.93, no. 2139 Uk93.

[21] R. Phillips, *Dissipative opertors and hyperbolic systems of partial differential equations*, Trans. Amer. Math. Soc. **90** (1959), 192–254.

[22] R.S. Phillips, *On dissipative operators, in "Lectures in Differential Equations"*, vol. II, Van Nostrand-Reinhold, New York, 1969, 65–113.

[23] F.S. Rofe-Beketov, *Numerical range of linear relation and maximal relations*, Theory of Functions, Functional Anal. and Appl. **44** (1985), 103–112.
[24] O.G. Storozh, *The description of some classes of extensins of nonnegative symmetric operators*, Dokl. Acad. Science Ukraine, Ser. A (1987), no. 10, 15–17.
[25] O.G. Storozh, *Accretive operators contiguous to positive definite*, Ukrainian Math. J. **41** (1989), no. 6, 789–794.
[26] O.G. Storozh, *Extremal extensions of nonnegative operators and accretive boundary value problems*, Ukrainian Math. J. **42** (1990), no. 6, 858–860.
[27] M.I. Vishik, *On general boundary conditions for elliptic differential equations*, Trudy Moskov. Mat. Obsc. **1** (1952), 187–246.

Department of Higher and Applied Mathematics
East Ukrainian University
Kvartal Molodyozhny, 20-A Lugansk, 348034
Ukraine
yma@vugu.lugansk.ua

AMS Classification: 47A20, 47B44

Operator Theory:
Advances and Applications, Vol. 118

A Proof of the Continuous Commutant Lifting Theorem

Ramon Bruzual and Marisela Dominguez**

We give a proof of the continuous commutant lifting theorem using an extension of a Krein theorem.

Introduction

Let a be such that $0 < a \leq +\infty$ and let $I = (-a, a)$. A kernel on I is a function $K : I \times I \to \mathbf{C}$. The kernel K is said to be positive definite if for any positive integer n and any $x_1, \ldots, x_n$ in I, $\lambda_1, \ldots, \lambda_n$ in $\mathbf{C}$ we have

$$\sum_{i,j=1}^{n} K(x_i, x_j)\lambda_i\overline{\lambda_j} \geq 0,$$

and K is said to be a Toeplitz kernel if there exists a function $k : I - I \to \mathbf{C}$ such that $K(x, y) = k(x - y)$ for all x, y in I.

M.G. Krein [8] proved that every continuous positive definite Toeplitz kernel on $I = (-a, a)$ can be extended to a continuous positive definite Toeplitz kernel on the whole line.

The concept of positive definite kernel on I can be extended to a more general context in the following natural way: Let $\mathcal{H}$ be a Hilbert space. A kernel $K : I \times I \to \mathcal{L}(\mathcal{H})$ is said to be positive definite if

$$\sum_{x,y \in I} \langle K(x, y)h(x), h(y)\rangle_{\mathcal{H}} \geq 0$$

for each function $h : I \to \mathcal{H}$ of finite support.

By the Naimark dilation theorem [11] every positive definite Toeplitz kernel K on $\mathbf{R}$ is of the form

$$K(x) = \tau^* U_x \tau \quad \text{for all} \quad x \in \mathbf{R}$$

where $\{U_x\}_{x \in \mathbf{R}}$ is a unitary representation of $\mathbf{R}$ on a larger Hilbert space $\mathcal{G}$ and $\tau : \mathcal{H} \to \mathcal{G}$ is a bounded operator.

M.L. Gorbachuk [7] extended the result of Krein for operator valued continuous Toeplitz kernels defined on an interval $I = (-a, a)$. So we have that a continuous

*Both authors were partially supported by the CDCH of the Universidad Central de Venezuela.

positive definite Toeplitz kernel on an interval I is of the form

$$K(x) = \tau^* U_x \tau \quad \text{for all} \quad x \in I$$

where $\{U_x\}_{x\in\mathbf{R}}$ is a strongly continuous unitary representation of $\mathbf{R}$ on a larger Hilbert space $\mathcal{G}$ and $\tau : \mathcal{H} \to \mathcal{G}$ is a bounded operator.

Several problems in analysis led Cotlar and Sadosky to introduce the so-called generalized Toeplitz kernels, as kernels defined in $\mathbf{Z} \times \mathbf{Z}$ or in $\mathbf{R} \times \mathbf{R}$ [5]. These kernels satisfy a condition more general than being Toeplitz in their domain. Also generalized Toeplitz kernels with domain a product of intervals on the real line have been considered. More precisely, for a such that $0 < a \leq +\infty$ set $I = (-a, a)$, $I_1 = I \cap [0, +\infty)$ and $I_2 = I \cap (-\infty, 0)$.

Definition: Let (H_1, H_2) be a pair of Hilbert spaces. An operator valued generalized Toeplitz kernel on (I, H_1, H_2) is a kernel K such that there exist four functions $C_{\alpha\beta} : I_\alpha - I_\beta \to L(H_\alpha, H_\beta)$ for $\alpha, \beta = 1, 2$ such that

$$K(x, y) = C_{\alpha\beta}(x - y) \quad \text{if} \quad (x, y) \in I_\alpha \times I_\beta$$

and $C_{12}(r) = C_{21}(-r)^*$ for all $r \in I_1 - I_2 = [0, 2a)$.

A generalized Toeplitz kernel K on (I, H_1, H_2) is positive definite if

$$\sum_{\alpha,\beta=1}^{2} \sum_{(x,y)\in I_\alpha\times I_\beta} \langle C_{\alpha\beta}(x-y)h_\alpha(x), h_\beta(y)\rangle_{H_\beta} \geq 0$$

for each pair of functions $h_1 : I_1 \to H_1$, $h_2 : I_2 \to H_2$ with finite support.

An approach to generalized Toeplitz kernels with values on $L(H_\alpha, H_\beta)$ through the notion of local semigroups of isometric operators was given in [4]. There it is proved the following result.

Theorem 1 ([4], pag. 795). *Let K be a positive definite generalized Toeplitz kernel on (I, H_1, H_2) given by four weakly continuous functions such that $C_{\alpha\alpha}(0) = I_{H_\alpha}$ for $\alpha = 1, 2$. Then there exist a Hilbert space G containing H_1 and H_2 as closed subspaces and a strongly continuous unitary representation $(U_r)_{r\in\mathbf{R}}$ of $\mathbf{R}$ on G such that:*

$$C_{\alpha\beta}(r) = P^G_{H_\beta} U_r \mid_{H_\alpha} \quad \text{for all } r \in I_\alpha - I_\beta \text{ for } \alpha, \beta = 1, 2.$$

$$G = \bigvee_{r\in\mathbf{R}} U_r H_1 \vee \bigvee_{r\in\mathbf{R}} U_r H_2.$$

(P^G_H denotes the orthogonal projection of the Hilbert space G on the Hilbert space H when $H \subset G$).

In [1] Arocena used the representation theorem for generalized Toeplitz kernels on $\mathbf{Z}$ to give a new proof of the commutant lifting theorem. According to a result of

Sz.-Nagy [10] every contraction semigroup $(W(r))_{r\in\mathbf{R}_1} \subset L(H)$ with parameter in $\mathbf{R}_1 = [0, +\infty)$ has a unique minimal unitary dilation $(U(r))_{r\in\mathbf{R}} \subset L(G)$ and then a unique minimal isometric dilation, given by $V(r) = U(r) \mid_F$ for $r \in \mathbf{R}_1$ where $F = \bigvee_{r\in\mathbf{R}_1} U(r)H$, so it is natural to ask if the commutant lifting theorem can be extended to $\mathbf{R}$. In [2] Arocena proved a version of the commutant lifting theorem (Theorem 2 below) reducing the problem in $\mathbf{R}$ to that in $\mathbf{Z}$ considering the cogenerators of the semigroups, see also [6]. A more natural proof can be obtained through Theorem 1 (with $a = +\infty$), as shown below.

Theorem 2 ([2]). *For $\alpha = 1, 2$ let H_α be Hilbert spaces and $(W_\alpha(r))_{r\in\mathbf{R}_1} \subset L(H_\alpha)$ be two strongly continuous semigroups of contractions. Let $(V_\alpha(r))_{r\in\mathbf{R}_1} \subset L(F_\alpha)$ be the minimal isometric dilation of $(W_\alpha(r))_{r\in\mathbf{R}_1}$ and let $(U_\alpha(r))_{r\in\mathbf{R}_1} \subset L(G_\alpha)$ be the minimal unitary dilation of $(W_\alpha(r))_{r\in\mathbf{R}_1}$. Let $X \in L(H_1, H_2)$ be such that $XW_1(r) = W_2(r)X$ for all $r \in \mathbf{R}_1$. Then*

(a) *There exists $Y \in L(F_1, F_2)$ such that:*

  (i) *$YV_1(r) = V_2(r)Y$ for all $r \in \mathbf{R}_1$.*
  (ii) $P^{F_2}_{H_2}Y = XP^{F_1}_{H_1}$.
  (iii) $\|Y\| = \|X\|$.

(b) *There exists $Z \in L(G_1, G_2)$ such that:*

  (i) *$ZU_1(r) = U_2(r)Z$ for all $r \in \mathbf{R}$.*
  (ii) $Z(F_1) \subset F_2$.
  (iii) $P^{G_2}_{H_2}Z \mid_{F_1} = XP^{G_1}_{H_1}$.
  (iv) $\|Z\| = \|X\|$.

A Proof of the Continuous Commutant Lifting Theorem

Our proof is as follows:

Step 1: By homogeneity we can suppose that $\|X\| = 1$.

Step 2: We consider the following generalized Toeplitz kernel on $(\mathbf{R}, H_1, H_2)$:

$$C_{11}(r) = \begin{cases} W_1(r) & r \in \mathbf{R}_1 \\ W_1(-r)^* & r \in \mathbf{R}_2 \end{cases} \qquad C_{22}(r) = \begin{cases} W_2(r) & r \in \mathbf{R}_1 \\ W_2(-r)^* & r \in \mathbf{R}_2 \end{cases}$$

$$C_{12}(r) = XW_1(r) = W_2(r)X \text{ if } r \in \mathbf{R}_1$$

$$C_{21}(r) = C_{12}(-r)^* \text{ if } r \in \mathbf{R}_2.$$

We prove that C is positive definite on $\mathbf{R}$:

Let $h_\alpha : \mathbf{R}_\alpha \to H_\alpha$, $\alpha = 1, 2$ be two functions with finite support. An easy calculation shows that

$$\sum_{(x,y)\in\mathbf{R}_2\times\mathbf{R}_1} \langle C_{21}(x-y)h_2(x), h_1(y)\rangle_{H_1}$$
$$= \sum_{(x,y)\in\mathbf{R}_1\times\mathbf{R}_2} \overline{\langle C_{12}(x-y)h_1(x), h_2(y)\rangle_{H_2}}.$$

Let $(U_1(r))_{r\in\mathbf{R}} \subset L(G_1)$ and $(U_2(r))_{r\in\mathbf{R}} \subset L(G_2)$ be the minimal unitary dilations of $(W_1(r))_{r\in\mathbf{R}_1}$ and $(W_2(r))_{r\in\mathbf{R}_2}$.

If $(x, y) \in \mathbf{R}_1 \times \mathbf{R}_2$ then

$$\begin{aligned}
\langle C_{12}(x-y)h_1(x), h_2(y)\rangle_{H_2} &= \langle XW_1(x-y)h_1(x), h_2(y)\rangle_{H_2} \\
&= \langle XW_1(-y)W_1(x)h_1(x), h_2(y)\rangle_{H_2} \\
&= \langle W_2(-y)XW_1(x)h_1(x), h_2(y)\rangle_{H_2} \\
&= \langle XW_1(x)h_1(x), W_2(-y)^*h_2(y)\rangle_{H_2} \\
&= \langle XP_{H_1}^{G_1}U_1(x)h_1(x), P_{H_2}^{G_2}U_2(-y)^*h_2(y)\rangle_{H_2}.
\end{aligned}$$

If $(x, y) \in \mathbf{R}_1 \times \mathbf{R}_1$ then

$$\begin{aligned}
\langle C_{11}(x-y)h_1(x), h_1(y)\rangle_{H_1} &= \langle W_1(x-y)h_1(x), h_1(y)\rangle_{H_1} \\
&= \langle P_{H_1}^{G_1}U_1(x-y)h_1(x), h_1(y)\rangle_{H_1} \\
&= \langle U_1(-y)U_1(x)h_1(x), h_1(y)\rangle_{G_1} \\
&= \langle U_1(x)h_1(x), U_1(y)h_1(y)\rangle_{G_1}.
\end{aligned}$$

An analogous result holds for C_{22}.

Therefore, if we put $f_\alpha = \sum_{r\in\mathbf{R}_\alpha} U_\alpha(r)h_\alpha(r)$, then we have

$$\sum_{\alpha,\beta=1}^{2} \sum_{(x,y)\in\mathbf{R}_\alpha\times\mathbf{R}_\beta} \langle C_{\alpha\beta}(x-y)h_\alpha(x), h_\beta(y)\rangle_{H_\beta}$$
$$= \|f_1\|_{G_1}^2 + \|f_2\|_{G_2}^2 + 2Re\langle XP_{H_1}^{G_1}f_1, P_{H_2}^{G_2}f_2\rangle_{G_2} \geq 0$$

because $\|X\| = 1$.

Step 3: By Theorem 1 there exits a Hilbert space G containing H_1 and H_2 as closed subspaces and a group of unitary operators $(U(r))_{r\in\mathbf{R}} \subset L(G)$ such that

$$C_{\alpha\beta}(r) = P_{H_\beta}^{G}U(r)\mid_{H_\alpha} \quad \text{for all} \quad r \in \mathbf{R}_\alpha - \mathbf{R}_\beta \quad \text{for} \quad \alpha, \beta = 1, 2$$

and

$$G = \bigvee_{r\in\mathbf{R}} U(r)H_1 \vee \bigvee_{r\in\mathbf{R}} U(r)H_2.$$

Step 4: For $\alpha = 1, 2$ we define

$$F_\alpha = \bigvee_{r \in \mathbf{R}_1} U(r) H_\alpha$$

and $V_\alpha(r) : F_\alpha \to F_\alpha$ by $V_\alpha(r) = U(r) \mid_{F_\alpha}$.

It is clear that $(V_\alpha(r))_{r \in \mathbf{R}_1}$ is a semigroup of isometric operators and it is easy to check, using that the minimal isometric dilation is determined up to isomorphism, that $(V_\alpha(r))_{r \in \mathbf{R}_1}$ is the minimal isometric dilation of $(W_\alpha(r))_{r \in \mathbf{R}_1}$.

Step 5: Define $Y \in L(F_1, F_2)$ by $Y = P^G_{F_2} \mid_{F_1}$.

Then it is clear that

$$\langle Y f_1, f_2 \rangle_{F_2} = \langle f_1, f_2 \rangle_G \quad \text{for all} \quad f_1 \in F_1, f_2 \in F_2$$

Let $r \in \mathbf{R}_1$, $h_1 \in H_1$ and $h_2 \in H_2$ then

$$\begin{aligned}
\langle P^{F_2}_{H_2} Y V_1(r) h_1, h_2 \rangle_{H_2} &= \langle P^{F_2}_{H_2} Y V_1(r) h_1, h_2 \rangle_{F_2} = \langle Y V_1(r) h_1, h_2 \rangle_{F_2} \\
&= \langle V_1(r) h_1, h_2 \rangle_G = \langle U(r) h_1, h_2 \rangle_G \\
&= \langle P^G_{H_2} U(r) h_1, h_2 \rangle_G = \langle P^G_{H_2} U(r) h_1, h_2 \rangle_{H_2} \\
&= \langle P^G_{H_2} (U(r) \mid_{H_1}) h_1, h_2 \rangle_{H_2} = \langle C_{12}(r) h_1, h_2 \rangle_{H_2} \\
&= \langle X W_1(r) h_1, h_2 \rangle_{H_2} = \langle X P^{F_1}_{H_1} (V_1(r) h_1), h_2 \rangle_{H_2}.
\end{aligned}$$

Since H_1 is generated by the elements of the form $V_1(r) h_1$ for $r \in \mathbf{R}_1$ and $h_1 \in H_1$ we have that

$$P^{F_2}_{H_2} Y = X P^{F_1}_{H_1}.$$

Step 6: Proof of $\|Y\| = \|X\|$.

We only have to show that $\|Y\| = 1$.

$\|Y\| \leq 1$ is clear and $1 = \|X\| = \|X P^{F_1}_{H_1}\| = \|P^{F_2}_{H_2} Y\| \leq \|Y\|$.

Step 7: Proof of $Y V_1(r) = V_2(r) Y$ for all $r \in \mathbf{R}_1$.

It is enough to show that given $r \in \mathbf{R}_1 - \{0\}$ then

$$\langle Y V_1(r) V_1(x) h_1, V_2(y) h_2 \rangle_{F_2} = \langle V_2(r) Y V_1(x) h_1, V_2(y) h_2 \rangle_{F_2}$$

for all $x, y \in \mathbf{R}_1$, $h_1 \in H_1$ and $h_2 \in H_2$.

If $y \geq r$ then

$$\begin{aligned}
&\langle Y V_1(r) V_1(x) h_1, V_2(y) h_2 \rangle_{F_2} = \langle V_1(r + x) h_1, V_2(y) h_2 \rangle_G \\
&= \langle U(r + x) h_1, U(y) h_2 \rangle_G \\
&= \langle U(x) h_1, U(y - r) h_2 \rangle_G = \langle V_1(x) h_1, V_2(y - r) h_2 \rangle_G \\
&= \langle Y V_1(x) h_1, V_2(y - r) h_2 \rangle_{F_2} = \langle V_2(r) Y V_1(x) h_1, V_2(y) h_2 \rangle_{F_2}.
\end{aligned}$$

If $y < r$ then, using $W_2(r)^* = V_2(r)^* \mid_{H_2}$, we obtain:

$$\begin{aligned}
&\langle Y V_1(r) V_1(x) h_1, V_2(y) h_2 \rangle_{F_2} \\
&= \langle V_1(r+x) h_1, V_2(y) h_2 \rangle_G \\
&= \langle U(r+x) h_1, U(y) h_2 \rangle_G \\
&= \langle U(r+x-y) h_1, h_2 \rangle_G = \langle P^G_{H_2} U(r+x-y) h_1, h_2 \rangle_G \\
&= \langle P^G_{H_2} U(r+x-y) \mid_{H_1} h_1, h_2 \rangle_{H_2} = \langle C_{12}(r+x-y) h_1, h_2 \rangle_{H_2} \\
&= \langle W_2(r+x-y) X h_1, h_2 \rangle_{H_2} = \langle W_2(r-y) W_2(x) X h_1, h_2 \rangle_{H_2} \\
&= \langle W_2(r-y) X W_1(x) h_1, h_2 \rangle_{H_2} \\
&= \langle X W_1(x) h_1, W_2(r-y)^* h_2 \rangle_{H_2} \\
&= \langle X P^{F_1}_{H_1} V_1(x) h_1, W_2(r-y)^* h_2 \rangle_{H_2} \\
&= \langle P^{F_2}_{H_2} Y V_1(x) h_1, W_2(r-y)^* h_2 \rangle_{F_2} \\
&= \langle Y V_1(x) h_1, V_2(r-y)^* h_2 \rangle_{F_2} = \langle V_2(r-y) Y V_1(x) h_1, h_2 \rangle_{F_2} \\
&= \langle V_2(r) Y V_1(x) h_1, V_2(y) h_2 \rangle_{F_2}.
\end{aligned}$$

In order to prove (b) we define $Z = i^{G_2}_{F_2} Y P^{G_1}_{F_1}$ where $i^{G_2}_{F_2}$ is the inclusion of F_2 in G_2 and the proof follows straightforward.

References

[1] R. Arocena, *Generalized Toeplitz kernels and dilations of intertwining operators.* Integral Equations and Operator Theory **6** (1983), pp. 759–778.

[2] R. Arocena, *Generalized Toeplitz kernels and dilations of intertwining operators II: the continuous case.* Acta Sci. Math. (Szeged) **53** (1989), pp. 123–137.

[3] R. Arocena and M. Cotlar, *Dilation of generalized Toeplitz kernels and some vectorial moment and weighted problems.* Lecture Notes in Math. **908**, Springer (1982), pp. 169–188.

[4] R. Bruzual, *Local semigroups of contractions and some applications to Fourier representation theorems.* Int. Eq. and Op. Theory **10** (1987), pp. 780–801.

[5] M. Cotlar and C. Sadosky, *On the Helson-Szegö theorem and a related class of modified Toeplitz kernels.* Proc. Symp. Pure Math. AMS. **35-I** (1979), pp. 383–407.

[6] M. Cotlar and C. Sadosky, *Prolongements des formes de Hankel généralisées en formes de Toeplitz.* C. R. Acad. Sci. Paris **305** (1987), pp. 167–170.

[7] M.L. Gorbachuk, *Representation of positive definite operator functions.* Ukrain. Math. J. **17**, no. 2 (1965), pp. 29–46.

[8] M.G. Krein, *Sur le problème du prolongement des fonctions hermitiennes positives el continues.* Dokl. Akad. Nauk. SSSR **26** (1940), pp. 17–22.

[9] M.G. Krein, *On the representation of functions by Fourier Stieltjes integrals.* (Russian). Učenije Zapiski Kuibishevskogo Gosud. Pedag. i Učitelskogo Inst. **7** (1943), pp. 123–148.

[10] B. Sz.-Nagy, *Transformations de l'espace de Hilbert, fonctions de type positif sur un groupe.* Acta. Sc. Math. Szeged. **15** (1954), pp. 104–114.

[11] M. Naimark, *Positive definite operator functions on a commutative group.* Bulletin (Izvestiya) Acad. Sci. URSS. **7** (1943), pp. 237–244.

[12] B. Sz.-Nagy and C. Foias, *Harmonic analysis of operators on Hilbert space.* North Holland Publishing Co., 1970.

Ramón Bruzual
Dpto de Matemáticas. Fac. Ciencias.
Universidad Central de Venezuela
Mailing address:
Apartado Postal 47686
Los Chaguaramos, Caracas 1041-A
Venezuela
rbruzual@euler.ciens.ucv.ve

Marisela Domínguez
Dpto de Matemáticas. Fac. Ciencias.
Universidad Central de Venezuela
Mailing address:
Apartado Postal 47159
Los Chaguaramos, Caracas 1041-A
Venezuela
mdomin@euler.ciens.ucv.ve

MSC numbers: Primary 47A20, Secondary 47D03

Operator Theory:
Advances and Applications, Vol. 118

An Outer Derivation Construction on the Algebra of Singular Integral Operators with General Coefficients in Weighted Spaces and its Applications

V.M. Deundyak and K.A. Georgiev

We construct an outer derivation on the non-closed algebra generated by the singular Cauchy operator $S_{p,\varrho}$ on the weighted space $L_{p,\varrho}(\mathbf{T})$ and operators of multiplication by essentially bounded functions utilizing properties of the commutator $[M_b; S_{p,\varrho}]$ where $\mathbf{T}$ is the unit circle and $b \in BMO(\mathbf{T})$. By the use of this derivation some algebraic properties of the Banach algebra of singular integral operators $\mathfrak{R}(S_{p,\varrho}, \mathcal{A})$ are investigated on the spaces $L_{p,\varrho}(\mathbf{T})$ for wide classes of coefficient algebras $\mathcal{A}$ and weights ϱ. In particular, the algebra $\mathfrak{R}(S_{p,\varrho}, \mathcal{A})$ modulo its commutator ideal is considered, a presymbol for the algebra $\mathfrak{R}(S_{p,\varrho}, \mathcal{A})$ is constructed and some necessary conditions for operators in $\mathfrak{R}(S_{p,\varrho}, \mathcal{A})$ to be Fredholm are obtained.

For $p \in (1, \infty)$ let ϱ be a Hunt-Muckenhoupt-Wheeden weight of the class A_p, $L_{p,\varrho}$ be the weighted L_p-space on the unit circle $\mathbf{T}$ with the norm $\|f\|_{p,\varrho} = (\int |f(t)|^p \varrho dm(t))^{1/p}$ where m is the Lebesgue measure, End $L_{p,\varrho}$ be the Banach algebra of all bounded linear operators on $L_{p,\varrho}$, Comp $L_{p,\varrho}$ be its ideal of all compact operators, $S_{p,\varrho}$ be the singular Cauchy operator

$$(S_{p,\varrho} f)(t) = \frac{1}{\pi i} \int \frac{f(\tau)}{\tau - t} d\tau,$$

belonging to End $L_{p,\varrho}$ since $\varrho \in A_p$, $P^{\pm}_{p,\varrho} = \frac{1}{2}(I_{p,\varrho} \pm S_{p,\varrho})$ where $I_{p,\varrho}$ is the unit of End $L_{p,\varrho}$. In the case $\varrho = 1$ we put for brevity $L_p = S_{p,1}$, $S_p = S_{p,1}$, $I_p = I_{p,1}$, $P^{\pm}_p = P^{\pm}_{p,1}$. The operator of multiplication by a function a will be denoted by M_a. Let $\mathcal{A}$ be a closed subalgebra of the Banach algebra L_∞ of all essentially bounded functions on $\mathbf{T}$. We denote by $\mathfrak{R}_0(S_{p,\varrho}, \mathcal{A})$ the non-closed subalgebra of End $L_{p,\varrho}$ generated by $S_{p,\varrho}$ and $\{M_a | a \in \mathcal{A}\}$ and let $\mathfrak{R}(S_{p,\varrho}, \mathcal{A})$ be the closure of $\mathfrak{R}_0(S_{p,\varrho}, \mathcal{A})$ in End $L_{p,\varrho}$. Let Com($\mathfrak{R}_0(S_{p,\varrho}, \mathcal{A})$) (Com($\mathfrak{R}(S_{p,\varrho}, \mathcal{A})$)) denote the commutator ideal of $\mathfrak{R}_0(S_{p,\varrho}, \mathcal{A})$ ($\mathfrak{R}(S_{p,\varrho}, \mathcal{A})$), that is the smallest ideal (the smallest closed ideal) containing commutators $[A, B] = AB - BA$ for all $A, B \in \mathfrak{R}_0(S_{p,\varrho}, \mathcal{A})$ ($A, B \in \mathfrak{R}(S_{p,\varrho}, \mathcal{A})$).

Let BMO be the space of functions of bounded mean oscillation on $\mathbf{T}$. Recall that by the Fefferman theorem a function $b (\in L_1)$ belongs to BMO if and only if $b = u + \tilde{v}$ where u and v are in L_∞ and $\tilde{v}$ is the harmonic conjugate of v. Generally speaking, the operator M_b is not bounded on $L_{p,\varrho}$ for $b \in BMO$. Since the article by R.R. Coifman, R. Rochberg and G. Weiss [2] the problem of the boundedness of the commutator of M_b and a singular Calderon-Zygmund operator on diverse funcional spaces has being studied. In the present article some algebraic properties of the commutator are investigated with the use of the notion of derivation on an algebra. Recall that a linear operator δ on an algebra $\mathfrak{R}$ is said to be a derivation if $\delta(a\cdot b) = \delta(a)\cdot b + a\cdot\delta(b)$ for all $a, b \in \mathfrak{R}$. A derivation δ on $\mathfrak{R}$ is called inner if there exists $a \in \mathfrak{R}$ such that $\delta(b) = [a, b]$ for all $b \in \mathfrak{R}$. In Section 1, some derivation $\delta_{p,\varrho;b}$ on $\mathfrak{R}_0(S_{p,\varrho}, \mathcal{A})$, where $u, v \in \mathcal{A}$, is constructed; this derivation has two important properties: 1) for each $F \in \mathfrak{R}_0(S_{p,\varrho}, \mathcal{A})$, $\delta_{p,\varrho;b}(F)$ coincides with $[M_b, F]$ on some set dense in $L_{p,\varrho}$, 2) $\delta_{p,\varrho;b}(\mathfrak{R}_0(S_{p,\varrho}, \mathcal{A})) \subset \mathrm{Com}(\mathfrak{R}_0(S_{p,\varrho}, \mathcal{A}))$ (Theorem 1). Since in general $M_b \notin \mathrm{End}\, L_{p,\varrho}$, the derivation $\delta_{p,\varrho;b}$ is not inner and mentioned in the sequel as an outer one.

The properties of the derivation are applied in Section 2 to the investigation of weighted singular integral operators. Namely, let $\omega \in A_p$, let α_ω be the similarity isomorphism of $\mathrm{End}(L_{p,\omega})$ onto $\mathrm{End}(L_p)$ defined by

$$\alpha_\omega : F(\in \mathrm{End}\, L_{p,\omega}) \longmapsto M_{\omega^{1/p}} F M_{\omega^{-1/p}}.$$

The family $A(z) = M_{\omega^{z/p}} S_{p,\omega_z} M_{\omega^{-z/p}}$ of weighted Cauchy singular operators is introduced, where $\omega_z = w^{\mathrm{Re}\, z}$, z ranges over some strip region Δ of the complex plane. Note that S_p and $\alpha_\omega(S_{p,\omega})$ are members of $A(z)$ for $z = 0$ and $z = 1$ respectively. It is proved in Theorem 2 that $A(z)$ is a holomorphic function with derivatives $A^{(m)}(z) = M_{\omega^{z/p}}((\delta_{p,\omega_z;b})^m(S_{p,\omega_z}))M_{\omega^{-z/p}}$, where $b = (1/p)\ln\omega\,(\in BMO)$. Under the condition that the functions u and v from the Fefferman representation of b belong to the coefficient algebra $\mathcal{A}$ it is shown by using the properties of the derivation from Theorem 1 and the results of Theorem 2 that α_ω isomorphically maps $\mathfrak{R}(S_{p,\omega}, \mathcal{A})$ onto $\mathfrak{R}(S_p, \mathcal{A})$ and that

$$\alpha_\omega(S_{p,\omega}) - S_p \;\in\; \mathrm{Com}(\mathfrak{R}(S_{p,\omega}, \mathcal{A}))$$

(Theorems 3 and 4). The corresponding results in the case where ω is a Khvedelidze power weight, $\mathcal{A}$ is the algebra of piecewise continuous functions and the curve is piecewise Lyapunov were obtained in [3], [9], [12]. Namely, it was shown that the algebras $\mathfrak{R}(S_{p,\omega}, \mathcal{A})$ and $\mathfrak{R}(S_p, \mathcal{A})$ are isomorphic, the isomorphism being some analog of α_ω.

In Section 3 the results of Sections 1 and 2 are applied to the investigation of Fredholmness of operators in $\mathfrak{R}(S_{p,\omega}, \mathcal{A})$ where, as above, $\omega \in A_p$, $\mathcal{A}$ is an arbitrary closed subalgebra of L_∞ containing the functions u and v from the Fefferman representation of the function $(1/p)\ln\omega$. More precisely, a presymbol $\mu_{\mathcal{A},\omega}$ for the algebra $\mathfrak{R}(S_{p,\omega}, \mathcal{A})$ is constructed, which is an epimorphism from $\mathfrak{R}(S_{p,\omega}, \mathcal{A})$ onto $\mathcal{A} \oplus \mathcal{A}$ with the kernel $\mathrm{Com}(\mathfrak{R}(S_{p,\omega}, \mathcal{A}))$. This allows us to obtain necessary conditions for operators in $\mathfrak{R}(S_{p,\omega}, \mathcal{A})$ to be Fredholm in terms of the presymbol. The unweighted case has been considered in [8]. It is obviously difficult to introduce a symbol calculus modulo the ideal $\mathrm{Comp}\, L_{p,\omega}$ and to obtain necessary and sufficient conditions of Fredholmness when the coefficient algebra $\mathcal{A}$ is subject of such weak restrictions. However, one should mention that for less general coefficient algebras (the algebras of piecewise continuous and piecewise quasicontinuous functions) it is possible to introduce a symbol calculus and to obtain necessary and sufficient conditions of Fredholmness even when more complicated curves are considered. Results of that type are contained in the papers of A. Böttcher, R. Douglas, I. Gohberg, Y. Karlovich, N. Krupnik, S. Roch, D. Sarason, B. Silbermann, I. Simonenko, I. Spitkovsky etc (one can find the bibliography in [1], [10]).

1 The Construction of the Derivation

We will use the following properties of BMO which can easily be obtained on the basis of [4] (see [6]).

Lemma 1 *Let $b \in BMO$ Then 1) for $1 < p < r < \infty$ and $v \in L_\infty$ we have the embeddings*

$$M_b(L_r) \subset L_p, \qquad (M_{v \pm i\tilde{v}} P_p^{\pm})(L_r) \subset P_p^{\pm}(L_p),$$

2) for $1 < p < \infty$ and $\varrho \in A_p$ the set $B_b = \cup_{\lambda > 0} (\chi_\lambda L_\infty)$ is densly embedded into $L_{p,\varrho}$ where χ_λ is charactiristic function of the set $\{t \in \mathbf{T} | |b(t)| \leq \lambda\}$.

Consider the function $b \in BMO$, fix a Fefferman representation $b = u + \tilde{v}$ for it and suppose untill the end of the section that the coefficient algebra $\mathcal{A}$ contains the functions u and v. Let us fix p and r, $1 < p < r < \infty$. Since $F(L_r) \subset L_r$ for each $F \in \mathfrak{R}_0(S_p, \mathcal{A})$, we deduce from Lemma 1 that $[M_b, F](L_r) \subset L_p$. In view of this fact the following equality holds for every $F, G \in \mathfrak{R}_0(S_p, \mathcal{A})$ and $f \in L_r$:

$$(1) \qquad [M_b, FG]f = [M_b, F]Gf + F[M_b, G]f \quad (\in L_p).$$

The following lemma is crucial in this section and allows together with Lemma 3 to construct the derivation in Theorem 1.

Lemma 2 *For each $f \in L_r$,*

$$[M_b, S_p]f = [M_u, S_p]f - iS_p[M_v, S_p]f.$$

Proof: According to Lemma 1,

$$P_p^- M_{v+i\tilde{v}} P_p^+ f = 0, \quad P_p^+ M_{v-i\tilde{v}} P_p^- f = 0.$$

Adding together these equalities and substituting $\frac{1}{2}(I_p \pm S_p)$ for $P_p^{\pm}$ we arrive at

$$(2M_v + 2iM_{\tilde{v}} S_p - 2iS_p M_{\tilde{v}} - S_p M_v S_p) f = 0$$

or

$$[M_{\tilde{v}}, S_p]f = -iS_p[M_v, S_p]f.$$

Due to the latter equality,

$$\begin{aligned}[M_b, S_p]f &= [M_{u+\tilde{v}}, S_p]f = [M_u, S_p]f + [M_{\tilde{v}}, S_p]f \\ &= [M_u, S_p]f - iS_p[M_v, S_p]f.\end{aligned}$$

□

Lemma 3 *For each $F \in \Re_0(S_p, \mathcal{A})$, there exists a unique operator $\delta_{p;b}(F)$ in* $\mathrm{Com}(\Re_0(S_p, \mathcal{A}))$ *coinciding on L_r with $[M_b, F]$.*

Proof: Provided $\delta_{p;b}(F)$ exists, it is unique by the density of L_r in L_p.

It follows from Lemma 2 that $\delta_{p;b}(S_p) = [M_u, S_p] - iS_p[M_v, S_p] \in \mathrm{Com}(\Re_0(S_p, \mathcal{A}))$. If

(2) $$F = \prod_{k=1}^{m} M_{a_k} S_p M_{b_k} \qquad (a_k, b_k \in \mathcal{A}),$$

then by (1)

$$\begin{aligned}[M_b, F]f &= \sum_{k=1}^{m} \left(\prod_{l=1}^{k-1} M_{a_l} S_p M_{b_l} \right) M_{a_k} [M_b, S_p] \\ & M_{b_k} \left(\prod_{l=k+1}^{m} M_{a_l} S_p M_{b_l} \right) f \;\; (f \in L_r).\end{aligned}$$

Hence,

$$\begin{aligned}\delta_{p;b}(F) &= \sum_{k=1}^{m} \left(\prod_{l=1}^{k-1} M_{a_l} S_p M_{b_l} \right) M_{a_k} \delta_{p;b}(S_p) M_{b_k} \left(\prod_{l=k+1}^{m} M_{a_l} S_p M_{b_l} \right) \\ &\in \mathrm{Com}(\Re_0(S_p, \mathcal{A})).\end{aligned}$$

An arbitrary operator $F \in \Re_0(S_p, \mathcal{A})$ can be written as a sum $\sum_{j=1}^{n} F_j$ where F_j has the form (2) for $j = 1, \dots, n$. Therefore, $\delta_{p;b}(F) = \sum_{j=1}^{n} \delta_{p;b}(F_j)$. □

Thus, the correspondence $F (\in \Re_0(S_p, \mathcal{A})) \mapsto \delta_{p;b}(F)$ determines a mapping $\delta_{p;b} : \Re_0(S_p, \mathcal{A}) \to \Re_0(S_p, \mathcal{A})$ acting on the generators of $\Re_0(S_p, \mathcal{A})$ as follows:

$$\delta_{p;b}(S_p) = [M_u, S_p] - iS_p[M_v, S_p]; \qquad \delta_{p;b}(M_a) = 0 \;\; (a \in \mathcal{A}).$$

Moreover,

$$\delta_{p;b}(\Re_0(S_p, \mathcal{A})) \subset \mathrm{Com}(\Re_0(S_p, \mathcal{A})).$$

Since for each $F \in \Re_0(S_p, \mathcal{A})$ the operator $\delta_{p;b}(F)$ and the commutator $[M_b, F]$ coincide on the set L_r, which is dense in L_p, it follows that for $\delta_{p;b}$ the analog of (1) holds:

$$\delta_{p;b}(FG) = \delta_{p;b}(F)\, G + F\, \delta_{p;b}(G).$$

Thus, $\delta_{p;b}$ is a derivation.

In the following theorem we extend the results obtained above to the case of an arbitrary weight $\varrho \in A_p$.

Theorem 1 *Let $\varrho \in A_p$. There exists a unique derivation $\delta_{p,\varrho;b}$ on the algebra $\Re_0(S_{p,\varrho}, \mathcal{A})$ acting on the algebra generators as follows:*

$$\delta_{p,\varrho;b}(S_{p,\varrho}) = [M_u, S_{p,\varrho}] - iS_{p,\varrho}[M_v, S_{p,\varrho}]; \quad \delta_{p,\varrho;b}(M_a) = 0 \quad (a \in \mathcal{A}). \tag{3}$$

Moreover,

$$\delta_{p,\varrho;b}(\Re_0(S_{p,\varrho}, \mathcal{A})) \subset \mathrm{Com}(\Re_0(S_{p,\varrho}, \mathcal{A})), \tag{4}$$

$$(\delta_{p,\varrho;b}(F))f = [M_b, F]f \qquad (f \in B_b). \tag{5}$$

Proof: It is well known that an arbitrary derivation on an algebra with a finite set of generators is uniquely determined by its action on the generators. Therefore, if a derivation $\delta_{p,\varrho;b}$ with the property (3) exists then it is unique.

Let us consider the natural isomorphism

$$\beta_\varrho : \Re_0(S_{p,\varrho}, \mathcal{A}) \longrightarrow \Re_0(S_p, \mathcal{A})$$

that takes the generators $S_{p,\varrho}$ and $M_a \in \Re_0(S_{p,\varrho}, \mathcal{A})$ to the generators S_p and $M_a \in \Re_0(S_p, \mathcal{A})$ respectively. Namely, select some representation for an arbitrary operator $F \in \Re_0(S_{p,\varrho}, \mathcal{A})$ via $S_{p,\varrho}$ and M_a and define $\beta_\varrho(F)$ by the substitution

$$S_{p,\varrho} \longmapsto S_p; \quad M_a(\in \Re_0(S_{p,\varrho}, \mathcal{A})) \longmapsto M_a(\in \Re_0(S_p, \mathcal{A}))$$

in the representation. Since $L_p \cap L_{p,\varrho}$ is dense in both L_p and $L_{p,\varrho}$, we have

$$Ff = (\beta_\varrho(F))f \qquad (f \in L_p \cap L_{p,\varrho}). \tag{6}$$

This equality implies that $\beta_\varrho(F)$ does not depend on the choice of the representation for F. Moreover, it follows from the definition that β_ϱ is an isomorphism.

It is easy to verify that the mapping

$$\delta_{p,\varrho;b} = \beta_\varrho^{-1}\delta_{p;b}\beta_\varrho$$

is a derivation and possesses the property (3).

By the definition of $\delta_{p,\varrho;b}$ the relation (4) follows from the analogous relation for $\delta_{p;b}$.

Due to (6), $(\delta_{p,\varrho;b}(F))f = (\delta_{p,;b}(F))f$ for every $f \in B_b \subset L_p \cap L_{p,\varrho}$. On the other hand, $(\delta_{p,;b}(F))f = [M_b, F]f$ for every $f \in B_b \subset L_r$. The last two equalities imply (5). □

Note that $\delta_{p,1;b} = \delta_{p;b}$. The equality (5) shows that the derivation $\delta_{p,\varrho;b}$ is in some sense the operation of taking the commutator with the operator M_b. As mentioned above, since in general $M_b \notin \mathrm{End}\, L_{p,\varrho}$, the derivation $\delta_{p,\varrho;b}$ is an outer one. Roughly speaking, Theorem 1 states that the commutator $[M_b, F]$, where $F \in \Re_0(S_{p,\varrho}, \mathcal{A})$, is a bounded operator and, moreover, belongs to $\mathrm{Com}(\Re_0(S_{p,\varrho}, \mathcal{A}))$.

2 The Description of the Weighted Cauchy Singular Operator Based on the Derivation $\delta_{p,\varrho;b}$

The following lemma contains some properties of Hunt-Muckenhoupt-Wheeden weights we use in what follows.

Lemma 4 *Let* $p \in (1, \infty)$, $\omega \in A_p$. *Then*

1. $\ln \omega \in BMO$.
2. *There are numbers* $\sigma_1 \langle 0$ *and* $\sigma_2 \rangle 1$ *such that* $\omega^{\mu} \in A_p$ *for every* $\mu \in (\sigma_1, \sigma_2)$; *moreover* $\|S_{p,\omega^{\mu}}\|$ *is a locally bounded function of the variable* $\mu \in (\sigma_1, \sigma_2)$.

Proof: Statement 1 is contained in Lemma VI.6.5 in [4]. Statement 2 can be deduced from Lemma VI.6.8 and Corollary VI.6.10 of [4]. □

Fix $p \in (1, \infty)$, $\omega \in A_p$ and let us suppose in what follows that σ_1 is the smallest number satisfying the condition of statement 2 of Lemma 4 while σ_2 is the biggest one. (The case where $\sigma_1 = -\infty$ and/or $\sigma_2 = \infty$ is not excluded.)

Let us denote by Δ the strip region $\sigma_1 < \text{Re}z < \sigma_2$ of the complex plane, and for every $z \in \Delta$ consider the weight $\omega_z = \omega^{\text{Re}z}$. Then, according to Lemma 4, $\omega_z \in A_p$ for every $z \in \Delta$. Since the operator $M_{\omega^{z/p}}$ maps L_{p,ω_z} isometrically and isomorphically onto L_p, we see from the equality

$$A(z) = M_{\omega^{z/p}} S_{p,\omega_z} M_{\omega^{-z/p}} \qquad (z \in \Delta)$$

that the operator-valued function

$$A : \Delta \longrightarrow \text{End}(L_p)$$

is well-defined. Note that $A(0) = S_p$, $A(1) = \alpha_{\omega}(S_{p,\omega})$.

Consider the function $b = (1/p) \ln \omega$, which is in BMO by Lemma 4, fix a Fefferman representation $b = u + \tilde{v}$ for it and suppose untill the end of the section that the coefficient algebra $\mathcal{A}$ contains the functions u and v. For every $z \in \Delta$ let us consider on $\mathfrak{R}_0(S_{p,\omega_z}, \mathcal{A})$ the outer derivation

$$\delta_z = \delta_{p,\omega_z;b} (= \beta_{\omega_z}^{-1} \delta_{p;b} \beta_{\omega_z})$$

where $\delta_{p,\omega_z;b}$ is $\delta_{p,\varrho;b}$ from Theorem 1 for $\varrho = \omega_z$. It follows from the definition that

$$\beta_{\omega_z} \delta_z = \delta_0 \beta_{\omega_z}. \tag{7}$$

In the following theorem the connection between the function $A(z)$ and the family of derivations $\{\delta_z\}_{z \in \Delta}$ is shown. Its proof is essentially based on the results of Theorem 1, on the equality (5) in particular.

Theorem 2 *The function $A(z)$ is holomorphic on Δ. Moreover, for every $z \in \Delta$ and for every positive integer m the following equality is valid:*

$$A^{(m)}(z) = M_{\omega^{z/p}}((\delta_z)^m(S_{p,\omega_z}))M_{\omega^{-z/p}}.$$

Let $q = p/(p-1)$. We shall denote the bilinear form $\int a(t)b(t)\,dm(t)$ by $\langle a, b\rangle$ where $a \in L_p$ and $b \in L_q$. For the proof of Theorem 2 we shall need

Lemma 5 *Let $\lambda > 0$, let χ_λ be the characteristic function of the set $\{t \in \mathbf{T} : |b(t)| \leq \lambda\}$ and ∂ is an inner derivation on* End L_p *defined by*

$$\partial(G) = [M_{\chi_\lambda b}, G] \qquad (G \in \text{End}\, L_p).$$

Then for every operator $F \in \mathfrak{R}_0(S_{p,\omega_z}, \mathcal{A})$, for every functions $f, g \in \chi_\lambda L_\infty$, every $z \in \Delta$ and every positive integer m the following equality is valid:

$$\langle M_{\omega^{z/p}}(\delta_z{}^m(F))M_{\omega^{-z/p}}f, g\rangle = \langle M_{\exp(z\chi_\lambda b)}(\partial^m(\beta_{\omega_z}(F)))M_{\exp(-z\chi_\lambda b)}f, g\rangle.$$
(8)

In particular,

$$\langle M_{\omega^{z/p}}(\delta_z{}^m(S_{p,\omega_z}))M_{\omega^{-z/p}}f, g\rangle = \langle M_{\exp(z\chi_\lambda b)}(\partial^m(S_p))M_{\exp(-z\chi_\lambda b)}f, g\rangle.$$
(9)

Proof: We proceed by induction on m.

1) Since $M_{\omega^{-z/p}}f \in \chi_\lambda L_\infty$, for $m = 1$ equality (8) follows from (5):

$$\begin{aligned}\langle M_{\omega^{z/p}}(\delta_z(F))M_{\omega^{-z/p}}f, g\rangle &= \langle M_{\omega^{z/p}}[M_b, \beta_{\omega_z}(F)]M_{\omega^{-z/p}}f, g\rangle \\ &= \langle M_{\exp(z\chi_\lambda b)}(\partial(\beta_{\omega_z}(F)))M_{\exp(-z\chi_\lambda b)}f, g\rangle.\end{aligned}$$

2) We assume that (8) holds for $m-1$ and verify it for m.

Let us first consider the case $z = 0$, for which desired equality (8) takes the form

(10) $$\langle(\delta_0{}^m(F))f, g\rangle = \langle(\partial^m(F))f, g\rangle.$$

By virtue of the induction assumption

$$\langle\partial^{m-1}(F)f, M_{\chi_\lambda b}g\rangle = \langle\delta_0{}^{m-1}(F)f, M_{\chi_\lambda b}g\rangle,$$

$$\langle(\partial^{m-1}(F))(M_{\chi_\lambda b}f), g\rangle = \langle(\delta_0{}^{m-1}(F))(M_{\chi_\lambda b}f), g\rangle.$$

In view of these equalities we arrive at

$$\begin{aligned}\langle(\partial^m(F))f, g\rangle &= \langle M_{\chi_\lambda b}(\partial^{m-1}(F))f, g\rangle - \langle(\partial^{m-1}(F))(M_{\chi_\lambda b}f), g\rangle \\ &= \langle M_{\chi_\lambda b}(\delta_0{}^{m-1}(F))f, g\rangle - \langle(\delta_0{}^{m-1}(F))(M_{\chi_\lambda b}f), g\rangle \\ &= \langle\partial(\delta_0{}^{m-1}(F))f, g\rangle = \langle(\delta_0{}^m(F))f, g\rangle\end{aligned}$$

and thus equality (10) is proved.

From (10) it is easy to obtain the equality

$$\langle(\partial^{m-1}(\delta_0(F)))f, g\rangle = \langle(\partial^m(F))f, g\rangle.$$

This equality together with the inclusions

$$\exp(-z\chi_\lambda b)\cdot f,\ \exp(z\chi_\lambda b)\cdot g \in \chi_\lambda L_\infty,$$

yields

$$\begin{aligned}(11)\qquad &\langle M_{\exp(z\chi_\lambda b)}(\partial^{m-1}(\delta_0(F)))M_{\exp(-z\chi_\lambda b)}f, g\rangle\\ &= \langle M_{\exp(z\chi_\lambda b)}(\partial^m(F))M_{\exp(-z\chi_\lambda b)}f, g\rangle.\end{aligned}$$

Let us now consider the general case $z \in \Delta$. Making use of the inductive assumption with the operator $\delta_z(F)$ instead of F we obtain, due to (7) and (11),

$$\begin{aligned}&\langle M_{\omega^{z/p}}(\delta_z{}^m(F))M_{\omega^{-z/p}}f, g\rangle = \langle M_{\exp(z\chi_\lambda b)}(\partial^{m-1}(\beta_{\omega_z}(\delta_z(F))))\\ &M_{\exp(-z\chi_\lambda b)}f, g\rangle = \langle M_{\exp(z\chi_\lambda b)}(\partial^{m-1}(\delta_0(\beta_{\omega_z}(F))))M_{\exp(-z\chi_\lambda b)}f, g\rangle\\ &= \langle M_{\exp(z\chi_\lambda b)}(\partial^m(\beta_{\omega_z}(F)))M_{\exp(-z\chi_\lambda b)}f, g\rangle.\end{aligned}$$

Thus, equality (8) is proved. Equality (9) results from (8) for $F = S_{p,\omega_z}$. □

Proof of Theorem 2: By statement 2 of Lemma 1 the set B_b is dense in both L_p and L_q where $q = p/(p-1)$. Furthermore, the function $A(z)$ is locally bounded on Δ. It follows from these facts and Theorem III.3.12 from [11] that if for every $f, g \in B_b$ the scalar function $\langle A(z)f, g\rangle$ is holomorphic on Δ then the operator-valued function $A(z)$ is holomorphic on Δ as well.

Let us fix arbitrary $f, g \in B_b$ and verify the holomorphy of the function $\langle A(z)f, g\rangle$.

In accordance with the definition of B_b, we have $f, g \in \chi_\lambda L_\infty$ for some $\lambda > 0$. We consider the operator-valued function

$$B(z) = M_{\exp(z\chi_\lambda b)}S_pM_{\exp(-z\chi_\lambda b)} \qquad (z \in \Delta)$$

and show that

$$(12)\qquad \langle A(z)f, g\rangle = \langle B(z)f, g\rangle.$$

Note that by the choice of f and g the following equalities are valid:

$$\exp(zb)\cdot g = \exp(z\chi_\lambda b)\cdot g,\quad \exp(-zb)\cdot f = \exp(-z\chi_\lambda b)\cdot f,$$

$$S_{p,\omega_z}(\exp(-zb)\cdot f) = S_p(\exp(-z\chi_\lambda b)\cdot f).$$

Hence

$$\begin{aligned}\langle A(z)f, g\rangle &= \int (A(z)f)g\,dm\\ &= \int (g\cdot\exp(zb))S_{p,\omega_z}(\exp(-zb)\cdot f)\,dm\\ &= \int (g\cdot\exp(z\chi_\lambda b))S_{p,\omega_z}(\exp(-z\chi_\lambda b)\cdot f)\,dm\\ &= \int (B(z)f)g\,dm = \langle B(z)f, g\rangle.\end{aligned}$$

Since $\chi_\lambda b \in L_\infty$, the function $B(z)$ is holomorphic on Δ. It is immediately verified that its derivatives satisfy the equalities

$$B^{(m)}(z) = M_{\exp(z\chi_\lambda b)}((\partial)^m(S_p))M_{\exp(-z\chi_\lambda b)} \tag{13}$$

where ∂ is the inner derivation defined in Lemma 5.

Equality (12) and the holomorphy of $B(z)$ imply the holomorphy of the scalar function $\langle A(z)f, g\rangle$ and consequently the holomorphy of $A(z)$. Therefore

$$\langle A^{(m)}(z)f, g\rangle = \langle B^{(m)}(z)f, g\rangle,\ m = 1, 2, \ldots,$$

which in combination with (13) and (9) yields the desired formulas for $A^{(m)}(z)$ in weak form:

$$\langle A^{(m)}(z)f, g\rangle = \langle M_{\omega^{z/p}}((\delta_z)^m(S_{p,\omega_z}))M_{\omega^{-z/p}}f, g\rangle \quad (f, g \in B_b).$$

By the density of B_b in both L_p and L_q the "weak coincidence" of the operators $A^{(m)}(z)$ and $M_{\omega^{z/p}}((\delta_z)^m(S_{p,\omega_z}))M_{\omega^{-z/p}}$ implies their equality. □

By $K(a, \tau)$ we shall denote the open disk of the radius τ and the center a.

Corollary 1 *Let $K(\zeta, \tau) \subset \Delta$. At every point $z \in K(\zeta, \tau)$ the function $A(z)$ can be represented in the form*

$$A(z) = M_{\omega^{\zeta/p}}(\exp((z-\zeta)\delta_\zeta)(S_{p,\omega_\zeta}))M_{\omega^{-\zeta/p}}.$$

Proof: This follows from Taylor series expansion of $A(z)$ on the disk $K(\zeta, \tau)$ by powers of $z - \zeta$. □

Theorem 3 $A(z) - A(\zeta) \in \mathrm{Com}(\mathfrak{R}(S_p, \mathcal{A}))$ *for every $z, \zeta \in \Delta$. In particular,*

$$\alpha_\omega(S_{p,\omega}) - S_p \in \mathrm{Com}(\mathfrak{R}(S_p, \mathcal{A})). \tag{14}$$

Proof: Let us first show that, for every $\zeta \in \Delta$,

$$A(\zeta) - A(0) \in \mathrm{Com}(\mathfrak{R}(S_p, \mathcal{A})). \tag{15}$$

Consider the analytic continuation of the function $A(z)$ from a neighborhood of the point $z = 0$ into a neighborhood of the point $z = \zeta$ by disks $K(z_k, \tau_k) \subset \Delta,\ k = 0, 1, \dots, l$ possessing the following properties:

$$z_0 = 0,\ \ z_l = \zeta,\ \ z_k \in K(z_{k-1}, \tau_{k-1}),\ k = 1, \dots, l.$$

Then for $k = 1, \dots, l$ the following equalities are valid:

$$A^{(n)}(z_k) = \sum_{m=n}^{\infty} \frac{A^{(m)}(z_{k-1})}{(m-n)!}(z_k - z_{k-1})^{m-n}, \qquad n = 1, 2, \dots \tag{16}$$

Theorem 2 and relation (4) from Theorem 1 imply

$$A^{(m)}(z_0) = (\delta_{p;b})^m(S_p) \in \mathrm{Com}(\mathfrak{R}(S_p, \mathcal{A})), \quad m = 1, 2, \dots,$$

which, due to (16), yields

$$A^{(n)}(z_k) \in \mathrm{Com}(\mathfrak{R}(S_p, \mathcal{A})), \quad n = 1, 2, \dots$$

for $k = 1, \dots, l$. These relations imply

$$A(z_k) - A(z_{k-1}) = \sum_{m=1}^{\infty} \frac{A^{(m)}(z_{k-1})}{m!}(z_k - z_{k-1})^m \ \in \mathrm{Com}(\mathfrak{R}(S_p, \mathcal{A})).$$

The sum of the left sides of the obtained equalities gives $A(z_l) - A(z_0) = A(\zeta) - A(0)$, whence (15) follows. From (15) we obtain

$$A(z) - A(\zeta) = (A(z) - A(0)) - (A(\zeta) - A(0)) \in \mathrm{Com}(\mathfrak{R}(S_p, \mathcal{A})).$$

The relations (14) result from (15) for $\zeta = 1$. □

Let us note that the proof of important relation (14) is based on relation (4) from Theorem 1.

By arguments similar to those of the proof of Theorem 3 one can show that

$$\alpha_\omega^{-1}(S_p) - S_{p,\omega} \ \in\ \mathrm{Com}(\mathfrak{R}(S_{p,\omega}\mathcal{A})). \tag{17}$$

The following theorem is the main result of Section 2.

Theorem 4 *The mapping α_ω establishes a similarity isomorphism of the Banach algebra $\mathfrak{R}(S_{p,\omega}, \mathcal{A})$ onto the Banach algebra $\mathfrak{R}(S_p, \mathcal{A})$. Moreover, α_ω(Com $(\mathfrak{R}(S_{p,\omega}\mathcal{A}))) = \mathrm{Com}(\mathfrak{R}(S_p, \mathcal{A}))$.*

Proof: Recall that α_ω is a similarity isomorphism of End $L_{p,\omega}$ onto End L_p. This isomorphism transfers the operator M_a acting on $L_{p,\omega}$ to the operator M_a acting on L_p. Hence, by virtue of (14), α_ω monomorphically maps $\mathfrak{R}(S_{p,\omega}, \mathcal{A})$ onto $\mathfrak{R}(S_p, \mathcal{A})$. On the other hand, by virtue of (17), α_ω^{-1} monomorphically maps $\mathfrak{R}(S_p, \mathcal{A})$ onto $\mathfrak{R}(S_{p,\omega}, \mathcal{A})$. This gives the first statement of the theorem. The second one is a consequence of the fact that α_ω, being a similarity isomorphism, transfers commutators to commutators. □

Theorems 3 and 4 as well as results close to them were announced in [5], [7] in a slightly other form.

3 Presymbol Construction

For a closed subalgebra $\mathcal{A}$ of the Banach algebra L_∞ let us denote by $G(\mathcal{A})$ the group of invertible elements of $\mathcal{A}$, by $\mathcal{A} \oplus \mathcal{A}$ the direct sum with the norm $\|(\varphi; \psi)\| = \max\{\|\varphi\|_\infty; \|\psi\|_\infty\}$ and by $G(\mathcal{A} \oplus \mathcal{A}) = G(\mathcal{A}) \oplus G(\mathcal{A})$ the group of invertible elements of $\mathcal{A} \oplus \mathcal{A}$. By γ_p we denote the norm $\|P_p^\pm\|$ (see [10] for estimates). In the following theorem a presymbol $\mu_{\mathcal{A},\omega}$ of the algebra $\mathfrak{R}(S_{p,\omega}, \mathcal{A})$ is defined and necessary conditions for an operator $F \in \mathfrak{R}(S_{p,\omega}, \mathcal{A})$ to be Fredholm are established.

Theorem 5 *Let $p \in (1, \infty)$, $\omega \in A_p$, and suppose the coefficient algebra $\mathcal{A}$ contains the functions u and v from the Fefferman representation for the function $(1/p)\ln\omega$. For every operator F in $\mathfrak{R}(S_{p,\omega}, \mathcal{A})$ the following statements are valid:*

1. *There exists a unique representation of F in the form*

$$F = M_\varphi P_{p,\omega}^+ + M_\psi P_{p,\omega}^- + K \tag{18}$$

   where $\varphi, \psi \in \mathcal{A}$ and $K \in \mathrm{Com}(\mathfrak{R}(S_{p,\omega}, \mathcal{A}))$. The mapping $\mu_{\mathcal{A},\omega}$ defined by the correspondence $F \mapsto (\varphi, \psi)$ is an epimorphism of the Banach algebras

$$\mu_{\mathcal{A},\omega} : \mathfrak{R}(S_{p,\omega}, \mathcal{A}) \longrightarrow \mathcal{A} \oplus \mathcal{A}$$

   with the kernel $ker(\mu_{\mathcal{A},\omega}) = \mathrm{Com}(\mathfrak{R}(S_{p,\omega}, \mathcal{A}))$ and the norm $\|\mu_{\mathcal{A},\omega}\| \leq (\gamma_p)^{-1}$.
2. *If the operator F is Fredholm and for the algebra $\mathcal{A}$ the condition $G(\mathcal{A}) = \mathcal{A} \cap G(L_\infty)$ holds then*

$$\mu_{\mathcal{A},\omega}(F) \in G(\mathcal{A} \oplus \mathcal{A}).$$

Proof: Theorem 5 was proved by the authors in [8] for the case $\omega = 1$. The general case will be deduced from that particular one on the basis of the results obtained in Sections 1 and 2. Put for brevity $\mu_{\mathcal{A}} = \mu_{\mathcal{A},1}$.

Let us prove the first statement of the theorem. Let F be an arbitrary operator in $\mathfrak{R}(S_{p,\omega}, \mathcal{A})$. By Theorem 4 the operator $\alpha_\omega(F)$ belongs to the unweighted algebra $\mathfrak{R}(S_p, \mathcal{A})$ and, hence, due to the results of [8], there exists a unique representation

$$\alpha_\omega(F) = M_\varphi P_p^+ + M_\psi P_p^- + K_1 \tag{19}$$

where $\varphi, \psi \in \mathcal{A}$ and $K_1 \in \mathrm{Com}(\mathfrak{R}(S_p, \mathcal{A}))$. Clearly,

$$F = M_\varphi \alpha_\omega^{-1}(P_p^+) + M_\psi \alpha_\omega^{-1}(P_p^-) + \alpha_\omega^{-1}(K_1).$$

By Theorem 4

$$\alpha_\omega^{-1}(K_1) \in \mathrm{Com}(\mathfrak{R}(S_{p,\omega}, \mathcal{A}))$$

while due to (17),

$$\alpha_\omega^{-1}(P_p^\pm) - P_{p,\omega}^\pm \in \mathrm{Com}(\mathfrak{R}(S_{p,\omega}, \mathcal{A})).$$

Thus, for F we have obtained the desired equality (18) with the operator

$$\begin{aligned} K &= M_\varphi(\alpha_\omega^{-1}(P_p^+) - P_{p,\omega}^+) + M_\psi(\alpha_\omega^{-1}(P_p^-) - P_{p,\omega}^-) + \alpha_\omega^{-1}(K_1) \\ &\in \mathrm{Com}(\mathfrak{R}(S_{p,\omega}, \mathcal{A})) \end{aligned}$$

and the functions φ, ψ from (19). Notice that

$$(\varphi, \psi) = \mu_\mathcal{A}(\alpha_\omega(F)). \tag{20}$$

The uniqueness of the obtained representation (18) follows from that of the representation (19). For the mapping $\mu_{\mathcal{A},\omega}$, due to (20), the following equality holds:

$$\mu_{\mathcal{A},\omega} = \mu_\mathcal{A} \circ \alpha_\omega. \tag{21}$$

Thus, it results that $\mu_{\mathcal{A},\omega}$ is an epimorphism and

$$\|\mu_{\mathcal{A},\omega}\| \le \|\mu_\mathcal{A}\| \|\alpha_\omega\| \le (\gamma_p)^{-1}.$$

By Theorem 4, $ker(\mu_{\mathcal{A},\omega}) = \mathrm{Com}(\mathfrak{R}(S_{p,\omega}, \mathcal{A}))$.

Let us now prove the second statement of the theorem. Let F ($\in \mathfrak{R}(S_{p,\omega}, \mathcal{A})$) be a Fredholm operator and suppose the condition $G(\mathcal{A}) = \mathcal{A} \cap G(L_\infty)$ holds. Since α_ω is similarity isomorphism, the operator $\alpha_\omega(F)$ is Fredholm. Hence, it follows from the results of [8] that $\mu_\mathcal{A}(\alpha_\omega(F)) \in G(\mathcal{A} \oplus \mathcal{A})$. Consequently, due to (21), $\mu_{\mathcal{A},\omega}(F) \in G(\mathcal{A} \oplus \mathcal{A})$. □

References

[1] A. Böttcher and Yu.I. Karlovich, Carleson curves, Muckenhoupt weights, and Toeplitz operators, Birkhäuser Verlag, Basel, Boston, Berlin, 1997.

[2] R.R. Coifman, R. Rochberg and G. Weiss, *Factorization theorems for Hardy spaces in several complex variables*, Ann. Math. **103** (1976), pp. 611–635.

[3] V.M. Deundyak, I.B. Simonenko and V.A. Stukopin, *Homotopy classification of families of Noetherian singular operators with piecewise continuous coefficients on a composite contour*, Dokl. Akad. Nauk (Russia) **329** (1993), pp. 305–309 (Russian).

[4] J.B. Garnett, Bounded analytic functions, Academic-Press, 1981.

[5] K. Georgiev and V. Deundyak, *The structure of algebras of singular integral operators on weighted L_p-spaces on the circle*, Izv. vuzov. Matematika **2** (1993), pp. 84–87 (Russian).

[6] K. Georgiev and V. Deundyak, *On the membership of the weighted Cauchy singular operator to some operator algebra*, Manuscript No. 574–B93, deposited at VINITI, 1993 (Russian).

[7] K. Georgiev and V. Deundyak, *A criterion for the weighted Cauchy singular operator to belong to the algebra of singular integral operators with coefficients in a Sarason algebra*, Funkt. analiz i ego prilozh. **28** (1994), pp. 80–82 (Russian).

[8] K. Georgiev and V. Deundyak, The presymbol of the algebra of singular integral operators, Manuscript no. 140–B97, deposited at VINITI, 1997 (Russian).

[9] I. Gohberg and N. Krupnik, *On singular integral equations with unbounded coefficients*, Matem. issled.-Kishinev **5** (1970), pp. 46–57 (Russian).

[10] I. Gohberg and N. Krupnik, One-dimensional linear singular integral equations, vol. **1**, **2**, Birkhäuser Verlag, Basel, Boston, Berlin, 1992.

[11] Kato, *Perturbation theory for linear operators*, Springer-Verlag, Berlin, Heidelberg, New York, 1966.

[12] I.B. Simonenko, Chin' Ngoc Min' (Chinh Ngoc Minh), *The local method in the theory of one-dimensional singular integral equations with piecewise continuous coefficients*. Noethericity, Rostov-on-Don, 1986 (Russian).

Department of Mathematics
Don State Technical University
Gagarin sq., 1, Rostov-on-Don, 344010
Russia

AMS Subject Classification: 47B47, 45E05

Operator Theory:
Advances and Applications, Vol. 118

Self-adjoint Operators with Inner Singularities and Pontryagin Spaces

Aad Dijksma, Heinz Langer, Yuri Shondin† and Chris Zeinstra*

Let A_0 be an unbounded self-adjoint operator in a Hilbert space $\mathcal{H}_0$ and let χ be a generalized element of order $-m-1$ in the rigging associated with A_0 and the inner product $\langle\cdot,\cdot\rangle_0$ of $\mathcal{H}_0$. In [S1, S2, S3] operators H^t, $t \in \mathbf{R}\cup\{\infty\}$, are defined which serve as an interpretation for the family of operators $A_0+t^{-1}\langle\,\cdot\,,\chi\rangle_0\,\chi$. The second summand here contains the inner singularity mentioned in the title. The operators H^t act in Pontryagin spaces of the form $\Pi_m = \mathcal{H}_0\oplus\mathbf{C}^m\oplus\mathbf{C}^m$ where the direct summand space $\mathbf{C}^m \oplus \mathbf{C}^m$ is provided with an indefinite inner product. They can be interpreted both as a canonical extension of some symmetric operator S in Π_m and also as extensions of a one-dimensional restriction S_0 of A_0 in $\mathcal{H}_0$ and hence they can be characterized by a class of Straus extensions of S_0 as well as via M.G. Krein's formulas for (generalized) resolvents. In this paper we describe both these realizations explicitly and study their spectral properties. A main role is played by a special class of Q-functions. Factorizations of these functions correspond to the separation of the nonpositive type spectrum from the positive spectrum of H^t. As a consequence, in Subsection 7.3 a family of self-adjoint Hilbert space operators is obtained which can serve as a nontrivial quantum model associated with the operators $A_0+t^{-1}\langle\,\cdot\,,\chi\rangle_0\,\chi$.

1 Introduction

1.1

In this paper we study differential operators with singular potentials by the method which was developed in [S1], [DT], [S3]. Recall that this method proceeds as follows. Given an expression of the form

$$A_0+\frac{1}{t}\langle\,\cdot\,,\chi\rangle_0\chi, \tag{1.1}$$

where A_0 is a self-adjoint operator in some Hilbert space $\mathcal{H}_0$, for example a differential operator, χ is some generalized element, for example a distribution, and $t \in \mathbf{R}$ is a parameter. In fact, χ is supposed to be an element of a space $\mathcal{H}_{-m-1}$ with negative norm constructed by means of A_0, see Section 1. Then the operator A_0 has the property, that for some fixed nonreal λ_0 the resolvent $R_0(\lambda_0) = (A_0-\lambda_0)^{-1}$ and its powers can be applied to χ and this 'smoothes' χ,

*The research of Heinz Langer was supported by the Fonds zur Förderung der wissenschaftlichen Forschung of Austria, Project P 12176 MAT.

†The research of Yuri Shondin was supported by the Netherlands Organization for Scientific Research NWO (NB 61-377) and by INTAS (93-0249-EXT).

which means $R_0(\lambda_0)^{m+1}\chi \in \mathcal{H}_0$ for some $m \in \mathbf{N}$. We introduce the elements

$$
\begin{aligned}
R_0(\lambda_0)\chi &= \chi_{-m},\ R_0(\lambda_0)^2\chi = \chi_{-m+1}, \ldots, R_0(\lambda_0)^m\chi = \chi_{-1}, \\
R_0(\lambda_0)^{m+1}\chi &= \chi_0, \ldots, R_0(\lambda_0)^{2m}\chi = \chi_{m-1}.
\end{aligned} \tag{1.2}
$$

Here $\chi_0, \chi_1, \ldots, \chi_{m-1} \in \mathcal{H}_0$, $\chi_{-m}, \ldots, \chi_{-1}$ are just formal elements. In this introduction we assume for simplicity that $m = 1$, hence the elements in (1.2) reduce to

$$R_0(\lambda_0)\chi = \chi_{-1},\ R_0(\lambda_0)^2\chi = \chi_0.$$

We introduce a new space by adding the improper element χ_{-1} and also χ_0 formally to $\mathcal{D}(A_0)$, that is, we consider the set $\mathcal{P}_1$ of all elements

$$\Phi = \varphi + a_1\chi_{-1} + a_2\chi_0,$$

where $\varphi \in \mathcal{D}(A_0)$ and $a_1, a_2 \in \mathbf{C}$. By the properties of the generalized element χ_{-1}, the inner product $\langle\chi_{-1}, \varphi\rangle_0$ and trivially also the inner products $\langle\chi_0, \varphi\rangle_0$, $\langle\chi_0, \chi_0\rangle_0$ make sense, whereas the inner products $\langle\chi_{-1}, \chi_0\rangle_0$ and $\langle\chi_{-1}, \chi_{-1}\rangle_0$ are not defined. We choose the first one to be a complex, the second one to be a real number, so far arbitrary but later they will be subjected to a condition (see (3.3) below), and set

$$
\begin{aligned}
\langle\chi_{-1}, \chi_{-1}\rangle_0 &=: g_{11},\ \langle\chi_{-1}, \chi_0\rangle_0 =: g_{21}, \\
\langle\chi_0, \chi_{-1}\rangle_0 &= g_{21}^* =: g_{12},\ \langle\chi_0, \chi_0\rangle_0 =: g_{22}.
\end{aligned}
$$

The space $\mathcal{P}_1$ is equipped with the inner product

$$\langle\Phi', \Phi\rangle_{\mathcal{P}_m} = \langle\varphi', \varphi\rangle_0 + \sum_{k=1}^{2}(a_k'\langle\chi_{-2+k}, \varphi\rangle_0 + a_k^*\langle\varphi', \chi_{-2+k}\rangle_0) + \sum_{i,j=1}^{2} a_i^* g_{ij} a_j',$$

where

$$\Phi = \varphi + \sum_{k=1}^{2} a_k\chi_{-2+k},\quad \Phi' = \varphi' + \sum_{k=1}^{2} a_k'\chi_{-2+k},\quad \varphi, \varphi' \in \mathcal{D}(A),\ a_k, a_k' \in \mathbf{C}.$$

It turns out (see [S1], Proposition 1, [JLT], Lemma 2.1, or (3.4) below) that this inner product has one negative square, hence the space $\mathcal{P}_1$ can be completed to a Pontryagin space Π_1.

The expression in (1.1) for the case $m = 1$ is given a meaning in the space Π_1 as follows. Let us first recall the formula for the resolvent of the one-dimensional perturbation

$$A_t := A_0 + \frac{1}{t}\langle\cdot, \omega\rangle_0\, \omega$$

of the operator A_0 if the element ω belongs to $\mathcal{H}_0$. This resolvent can be written as

$$(1.3)\qquad (A_t - z)^{-1} f = R_0(z) f - \frac{\langle R_0(z) f, \omega\rangle_0 R_0(z)\omega}{\langle R_0(z)\omega, \omega\rangle_0 + t}.$$

We shall give a meaning to (1.1) by giving a meaning in Π_1 to the right-hand side of (1.3) with ω replaced by the improper element χ. To this end we only have to give a meaning to the inner products in (1.3) with ω replaced by χ and to $R_0(z)\chi$. We start with the latter:

$$\begin{aligned} R_0(z)\chi &= R_0(\lambda_0)\chi + (z-\lambda_0) R_0(z) R_0(\lambda_0)\chi = \chi_{-1} + (z-\lambda_0) R_0(z)\chi_{-1} \\ &= \chi_{-1} + (z-\lambda_0)(R_0(\lambda_0)\chi_{-1} + (z-\lambda_0) R_0(z) R_0(\lambda_0)\chi_{-1}) \\ &= \chi_{-1} + (z-\lambda_0)(\chi_0 + (z-\lambda_0) R_0(z)\chi_0); \end{aligned}$$

here we have just used the resolvent relation. With this formula all the expressions in the numerator of (1.3) can be given a meaning in $\mathcal{P}_1$. The inner product $\langle R_0(z)\chi, \chi\rangle_0$ in the denominator of (1.3) is replaced by

$$\begin{aligned} &\langle R_0(z)\chi, \chi\rangle_0 - \operatorname{Re}\langle R_0(\lambda_0)\chi, \chi\rangle_0 \\ &= \frac{1}{2}\{\langle (R_0(z) - R_0(\lambda_0))\chi, \chi\rangle_0 + \langle (R_0(z) - R_0(\lambda_0^*))\chi, \chi\rangle_0\} \\ &= \frac{1}{2}\{(z-\lambda_0)\langle R_0(z) R_0(\lambda_0)\chi, \chi\rangle_0 + (z-\lambda_0^*)\langle R_0(z) R_0(\lambda_0^*)\chi, \chi\rangle_0\} \\ &= \frac{1}{2}\{(z-\lambda_0)\langle R_0(z)\chi, R_0(\lambda_0^*)\chi\rangle_{\mathcal{P}_1} + (z-\lambda_0^*)\langle R_0(z)\chi, R_0(\lambda_0)\chi\rangle_{\mathcal{P}_1}\}. \end{aligned}$$

Here we have subtracted $\operatorname{Re}\langle R_0(\lambda_0)\chi, \chi\rangle_0$, what formally is a real number and can be considered to be compensated by changing the parameter t.

We use these expressions in the right-hand side of (1.3) with ω replaced by χ. Instead of the operator A_0 in (1.3) we use a kind of lifting H^∞ of A_0 into the larger space Π_1, which is a self-adjoint relation with a one-dimensional multivalued part and the property that its resolvent set $\rho(H^\infty)$ coincides with the resolvent set $\rho(A_0)$ of A_0. Finally, the self-adjoint operators to be defined are denoted by H^t instead of A_t. Then the analogue of the formula (1.3) becomes (see (3.18) below)

$$(1.4)\qquad (H^t - z)^{-1} = (H^\infty - z)^{-1} - \Gamma_z\,(Q(z)+t)^{-1}\,\Gamma_{z^*}^*.$$

The operators $H^t, t\in\mathbf{R}$, are the canonical extensions of a symmetry S in Π_1 with defect (1, 1); they are also the self-adjoint extensions with exit into $\Pi_1 \supset \mathcal{H}_0$ of a one-dimensional restriction S_0 of A_0 in $\mathcal{H}_0$.

The Pontryagin space Π_1 is of the form

$$\Pi_1 = \mathcal{H}_0 \oplus \mathbf{C} \oplus \mathbf{C}$$

equipped with an inner product given by the Gram matrix

$$\mathbf{G} = \begin{pmatrix} I_0 & 0 & 0 \\ 0 & 0 & 1 \\ 0 & 1 & g_{11} \end{pmatrix}.$$

The function in the denominator of the formula (1.4) for the resolvent of the self-adjoint operator H^t is of the form

$$Q(z)+t=(z-\lambda_0)(z-\lambda_0^*)Q_0(z)+p_1(z)+t,$$

where

$$Q_0(z)=(z-\lambda_0^*)\langle R_0(z)\chi_{-1},\chi_0\rangle_0+g_{21},\quad p_1(z)=(z-\lambda_0)g_{11}-i\operatorname{Im}\lambda_0\,g_{11}.$$

The function $Q(z)$ belongs to the generalized Nevanlinna class N_1 and has a pole of nonpositive type at infinity. It is the so-called *Q-function*, corresponding to the symmetry S and the self-adjoint extension H^∞. The so-called *Γ-field* Γ_z, which also appears in (1.4), consists of linear mappings $\Gamma_z:\mathbf{C}\to\Pi_1$ defined by the relation

$$\Gamma_z=(I+(z-\lambda_0)(H^\infty-z)^{-1})\Gamma_{\lambda_0},$$

and takes the form

$$\Gamma_z 1=\begin{pmatrix}(z-\lambda_0)R_0(z)\chi_{-1}\\(z-\lambda_0)Q_0(z)\\1\end{pmatrix}.$$

Let $\lambda_0\in\mathbf{C}^+$ and $\lambda_0\in\sigma_p(H^t)$, the point spectrum of H^t. Then λ_0 is a complex zero of $Q(z)+t$ which implies $g_{11}=0$. The Γ-field Γ_z admits an expansion at λ_0 of the form

$$\Gamma_z 1=\begin{pmatrix}0\\0\\1\end{pmatrix}+(z-\lambda_0)\begin{pmatrix}R_0(z)\chi_{-1}\\Q_0(z)\\0\end{pmatrix}.$$

If we write this expansion as $\Gamma_z=\Gamma_{\lambda_0}+(z-\lambda_0)\Gamma_{0z}$ and denote by P_0 the orthogonal projection onto $\mathcal{H}_0$ then $P_0\Gamma_{0z}$ will also be a Γ-field which is in correspondence with $Q_0(z)$.

Now consider the function $-(Q(z)+t)^{-1}$. It belongs also to the class N_1 and has its poles of nonpositive type at the points λ_0 and λ_0^*. It is known [LT] that $-(Q(z)+t)^{-1}$ is the Q-function corresponding to H^t and S, and that

$$\Gamma_z^t=\frac{\alpha}{Q(z)+t}\Gamma_z$$

with some number α is the corresponding Γ-field. It admits an expansion at ∞, which is the generalized zero of nonpositive type of the function $-(Q(z)+t)^{-1}$:

$$\Gamma_z^t 1=\frac{\alpha}{z-\lambda_0^*}\begin{pmatrix}0\\1\\0\end{pmatrix}+\frac{1}{z-\lambda_0^*}\Gamma_{0z}^t 1,$$

where

$$\Gamma_{0z}^t 1=\frac{\alpha}{Q_0(z)}\begin{pmatrix}R_0(z)\chi_{-1}\\0\\(z-\lambda_0)^{-1}\end{pmatrix}.$$

We observe that the root space $\mathcal{L}_{\lambda_0}(H^t)$ of H^t at λ_0, which we denote by $\mathcal{M}$, is spanned by the leading term $\Gamma_{\lambda_0}1$ of the expansion of $\Gamma_z 1$ at the complex zero λ_0 of $Q(z)+t$. The root space of H^∞ at ∞, which is denoted by $\mathcal{L}$, is spanned by the leading term $\lim_{z\to\infty}(z-\lambda_0^*)\Gamma_z^t 1$ of the expansion of the dual Γ-field Γ_z^t at ∞. The subspaces $\mathcal{L}$ and $\mathcal{M}$ are in duality and Π_1 decomposes as

$$\Pi_1 = \mathcal{H}_0 \oplus (\mathcal{L} \dotplus \mathcal{M}).$$

By means of these constructions finally self-adjoint operators A^t are introduced which act in some Hilbert space $\widetilde{\mathcal{H}}_0$; this Hilbert space is either $\mathcal{H}_0$ or a one-dimensional extension of it. In examples, the operators A^t are objects of interest in quantum mechanics. They depend on some parameters which can be chosen appropriately.

In this paper we shall show that these constructions can be carried out in the general case of singular perturbations of order m as described at the beginnning. In a second part examples will be considered. A typical example is the Bessel equation on $(0,\infty)$ of order $\nu \in (0,\infty)$. Then the corresponding Q-function is

$$Q(z) = -\pi(-z)^\nu/(2\sin\pi\nu), \quad \nu \neq 1, 2, \dots.$$

It is a Nevanlinna function for $\nu < 1$, if $\nu > 1$ it belongs to the class N_m where $m = [\frac{\nu+1}{2}]$, and the method developed in this paper can be applied; see [DS].

1.2

We describe the contents of the sections of this paper. In Section 2 we recall basic notions and properties, which concern functions of the generalized Nevanlinna class N_m, and we describe their operator representation and review results about their generalized zeros of nonnegative type.

In Section 3 we introduce the symmetric nondensely defined operator S in the Pontryagin space Π_m with defect index $(1,1)$ and its self-adjoint extensions H^t; the integer $m \geq 0$ is determined by the regularity properties of the singular perturbation, that is, by the condition $\chi \in \mathcal{H}_{-m-1}\backslash\mathcal{H}_{-m}$. After the renormalization point λ_0 has been fixed the set of parameters which can be chosen freely in the description of the H^t's consists of $2m$ real numbers which determine the inner product of Π_m and H^∞, and a real number t which distinguishes the different self-adjoint extensions. The explicit expression (3.15) of the Q-function is fundamental for all the following considerations.

In Section 4 we consider the self-adjoint operators H^t as self-adjoint extensions of a symmetric operator S_0 in $\mathcal{H}_0$ ($\subset \Pi_m$) with exit into Π_m. In fact, the operators H^t are described in terms of the generalized resolvents and of the Straus extensions of S_0 in $\mathcal{H}_0$. We study the properties of the generalized Nevanlinna functions $\tau_t(z)$, which parametrize the generalized resolvents, and of the generalized the Schur functions $\theta_t(z)$, which parametrize the Straus extensions.

In Section 5 we introduce the subclass N_m^∞ of the class N_m of generalized Nevanlinna functions which is the class of Q-functions, which correspond to the singular perturbations considered in this paper. We characterize these functions $Q(z) \in N_m^\infty$ analytically and through their operator representation. For a given $\lambda_0 \in \mathbf{C}^+$ the Pontryagin space Π_m in this representation is decomposed as follows:

$$\Pi_m = \mathcal{H}_0 \oplus (\mathcal{L} \dot{+} \mathcal{M}).$$

Here $\mathcal{L}$ is the (m-dimensional neutral) root subspace at ∞ of the self-adjoint linear relation H^∞, and $\mathcal{M}$ is spanned by the vectors $\frac{1}{(j-1)!}\Gamma_{\lambda_0}^{(j-1)}1$, $j = 1, \ldots, m$, which are determined by the Γ-field Γ_z.

In Section 6 we generalize the construction of Section 3 to the case where the point λ_0 is replaced by a finite sequence $\Lambda = \{\lambda_1, \lambda_2, \ldots, \lambda_m\}$ of points $\lambda_j \in \rho(H^\infty)$ which leads to a decomposition

$$\Pi_m = \widehat{\mathcal{H}}_0 \oplus (\mathcal{L} \dot{+} \widehat{\mathcal{M}}),$$

where the subspace $\widehat{\mathcal{M}}$ is determined by Γ_z and the sequence Λ. Then we describe the operators H^t, some of its root subspaces and properties of the corresponding generalized resolvents and Straus extensions of a new symmetric operator $\widehat{S}_0$ in $\widehat{\mathcal{H}}_0$. At the end of this section also the case that Λ contains some points of the nonpositive type spectrum $\sigma^0(H^t)$ of H^t is considered. Recall that $\sigma^0(H^t)$ is the set of all eigenvalues of H^t in $\mathbf{C}^+ \cup \mathbf{R}$ with a nonpositive eigenvector.

In Section 7 we consider in more detail the special case that Λ is the set $\sigma^0(H^t)$ each point counted as often as the negative index of the algebraic eigenspace indicates. In this case we study the decomposition of Π_m from the point of view of the operator representations of the two functions $Q(z) + t$ and $-(Q(z) + t)^{-1}$. Additionally to the above decomposition another decomposition appears in a natural way:

$$\Pi_m = \widetilde{\mathcal{H}}_0 \oplus (\mathcal{L}_0 \dot{+} \widehat{\mathcal{M}}^0) \oplus \widehat{\mathcal{M}}^-.$$

Here $\widehat{\mathcal{M}}^0$ is the isotropic and $\widehat{\mathcal{M}}^-$ is a negative subspace of $\widehat{\mathcal{M}}$, and $\mathcal{L}_0$ is a subspace in $\mathcal{L}$ which is skewly linked with $\widehat{\mathcal{M}}^0$. The two decompositions of Π_m are related with two representations of $Q(z)$ and two functions $\widehat{Q}_0(z)$, $\widetilde{Q}_0(z) \in N_0$:

$$Q(z) = \widehat{b}_m(z)\widehat{b}_m^*(z)\widehat{Q}_0(z) + p_{2m-1}(z), \quad Q(z) = \widehat{b}_m(z)\widehat{b}_m^*(z)(-\widetilde{Q}_0(z)^{-1}),$$

where $\widehat{b}_m$ is the monic polynomial of degree m with zeros being the points of Λ. Further, H^t induces a self-adjoint operator A^t in the Hilbert space $\widetilde{\mathcal{H}}_0$ which is a self-adjoint extension of the symmetric operator $\widehat{S}_0$ in $\widehat{\mathcal{H}}_0$. This extension is canonical or with exit depending on the fact if $\widehat{M}$ is neutral or contains a negative part.

1.3

In this paper we always suppose that $m \geq 1$. We mention that singular perturbations (1.1) with $m = 0$, that is $\chi \in \mathcal{H}_{-1}$, are widely used in mathematical physics as quantum models of point-like interactions, see for example [AGHH]. In this case H^t can be interpreted entirely within the framework of Hilbert space extension theory as the one-parameter family of the canonical self-adjoint extensions of the one-dimensional restriction S_0 of A_0 to the domain $\{f \in \operatorname{dom} A_0 \mid \langle f, \chi\rangle_0 = 0\}$. More general quantum models based on self-adjoint extensions of S_0 with exit into a larger Hilbert space were also considered in mathematical physics, see for example [P].

Quantum models with singular perturbations with $\chi \in \mathcal{H}_{-m-1}$, $m > 0$, were treated also in an indefinite metric space in [Be], [S1], [DT], [S3], [C]. In this case H^t is not an appropriate candidate for the description of the dynamics in the quantum model because of the indefiniteness of the inner product. But if $\widehat{\mathcal{M}}$ is the m-dimensional nonpositive invariant subspace of H^t then the self-adjoint operator A^t in the Hilbert space $\widetilde{\mathcal{H}}_0 = \widehat{\mathcal{M}}^{\perp}/\widehat{\mathcal{M}}^0$ can be considered as such a candidate.

Summing up the result of this paper, a construction of a nontrivial quantum model is presented which is based on the following scheme:

$$\{\mathcal{H}_0, A_0, \chi\} \longrightarrow \{\Pi_m, H^t, S\} \longrightarrow \{\widehat{\mathcal{H}}_0, \widehat{A}^t, \widehat{S}_0\}.$$

And we observe that the triple $\{\widehat{\mathcal{H}}_0, \widehat{A}^t, \widehat{S}_0\}$ is isomorhic to a triple $\{\mathcal{H}_0, A^{\wedge}, S_0^{\wedge}\}$, where $S_0^{\wedge}$ is the one-dimensional restriction of A_0 to the domain

$$\operatorname{dom} S_0^{\wedge} = \{f \in \operatorname{dom} A_0 \mid \langle f, \chi_{-1}^{\wedge}\rangle_0 = 0\}$$

with $\chi_{-1}^{\wedge} = (\widehat{b}_m(A_0))^{-1}\chi \in \mathcal{H}_{-1}$, and $A^{\wedge}$ is a self-adjoint extension of $S_0^{\wedge}$ with exit into the Hilbert space $\mathcal{H} = \mathcal{H}_0 \oplus \mathbf{C}^r$, $r = \dim \widehat{\mathcal{M}}^{-}$. In a special case, when the invariant subspace $\widehat{\mathcal{M}}$ is neutral, a singular perturbation of the type (1.1) is obtained but with the generalized element $\chi_{-1}^{\wedge}$ instead of χ.

2 Preliminaries

2.1

In the following, Π_m is a Pontryagin space with m negative squares. Let A be a self-adjoint operator in Π_m. According to a well-known theorem of L.S. Pontryagin it has an m-dimensional nonpositive invariant subspace $\mathcal{L}^0(A)$ such that the spectrum

$$\sigma^0(A) := \sigma(A|\mathcal{L}^0(A))$$

of the restriction of A to $\mathcal{L}^0(A)$ is in the closed upper half plane. In general, the invariant subspace $\mathcal{L}^0(A)$ is not unique, however the set $\sigma^0(A)$ is uniquely determined. If each eigenvalue of the operator A is geometrically simple then also the subspace $\mathcal{L}^0(A)$ is uniquely determined (see for example [AI], [IKL]).

An eigenvalue λ of A is said to be of *positive* (*nonpositive, neutral* etc., respectively) *type* if all the corresponding eigenvectors x are positive (nonpositive, neutral etc., respectively), that is, $\langle x, x\rangle_{\Pi_m} > 0$ ($\langle x, x\rangle_{\Pi_m} \geq 0$, $\langle x, x\rangle_{\Pi_m} = 0$ etc., respectively). If the eigenvalue λ of A is nonreal then it is of neutral type; if it is real and an isolated point of $\sigma(A)$, then the algebraic eigenspace is nondegenerated. If the real eigenvalue is not an isolated point of $\sigma(A)$, then the algebraic eigenspace can be degenerated.

Two subspaces $\mathcal{F}$ and $\mathcal{G}$ of the Pontryagin space Π_m are said to be *in duality* if no vector of $\mathcal{F}$ is orthogonal to $\mathcal{G}$ and no vector of $\mathcal{G}$ is orthogonal to $\mathcal{F}$.

2.2

Let m be a nonnegative integer. The complex function F belongs to the *class* N_m, if it has the following property: F is meromorphic on $\mathbf{C}\backslash\mathbf{R}$, $F(z^*) = F(z)^*$ for $z \in \rho(F)$, the largest open set in $\mathbf{C}$ on which F is locally holomorphic, and the Nevanlinna kernel

$$K_F(z, w) = \frac{F(z) - F(w)^*}{z - w^*}, \quad z, w \in \rho(F), \ z \neq w^*, \tag{2.1}$$

has m negative squares. The functions of the class N_m can also be characterized by their *operator representation*: $F \in N_m$ if and only if F admits the representation

$$F(z) = F(\lambda_0)^* + (z - \lambda_0^*)\Gamma^*(I + (z - \lambda_0)(A - z)^{-1})\Gamma 1, \quad z \in \rho(A), \tag{2.2}$$

where A is a self-adjoint operator or a self-adjoint relation with nonempty resolvent set $\rho(A)$ in a Pontryagin space Π_m with negative index m, $\lambda_0 \in \rho(F) = \rho(A)$, $\Gamma : \mathbf{C} \to \Pi_m$ is a linear mapping and the minimality condition

$$\Pi_m = \overline{\operatorname{span}}\,\{\operatorname{ran}\Gamma, \ \operatorname{ran}(A - z)^{-1}\Gamma \mid z \in \rho(A)\}, \tag{2.3}$$

is satisfied. The condition (2.3) implies that the operator representation is unique up to an isomorphism, and we write sometimes $A(F)$ and $\Pi_m(F)$ to denote the dependence of A and Π_m on the function F. The functions in N_m are called *(generalized) Nevanlinna functions with m negative squares.* If $m = 0$ they are simply called *Nevanlinna functions.* If $F \in N_0$, then $\mathbf{C}\backslash\mathbf{R} \subset \rho(F)$ and the condition that the kernel K_F is positive definite is equivalent to $\operatorname{Im} F(z)/\operatorname{Im} z \geq 0$, $z \in \mathbf{C}\backslash\mathbf{R}$. Moreover, the function F belongs to N_0 if and only if it admits the integral representation

$$F(z) = \alpha + \beta z + \int_{-\infty}^{\infty} \frac{zt + 1}{t - z}\, d\sigma(t), \tag{2.4}$$

with $\alpha \in \mathbf{R}$, $\beta \geq 0$, and a left continuous, nondecreasing, and bounded function σ on $\mathbf{R}$.

Besides the Nevanlinna class N_m also the *Schur class* S_m plays a role in the sequel: It consists of all functions which are defined and meromorphic on the open upper half-plane $\mathbf{C}^+$ and have the property that the Schur kernel

$$S_\theta(z, w) := \frac{1 - \theta(w)^*\theta(z)}{i(w^* - z)}, \quad z, w \in \rho(\theta),$$

has m negative squares. Usually these functions are defined on the open unit disc $\mathbf{D}$ and then the denominator of the kernel function has to be replaced by $1 - w^*z$. For the properties of such Schur functions we refer to [ADRS]. In connection with the so-called Straus extensions (see (2.12) below) it is more convenient to work on $\mathbf{C}^+$ than on $\mathbf{D}$. Evidently, the two classes are related via a fractional linear transformation of the independent variable.

2.3

The notion of a Q-function is of fundamental importance in this paper. Given is a pair S, A, where S is a closed symmetric relation with defect index $(1, 1)$ in the Pontryagin space Π_m and A is a self-adjoint extension of S in Π_m with nonempty resolvent set $\rho(A)$. Choose $\lambda_0 \in \rho(A)$ and a nonzero linear mapping $\Gamma : \mathbf{C} \to \ker(S^* - \lambda_0)$, and define for $z \in \rho(A)$

$$\Gamma_z = (I + (z - \lambda_0)(A - z)^{-1})\Gamma,$$

which maps $\mathbf{C}$ into $\ker(S^* - z)$; evidently, $\Gamma_{\lambda_0} = \Gamma$ and

$$\Gamma_z = (I + (z - \zeta)(A - z)^{-1})\Gamma_\zeta, \quad (A - \zeta)^{-1}\Gamma_z = (A - z)^{-1}\Gamma_\zeta. \tag{2.5}$$

The family of operators Γ_z, $z \in \rho(A)$, is called the *Γ-field corresponding to $\Gamma = \Gamma_{\lambda_0}$ and A*, or, if Γ is fixed, the *Γ-field corresponding to A*. A *Q-function* associated with the pair S, A is any function $Q(z)$ that satisfies

$$\frac{Q(z) - Q(w)^*}{z - w^*} = \Gamma_w^*\Gamma_z 1, \quad z, w \in \rho(A). \tag{2.6}$$

This requirement determines $Q(z)$ up to a real constant c:

$$\begin{aligned} Q(z) &= c - i\mathrm{Im}\,\lambda_0\,\Gamma_{\lambda_0}^*\Gamma_{\lambda_0}1 + (z - \lambda_0^*)\Gamma_{\lambda_0}^* \\ &\qquad (I + (z - \lambda_0)(A - z)^{-1})\Gamma_{\lambda_0}1. \end{aligned} \tag{2.7}$$

It also implies that $Q(z)$ belongs to a class $N_{m'}$ for some $m' \le m$; if the minimality condition (2.3) holds then $m' = m$. Observe that a Q-function is never a real constant on $\mathbf{C}$. Conversely, any generalized Nevanlinna function $F(z)$ which is not a real constant is a Q-function, namely for the self-adjoint relation $A(F)$ in its operator representation and the symmetric relation:

$$S = \{\{f, g\} \in A(F) \mid \Gamma^*(g - \lambda_0^* f) = 0\}.$$

Of interest below is also the case that $m = 0$, that is, S is a symmetric relation with defect index $(1, 1)$ and A is one of its self-adjoint extensions in a Hilbert space $\mathcal{H}$. This case was studied for example in [LT]. Then the assumption $\rho(A) \neq \emptyset$ is automatically satisfied. It was shown in [LT] (see also [KL2]) that the Q-function $Q(z)$ satisfies the relation

$$\lim_{y\to\infty} y\mathrm{Im}\,Q(iy) = \infty \tag{2.8}$$

if and only if $\overline{\mathrm{dom}\, S} = \overline{\mathrm{dom}\, A}$, and this is equivalent to $\mathrm{ran}\,\Gamma \cap \mathrm{dom}\, A = \{0\}$; $Q(z)$ satisfies the relation

$$\lim_{y\to\infty} \frac{1}{y} Q(iy) = 0 \tag{2.9}$$

if and only if A is an operator, or equivalently, if A is densely defined. Thus, the relations (2.8) and (2.9) together are equivalent to the fact that $Q(z)$ is the Q-function of a densely defined symmetric operator with defect index (1, 1) in a Hilbert space.

The Q-function apears in M.G. Krein's formula for the description of the generalized resolvents of a closed symmetric relation S with defect index (1, 1) in a Pontryagin space Π_κ ([KL], [KL1], [LT]). Recall that a *generalized resolvent of* S is a function $R(z)$ of the form

$$R(z) = \widetilde{P}_{\Pi_\kappa}(\widetilde{A} - z)^{-1}|_{\Pi_\kappa}, \tag{2.10}$$

where $\widetilde{A}$ is a self-adjoint extension of S in some Pontryagin space $\widetilde{\Pi}_{\widetilde{\kappa}}$ which contains Π_κ, and $\widetilde{P}_{\Pi_\kappa}$ is the orthogonal projection in $\widetilde{\Pi}_{\widetilde{\kappa}}$ onto Π_κ. We fix a self-adjoint extension A of S in Π_κ with nonempty resolvent set, a complex number $\lambda_0 \in \mathbf{C}^+ \cap \rho(A)$ and a nonzero mapping $\Gamma_{\lambda_0} : \mathbf{C} \to \ker(S^* - \lambda_0)$. The Γ-field corresponding to Γ_{λ_0} and A is again denoted by Γ_z. Then the following formula establishes a bijective correspondence between all generalized resolvents $R(z)$ of S, corresponding to a self-adjoint extension of S in some Pontryagin space $\widetilde{\Pi}_{\widetilde{\kappa}}$, and all functions $t(z) \in \cup_{j=0}^{\infty} N_j$ with the property $t(z) + Q(z) \not\equiv 0$:

$$R(z) = (A - z)^{-1} - \Gamma_z(Q(z) + t(z))^{-1}\Gamma_{z^*}^*. \tag{2.11}$$

The formula (2.11) holds for all $z \in \rho(A) \cap \rho(\widetilde{A})$, if $\widetilde{A}$ denotes the self-adjoint extension in (2.10); in particular, it holds for all but possibly a finite number of nonreal z. If in (2.11) the parameter function $t(z)$ belongs to N_j then $\widetilde{\kappa} = \kappa + j$. If t in (2.11) is a (real) constant then the corresponding (minimal) self-adjoint extension $\widetilde{A}$ acts in the original space Π_κ; such extensions $\widetilde{A}$ are called *canonical.* In this case, evidently, the generalized resolvent $R(z)$ is just the resolvent of the extension $\widetilde{A}$, and the formula (2.11) describes the extension $\widetilde{A}$ explicitly through its resolvent. If t is not a real constant, the dimension of the *exit space* $\widetilde{\Pi}_{\widetilde{\kappa}} \ominus \Pi_\kappa$ coincides with the sum $n_+(t) + n_-(t)$, where $n_+(t)$ ($n_-(t)$, respectively) denotes the number of positive (negative, respectively) squares of the kernel $K_t(z, w)$ from (2.1), and the signature of the inner product on the exit space is given by these numbers ([DLS3]).

As an operator function in the originally given space Π_κ, the generalized resolvent $R(z)$ contains all the information about the corresponding (minimal) extension $\widetilde{A}$ of S. The same is true for the so-called *Straus relation* or *Straus extension* $T_t(z)$ *of* S in Π_κ, associated with $\widetilde{A}$ in $\widetilde{\Pi}_{\widetilde{\kappa}}$. It is defined as follows:

$$T_t(z) = \{\{\widetilde{P}_{\Pi_\kappa} f, \widetilde{P}_{\Pi_\kappa} g\} | \{f, g\} \in \widetilde{A},\ g - zf \in \Pi_\kappa\}, \quad z \in \mathbf{C}. \tag{2.12}$$

Evidently, if $\widetilde{A}$ is a canonical extension (that is, t is a real constant), then $T_t(z) = \widetilde{A}$ for all $z \in \mathbf{C}$. In this paper, Straus relations appear only in the Hilbert space $\mathcal{H}_0$, that is, $\kappa = 0$, and for $\widetilde{\kappa} = m$.

2.4

If $F(z) \in N_m$, the point $x_0 \in \mathbf{R} \cup \{\infty\}$ is called a *generalized zero of* $F(z)$ *of nonpositive type of multiplicity* π_0 if for each sufficiently small neighbourhood $\mathcal{U}$ of x_0 there exists a $\delta_{\mathcal{U}} > 0$ such that $0 < \alpha < \delta_{\mathcal{U}}$ implies that the equation $F(z) = -i\alpha$ has π_0 solutions in $\mathcal{U} \cap \mathbf{C}^+$. This is equivalent to the following. If we consider an operator representation (2.2) of the function $\widetilde{F}(z) := -F(z)^{-1}$:

$$\widetilde{F}(z) = \widetilde{F}(\lambda_0)^* + (z - \lambda_0^*)\Gamma^*(I + (z - \lambda_0)(\widetilde{A} - z)^{-1})\Gamma,$$

then x_0 is an eigenvalue of nonpositive type of the self-adjoint relation $\widetilde{A}$, and the indefinite inner product has π_0 negative squares on the root space of $\widetilde{A}$ at x_0. The generalized zeros of nonpositive type in $\mathbf{R}$ can be characterized analytically as follows: $x_0 \in \mathbf{R}$ is a generalized zero of nonpositive type of $F(z)$ of multiplicity $\pi_0 > 0$ if and only if

1. $\lim_{z \widehat{\to} x_0} \frac{F(z)}{(z-x_0)^{2\pi_0-1}}$ is finite and nonpositive,
2. $\lim_{z \widehat{\to} x_0} \frac{F(z)}{(z-x_0)^{2\pi_0+1}}$ is either finite and positive, or it is ∞.

Here $\lim_{z \widehat{\to} x_0}$ denotes the nontangential limit.

Lemma 2.1 *Consider the function* $F(z) \in N_m$ *with the operator representation (2.2), and let* Γ_z *be the* Γ*-field corresponding to* A *and* Γ. *If* x_0 *is a generalized zero of nonpositive type and of multiplicity* π_0 *of* $F(z)$, *then the limit* $\Gamma_{x_0}1 := \lim_{z \widehat{\to} x_0} \Gamma_z 1$ *and, if* $\pi_0 > 1$, *the limits*

$$\Gamma_{x_0}^{(j)}1 := \lim_{z \widehat{\to} x_0} \Gamma_z^{(j)}1, \quad j = 1, 2, \ldots, \pi_0 - 1, \tag{2.13}$$

exist in the norm topology of Π_m. *Moreover,*

$$\begin{cases} \langle \Gamma_{x_0}^{(j)}1, \Gamma_{x_0}^{(k)}1 \rangle_{\Pi_m} = 0, \quad 0 \le j, k \le \pi_0 - 1, \quad j + k \ne 2\pi_0 - 2, \\ \langle \Gamma_{x_0}^{(\pi_0-1)}1, \Gamma_{x_0}^{(\pi_0-1)}1 \rangle_{\Pi_m} \le 0, \end{cases} \tag{2.14}$$

hence the subspace spanned by the elements $\Gamma_{x_0}^{(j)}1$, $j = 0, 1, \ldots, \pi_0 - 1$, *is nonpositive.*

Proof: In [DLLS] it was shown that if $x_0 \in \mathbf{R}$ is a generalized zero of nonpositive type of multiplicity π_0 of $F(z)$ then $F(z)$ admits the representation

$$F(z) = (z - x_0)^{2\pi_0} \frac{c(z)c^*(z)}{d(z)d^*(z)} F_0(z).$$

Here $c(z)$ is the monic polynomial the zeros of which are the zeros in $\mathbf{C}^+$ or generalized zeros of nonpositive type of $F(z)$ except x_0, counted according to their multiplicities, $d(z)$ is the monic polynomial the zeros of which are the poles in $\mathbf{C}^+$ or generalized poles of nonpositive type of $F(z)$, also counted according to their multiplicities, and $F_0(z)$ is a Nevanlinna function. Denote by $\sigma_0(t)$ the nondecreasing function in the integral representation (2.4) of $F_0(z)$. We choose an open bounded interval Δ containing x_0 but no other generalized zero or generalized pole of $F(z)$, introduce the function

$$F_\Delta(z) := \int_\Delta \frac{1}{t-z}\, d\sigma(t) \text{ with } d\sigma(t) := \frac{(t-x_0)^{2\pi_0} c(t) c^*(t)}{d(t) d^*(t)}\, d\sigma_0(t),$$

and write $F(z) = F_\Delta(z) + \widehat{F}(z)$. Then the function $\widehat{F}(z)$ is holomorphic in Δ and $\int_\Delta \frac{d\sigma(t)}{|t-x_0|^{2\pi_0}} < \infty$. The last inequality yields for $k = 0, 1, \ldots, 2\pi_0 - 1$

$$\lim_{z \widehat{\to} x_0} \left(\int_\Delta \frac{d\sigma(t)}{(t-z)^{k+1}} - \int_\Delta \frac{d\sigma(t)}{(t-x_0)^{k+1}} \right)$$

$$= \lim_{z \widehat{\to} x_0} \int_\Delta \left(\left(\frac{t-x_0}{t-z} \right)^{k+1} - 1 \right) \frac{d\sigma(t)}{(t-x_0)^{k+1}} = 0$$

(observe that the term $(\frac{t-x_0}{t-z})^{k+1}$ remains bounded when $z \widehat{\to} x_0$), and therefore the limit

$$F^{(k)}(x_0) = \lim_{z \widehat{\to} x_0} F^{(k)}(z) = k! \int_\Delta \frac{d\sigma(t)}{(t-x_0)^{k+1}} + \widehat{F}^{(k)}(x_0) = \lim_{z \widehat{\to} x_0} \frac{F(z)}{(z-x_0)^k}$$

exists and according to 1. it is zero for $k = 0, 1, \ldots, 2\pi_0 - 2$.

In order to prove the existence of the limits in (2.13) we use the following criterion for the convergence of a sequence of elements in Π_m, see [IKL]: The sequence $(f_n) \subset \Pi_m$ converges in norm if and only if the sequences $(\langle f_n, g \rangle_{\Pi_m})$ converge for all g from a dense subset of Π_m and the limit $\lim_{n,m\to\infty} \langle f_m, f_n \rangle_{\Pi_m}$ exists. Now we observe for $j = 0, 1, \ldots, \pi_0 - 1$ the relations

$$\lim_{z \widehat{\to} x_0} \langle \Gamma_z^{(j)} 1, \Gamma_u 1 \rangle_{\Pi_m} = \lim_{z \widehat{\to} x_0} \frac{d^j}{dz^j} \frac{F(z) - F(u)^*}{z - u^*} = (-1)^{j+1} j! \frac{F(u)^*}{(x_0 - u^*)^{j+1}}$$

and

$$\begin{aligned} \lim_{z, \zeta^* \widehat{\to} x_0} \langle \Gamma_z^{(j)} 1, \Gamma_\zeta^{(j)} 1 \rangle_{\Pi_m} &= \lim_{z, \zeta^* \widehat{\to} x_0} \frac{d^j}{dz^j} \frac{d^j}{d\zeta^{*j}} \frac{F(z) - F(\zeta)^*}{z - \zeta^*} \\ &= \frac{(j!)^2}{(2j+1)!} \widehat{F}^{(2j+1)}(x_0) + (j!)^2 \int_\Delta \frac{d\sigma(t)}{|t-x_0|^{2j+2}} \\ &= \frac{(j!)^2}{(2j+1)!} F^{(2j+1)}(x_0), \end{aligned}$$

and the relations (2.13) follow. In the same way,

$$\begin{aligned}\langle \Gamma_{x_0}^{(j)} 1, \Gamma_{x_0}^{(k)} 1\rangle_{\Pi_m} &= \frac{j!k!}{(j+k+1)!} F^{(j+k+1)}(x_0) \\ &= \frac{j!k!}{(j+k+1)!} \lim_{z \hat{\to} x_0} \frac{F(z)}{(z-x_0)^{j+k+1}},\end{aligned}$$

and the relations (2.14) follow from 1. □

3 The Family of Self-adjoint Extensions H^t

3.1

We consider an unbounded self-adjoint operator A_0 in a Hilbert space $\mathcal{H}_0$ with inner product $\langle \cdot, \cdot \rangle_0$. If m is a nonnegative integer, A_0 generates a rigging (see [B1], [B2])

$$\mathcal{H}_{m+1} \hookrightarrow \mathcal{H}_m \hookrightarrow \cdots \hookrightarrow \mathcal{H}_1 \hookrightarrow \mathcal{H}_0 \hookrightarrow \mathcal{H}_{-1} \hookrightarrow \cdots \hookrightarrow \mathcal{H}_{-m} \hookrightarrow \mathcal{H}_{-m-1}. \tag{3.1}$$

Here the Hilbert spaces $\mathcal{H}_k$ with positive index $k = 1, \ldots, m+1$ are defined as follows:

$\mathcal{H}_k$ is the linear space dom A_0^k equipped with the inner product and norm

$$\langle u, v \rangle_k = \sum_{j=0}^{k} \langle A_0^j u, A_0^j v \rangle_0, \quad \|u\|_k = \sqrt{\langle u, u \rangle_k}.$$

The inclusion mappings $\hookrightarrow$ between these spaces are contractions and have dense ranges, and also the products of these mappings have the same properties. Denote by J_k the inclusion map $\mathcal{H}_k \hookrightarrow \mathcal{H}_0$. For $f \in \mathcal{H}_0$, the antilinear mapping

$$u \in \mathcal{H}_k \longmapsto \langle f, J_k u \rangle_0$$

is bounded on $\mathcal{H}_k$ and we set

$$\|f\|_{-k} = \sup \frac{|\langle f, J_k u \rangle_0|}{\|u\|_k}.$$

Then $\|\cdot\|_{-k}$ defines a norm on $\mathcal{H}_0$ and the completion of $\mathcal{H}_0$ with respect to this norm is the so-called negative space $\mathcal{H}_{-k}$, $k = 1, \ldots, m+1$. It is a Hilbert space, which can be identified as the space of all continuous antilinear functionals on $\mathcal{H}_k$: For $\psi \in \mathcal{H}_{-k}$ there is a sequence of elements $f_n \in \mathcal{H}_0$ which converges in $\mathcal{H}_{-k}$ to ψ such that the limit

$$\langle \psi, u \rangle_0 := \lim_{n \to \infty} \langle f_n, E_k u \rangle_0, \quad u \in \mathcal{H}_k,$$

exists, and $|\langle\psi, u\rangle_0| \leq \|\psi\|_{-k}\|u\|_k$. The form $\langle\psi, u\rangle_0$ is viewed as a sesquilinear form on $\mathcal{H}_{-k} \times \mathcal{H}_k$, and we set for $\psi \in \mathcal{H}_{-k}$, $u \in \mathcal{H}_k$

$$\langle u, \psi\rangle_0 = \langle\psi, u\rangle_0^*.$$

It is well-known that the inclusion mappings in (3.1) to the right of the space $\mathcal{H}_0$ make sense and are also contractions with dense ranges. The proofs and more information about riggings can be found in [B1], [B2].

We consider an element $\chi \in \mathcal{H}_{-m-1}\backslash\mathcal{H}_{-m}$, fix a complex number $\lambda_0 \in \mathbf{C}^+$ and introduce the elements χ_{-1}, χ_0:

$$\chi_{-1} = R_0(\lambda_0)^m\chi, \quad \chi_0 = R_0(\lambda_0)^{m+1}\chi$$

and

$$\varepsilon_j = R_0(\lambda_0)^j\chi, \quad j = 1, 2, \ldots, 2m. \tag{3.2}$$

Note that $\varepsilon_j \in \mathcal{H}_{j-m-1}\backslash\mathcal{H}_{j-m}$, hence the $2m$ elements $\varepsilon_1, \varepsilon_2, \ldots, \varepsilon_{2m}$ are linearly independent in the space $\mathcal{H}_{-m}$. The elements $\varepsilon_m = \chi_{-1} \in \mathcal{H}_{-1}\backslash\mathcal{H}_0$, $\varepsilon_{m+1} = \chi_0 \in \mathcal{H}_0\backslash\mathcal{H}_1$ will play a special role in the sequel.

3.2

The Pontryagin space Π_m. By $\mathcal{P}_m$ we denote the linear space of all elements

$$\Phi = \varphi + \sum_{j=1}^{2m} a_j\varepsilon_j, \quad \varphi \in \mathcal{H}_m, \ a_j \in \mathbf{C}, \ j = 1, 2, \ldots, 2m.$$

The sum is considered either as a formal one or as a sum in $\mathcal{H}_{-m}$, which amounts to the same since the elements ε_j, $j = 1, 2, \ldots, 2m$, are linearly independent in $\mathcal{H}_{-m}$. An inner product $\langle\Phi', \Phi\rangle_{\mathcal{P}_m}$ between two such elements Φ, Φ' is defined as follows (with evident notation):

$$\langle\Phi', \Phi\rangle_{\mathcal{P}_m} = \langle\varphi', \varphi\rangle_0 + \sum_{j=1}^{2m}(a'_j\langle\varepsilon_j, \varphi\rangle_0 + a_j^*\langle\varphi', \varepsilon_j\rangle_0) + \sum_{i,j=1}^{2m} a_i^* g_{ij} a'_j.$$

Here the numbers g_{ij}, $i, j = 1, \ldots, 2m$, are formally

$$g_{ij} = \langle\varepsilon_j, \varepsilon_i\rangle_0,$$

however, the expression on the right-hand side makes sense only if $i + j \geq 2m+2$, if $i + j \leq 2m + 1$ then g_{ij} is chosen arbitrarily but subject to the conditions

$$\begin{aligned} g_{ij}^* &= g_{ji}, \quad i, j = 1, 2, \ldots, 2m; \\ g_{ij} &= \frac{g_{i-1,j} - g_{i,j-1}}{\lambda_0 - \lambda_0^*}, \quad i, j = 2, 3, \ldots, 2m. \end{aligned} \tag{3.3}$$

The first condition implies that the inner product $\langle \cdot, \cdot \rangle_{\mathcal{P}_m}$ is symmetric, both conditions are needed in order to prove the self-adjointness of the operator H^∞ defined below, see [DT, S3]. The g_{ij} are not uniquely determined by the relations (3.3), there remains a free choice of $2m$ real constants. For example, the real numbers g_{jj} and $\operatorname{Re} g_{j,j+1}$, $j = 1, \ldots, m$, can be chosen arbitrarily. Then, by the relation (3.3), $\operatorname{Im} g_{j,j+1} = \operatorname{Im} \lambda_0 \, g_{j+1,j+1}$.

We form the matrix $G = (g_{ij})_{i,j=1}^m$ and consider the space

$$\Pi_m = \mathcal{H}_0 \oplus \mathbf{C}^m \oplus \mathbf{C}^m \tag{3.4}$$

provided with the inner product

$$\left\langle \begin{pmatrix} f \\ b \\ c \end{pmatrix}, \begin{pmatrix} f' \\ b' \\ c' \end{pmatrix} \right\rangle_{\Pi_m} = \langle f, f' \rangle_0 + \left(\begin{pmatrix} 0 & I_m \\ I_m & G \end{pmatrix} \begin{pmatrix} b \\ c \end{pmatrix}, \begin{pmatrix} b' \\ c' \end{pmatrix} \right)_{\mathbf{C}^{2m}},$$

$$f, f' \in \mathcal{H}_0, \quad \begin{pmatrix} b \\ c \end{pmatrix}, \begin{pmatrix} b' \\ c' \end{pmatrix} \in \mathbf{C}^m \oplus \mathbf{C}^m.$$

It can be checked easily (see [S3]) that the mapping τ defined by

$$\tau \left(\varphi + \sum_{k=1}^{2m} a_k \varepsilon_k \right) = \begin{pmatrix} \varphi + \sum_{k=m+1}^{2m} a_k \varepsilon_k \\ \left(\langle \varphi, \varepsilon_i \rangle_0 + \sum_{k=m+1}^{2m} a_k g_{ik} \right)_{i=1}^m \\ (a_k)_{k=1}^m \end{pmatrix}$$

defines an isometry from $\mathcal{P}_m$ to Π_m. It has a dense range, and so the Pontryagin space Π_m can be viewed as the completion of the pre-Pontryagin space $\mathcal{P}_m$.

3.3

The self-adjoint relation H^∞ in Π_m. The elements of the space $\mathcal{P}_m$ can also be considered as elements of $\mathcal{H}_{-m}$. The resolvent identity implies that for $z \in \mathbf{C}\backslash\mathbf{R}$ the extended resolvent operator $R_0(z)$ maps $\mathcal{P}_m$ into itself and the mapping $\tau R_0(z) \tau^{-1}$ defined on the dense set $\tau \mathcal{P}_m$ is closable. Its closure is the bounded operator $R^\infty(z)$ on Π_m given by the relation

$$\begin{aligned} R^\infty(z) \begin{pmatrix} f \\ (b_j)_{j=1}^m \\ (c_j)_{j=1}^m \end{pmatrix} & \\ = \begin{pmatrix} R_0(z) f + c_{m+1}(z) R_0(z) \chi_{-1} \\ \left((z - \lambda_0^*)^{m-j} \langle R_0(z) f, \chi_{-1} \rangle_0 + q_j(z) c_{m+1}(z) + b_j(z) \right)_{j=1}^m \\ (c_j(z))_{j=1}^m \end{pmatrix}&, \end{aligned} \tag{3.5}$$

where

$$
(3.6)\qquad
\begin{cases}
c_j(z) = \displaystyle\sum_{k=1}^{j-1}(z-\lambda_0)^{j-k-1}c_k, & j = 1,\dots,m+1,\\
b_j(z) = \displaystyle\sum_{k=j+1}^{m}(z-\lambda_0^*)^{k-1-j}b_k, & j = 1,\dots,m,\\
q_j(z) = (z-\lambda_0^*)^{m-j}Q_0(z) & \\
\qquad\qquad + \displaystyle\sum_{k=1}^{m-j}(z-\lambda_0^*)^{k-1}g_{j+k,m}, & j = 1,\dots,m,
\end{cases}
$$

and

$$
(3.7)\qquad Q_0(z) = (z-\lambda_0)\langle R_0(z)\chi_0, \chi_{-1}\rangle_0 + g_{m,m+1}.
$$

The operator $R^\infty(z)$ has the following properties for $z, w \in \mathbf{C}\backslash\mathbf{R}$; here e_j stands for the element of $\mathbf{C}^m$ with the j-th entry equal to 1 and all the other entries 0.

(a) $R^\infty(z)$ *is a bounded operator in* Π_m;

(b) $R^\infty(z) - R^\infty(w) = (z-w)R^\infty(z)R^\infty(w)$;

(c) $R^\infty(z)^* = R^\infty(z^*)$, *the adjoint is with respect to the indefinite inner product in* Π_m;

(d) $\ker R^\infty(z) = \{0\} \oplus \operatorname{span}\{e_1\} \oplus \{0\}$.

The properties (a)–(c) imply that $R^\infty(z)$ is the resolvent of a self-adjoint relation H^∞ in Π_m:

$$
(3.8)\qquad H^\infty = \{\{R^\infty(z)f, (I + zR^\infty(z))f\} \mid f \in \Pi_m\}.
$$

The set on the right-hand side is independent of $z \in \rho(A_0)$. The multivalued part (or the geometric eigenspace corresponding to the eigenvalue ∞) of H^∞ is given by the expression on the right-hand side of (d), and $\rho(H^\infty) = \rho(A_0)$ since the resolvent of H^∞ is holomorphic there. In particular, the set $\mathbf{C}\backslash\mathbf{R}$ belongs to $\rho(H^\infty)$.

We explain the relations between the spaces and the operators with the following diagram:

$$
\begin{array}{ccccc}
\mathcal{P}_m & \xleftarrow{\ \tau^{-1}\ } & \tau(\mathcal{P}_m) \subset \Pi_m & \qquad & \Pi_m \\
R_0(z)\Big\downarrow & & \Big\downarrow \tau R_0(z)\tau^{-1} & & \Big\downarrow R^\infty(z) = (H^\infty - z)^{-1} \\
\mathcal{P}_m & \xrightarrow[\ \tau\ ]{} & \Pi_m & & \Pi_m
\end{array}
$$

If we choose $z = \lambda_0$ in (3.8) we obtain the following explicit representation of H^∞:

$$H^\infty = \left\{ \left\{ \begin{pmatrix} h + c_{m+1}\chi_0 \\ (b_j)_{j=1}^m \\ (c_j)_{j=1}^m \end{pmatrix}, \begin{pmatrix} A_0 h + \lambda_0 c_{m+1}\chi_0 \\ \begin{pmatrix} \lambda_0^* b_1 + b_0 \\ (\lambda_0^* b_j + b_{j-1} - g_{jm} c_{m+1})_{j=2}^m \end{pmatrix} \\ (\lambda_0 c_j + c_{j+1})_{j=1}^m \end{pmatrix} \right\} \right.$$
$$\left| \begin{array}{l} h \in \operatorname{dom} A_0,\ c_1 = 0, \\ c_{j+1}, b_{j-1} \in \mathbf{C},\ j = 1, \dots, m \\ b_m = \langle h, \chi_{-1} \rangle + g_{m,m+1} c_{m+1}, \end{array} \right\}.$$

The relation H^∞ is a kind of "lifting" of A_0 from $\mathcal{H}_0$ to the Pontryagin space Π_m.

3.4

The symmetry S in Π_m. First we define a one-dimensional restriction S of H^∞ in the space Π_m by the condition, that its defect subspace $\ker(S^* - \lambda_0) = \operatorname{ran}(S - \lambda_0^*)^\perp$ is spanned by the vector

$$\tau\varepsilon_1 = \begin{pmatrix} 0 \\ 0 \\ e_1 \end{pmatrix}.$$

Therefore S and its adjoint S^* in Π_m are given by the relations

$$S = H^\infty \cap \{\{\tau\varepsilon_1, \lambda_0\tau\varepsilon_1\}\}^*, \quad S^* = H^\infty \dot{+} \operatorname{span}\{\{\tau\varepsilon_1, \lambda_0\tau\varepsilon_1\}\}. \tag{3.9}$$

Since

$$\left\langle \begin{pmatrix} \varphi \\ (b_j)_{j=1}^m \\ (c_j)_{j=1}^m \end{pmatrix}, \begin{pmatrix} 0 \\ 0 \\ e_1 \end{pmatrix} \right\rangle_{\Pi_m} = b_1 + \sum_{j=1}^m c_j g_{1j},$$

S is the set of elements

$$\left\{ \begin{pmatrix} h + c_{m+1}\chi_0 \\ (b_j)_{j=1}^m \\ (c_j)_{j=1}^m \end{pmatrix}, \begin{pmatrix} A_0 h + \lambda_0 c_{m+1}\chi_0 \\ \begin{pmatrix} \lambda_0^* b_1 + b_0 \\ (\lambda_0^* b_j + b_{j-1} - g_{jm} c_{m+1})_{j=2}^m \end{pmatrix} \\ (\lambda_0 c_j + c_{j+1})_{j=1}^m \end{pmatrix} \right\} \in H^\infty$$

for which

$$b_0 + (\lambda_0 - \lambda_0^*) \sum_{j=2}^m g_{1j} c_j + \sum_{j=1}^m g_{1j} c_{j+1} = 0. \tag{3.10}$$

Note that S is a nondensely defined operator since it is a restriction of the relation H^∞ and the multivalued part $H^\infty(0)$ is given by b_0, which must vanish because of (3.10). Without loss of generality we may and shall assume that S is simple (see Subsection 3.6 below).

The family of self-adjoint operators H^t, $t \in \mathbf{R}$, will be defined by M.G. Krein's formula (2.11) as the set of self-adjoint extensions of the symmetric operator S with $A = H^\infty$ as fixed canonical self-adjoint extension. In order to make this explicit, we have to introduce the Q-function corresponding to S and $A = H^\infty$ and the corresponding Γ-field. Define for nonreal z the Γ-field $\Gamma_z : \mathbf{C} \to \ker(S^* - z)$:

$$\text{(3.11)} \qquad \begin{aligned} \Gamma_{\lambda_0} 1 &= \begin{pmatrix} 0 \\ 0 \\ e_1 \end{pmatrix}, \\ \Gamma_z 1 &= (I + (z - \lambda_0) R^\infty(z)) \Gamma_{\lambda_0} 1 = \begin{pmatrix} (z - \lambda_0)^m R_0(z) \chi_{-1} \\ ((z - \lambda_0)^m q_j(z))_{j=1}^m \\ ((z - \lambda_0)^{j-1})_{j=1}^m \end{pmatrix}. \end{aligned}$$

Note that $\Gamma_{\lambda_0} 1 = \tau \varepsilon_1$. As a Q-function associated with S and H^∞ we can choose

$$\text{(3.12)} \qquad \begin{aligned} Q(z) &= (z - \lambda_0^*) \Gamma_{\lambda_0}^* \Gamma_z 1 - i g_{11} \operatorname{Im} \lambda_0 \\ &= (z - \lambda_0) \Gamma_{z^*}^* \Gamma_{\lambda_0} 1 + i g_{11} \operatorname{Im} \lambda_0. \end{aligned}$$

By formula (3.7) we introduced the function $Q_0(z)$:

$$Q_0(z) = (z - \lambda_0) \langle R_0(z) \chi_0, \chi_{-1} \rangle_0 + g_{m,m+1}.$$

We recall that $\operatorname{Im} g_{m,m+1} = \operatorname{Im} \lambda_0 \langle \chi_0, \chi_0 \rangle_0$. It follows from this representation that $Q_0(z)$ is the Q-function of a symmetric densely defined operator S_0 and the operator A_0, where S_0 is the one-dimensional restriction of A_0 to the domain

$$\text{(3.13)} \qquad \operatorname{dom} S_0 = \{ f \in \operatorname{dom} A_0 \mid \langle f, \chi_{-1} \rangle_0 = \langle f, R_0(\lambda_0)^m \chi \rangle_0 = 0 \}.$$

If $F(z)$ is an analytic function, we denote by $F^*(z)$ the function $F^*(z) := F(z^*)^*$; the function $F(z)$ is called *self-adjoint* if $F(z) = F^*(z)$. In the sequel we denote by $p_{2m-1}(z)$ the following polynomial of degree at most $2m - 1$:

$$\text{(3.14)} \qquad \begin{aligned} p_{2m-1}(z) &= \sum_{j=2}^{m} (z - \lambda_0)^m (z - \lambda_0^*)^{j-1} g_{jm} \\ &\quad + \sum_{j=1}^{m} (z - \lambda_0)^{j-1} (z - \lambda_0^*) g_{1j} - i \operatorname{Im} \lambda_0 \, g_{11}. \end{aligned}$$

The next proposition shows that this polynomial is self-adjoint.

Proposition 3.1 *The Q-function $Q(z)$ given by* (3.12) *is analytic in $\rho(A_0) = \rho(H^\infty)$ and it can be represented as*

$$\text{(3.15)} \qquad Q(z) = (z - \lambda_0)^m (z - \lambda_0^*)^m Q_0(z) + p_{2m-1}(z),$$

where $p_{2m-1}(z)$ is the self-adjoint polynomial given by (3.14).

Proof: On account of (3.11) we have

$$\Gamma_{\lambda_0}^*\Gamma_z 1 = (z-\lambda_0)^m q_1(z) + e_1^* G \begin{pmatrix} 1 \\ (z-\lambda_0) \\ \vdots \\ (z-\lambda_0)^{m-1} \end{pmatrix}.$$

We insert this expression in (3.12) and obtain

$$Q(z) = (z-\lambda_0)^m (z-\lambda_0^*)^m Q_0(z) + \widehat{p}_{2m-1}(z),$$

where

$$\begin{aligned}\widehat{p}_{2m-1}(z) &= \sum_{j=2}^{m} (z-\lambda_0)^m (z-\lambda_0^*)^{j-1} g_{jm} \\ &\quad + \sum_{j=1}^{m} (z-\lambda_0)^{j-1}(z-\lambda_0^*) g_{1j} - i\operatorname{Im}\lambda_0 g_{11}.\end{aligned}$$

Thus the polynomial $\widehat{p}_{2m-1}(z)$ coincides with the polynomial (3.14). This proves the desired representation of the Q-function. The self-adjointness of the polynomial $p_{2m-1}(z)$ follows from the self-adjointness of $Q(z)$ and of $Q_0(z)$. □

Proposition 3.2 *The polynomial* $p_{2m-1}(z)$ *in* (3.16) *can also be written as*

$$\begin{aligned}p_{2m-1}(z) &= \sum_{j=1}^{m-1} (z-\lambda_0)^j (z-\lambda_0^*)^j g_{j,j+1} \\ &\quad + \sum_{j=1}^{m} (z-\lambda_0)^j (z-\lambda_0^*)^{j-1} g_{jj} + i\operatorname{Im}\lambda_0 g_{11}.\end{aligned} \tag{3.16}$$

Proof: For $k = 1, \dots, m$, we introduce the polynomials

$$\begin{aligned} r_{2k-1}(z) &:= \sum_{j=2}^{k} (z-\lambda_0)^k (z-\lambda_0^*)^{j-1} g_{jk}, \\ s_k(z) &:= \sum_{j=1}^{k} (z-\lambda_0)^{j-1}(z-\lambda_0^*) g_{1j} - i\operatorname{Im}\lambda_0 g_{11}\end{aligned}$$

and $p_{2k-1}(z) := r_{2k-1}(z) + s_k(z)$; for $k = m$ the last polynomial $p_{2m-1}(z)$ coincides with the one in (3.14). We shall show that the polynomials $p_{2k-1}(z)$ satisfy the following recurrence relation for $k = 1, 2, \dots, m-1$:

$$\begin{aligned} p_{2(k+1)-1}(z) &= p_{2k-1}(z) + (z-\lambda_0)^{k+1}(z-\lambda_0^*)^k g_{k+1,k+1} \\ &\quad + (z-\lambda_0)^k (z-\lambda_0^*)^k g_{k,k+1}.\end{aligned} \tag{3.17}$$

Since $p_1(z) = zg_{11} - \operatorname{Re}\lambda_0\, g_{11}$, the proposition is an easy consequence of (3.17).

In order to prove the relations (3.17) we express $r_{2(k+1)-1}(z)$ through $r_{2k-1}(z)$. To this end we transform a generic summand in $r_{2(k+1)-1}(z)$ as follows:

$$\begin{aligned}(z &- \lambda_0)^{k+1}(z-\lambda_0^*)^{j-1}g_{j,k+1}\\ &= (z-\lambda_0)^k(z-\lambda_0^*)^j g_{j+1,k} + (z-\lambda_0)^k(z-\lambda_0^*)^j(g_{j,k+1}-g_{j+1,k})\\ &\quad -(\lambda_0-\lambda_0^*)(z-\lambda_0)^k(z-\lambda_0^*)^{j-1}g_{j,k+1}\\ &= (z-\lambda_0)^k(z-\lambda_0^*)^j g_{j+1,k} + (z-\lambda_0)^k(\lambda_0-\lambda_0^*)\\ &\quad ((z-\lambda_0^*)^j g_{j+1,k+1} - (z-\lambda_0^*)^{j-1}g_{j,k+1}).\end{aligned}$$

Here in the last step we have used the relation (3.3). Now, substituting the last expression into the sum for $r_{2(k+1)-1}(z)$ and using again (3.3) we obtain

$$\begin{aligned}r_{2(k+1)-1}(z) &= r_{2k-1}(z) + (z-\lambda_0)^{k+1}(z-\lambda_0^*)^k g_{k+1,k+1}\\ &\quad + (z-\lambda_0)^k(z-\lambda_0^*)^k g_{k,k+1}\\ &\quad - (z-\lambda_0)^k(z-\lambda_0^*)(g_{2k}+(\lambda_0-\lambda_0^*)g_{2,k+1})\\ &= r_{2k-1}(z) - (z-\lambda_0)^k(z-\lambda_0^*)g_{1,k+1}.\end{aligned}$$

The polynomials s_k satisfy the relations

$$s_{k+1}(z) = s_k(z) + (z-\lambda_0)^k(z-\lambda_0^*)g_{1,k+1},$$

and we obtain finally

$$\begin{aligned}p_{2(k+1)-1}(z) &= r_{2(k+1)-1}(z) + s_{k+1}(z)\\ &= r_{2k-1}(z) + s_k(z) + (z-\lambda_0)^{k+1}(z-\lambda_0^*)^k g_{k+1,k+1}\\ &\quad + (z-\lambda_0)^k(z-\lambda_0^*)^k g_{k,k+1}.\end{aligned}$$

These relations coincide with (3.17) since $p_{2k-1}(z) = r_{2k-1}(z) + s_k(z)$ by definition. □

It is easy to see that if $p_{2m-1}(z)$ is a polynomial of degree $2m-1$ then it is self-adjoint if and only if it is of the form

$$p_{2m-1}(z) = \sum_{j=0}^{m-1}(z-\lambda_0)^j(z-\lambda_0^*)^j b_j + \sum_{j=1}^{m}(z-\lambda_0)^j(z-\lambda_0^*)^{j-1}c_j,$$

where the c_j's are real and $\operatorname{Im} b_j = c_{j+1}\operatorname{Im}\lambda_0$.

We have mentioned already that the real numbers g_{jj}, $\operatorname{Re} g_{j,j+1}$, $j=1,\dots,m$, can be choosen as parameters. These numbers except for $\operatorname{Re} g_{m,m+1}$ determine the inner product on Π_m; the representation of the polynomial $p_{2m-1}(z)$ in the form (3.16) shows that this amounts to the choice of the polynomial $p_{2m-1}(z)$. The number $\operatorname{Re} g_{m,m+1}$ enters into H^∞ and S, and therefore in the polynomial $p_{2m}(z)$ in (4.2) below which is used to describe the Straus extensions of S corresponding to H^t. In the following $p_{2m-1}(z)$ is always the self-adjoint polynomial given by (3.14) or (3.16).

3.5

Now M.G. Krein's formula (2.11) describes the canonical self-adjoint extensions H^t of S through their resolvents:

$$(H^t - z)^{-1} = (H^\infty - z)^{-1} - \Gamma_z (Q(z) + t)^{-1} \Gamma_{z^*}^*, \tag{3.18}$$

which holds for all $z \in \rho(H^\infty)$ such that $(Q(z)+t)^{-1} \neq 0$; t is a real parameter, and formally for $t = \infty$ we obtain the self-adjoint extension H^∞. A straightforward calculation (see [S3]) gives the following explicit expression for the self-adjoint operator extension H^t, $t \in \mathbf{R}$:

$$H^t = S \dot{+} \operatorname{span}\left\{\left\{\begin{pmatrix} 0 \\ 0 \\ e_1 \end{pmatrix}, \begin{pmatrix} 0 \\ -(t + i\, g_{11} \operatorname{Im} \lambda_0) e_1 \\ \lambda_0 e_1 \end{pmatrix}\right\}\right\}$$

$$= \left\{\left\{\begin{pmatrix} h + c_{m+1}\chi_0 \\ (b_j)_{j=1}^m \\ (c_j)_{j=1}^m \end{pmatrix}, \begin{pmatrix} A_0 h + \lambda_0 c_{m+1}\chi_0 \\ \begin{pmatrix} \lambda_0^* b_1 + b_0 \\ (\lambda_0^* b_j + b_{j-1} - g_{jm} c_{m+1})_{j=2}^m \end{pmatrix} \\ (\lambda_0 c_j + c_{j+1})_{j=1}^m \end{pmatrix}\right\} \right.$$

$$\left. \begin{array}{|l} h \in \operatorname{dom} A_0,\ c_1, \dots, c_{m+1}, b_0, \dots b_m \in \mathbf{C}, \\ b_m = \langle h, \chi_{-1}\rangle + g_{m,m+1} c_{m+1}, \\ b_0 + (\lambda_0 - \lambda_0^*) \sum_{j=2}^m g_{1j} c_j + \sum_{j=1}^m g_{1j} c_{j+1} = -(t + i \operatorname{Im} \lambda_0 g_{11}) c_1 \end{array} \right\}.$$

(3.19)

If we set

$$u = \tau\varepsilon_1 = \begin{pmatrix} 0 \\ 0 \\ e_1 \end{pmatrix}, \quad v = \lambda_0 u = \begin{pmatrix} 0 \\ 0 \\ \lambda_0 e_1 \end{pmatrix}, \quad w = \begin{pmatrix} 0 \\ e_1 \\ 0 \end{pmatrix},$$

then H^t, $t \in \mathbf{R}$, can be written as

$$H^t = S \dot{+} \operatorname{span}\{\{u, v - (t + Q(\lambda_0))w\}\}, \quad t \in \mathbf{R}, \tag{3.20}$$

or as

$$H^t = S^* \cap \{\{u, v - (t + Q(\lambda_0))w\}\}^*, \quad S^* = H^\infty \dot{+} \operatorname{span}\{\{u, v\}\}.$$

Note that $Q(\lambda_0) = i \operatorname{Im} \lambda_0\, g_{11}$. Hence the subspace H^t is the set of all elements $\{h_\infty + cu, f_\infty + cv\} \in S^*$, where $\{h_\infty, f_\infty\} \in H^\infty$ and $c \in \mathbf{C}$ satisfy

$$\Gamma_{\lambda_0}^* (f_\infty - \lambda_0^* h_\infty) = -(t + Q(\lambda_0))c. \tag{3.21}$$

Finally, we prove some statements about the spectrum of the operators H^t, $t \in \mathbf{R}$. Evidently, since S is a simple symmetric operator with defect index (1, 1) and the set $\rho(A_0)$ consists only of points of regular type of S, all the eigenvalues of H^t in $\rho(A_0)$ have geometric multiplicity one. Recall that $\sigma^0(H^t)$ denotes the spectrum of the restriction $H^t|_{\mathcal{L}^0(H^t)}$ of H^t to its m-dimensional nonpositive invariant subspace $\mathcal{L}^0(H^t)$, $\sigma^0(H^t) \subset \mathbf{R} \cup \mathbf{C}^+$. Further, we set

$$\mathcal{L}^0_\mu(H^t) := \mathcal{L}^0(H^t) \cap \mathcal{L}_\mu(H^t).$$

In the sequel a chain of root vectors of an operator or a linear relation at an eigenvalue always begins with the eigenvector corresponding to this eigenvalue.

Theorem 3.3 *Let Γ_z be the Γ-field associated with H^∞ and S, and let H^t, $t \in \mathbf{R}$, be the self-adjoint extension of S given by* (3.18). *Then the following statements hold:*

(i) *Suppose that $\lambda \in \rho(A_0)$. Then $\lambda \in \sigma_p(H^t)$ and the corresponding root space $\mathcal{L}_\lambda(H^t)$ has dimension k if and only if λ is a zero of order k of the function $Q(z) + t$. The elements*

$$\Gamma_\lambda 1,\ \frac{1}{1!}\Gamma_\lambda^{(1)} 1,\ \ldots,\ \frac{1}{(k-1)!}\Gamma_\lambda^{(k-1)} 1 \tag{3.22}$$

form a chain of root vectors of H^t at λ.

(ii) *$\mu \in \mathbf{R} \cap \sigma^0(H^t)$, that is, μ is a real eigenvalue of nonpositive type of H^t, if and only if it is a generalized zero of nonpositive type of the function $Q(z) + t$; in this case, if (π_+, π_0, π_-) is the signature of the inner product on the corresponding root space $\mathcal{L}_\mu(H^t)$ and π is the multiplicity of the generalized zero μ of nonpositive type of the function $Q(z) + t$, then $\pi = \pi_- + \pi_0$ and the subspace $\mathcal{L}^0_\mu(H^t)$ is spanned by the elements*

$$\Gamma_\mu 1,\ \frac{1}{1!}\Gamma_\mu^{(1)} 1,\ \ldots,\ \frac{1}{(\pi-1)!}\Gamma_\mu^{(\pi-1)} 1 \tag{3.23}$$

which form a chain of root vectors of H^t at μ.

(iii) *If the interval (a, b) is contained in $\rho(A_0)$, then the total multiplicity of the eigenvalues of H^t in (a, b) is at most $2m+1$. If two consecutive eigenvalues are algebraically simple then their types are alternating.*

Before we prove Theorem 3.3 we make some simple observations. Here we write $A := H^\infty$. Firstly, it is easy to check that the elements $h_0, h_1, \ldots, h_{k-1}$ form a chain of root vectors of H^t at λ if and only if for $z \in \rho(H^t)$ the relation

$$(H^t - z)^{-1} h_{k-1} = -\frac{h_{k-1}}{z-\lambda} - \frac{h_{k-2}}{(z-\lambda)^2} - \cdots - \frac{h_0}{(z-\lambda)^k} \tag{3.24}$$

holds. Secondly, if $z, \zeta \in \mathbf{C}$, $z \neq \zeta$, and $s \in \mathbf{R}$ the identity

$$\frac{1}{(s-z)(s-\zeta)^{k-1}} - \frac{s-\zeta}{(s-z)(z-\zeta)^k}$$
$$= -\frac{1}{(s-\zeta)^{k-1}(z-\zeta)} - \frac{1}{(s-\zeta)^{k-2}(z-\zeta)^2} - \cdots - \frac{1}{(z-\zeta)^k}$$

can easily be checked. If Γ_ζ denotes again the Γ-field corresponding to $A = H^\infty$, it follows for $z, \zeta \in \rho(A)$ that

$$\begin{aligned} &(A-z)^{-1}(A-\zeta)^{-(k-1)}\Gamma_\zeta \\ (3.25) \quad &- \frac{1}{(z-\zeta)^k}(I + (z-\zeta)(A-z)^{-1})\Gamma_\zeta = -\frac{1}{z-\zeta}(A-\zeta)^{-(k-1)}\Gamma_\zeta \\ &- \frac{1}{(z-\zeta)^2}(A-\zeta)^{-(k-2)}\Gamma_\zeta - \cdots - \frac{1}{(z-\zeta)^k}\Gamma_\zeta. \end{aligned}$$

Observing (2.5) and that $\Gamma_\zeta^{(j)} = j!(A-\zeta)^{-j}\Gamma_\zeta$ we obtain

$$\begin{aligned} &(A-z)^{-1}\frac{1}{(k-1)!}\Gamma_\zeta^{(k-1)} - \frac{1}{(z-\zeta)^k}\Gamma_z = -\frac{1}{z-\zeta}\frac{1}{(k-1)!}\Gamma_\zeta^{(k-1)} \\ (3.26) \quad & \\ &- \frac{1}{(z-\zeta)^2}\frac{1}{(k-2)!}\Gamma_\zeta^{(k-2)} - \cdots - \frac{1}{(z-\zeta)^k}\Gamma_\zeta. \end{aligned}$$

Finally, the relation between $Q(z)$ and Γ_z implies

$$\begin{aligned} \Gamma_{z^*}^* \frac{\Gamma_\zeta^{(k-1)}1}{(k-1)!} &= \frac{1}{(z-\zeta)^k}\Big(Q(z) - Q(\zeta) - (z-\zeta)\frac{1}{1!}Q'(\zeta) \\ (3.27) \quad & \\ &- \cdots - (z-\zeta)^{k-1}\frac{1}{(k-1)!}Q^{(k-1)}(\zeta)\Big). \end{aligned}$$

Proof of Theorem 3.3: (i) Suppose that $\lambda \in \rho(Q)$ is a zero of the function $Q(z)+t$ of order k. The elements $\Gamma_\lambda 1, \Gamma_\lambda^{(1)}1, \ldots, \Gamma_\lambda^{(k-1)}1$ span a k-dimensional subspace. The relation (3.26) implies

$$\begin{aligned} &(A-z)^{-1}\frac{\Gamma_\lambda^{(k-1)}1}{(k-1)!} - \frac{\Gamma_z 1}{(z-\lambda)^k} \\ (3.28) \quad & \\ &= -\frac{1}{z-\lambda}\frac{\Gamma_\lambda^{(k-1)}1}{(k-1)!} - \frac{1}{(z-\lambda)^2}\frac{\Gamma_\lambda^{(k-2)}1}{(k-2)!} - \cdots - \frac{1}{(z-\lambda)^k}\Gamma_\lambda 1. \end{aligned}$$

Further, observing that $Q'(\lambda) = \cdots = Q^{(k-1)}(\lambda) = 0$, from (3.27) we obtain

$$\Gamma_{z^*}^*\Gamma_\lambda^{(k-1)}1 = (k-1)!\frac{Q(z)-Q(\lambda)}{(z-\lambda)^k}.$$

Now we apply M.G. Krein's formula (3.18) and obtain with $t = -Q(\lambda)$ and (3.26)

$$(H^t - z)^{-1}\frac{\Gamma_\lambda^{(k-1)}1}{(k-1)!} = (A-z)^{-1}\frac{\Gamma_\lambda^{(k-1)}1}{(k-1)!} - \Gamma_z(Q(z)+t)^{-1}\frac{\Gamma_{z^*}^*\Gamma_\lambda^{(k-1)}1}{(k-1)!}$$

$$= (A-z)^{-1}\frac{\Gamma_\lambda^{(k-1)}1}{(k-1)!} - \frac{1}{(z-\lambda)^k}\Gamma_z 1 = -\frac{1}{z-\lambda}\frac{\Gamma_\lambda^{(k-1)}1}{(k-1)!}$$

$$-\frac{1}{(z-\lambda)^2}\frac{\Gamma_\lambda^{(k-2)}1}{(k-2)!} - \cdots - \frac{1}{(z-\lambda)^k}\Gamma_\lambda 1.$$

If we compare this relation with (3.24) it follows that the elements

$$\Gamma_\lambda 1,\ \frac{1}{1!}\Gamma_\lambda^{(1)}1,\ \ldots,\ \frac{1}{(k-1)!}\Gamma_\lambda^{(k-1)}1$$

form a chain of root vectors of H^t at λ.

(ii) Since $Q(z) \in N_m$, for $t \in \mathbf{R}$ also the function $-(Q(z)+t)^{-1}$ belongs to the class N_m. Without loss of generality we can suppose that Re $Q(\lambda_0) = 0$ (otherwise the parameter t is changed accordingly). Now we consider the Q-function $Q_t(z)$ which corresponds to the symmetric operator S and its self-adjoint extension H^t, again chosen such that Re $Q_t(\lambda_0) = 0$:

$$Q_t(z) = -i\,\mathrm{Im}\,\lambda_0\,\Gamma_{\lambda_0}^*\Gamma_{\lambda_0} + (z-\lambda_0^*)\Gamma_{\lambda_0}^*(I + (z-\lambda_0)(H^t-z)^{-1})\Gamma_{\lambda_0}.$$

It was shown in [HLS], Proposition 3.2, that the relation

$$-\frac{1}{Q(z)+t} = -\frac{t}{t^2 + (\mathrm{Im}\,Q(\lambda_0))^2} + \frac{1}{t^2+(\mathrm{Im}\,Q(\lambda_0))^2}Q_t(z)$$

holds. This shows that in the operator representation of the function $-\frac{1}{Q(z)+t}$ as a self-adjoint operator the operator H^t can be chosen. The generalized zeros of nonpositive type of $-\frac{1}{Q(z)+t}$ are the real eigenvalues of H^t with a nonpositive eigenvector and the multiplicity of such a generalized zero μ of nonpositive type coincides with the dimension of $\mathcal{L}_\mu^0(H^t)$ (see [KL3]). The elements in (3.23) exist according to Lemma 2.1 as nontangential limits, and their span is a nonpositive subspace. The proof that these elements form a Jordan chain of H^t at μ is the same as in (i): Develop $\Gamma_{z^*}^*\Gamma_\lambda$ for Im $\lambda > 0$ and then take the nontangential limit $\lambda \hat{\to} \mu$ to obtain the formula

$$(H^t-z)^{-1}\frac{\Gamma_\mu^{(k-1)}1}{(k-1)!}$$

$$= -\frac{1}{z-\mu}\frac{\Gamma_\mu^{(k-1)}1}{(k-1)!} - \frac{1}{(z-\mu)^2}\frac{\Gamma_\mu^{(k-2)}1}{(k-2)!} - \cdots - \frac{1}{(z-\mu)^k}\Gamma_\mu 1.$$

Hence for z in a sector $0 < \frac{\pi}{2} - \omega < \arg(z-\mu) < \frac{\pi}{2} + \omega$,

$$(H^t-z)^{-1}h_{k-1} = -\frac{h_{k-1}}{z-\mu} - \frac{h_{k-2}}{(z-\mu)^2} - \cdots - \frac{h_0}{(z-\mu)^k},$$

which is the analogue of (3.24).

(iii) By Proposition 3.1, the Q-function is of the form $Q(z) = (z-\lambda_0)^m (z-\lambda_0^*)^m Q_0(z) + p_{2m-1}(z)$ with a self-adjoint polynomial $p_{2m-1}(z)$ of degree at most $2m-1$ and $Q_0(z) \in N_0$. If we use the integral representation (2.4) for $Q_0(z)$, the relation

$$\frac{d^{2m+1}}{dz^{2m+1}}\left(\frac{(z-\lambda_0)^m(z-\lambda_0^*)^m}{s-z}\right) = (2m+1)!\frac{|s-\lambda_0|^{2m}}{(s-z)^{2m+2}}$$

and observe that $\widehat{Q}_0(z)$ is holomorphic on the interval (a, b) it follows that

$$(Q(z)+t)^{(2m+1)} = (2m+1)!\left(\beta + \int_{-\infty}^{+\infty} \frac{|s-\lambda_0|^{2m}(1+s^2)}{(s-z)^{2m+2}}\, d\sigma(s)\right)$$

which is positive on (a, b). Therefore the function $Q(z)+t$ can have at most $2m+1$ zeros in (a, b), and if these zeros are simple the signs of their derivatives are alternating. On the other hand, if λ is a real simple eigenvalue of H^t in $\rho(A_0)$, it is a zero of $Q(z)+t$, $\Gamma_\lambda 1$ is the corresponding eigenvector and the definition (2.6) of the Q-function implies

$$Q'(\lambda) = \langle \Gamma_\lambda 1, \Gamma_\lambda 1\rangle_{\Pi_m}.$$

□

Remark 3.4 We mention, that the last statement about the alteration of the signs of the eigenvectors corresponding to eigenvalues in (a, b) can be generalized to the case if the eigenvalues are not simple. We do not formulate this statement, but mention that it can be explained by a perturbation type argument starting from the situation that all the eigenvalues in (a, b) are simple and that under the perturbation some of the eigenvalues meet, say at λ. Then the number of positive (negative, respectively) squares of the algebraic eigenspace at λ is given by the number of positive (negative, respectively) type eigenvalues which have met.

3.6

In the previous subsections we have constructed Π_m, H^∞, S, H^t and the Γ-field Γ_z from the data $\{\mathcal{H}_0, A_0, \chi\}$. Recall $\chi_{-1} = (A_0-\lambda_0)^{-m}\chi$.

Proposition 3.5 *If* $\{\mathcal{H}_0, A_0, \chi\}$ *is minimal in the sense that*

$$\mathcal{H}_0 = \overline{\operatorname{span}}\,\{(A_0-z)^{-1}\chi_{-1} \mid z \in \rho(A_0)\},$$

then S is simple in Π_m in the sense that

$$\Pi_m = \overline{\operatorname{span}}\,\{\ker(S^*-z) \mid z \in \rho(A_0)\}.$$

In particular, S has no nonreal eigenvalue.

Proof: Assume $\begin{pmatrix} u \\ \beta \\ \gamma \end{pmatrix}$ with $u \in \mathcal{H}_0$, $\beta, \gamma \in \mathbf{C}^m$, is orthogonal in Π_m to all $\Gamma_z 1$, $z \in \rho(A_0)$, given by (3.11):

$$\frac{1}{(z-\lambda_0)^m}\left\langle \Gamma_z 1, \begin{pmatrix} u \\ \beta \\ \gamma \end{pmatrix}\right\rangle_{\Pi_m} = 0, \; z \in \rho(A_0)\backslash\{\lambda_0\}.$$

Expanding the left-hand side around $z = \lambda_0$ we find $\beta = -G^*\gamma$ and with $\gamma = (\gamma_j)_{j=1}^m$

$$(3.29) \qquad \langle R_0(z)\chi_{-1}, u\rangle_0 + \sum_{j=1}^m \gamma_j^* q_j(z) = 0, \; z \in \rho(A_0).$$

Choose m positive numbers $0 < y_1 < \ldots < y_l < \ldots < y_m$. Then the Vandermonde matrix $M := ((-iy_l)^{m-j})_{j,l=1}^m$ is invertible and for $1 \le k \le m$ we define the m-vector $b^k = (b_l^k)_{l=1}^m$ by $b^k = M^{-1}e_k$. Further, we set for all natural numbers $n \ge n_0$,

$$z_{ln} = \lambda_0^* - iny_l, \qquad c_{ln}^k = \frac{b_l^k(z_{ln}-\lambda_0^*)^{m-1}}{n^{m-k}q_1(z_{ln})}, \quad l = 1, 2, \ldots, m,$$

where the natural number n_0 is chosen such that the zeros of $q_1(z)$ lie within the circle with radius $|\lambda_0^* - in_0y_1|$ around the origin. We claim that in $\mathcal{H}_0 \oplus \mathbf{C}^m$ as $n \to \infty$,

$$\sum_{l=1}^m c_{ln}^k \begin{pmatrix} R_0(z_{ln})\chi_{-1} \\ (q_j(z_{ln}))_{j=1}^m \end{pmatrix} \longrightarrow \begin{pmatrix} 0 \\ e_k \end{pmatrix}, \quad k = 1, 2, \ldots, m.$$

The claim and (3.29) imply that $\gamma_k = 0$, $k = 1, 2, \ldots, m$, and $\langle R_0(z)\chi_{-1}, u\rangle_{\mathcal{H}_0} = 0$ for all $z \in \mathbf{C}\backslash\mathbf{R}$. Hence $\gamma = 0$ and $\beta = -G^*\gamma = 0$, and the minimality hypothesis implies that $u = 0$, and so the proposition holds true.

It remains to check the claim. We have for $c_r \in \mathbf{C}$ and $z_r \in \rho(A_0)\backslash\mathbf{R}$

$$\left\|\sum_r c_r R_0(z_r)\chi_{-1}\right\|_0^2 = \sum_{r,s} c_r c_s^* \frac{Q_0(z_r) - Q_0(z_s)^*}{z_r - z_s^*} = \left\|\sum_r c_r \Gamma_{z_r} 1\right\|^2$$

$$\le \left(\sum_r |c_r| \|\Gamma_{z_r} 1\|\right)^2 = \left(\sum_r |c_r| \sqrt{\frac{\operatorname{Im} Q(z_r)}{\operatorname{Im} z_r}}\right)^2$$

and hence with the above notation

$$\left\|\sum_{l=1}^m c_{ln}^k R_0(z_{ln})\chi_{-1}\right\|_0 \le \sum_{l=1}^m |c_{ln}^k| \sqrt{\frac{\operatorname{Im} Q_0(z_{ln})}{\operatorname{Im} z_{ln}}}$$

$$\le \sum_{l=1}^m \frac{|b_l^k|}{n^{m-k}\,|Q_0(z_{ln}) + o(\frac{1}{n})|} \sqrt{\frac{\operatorname{Im} Q_0(z_{ln})}{\operatorname{Im} z_{ln}}}.$$

For $k = 1, 2, \dots m-1$, the right-hand side tends to 0 as $n \to \infty$ because of (2.8) and (2.9). (Indeed, they imply

$$\lim_{z \widehat{\to} \infty} z \operatorname{Im} Q_0(z) = \infty, \quad \lim_{z \widehat{\to} \infty} \frac{1}{z} Q_0(z) = 0,$$

where $z \widehat{\to} \infty$ means that $z \to \infty$ within a sector $0 < \frac{\pi}{2} - \omega < \arg z < \frac{\pi}{2} + \omega$.)
For $k = m$ we rewrite the right-hand side as

$$\sum_{l=1}^{m} \frac{|b_l^m|}{|1 + o(\frac{1}{nQ_0(z_{ln})})|} \sqrt{\frac{\operatorname{Im}\,(-Q_0(z_{ln}))^{-1}}{\operatorname{Im}\, z_{ln}}}$$

and this expression also tends to 0 as $n \to \infty$. It follows that for $k = 1, 2, \dots m$,

$$\sum_{l=1}^{m} c_{ln}^k R_0(z_{ln}) \chi_{-1} \longrightarrow 0$$

in $\mathcal{H}_0$ as $n \to \infty$. From

$$\begin{aligned} \frac{q_j(z)}{(z - \lambda_0^*)^{m-j}} &= Q_0(z) + \sum_{p=j+1}^{m} \frac{g_{pm}}{(z - \lambda_0^*)^{m-p+1}} \\ &= \frac{q_1(z)}{(z - \lambda_0^*)^{m-1}} - \sum_{p=2}^{j} \frac{g_{pm}}{(z - \lambda_0^*)^{m-p+1}} \end{aligned}$$

we obtain that

$$\frac{q_j(z)}{q_1(z)} = \frac{1}{(z - \lambda_0^*)^{j-1}} - \frac{1}{q_1(z)} \sum_{p=2}^{j} \frac{g_{pm}}{(z - \lambda_0^*)^{j-p+1}}.$$

Hence for $j, k = 1, 2, \dots m$,

$$\begin{aligned} \sum_{l=1}^{m} c_{ln}^k q_j(z_{ln}) &= \sum_{l=1}^{m} \frac{b_l^k (z_{ln} - \lambda_0^*)^{m-1}}{n^{m-k}} \frac{q_j(z_{ln})}{q_1(z_{ln})} \\ &= n^{k-j} \sum_{l=1}^{m} b_l^k (-iy_l)^{m-j} \\ &\quad - \frac{1}{n^{m-k}} \sum_{l=1}^{m} \frac{b_l^k}{Q_0(z_{ln}) + o(\frac{1}{n})} \\ \left(\sum_{p=2}^{j} \frac{g_{pm}}{(z_{ln} - \lambda_0^*)^{j-p+1}} \right) &= \delta_{jk} - o\left(\frac{1}{n^{m-k}}\right) \longrightarrow \delta_{jk}, \end{aligned}$$

as $n \to \infty$. This completes the proof of the claim. □

Now assume that $\{\mathcal{H}_0, A_0, \chi\}$ is not minimal and define a subspace $\mathcal{H}_0^\chi$ in $\mathcal{H}_0$ by

$$\mathcal{H}_0^\chi = \overline{\text{span}}\,\{(A_0 - z)^{-1}\chi_{-1} \mid z \in \rho(A_0)\}. \tag{3.30}$$

Then $\mathcal{H}_0$ can be decomposed into the orthogonal sum of two subspaces

$$\mathcal{H}_0 = \mathcal{H}_0^\chi \oplus \mathcal{H}_0',$$

and this decomposition reduces the operator A_0: $A_0 = A_0^\chi \oplus A_0'$, where A_0^χ is a selfadjoint operator in $\mathcal{H}_0^\chi$ and A_0' is a selfadjoint operator in $\mathcal{H}_0'$. Clearly, A_0^χ is a cyclic operator in $\mathcal{H}_0^\chi$ with generating vector $\chi_0 = (A_0 - \lambda_0)^{-1}\chi_{-1}$.

If we choose as starting data the triple $\{\mathcal{H}_0^\chi, A_0^\chi, \chi\}$ and repeat the construction of the previous subsections, we obtain a realization $\{\Pi_m^\chi, H_\chi^\infty, S_\chi, H_\chi^t\}$. Moreover, we obtain the same Q-function $Q^\chi(z) \equiv Q(z)$ if we choose $g_{jj}^\chi \equiv g_{jj}$, $g_{j,j+1}^\chi \equiv g_{j,j+1}$, $j = 1, 2, \ldots, m$, and the same Γ-field $\Gamma_z^\chi \equiv \Gamma_z$ (see (3.11). Also, we find that

$$\Pi_m^\chi = \overline{\text{span}}\,\{\Gamma_z^\chi 1 \mid z \in \rho(A_0^\chi)\},$$

in other words, S_χ is simple in Π_m^χ, and

$$\Pi_m = \Pi_m^\chi \oplus \mathcal{H}_0'.$$

With respect to the latter decomposition H^∞, S, and H^t can be written as

$$H^\infty = H_\chi^\infty \oplus A_0', \quad S = S_\chi \oplus A_0', \quad H^t = H_\chi^t \oplus A_0'.$$

The new representations in Section 6 and the spectral factorization in Section 7 concern only the parts $\{\Pi_m^\chi, H_\chi^\infty, S_\chi, H_\chi^t\}$. Thus we have two possibilities: either we work with the whole space Π_m bearing in mind that the non-essential part $\mathcal{H}_0'$, A_0' is constantly present, or we work with the minimal realization $\{\Pi_m^\chi, H_\chi^\infty, S_\chi, H_\chi^t\}$ constructed from $\{\mathcal{H}_0^\chi, A_0^\chi, \chi\}$. We prefer the second possibility, partly because $\rho(Q) = \rho(A_0^\chi)$ (whereas, on account of Proposition 3.1, $\rho(A_0) \subset \rho(Q)$) and partly because it makes the formulation of some of the results less complicated. Rather than using χ as super- of subscript in the notation, without loss of generality we assume from now on, that the original data $\{\mathcal{H}_0, A_0, \chi\}$ satisfy the minimality condition (3.30), and hence that S is simple in Π_m.

4 The Operators H^t as Extensions with Exit

4.1

In the original Hilbert space $\mathcal{H}_0$ we consider the operator S_0 defined by (3.13). It can also be written as:

$$S_0 := \{\{f, A_0 f\} \in A_0 \mid (f, \chi_{-1})_0 = 0\} = A_0 \cap \{\{\chi_0, \lambda_0\chi_0\}\}^*.$$

This means, that the defect subspace of S_0 at λ_0^* is spanned by the element χ_0, or

$$\ker (S_0^* - \lambda_0) = \text{span}\, \{\chi_0\}.$$

As a one-dimensional restriction of the self-adjoint operator A_0 the operator S_0 is symmetric with defect index $(1, 1)$. Moreover,

$$S_0^* = A_0 \dot{+} \text{span}\, \{\{\chi_0, \lambda_0\chi_0\}\}.$$

Von Neumann's formula for S_0^* takes the form

$$S_0^* = S_0 \dot{+} \text{span}\, \{\{\chi_0, \lambda_0\chi_0\}\} \dot{+} \text{span}\, \{\{\widetilde{\chi}_0, \lambda_0^*\widetilde{\chi}_0\}\}; \tag{4.1}$$

here

$$\widetilde{\chi}_0 = (I + (\lambda_0^* - \lambda_0)R_0(\lambda_0^*))\chi_0 = R_0(\lambda_0^*)\chi_{-1},$$

and it holds $\langle\widetilde{\chi}_0, \widetilde{\chi}_0\rangle_0 = \langle\chi_0, \chi_0\rangle_0$. The canonical self-adjoint extensions A_θ of S_0 can be parametrized by the complex numbers θ of modulus one as follows:

$$A_\theta = S_0 \dot{+} \text{span}\, \{\{\widetilde{\chi}_0 - \theta\chi_0, \lambda_0^*\widetilde{\chi}_0 - \lambda_0\theta\chi_0\}\},$$

For example, $A_0 = A_\theta$ with $\theta = 1$, and $\{h, h'\} \in \mathcal{H}_0^2$ belongs to A_0 if and only if it can be written as

$$\{h, h'\} = \{f, S_0 f\} + \nu\{\widetilde{\chi}_0 - \chi_0, \lambda_0^*\,\widetilde{\chi}_0 - \lambda_0\,\chi_0\}$$

for some $f \in \text{dom}\, S_0$ and some $\nu \in \mathbf{C}$; since $g_{m+1,m+1} = \langle R_0(\lambda_0)\chi_{-1}, R_0(\lambda_0)\chi_{-1}\rangle_0$ it follows that

$$\nu = -\frac{\langle h, \chi_{-1}\rangle_0}{(\lambda_0 - \lambda_0^*)\, g_{m+1,m+1}}.$$

We consider the self-adjoint operators H^t in the space Π_m as self-adjoint extensions of S_0 in $\mathcal{H}_0$ with exit and characterize the Straus extension T_t of S_0 in $\mathcal{H}_0$ associated with H^t (compare (2.12)):

$$T_t(z) = \{\{P_0 f, P_0 g\} \mid \{f, g\} \in H^t, g - zf \in \mathcal{H}_0\}, \quad z \in \mathbf{C},$$

where P_0 is the projection in Π_m onto its first summand $\mathcal{H}_0$, see (3.4). To describe $T_t(z)$ explicitly, we have to introduce the polynomial $p_{2m}(z)$:

$$p_{2m}(z) = (z - \lambda_0)^m (z - \lambda_0^*)^m\, g_{m,m+1} + p_{2m-1}(z). \tag{4.2}$$

Note that $p_{2m}(z)$ is of degree $2m$ and not self-adjoint: $g_{m,m+1} \neq 0$, otherwise $g_{m+1,m+1} = 0$ and then $\chi_0 = 0$ and $\chi = 0$, and

$$p_{2m}^*(z) = p_{2m}(z) - (\lambda_0 - \lambda_0^*)(z - \lambda_0)^m (z - \lambda_0^*)^m g_{m+1,m+1}. \tag{4.3}$$

Theorem 4.1 *For the Straus extension $T_t(z)$ of S_0 associated with H^t, $t \in \mathbf{R}$, the following direct sum representation holds:*

$$T_t(z) = \begin{cases} S_0 \dotplus \operatorname{span}\{\{\widetilde{\chi}_0 - \theta_t(z)\chi_0,\ \lambda_0^*\widetilde{\chi}_0 - \lambda_0\theta_t(z)\chi_0\}\}, & \operatorname{Im} z > 0, \\ S_0 \dotplus \operatorname{span}\{\{\chi_0 - \theta_t^*(z)\widetilde{\chi}_0,\ \lambda_0\chi_0 - \lambda_0^*\theta_t^*(z)\widetilde{\chi}_0\}\}, & \operatorname{Im} z < 0, \end{cases}$$

where $\theta_t(z)$ is the generalized Schur function given by

$$\theta_t(z) = \frac{p_{2m}^*(z) + t}{p_{2m}(z) + t} \tag{4.4}$$

with the polynomial $p_{2m}(z)$ defined in (4.2). *In the case $t = \infty$: $\theta_\infty(z) \equiv 1$ and $T_\infty(z) \equiv A_0$.*

Proof: That $T_t(z)$ can be written as in the theorem for some Schur function $\theta_t(z)$ follows from for example [DLS2]. Therefore it suffices to prove (4.4) and the last statement. We first consider $t \in \mathbf{R}$. From the relation (3.19) it follows that $H^t = S_0 \dotplus S^t$, where S^t is the set of all elements of the form

$$\left\{\begin{pmatrix} \nu\left(\left(1 + \frac{c_{m+1}}{\nu}\right)\chi_0 - \widetilde{\chi}_0\right) \\ X_t \end{pmatrix}, \begin{pmatrix} \nu\left(\lambda_0\left(1 + \frac{c_{m+1}}{\nu}\right)\chi_0 - \lambda_0^*\widetilde{\chi}_0\right) \\ Y_t \end{pmatrix}\right\}, \tag{4.5}$$

with

$$X_t = \begin{pmatrix} (b_j)_{j=1}^m \\ (c_j)_{j=1}^m \end{pmatrix}, \quad Y_t = \begin{pmatrix} \begin{pmatrix} \lambda_0^* b_1 + b_0 \\ (\lambda_0^* b_j + b_{j-1} - g_{jm}c_{m+1})_{j=2}^m \end{pmatrix} \\ (\lambda_0 c_j + c_{j+1})_{j=1}^m \end{pmatrix},$$

$\nu, c_1, \dots, c_{m+1}, b_0, \dots b_m \in \mathbf{C}$, such that the relations

$$b_m = \nu\,(\lambda_0 - \lambda_0^*)\, g_{m+1,m+1} + g_{m,m+1}\, c_{m+1}, \tag{4.6}$$

$$b_0 = -(t + i\operatorname{Im}\lambda_0\, g_{11})\, c_1 - (\lambda_0 - \lambda_0^*) \textstyle\sum_{j=2}^m g_{1j}\, c_j - \sum_{j=1}^m g_{1j}\, c_{j+1}. \tag{4.7}$$

are satisfied. By (4.5), the Schur function $\theta_t(z)$ in the description of $T_t(z)$ is given by

$$\theta_t(z) = 1 + \frac{c_{m+1}}{\nu}, \tag{4.8}$$

where c_{m+1} must now be determined in dependence of z such that $Y_t - zX_t = 0$ holds, that is, from the system of equations

$$\begin{cases} (z - \lambda_0^*)\, b_1 - b_0 = 0, \\ (z - \lambda_0^*)\, b_j - b_{j-1} = -g_{jm}\, c_{m+1}, \quad j = 2, \dots, m, \end{cases} \tag{4.9}$$

$$(z - \lambda_0)c_j - c_{j+1} = 0, \quad j = 1, \dots, m. \tag{4.10}$$

The equations (4.10) imply

$$c_j = \frac{c_{m+1}}{(z-\lambda_0)^{m-j+1}}, \quad j = 1, \dots, m, \tag{4.11}$$

and hence b_0 in (4.7) can be expressed in terms of c_{m+1}. From (4.9) we obtain

$$(z-\lambda_0^*)^m \, b_m = b_0 - \left(\sum_{j=2}^{m} g_{jm}(z-\lambda_0^*)^{j-1} \right) c_{m+1}. \tag{4.12}$$

If we insert the expressions for b_m and b_0 given by (4.6) and (4.7) into (4.12) and observe (4.11), we get the equation

$$\nu(\lambda_0 - \lambda_0^*)(z-\lambda_0)^m (z-\lambda_0^*)^m g_{m+1,m+1} = -c_{m+1}(t + p_{2m}(z)),$$

where $p_{2m}(z)$ is the polynomial from (4.2). By (4.8) and (4.3),

$$\theta_t(z) = \frac{p_{2m}(z) - (z-\lambda_0)^m (z-\lambda_0^*)^m (\lambda_0 - \lambda_0^*) g_{00} + t}{p_{2m}(z) + t} = \frac{p_{2m}^*(z) + t}{p_{2m}(z) + t}.$$

The calculations for T_∞ are similar, but simpler: In this case $c_1 = 0$, so by (4.10), $c_{m+1} = 0$, and hence by (4.8), $\theta_\infty(z) = 1$. □

4.2

The self-adjoint extensions H^t, $t \in \mathbf{R}$, define generalized resolvents of S_0 in the originally given space $\mathcal{H}_0$, which we describe now. To this end we define for $z \in \rho(A_0)$

$$\varphi_0(z) := R_0(z)\chi_{-1}. \tag{4.13}$$

Then the corresponding Q-function $Q_0(z)$, which has to satisfy the relation

$$\frac{Q_0(z) - Q_0(w)^*}{z - w^*} = \langle \varphi_0(z), \varphi_0(w) \rangle_0, \quad z, w \in \rho(A_0),$$

can be chosen as in (3.7):

$$\begin{aligned} Q_0(z) &= (z-\lambda_0)\langle R_0(z)\chi_0, \chi_{-1}\rangle_0 + g_{m,m+1} \\ &= (z-\lambda_0^*)\langle R_0(z)\chi_{-1}, \chi_0\rangle_0 + g_{m+1,m}. \end{aligned} \tag{4.14}$$

Recall by P_0 we denote the orthogonal projection onto $\mathcal{H}_0$ in Π_m.

Theorem 4.2 *The generalized resolvent $R_0^t(z) = P_0(H^t - z)^{-1}|_{\mathcal{H}_0}$ of S_0 associated with H^t in Π_m, $t \in \mathbf{R}$, admits the representation*

$$R_0^t(z) = R_0(z) - \frac{\langle \,\cdot\,, \varphi_0(z^*)\rangle_0 \varphi_0(z)}{Q_0(z) + \tau_t(z)}, \tag{4.15}$$

with $\varphi_0(z)$ *given by the relation* (4.13), $Q_0(z)$ *given by* (4.14), *and as parameter the generalized Nevanlinna function*

$$\tau_t(z) = \frac{p_{2m-1}(z) + t}{(z-\lambda_0)^m (z-\lambda_0^*)^m}. \tag{4.16}$$

The generalized resolvent $P_0(H^\infty - z)^{-1}|_{\mathcal{H}_0}$ *of* S_0 *associated with* H^∞ *coincides with* $R_0(z)$.

Proof: For each $f \in \mathcal{H}_0$, we have to express $h = R_0^t(z)f$ in terms of $R_0(z)f$. We use that $R_0^t(z) = (T_t(z) - z)^{-1}$. If $h = R_0^t(z)f$, then $\{h, f + zh\} \in T_t(z)$. On account of Theorem 4.1 there exist $\{u, A_0 u\} \in S_0$ and $c \in \mathbf{C}$ such that

$$\{h, f + zh\} = \{u, A_0 u\} + c\{\widetilde{\chi}_0 - \theta_t(z)\chi_0, \lambda_0^*\widetilde{\chi}_0 - \lambda_0\theta_t(z)\chi_0\},$$

and hence

$$h = u + c(\widetilde{\chi}_0 - \theta_t(z)\chi_0), \tag{4.17}$$

$$f = (A_0 - z)u - c((z-\lambda_0^*)\widetilde{\chi}_0 - (z-\lambda_0)\theta_t(z)\chi_0). \tag{4.18}$$

From these equations we eliminate u and, using the identities

$$(I + (z-\lambda_0)R_0(z))\chi_0 = R_0(z)\chi_{-1} = (I + (z-\lambda_0^*)R_0(z))\widetilde{\chi}_0, \tag{4.19}$$

we obtain

$$h = R_0(z)f + c(1 - \theta_t(z))\varphi_0(z). \tag{4.20}$$

By (4.18),

$$R_0(z)f = u - c[(z-\lambda_0^*)R_0(z)\widetilde{\chi}_0 - \theta_t(z)(z-\lambda_0)R_0(z)\chi_0], \tag{4.21}$$

and hence, since $u \in \operatorname{dom} S_0$,

$$c = \frac{\langle f, \varphi_0(z^*)\rangle_0}{(z-\lambda_0^*)\langle R_0(z)\widetilde{\chi}_0, \chi_{-1}\rangle_0 - \theta_t(z)(z-\lambda_0)\langle R_0(z)\chi_0, \chi_{-1}\rangle_0}. \tag{4.22}$$

The two terms in the denominator can be written as

$$\theta_t(z)(z-\lambda_0)\langle R_0(z)\chi_0, \chi_{-1}\rangle_0 = \theta_t(z)(Q_0(z) - g_{m,m+1}),$$

and

$$\begin{aligned}
&(z-\lambda_0^*)\langle R_0(z)\widetilde{\chi}_0, \chi_{-1}\rangle_0 \\
&= (z-\lambda_0^*)\langle (I + (z-\lambda_0)R_0(z))R_0(\lambda_0)R_0(\lambda_0^*)\chi_{-1}, \chi_{-1}\rangle_0 \\
&= (z-\lambda_0^*)\langle (I + (z-\lambda_0)R_0(z))R_0(\lambda_0^*)\chi_0, \chi_{-1}\rangle_0 \\
&= (z-\lambda_0^*)\langle \chi_0, \chi_0\rangle_0 + (z-\lambda_0)\langle (R_0(z) - R_0(\lambda_0^*))\chi_0, \chi_{-1}\rangle_0 \\
&= (z-\lambda_0)\langle R_0(z)\chi_0, \chi_{-1}\rangle_0 + (\lambda_0 - \lambda_0^*)g_{m+1,m+1} \\
&= Q_0(z) - g_{m,m+1} + (\lambda_0 - \lambda_0^*)g_{m+1,m+1}.
\end{aligned}$$

From these identities, (4.22), and (4.20) we obtain (4.15) with

$$\tau_t(z) = \frac{g_{m,m+1} - \theta_t(z) g_{m+1,m}}{1 - \theta_t(z)},$$

which by (4.4) equals (4.16).

The last statement can be proved similarly but also follows from the relation $R(z) = (T(z) - z)^{-1}$ that always holds between generalized resolvent and Straus extension (corresponding to the same self-adjoint extension) and the fact that on account of Theorem 4.1, $T_\infty(z) = A_0$. □

4.3

Recall, that we have two M.G. Krein formulas related to the operators H^t: The formula (3.18) with the Q-function $Q(z)$ from (3.12) and the formula (4.15) with the Q-function $Q_0(z)$ from (3.7) or (4.14). Between these Q-functions the relation (3.15):

$$Q(z) = (z - \lambda_0)^m (z - \lambda_0^*)^m Q_0(z) + p_{2m-1}(z). \tag{4.23}$$

holds, which is fundamental for all the following considerations.

The original space of our considerations is the Hilbert space $\mathcal{H}_0$, the extended space Π_m is a Pontryagin space with m negative squares. According to the general extension theory (see [DLS1], [DLS3]) in the generic case the parameter $\tau_t(z)$ in the description of the generalized resolvents in Theorem 4.2 should belong to the class N_m and the parameter $\theta_t(z)$ in Theorem 4.1 to the class S_m. The precise statement in the following theorem is more complicated, since the self-adjoint operator H^t in the Pontrjagin space Π_m can have λ_0 as a normal eigenvalue; here *normal* means that it is isolated and that the corresponding Riesz subspace is finite dimensional.

Theorem 4.3 *For the parametrizations of the self-adjoint extensions $H^t, t \in \mathbf{R}$, in Theorems* 4.1 *and* 4.2 *the following statements hold:*

$$\lambda_0 \in \rho(H^t) \Longleftrightarrow \theta_t(z) \in S_m \Longleftrightarrow \tau_t(z) \in N_m, \tag{4.24}$$

and, if $\lambda_0 \in \sigma_p(H^t)$, for $k = 1, 2, \dots, m$

$$\dim \mathcal{L}_{\lambda_0}(H^t) = k \Longleftrightarrow \theta_t(z) \in S_{m-k} \Longleftrightarrow \tau_t(z) \in N_{m-k}. \tag{4.25}$$

In the first case, $p_{2m}(\lambda_0) + t \neq 0$, $p_{2m-1}(\lambda_0) + t \neq 0$, in the second case, $t = 0$ and λ_0 is a zero of order k of $p_{2m}(z)$.

In the proof of this theorem we use the following lemma.

Lemma 4.4 *The following statements are equivalent:*

(1) *H^t and Π_m are closely connected, that is,*

$$\Pi_m = \overline{\text{span}}\, \{\text{ran}\, (H^t - z)^{-1} f \mid z \in \rho(H^t),\ f \in \mathcal{H}_0\}. \tag{4.26}$$

(2) *The operator H^t does not have an eigenvalue with an eigenvector in $\Pi_m \ominus \mathcal{H}_0$.*

Proof: If $H^t u = z_0 u$ for some $u \in \Pi_m \ominus \mathcal{H}_0$, then u is orthogonal to the set on the right-hand side of (4.26), and hence H^t is not closely connected. Conversely, if H^t and Π_m are not closely connected, we denote the orthogonal complement in Π_m of the set on the right-hand side of (4.26) by $\mathcal{M}$. The relation

$$\lim_{y \to +\infty} -iy(H^t - iy)^{-1} = I,$$

implies that $\mathcal{M} \subset \Pi_m \ominus \mathcal{H}_0$. Let $z_0 \in \rho(H^t)$. It is easy to see that $(H^t - z_0)^{-1}\mathcal{M} \subset \mathcal{M}$. Since $\mathcal{M}$ is finite dimensional, $(H^t - z_0)^{-1}$ and hence also H^t have an eigenvalue with eigenvector in $\mathcal{M}$. □

Proof of Theorem 4.3: Recall that λ_0 is nonreal. Hence either $\lambda_0 \in \rho(H^t)$ or λ_0 is a normal eigenvalue of H^t.

Assume $\lambda_0 \in \rho(H^t)$. We shall show that statement (2) of Lemma 4.4 holds: Assume

$$(H^t - z)\begin{pmatrix} 0 \\ (b_j)_{j=1}^m \\ (c_j)_{j=1}^m \end{pmatrix} = 0.$$

From the formula (3.19) with $h = 0$ and $c_{m+1} = 0$, we see that the relations (4.7), (4.9), and (4.10) are valid. If $z \neq \lambda_0$, they imply that all c_j's and all b_j's are zero; this also holds if $z = \lambda_0$, because $\lambda_0 \in \rho(H^t)$. Hence H^t and Π_m are closely connected. On account of [DLS1, Theorem 2.1] this implies that $\theta_t(z) \in S_m$.

By Theorem 3.3, $\lambda_0 \in \sigma_p(H^t)$ if and only if

$$Q(\lambda_0) + t = 0. \tag{4.27}$$

The relation (3.15) implies that (4.27) is equivalent to $p_{2m-1}(\lambda_0) + t = 0$, and, finally, by (4.2), this is equivalent to $p_{2m}(\lambda_0) + t = 0$. Since $p_{2m-1}(z)$ is a self-adjoint polynomial, λ_0 and λ_0^* are zeros of $p_{2m}(z) + t = 0$ of the same order k, say, and therefore $0 \leq k \leq m$. It follows from (3.3), (3.16), and (4.2) that λ_0 is a zero of order $k > 0$ if and only if

$$t = 0, \quad g_{11} = 0, \quad g_{j,j+1} = 0 \; (j = 1, \ldots, k-1), \quad g_{k,k+1} \neq 0. \tag{4.28}$$

We shall use the following elements of Π_m:

$$u_j = \begin{pmatrix} 0 \\ 0 \\ e_j \end{pmatrix}, \quad w_j = \begin{pmatrix} 0 \\ e_j \\ 0 \end{pmatrix}, \quad j = 1, \ldots, m,$$

where as before e_j is the m-vector with j-th component $= 1$ and the other components $= 0$, $j = 1, 2, \ldots, m$. From the explicit formula (3.19) for H^t we find that $\lambda_0 \in \sigma_p(H^t)$ and $\dim \mathcal{L}_{\lambda_0}(H^t) = k > 0$ if and only if (4.28) holds and then $\mathcal{L}_{\lambda_0}(H^0)$ is spanned by $u_1, u_2, \ldots, u_k$. In fact (4.28) is equivalent to

$$H^t u_1 = \lambda_0 u_1, \; H^t u_j = \lambda_0 u_j + u_{j-1} \; (j = 2, 3, \ldots, k), \; u_k \notin \operatorname{ran}(H^t - \lambda_0).$$

From (4.28) and (3.3), we have $g_{ij} = 0$, $i, j = 1, 2, \ldots, k$, and it follows from the structure of the inner product on Π_m (see after (3.4)) that

$$\mathcal{L}_{\lambda_0}(H^0)^\perp = \mathcal{H}_0 \oplus \text{span}\,\{w_{k+1}, \ldots, w_m\} \oplus \text{span}\,\{u_1, \ldots, u_k\} \oplus (I - X)\ \text{span}\,\{u_{k+1}, \ldots, u_m\},$$

where $X : \text{span}\,\{u_{k+1}, \ldots, u_m\} \to \text{span}\,\{w_1, \ldots, w_k\}$ is the linear mapping such that

$$Xu_j = \sum_{i=1}^{k} g_{ij} w_i, \quad j = k+1, \ldots, m.$$

The root space $\mathcal{L}_{\lambda_0}(H^0)$ is neutral and is the isotropic part of $\mathcal{L}_{\lambda_0}(H^0)^\perp$. Therefore the quotient space $\mathcal{L}_{\lambda_0}(H^0)^\perp / \mathcal{L}_{\lambda_0}(H^0)$ is a Pontryagin space with negative index $m - k$ and of the form

$$\mathcal{L}_{\lambda_0}(H^0)^\perp / \mathcal{L}_{\lambda_0}(H^0) = \mathcal{H}_0 \oplus \ \text{span}\,\{w_{k+1}, \ldots, w_m\} \oplus (I - X)\ \text{span}\,\{u_{k+1}, \ldots, u_m\}.$$

The image of H^0 under the embedding of Π_m into this quotient space is a closely connected self-adjoint extension of S_0. The corresponding Schur function is given by the same formula but now $\theta_0(z) \in S_{m-k}$. We have proved the first "if and only if" statements in (4.24) and (4.25); the others follow from the relation (4.4) between the Schur function $\theta_t(z)$ and the Nevanlina function $\tau_t(z)$. □

5 The Class N_m^∞

5.1

By N_0^∞ we denote the set of all functions $Q_0(z) \in N_0$ with the properties

(a) $\lim_{y \to +\infty} y \,\text{Im}\, Q_0(iy) = \infty$,

(b) $\lim_{y \to +\infty} y^{-1} \,\text{Im}\, Q_0(iy) = 0$.

Recall (see Section 2 or [KL3]) that this is the set of all Q-functions of densely defined symmetric operators with defect index (1, 1) in a Hilbert space. The integral representation (2.4) of a function $Q_0(z) \in N_0^\infty$ is of the form

$$Q_0(z) = \alpha + \int_{-\infty}^{+\infty} \frac{1 + sz}{s - z}\, d\sigma(s), \tag{5.1}$$

with $\alpha \in \mathbf{R}$ and a finite measure σ on $\mathbf{R}$ such that $\int_{-\infty}^{+\infty} (1 + s^2) d\sigma(s) = \infty$.

If $m \in \mathbf{N}$, by N_m^∞ we denote the set of all functions $Q(z)$ which admit a representation

$$Q(z) = (z^2 + 1)^m\, Q_0(z) + q_{2m-1}(z), \tag{5.2}$$

where $q_{2m-1}(z)$ is a self-adjoint polynomial of degree at most $2m-1$ and $Q_0(z) \in N_0^\infty$. It follows easily, that $N_m^\infty \subset N_m$, and that for a function $Q(z) \in N_m^\infty$ the point ∞ is a generalized pole of $Q(z)$ of nonpositive type of multiplicity m; see [L]. Therefore, for $Q(z) \in N_m^\infty$ the total number of its zeros in $\mathbf{C}^+$ and its generalized zeros of nonpositive type on $\mathbf{R}$ equals m; here *total* means, that these zeros are counted according to their multiplicities.

We mention without proof two simple properties of the functions of the class N_m^∞.

(i) If $Q_1(z) \in N_{m_1}^\infty$, $Q_2(z) \in N_{m_2}^\infty$, then $Q_1(z) + Q_2(z) \in N_m^\infty$ with $m = \max\{m_1, m_2\}$.

(ii) If $Q_1(z) \in N_m^\infty$ with $m \geq 1$ and $Q_2(z) \in N_0$, then $Q_1(z) + Q_2(z) \in N_m^\infty$.

Theorem 5.1 *If $Q(z) \in N_m^\infty$ and $q(z)$ is a polynomial of degree m with zeros in the domain of holomorphy of $Q(z)$, then there exist a function $\widehat{Q}_0(z) \in N_0^\infty$ with $\rho(Q) = \rho(\widehat{Q}_0)$ and a self-adjoint polynomial $\widehat{p}_{2m-1}(z)$ of degree at most $2m-1$ such that*

$$Q(z) = q(z)q^*(z)\widehat{Q}_0(z) + \widehat{p}_{2m-1}(z); \tag{5.3}$$

here $\widehat{Q}_0(z)$ and $\widehat{p}_{2m-1}(z)$ are uniquely determined by $q(z)$ and $Q(z)$. Conversely, given a function $\widehat{Q}_0(z) \in N_0^\infty$, a polynomial $q(z)$ of degree m and a self-adjoint polynomial $\widehat{p}_{2m-1}(z)$ of degree at most $2m-1$, then the function $Q(z)$ in (5.3) belongs to the class N_m^∞ and $\rho(Q) = \rho(\widehat{Q}_0)$.

Proof: The function $\widehat{Q}_0(z)$ must be of the form

$$\widehat{Q}_0(z) = \widehat{\alpha} + \int_{-\infty}^{+\infty} \frac{1+sz}{s-z}\, d\widehat{\sigma}(s).$$

With the representation (5.1) of $Q(z)$ and (5.2) the Stieltjes-Livsic inversion formula implies

$$|q(s)|^2\, d\widehat{\sigma}(s) = (s^2+1)^m\, d\sigma(s).$$

Since by assumption the measure σ has no mass at the zeros of the polynomial $q(z)$ this equation determines $\widehat{\sigma}$ uniquely. The number α and the polynomial $\widehat{p}_{2m-1}(z)$ can easily be calculated if it is only observed that

$$\frac{|q(s)|^2(1+z^2)^m - q(z)q^*(z)(1+s^2)^m}{s-z}$$

is a polynomial in z of degree $\leq 2m-1$. □

Remark 5.2 If the condition $\rho(Q) = \rho(\widehat{Q}_0)$ is not satisfied then the decomposition (5.3) is not necessarily unique. Indeed, choose $Q_0(z) \in N_0^\infty$, $x_0 \in \mathbf{R} \cap \rho(Q_0)$ and $\gamma > 0$. Then the function $\widehat{Q}_0(z) := Q_0(z) + \gamma(x_0 - z)^{-1}$ belongs to N_0^∞. The function

$$Q(z) := (z-x_0)^2\widehat{Q}_0(z) = (z-x_0)^2 Q_0(z) + \gamma(x_0 - z) \tag{5.4}$$

belongs to N_1^∞ and both representations in (5.4) are of the form (5.3).

Remark 5.3 In Theorem 5.1 the assumption that the zeros of the function $q(z)$ are in the domain of holomorphy of $Q(z)$ can be weakened. If σ denotes the measure in the integral representation (5.1) of $Q_0(z)$, then, if $q(z)$ has a simple zero at x_0, it is for example sufficient that the following condition is satisfied:

$$\int_{x_0-\delta}^{x_0+\delta} \frac{d\sigma(s)}{(s-x_0)^2} < \infty \text{ for some } \delta > 0. \tag{5.5}$$

The condition (5.5) is satisfied if x_0 is a generalized zero of nonpositive type of $Q(z)$.

The following factorization of the function $Q(z) \in N_m^\infty$ is a special case of the results of [DLLS]. We prove it here for the convenience of the reader.

Theorem 5.4 *If $Q(z) \in N_m^\infty$ and z_0 is a nonreal zero or a generalized zero of nonpositive type of $Q(z)$, then*

$$Q(z) = (z - z_0)(z - z_0^*)Q_1(z) \tag{5.6}$$

with $Q_1(z) \in N_{m-1}^\infty$. The nonreal zeros or generalized zeros of nonpositive type of $Q_1(z)$, which are different from z_0, z_0^*, are the same (including multiplicities) as those of $Q(z)$, the multiplicity of z_0, z_0^* with respect to $Q_1(z)$ equals the multiplicity of z_0, z_0^* with respect to $Q(z)$ minus one.

Proof: We consider first the case of a nonreal pair of zeros z_0, z_0^* of $Q(z)$. Let

$$Q(z) = (z^2+1)^m \left(\alpha + \int_{-\infty}^{+\infty} \frac{1+sz}{s-z}\, d\sigma(s)\right) + p_{2m-1}(z)$$

be the representation (5.2) of $Q(z)$. We introduce the function

$$\widetilde{Q}(z) := (z^2+1)^{m-1} \left(\widetilde{\alpha} + \int_{-\infty}^{+\infty} \frac{1+sz}{s-z} \frac{s^2+1}{(s-z_0)(s-z_0^*)}\, d\sigma(s)\right) \tag{5.7}$$

with some $\widetilde{\alpha} \in \mathbf{R}$ and consider the difference $Q(z) - (z-z_0)(z-z_0^*)\widetilde{Q}(z)$. The difference of the two terms containing the integrals is of the form $(z^2+1)^{m-1}(az^2 + bz + c)$ with $a, b, c \in \mathbf{R}$, hence

$$\begin{aligned} &Q(z) - (z-z_0)(z-z_0^*)\widetilde{Q}(z) \\ &\quad = (z^2+1)^m\alpha - (z-z_0)(z-z_0^*)(z^2+1)^{m-1}\widetilde{\alpha} \\ &\qquad + (z^2+1)^{m-1}(az^2+bz+c) + p_{2m-1}(z). \end{aligned}$$

If we choose $\widetilde{\alpha} = \alpha + a$ then this difference becomes a polynomial of degree at most $2m-1$ and it vanishes at z_0 and z_0^*. Therefore it can be written as

$(z-z_0)(z-z_0^*)p_{2m-3}(z)$ with a polynomial $p_{2m-3}(z)$ of degree at most $2m-3$, and the representation

$$Q(z)=(z-z_0)(z-z_0^*)(\widetilde{Q}(z)+p_{2m-3}(z))$$

follows. If we observe that the function $Q_1(z)=\widetilde{Q}(z)+p_{2m-3}(z)$ belongs to the class N_{m-1}^{∞}, the theorem is proved for a nonreal zero z_0.

Let now $z_0=x_0\in\mathbf{R}$. Then the integral on the right-hand side of (5.7) exists according to Remark 5.3 and by a similar reasoning as in the case of a nonreal zero z_0 we come to the conclusion that the difference $Q(z)-(z-x_0)^2\widetilde{Q}(z)$ is of the form $(z-x_0)\gamma+(z-x_0)^2p_{2m-3}(z)$ with some self-adjoint polynomial $p_{2m-3}(z)$ of degree at most $2m-3$. Applying the criterion for x_0 being a generalized zero of nonpositive type we find $\gamma\le 0$, and the representation

$$Q(z)=(z-x_0)^2\left(\widetilde{Q}(z)+\frac{\gamma}{z-x_0}+p_{2m-3}(z)\right)$$

follows. □

In Section 7 the following special factorization will play an important role.

Corollary 5.5 *If $Q(z)\in N_m^{\infty}$ and $q_0(z)$ is the monic polynomial of degree m the zeros of which are the zeros of $Q(z)$ in $\mathbf{C}^+$ and the generalized zeros of $Q(z)$ of nonpositive type on* **R**, *including multiplicities, then*

$$Q(z)=q_0(z)q_0^*(z)\widehat{Q}_0(z) \tag{5.8}$$

with $\widehat{Q}_0(z)\in N_0^{\infty}$.

5.2

In the next theorem we characterize the functions $Q(z)\in N_m^{\infty}$ within the class N_m by the self-adjoint relation A in their operator representation. At the same time, a decomposition of the space Π_m is obtained which corresponds to the factorization (5.3) of the function $Q(z)$ with the choice $q(z)=(z-\lambda_0)^m$ where $\lambda_0\in\mathbf{C}^+$ is in the domain of holomorphy of $Q(z)$. We shall show how the operator representation of the function $Q(z)$ leads to a representation of the function $\widehat{Q}_0(z)$ which later also will be identified as its operator representation. In Sections 6 and 7 similar considerations will be repeated for other choices of the polynomial $q(z)$.

Let $Q(z)\in N_m^{\infty}$ have the following representation (5.3) with $q(z)=(z-\lambda_0)^m$:

$$Q(z)=(z-\lambda_0)^m(z-\lambda_0^*)^m\,Q_0(z)+\widehat{p}_{2m-1}(z); \tag{5.9}$$

here $Q_0(z)\in N_0$ and $\widehat{p}_{2m-1}(z)$ is a self-adjoint polynomial of degree at most $2m-1$. We consider the minimal operator representation of the function $Q(z)$ with a self-adjoint relation A in some Pontryagin space Π_m of negative index m:

$$\begin{aligned} Q(z) &= Q(\lambda_0^*)+(z-\lambda_0^*)\Gamma_{\lambda_0}^*(I+(z-\lambda_0)(A-z)^{-1})\Gamma_{\lambda_0}1 \\ &= Q(\lambda_0^*)+(z-\lambda_0^*)\Gamma_{\lambda_0}^*\Gamma_z 1,\quad z\in\rho(A), \end{aligned} \tag{5.10}$$

where

$$\Gamma_z = (I + (z - \lambda_0)(A - z)^{-1})\Gamma_{\lambda_0} \tag{5.11}$$

is the corresponding Γ-field. Expand the function Γ_z at λ_0:

$$\begin{aligned} \Gamma_z &= \Gamma_{\lambda_0} + (z - \lambda_0)(A - \lambda_0)^{-1}\Gamma_{\lambda_0} \\ &\quad + \cdots + (z - \lambda_0)^{m-1}(A - \lambda_0)^{-m+1}\Gamma_{\lambda_0} \\ &\quad + (z - \lambda_0)^m (A - \lambda_0)^{-m+1}(A - z)^{-1}\Gamma_{\lambda_0}. \end{aligned} \tag{5.12}$$

With the operators

$$\begin{aligned} E_j &:= (A - \lambda_0)^{-j+1}\,\Gamma_{\lambda_0}, \; j = 1, 2, \ldots m, \\ \Gamma_{0z} &:= (A - \lambda_0)^{-m+1}(A - z)^{-1}\Gamma_{\lambda_0} \end{aligned} \tag{5.13}$$

the expansion (5.12) becomes

$$\Gamma_z = \sum_{j=1}^{m} (z - \lambda_0)^{j-1} E_j + (z - \lambda_0)^m \Gamma_{0z}. \tag{5.14}$$

Then for any $c \in \mathbf{R}$ the function

$$Q_1(z) := \Gamma_{\lambda_0}^* (A - \lambda_0)^{-m+1}(A - \lambda_0^*)^{-m+1}(A - z)^{-1}\Gamma_{\lambda_0} 1 + c \tag{5.15}$$

satisfies the relation

$$\frac{Q_1(z) - Q_1(\zeta)^*}{z - \zeta^*} = \Gamma_{0\zeta}^* \Gamma_{0z} 1, \tag{5.16}$$

which follows immediately from the resolvent relation. We shall show that for $c = 0$ the relation

$$Q(z) - (z - \lambda_0)^m (z - \lambda_0^*)^m Q_1(z) = q_{2m-1}(z) - c(z - \lambda_0)^m (z - \lambda_0^*)^m$$

holds with a self-adjoint polynomial $q_{2m-1}(z)$ of degree at most $2m - 1$. Indeed, by (5.10) and (5.15), the difference on the left-hand side is up to the term $-c(z - \lambda_0)^m (z - \lambda_0^*)^m$ equal to

$$\begin{aligned} &\frac{1}{2}(Q(\lambda_0) - Q(\lambda_0^*)) + (z - \operatorname{Re}\lambda_0)\Gamma_{\lambda_0}^*\Gamma_{\lambda_0} 1 \\ &\quad + \Gamma_{\lambda_0}^*[(z - \lambda_0^*)(z - \lambda_0)(A - z)^{-1} \\ &\quad - (z - \lambda_0)^m (z - \lambda_0^*)^m (A - \lambda_0)^{-m+1}(A - \lambda_0^*)^{-m+1}(A - z)^{-1}]\Gamma_{\lambda_0} 1, \end{aligned}$$

and the expression in the square brackets is an operator polynomial in z of degree at most $2m - 1$. Therefore, the function $Q_1(z)$, apart from a real additive constant, coincides with the function $Q_0(z)$ in the representation (5.9). The function $Q_1(z)$ can also be written as:

$$Q_1(z) = \Gamma_{0\lambda_0}^* E_m 1 + (z - \lambda_0^*)\Gamma_{0\lambda_0}^* \Gamma_{0z} 1 + c = Q_1(\lambda_0^*) + (z - \lambda_0^*)\Gamma_{0\lambda_0}^* \Gamma_{0z} 1. \tag{5.17}$$

We introduce the subspaces

$$\mathcal{L} := \ker (A - z)^{-m}, \ z \in \rho(A), \quad \mathcal{M} := \operatorname{span} \{E_1 1, E_2 1, \ldots, E_m 1\}.$$

Observe that the subspace $\mathcal{L}$ is independent of $z \in \rho(A)$.

Lemma 5.6 *Suppose $Q(z) \in N_m$ and let $A = A(Q)$ be the self-adjoint linear relation in the Pontryagin space $\Pi_m = \Pi_m(Q)$ of the minimal operator representation (5.10) of $Q(z)$. Suppose further that the subspace $\mathcal{L}$ of Π_m is m-dimensional and neutral. Then:*

(i) *The subspaces $\mathcal{L}$ and $\mathcal{M}$ in Π_m form a dual pair.*

(ii) *There exists a unique decomposition*

$$\Pi_m = \mathcal{H}_0 \oplus (\mathcal{L} \dot{+} \mathcal{M}) \tag{5.18}$$

with some Hilbert space $(\mathcal{H}_0, \langle \cdot, \cdot \rangle_{\Pi_m})$ and $\mathcal{L}^\perp = \mathcal{H}_0 \oplus \mathcal{L}$.

(iii) *If P_0 denotes the orthogonal projection onto $\mathcal{H}_0$ in the decomposition* (5.18) *then*

$$A_0 := \{\{P_0 f, P_0 g\} \mid \{f, g\} \in A, \ g - zf \in \mathcal{H}_0\}, \quad z \in \rho(A), \tag{5.19}$$

is a self-adjoint relation in $\mathcal{H}_0$; here the set on the right-hand side is independent of $z \in \rho(A)$.

Proof: We set $R(z) := (A - z)^{-1}$. Evidently, the subspaces $\mathcal{L}$ and $\mathcal{L}^\perp$ are $R(z)$-invariant. Since $\operatorname{ran} R(z)^m \subset\subset \mathcal{L}^\perp$, the relations $\Gamma_{0\lambda_0} = R(\lambda_0)^{-m}\Gamma_{\lambda_0}$ and $\Gamma_{0z} = (1 + (z - \lambda_0)R(z))\Gamma_{0\lambda_0}$ imply that $\operatorname{ran} \Gamma_{0\lambda_0}$, $\operatorname{ran} \Gamma_{0z} \subset \mathcal{L}^\perp$ and hence also

$$\mathcal{M}' := \overline{\operatorname{span}} \{\Gamma_{0z} c_z \mid z \in \rho(A), \ c_z \in \mathbf{C}\} \subset \mathcal{L}^\perp.$$

Now we consider the expansion (5.14) of Γ_z. Since A is minimal, that is, Π_m is the closed linear span of the elements $\Gamma_z c_z$, $z \in \rho(A)$, $c_z \in \mathbf{C}$, it follows that $\Pi_m = \mathcal{M}' + \mathcal{M}$. Further, $\dim \mathcal{L} = \operatorname{codim} \mathcal{L}^\perp$ (see [Bo], Theorem 10.8, or [IKL], Corollary 1 after Theorem 3.4), therefore the inclusion $\mathcal{M}' \subset \mathcal{L}^\perp$ implies $\dim \mathcal{M} = m$, $\mathcal{M}' = \mathcal{L}^\perp$ and $\Pi_m = \mathcal{L}^\perp \dot{+} \mathcal{M}$. Moreover, $\mathcal{L} \cap \mathcal{M}^\perp = \{0\}$. Indeed, an element x belonging to this intersection is orthogonal to the whole space Π_m and hence $x = 0$. Therefore $\mathcal{M}$ and $\mathcal{L}$ are in duality, and (i) is proved. Now (ii) is a consequence of [IKL], Theorem 3.4.

In order to prove (iii) we observe the inclusions

$$R(z)\mathcal{L} \subset \mathcal{L}, \quad R(z)\mathcal{L}^{\perp} \subset \mathcal{L}^{\perp}$$

and the relations

$$\mathcal{L}^{\perp} = \mathcal{H}_0 \oplus \mathcal{L}, \quad R(z)^* = R(z^*).$$

They imply the following matrix representation of $R(z)$ with respect to (5.18):

$$R(z) = \begin{pmatrix} R_{00}(z) & 0 & R_{02}(z) \\ R_{10}(z) & R_{11}(z) & R_{12}(z) \\ 0 & 0 & R_{22}(z) \end{pmatrix}. \tag{5.20}$$

If we exchange the first and second rows and columns this matrix becomes triangular. Then it is easy to see that $R_{00}(z)$ satisfies to the resolvent relation. Since also $R_{00}(z)^* = R_{00}(z^*)$ for $z \in \rho(A)$, $R_{00}(z)$ is the resolvent of a self-adjoint linear relation

$$A_0 = \{\{R_{00} f_0, (I + zR_{00}) f_0\} \mid f_0 \in \mathcal{H}_0\},$$

which is just A_0 given by (5.19). □

Now we can characterize the functions from N_m^{∞} by their operator representation.

Theorem 5.7 *Suppose $Q(z) \in N_m$ for some $m \geq 1$ and let $A = A(Q)$ be the self-adjoint linear relation and $\Pi_m = \Pi_m(Q)$ be the Pontryagin space $\Pi_m = \Pi_m(Q)$ from the minimal operator representation* (5.10) *of $Q(z)$. Then $Q(z) \in N_m^{\infty}$ if and only if the subspace*

$$\mathcal{L} = \ker (A - z)^{-m}, \quad z \in \rho(A),$$

is m-dimensional, neutral and is the (full) root space of A at ∞.

Proof: Consider the minimal operator representation (5.10) of $Q(z) \in N_m$ with $\lambda_0 \in \rho(A) \cap \mathbf{C}^+$, and suppose that A has the property formulated in the theorem. Let the functions Γ_z and Γ_{0z} be defined as in (5.11) and (5.13), respectively, and let $Q_0(z)$ be the function $Q_1(z)$ from (5.15) for $c = 0$:

$$Q_0(z) \,(= Q_1(z)) = \Gamma_{\lambda_0}^* (A - \lambda_0)^{-m+1} (A - \lambda_0^*)^{-m+1} (A - z)^{-1} \Gamma_{\lambda_0}.$$

We show that $Q_0(z) \in N_0^{\infty}$. To this end we consider the corresponding kernel (see (5.16))

$$\frac{Q_0(z) - Q_0(\zeta)^*}{z - \zeta^*} = \Gamma_{0\zeta}^* \Gamma_{0z}.$$

As in the proof of Lemma 5.6 it follows that the subspace $\mathcal{L}^{\perp}$ is nonnegative and $R(z)$-invariant and also that $\operatorname{ran} \Gamma_{0z} \subset \mathcal{L}^{\perp}$, $z \in \rho(A)$. This implies

$$\operatorname{Im} Q_0(z) / \operatorname{Im} z = \Gamma_{0z}^* \Gamma_{0z} \geq 0 \text{ for all } z \in \mathbf{C}^+.$$

If for some $\zeta \in \mathbf{C}^+$ we would have $\Gamma_{0\zeta}^*\Gamma_{0\zeta} = 0$, then $\Gamma_{0\zeta}1$ would be neutral, hence $\Gamma_{0\zeta}1 \in \mathcal{L}$ and, consequently, $\Gamma_{0z}1 \in \mathcal{L}$ for all $z \in \rho(A)$ because of the relation between Γ_{0z} and $\Gamma_{0\zeta}$ (compare (2.5)):

$$\Gamma_{0z} = (I + (z-\zeta)(A-z)^{-1})\Gamma_{0\zeta} \tag{5.21}$$

and the $R(z)$-invariance of $\mathcal{L}$. In this case the decomposition (5.18) would be of the form $\Pi_m = \mathcal{L} \dot{+} \mathcal{M}$ which is impossible since the space Π_m is infinite dimensional.

Next we show that $Q_0(z) \in N_0^\infty$. According to Lemma 5.6 the space Π_m admits a decomposition (5.18) and A_0 from (5.19) is the self-adjoint relation in $\mathcal{H}_0$ such that $(A_0 - z)^{-1} = R_{00}(z)$. We shall show that $\ker R_{00}(z) = \{0\}$, that is, that A_0 is a self-adjoint operator. Suppose $R_{00}(z)x_0 = 0$ for some $x_0 \in \mathcal{H}_0$. Then $R(z)x_0 \in \mathcal{L}$ because of the matrix form (5.20), and this implies $x_0 \in \mathcal{L}$ by the condition of the theorem, hence $x_0 = 0$.

Define

$$\gamma(z) := P_0\Gamma_{0z}1 = P_0(A-\lambda_0)^{-m}\Gamma_z 1.$$

The inclusion $\operatorname{ran}(A-z)^{-1}\Gamma_z \subset \mathcal{L}^\perp$ and the matrix representation (5.20) imply

$$P_0(A-z)^{-1}\Gamma_z = P_0(A_0-z)^{-1}P_0\Gamma_z, \tag{5.22}$$

and hence, by (5.21),

$$\gamma(z) - \gamma(\zeta) = (z-\zeta)(A_0-z)^{-1}\gamma(\zeta).$$

In the proof of Lemma 5.6 it was shown that the subspace $\mathcal{L}^\perp = \mathcal{H}_0 \oplus \mathcal{L}$ is spanned by the elements $\Gamma_{0z}c_z$, $c_z \in \mathbf{C}$, $z \in \rho(A)$, hence $\rho(A) \subset \rho(A_0)$ implies

$$\mathcal{H}_0 = \overline{\operatorname{span}}\,\{\gamma_z c_z,\ c_z \in \mathbf{C},\ z \in \rho(A_0)\}.$$

Further, it is evident that in the representation (5.17) of $Q_0(z)$ we can insert $\gamma(\lambda_0)$ instead of $\Gamma_{0\lambda_0}1$ and $\gamma(z)$ instead of the $\Gamma_{0z}1$ and obtain

$$Q_0(z) = Q_0(\lambda_0^*) + (z-\lambda_0^*)\langle\gamma(z), \gamma(\lambda_0)\rangle_0.$$

From this new representation it follows that $Q_0(z)$ is the Q-function of the restriction S_0 of A_0 in the Hilbert space $\mathcal{H}_0$, defined on

$$\operatorname{dom} S_0 = \{f \in \operatorname{dom} A_0 | (A_0 - \lambda_0^*)f \perp \gamma(\lambda_0)\},$$

and the self-adjoint operator A_0.

We shall show that $\operatorname{dom} S_0$ is dense in $\mathcal{H}_0$. Since $\operatorname{dom} A_0$ is dense in $\mathcal{H}_0$, it is sufficient to show that $\gamma(\lambda_0) \notin \operatorname{dom} A_0$. Suppose that $\gamma(\lambda_0) \in \operatorname{dom} A_0$. Then, by the definition (5.19) of A_0, there exists a $y_0 \in \mathcal{H}_0$ such that $\gamma(\lambda_0) = P_0 R(\lambda_0)y_0$. On the other hand, $\gamma(\lambda_0) = P_0\Gamma_{0\lambda_0}1 = P_0 R(\lambda_0)E_m 1$. Taking into account that $R(\lambda_0)E_m 1 = \Gamma_{0\lambda_0}1$ and $R(\lambda_0)y_0$ belong to $\mathcal{H}_0 \oplus \mathcal{L}$ we obtain $R(\lambda_0)(x_0 - E_m 1) \in$

$\mathcal{L}$. But this implies $(x_0 - E_m 1) \in \mathcal{L}$ according to the condition of the theorem. The last inclusion is in contradiction with the decomposition (5.18) since $y_0 \in \mathcal{H}_0$ and $E_m 1 \in \mathcal{M}$. Therefore $\gamma(\lambda_0) \notin \operatorname{dom} A_0$. Thus, $Q_0(z)$ is the Q-function of the densely defined symmetric operator S_0 and hence $Q(z) \in N_0^\infty$.

In order to prove the necessity of the condition, we start from a representation (5.9) of $Q(z) \in N_m^\infty$ with a function $Q_0(z) \in N_0^\infty$. Consider the operator representation

$$Q_0(z) = Q_0(\lambda_0^*) + (z - \lambda_0^*)\langle \gamma(z), \gamma(\lambda_0)\rangle_0$$

of the function $Q_0(z)$ with a densely defined self-adjoint operator A_0 in some Hilbert space $\mathcal{H}_0$, an element $\gamma(\lambda_0) \in \mathcal{H}_0$ and $\gamma(z) = (1 + (z - \lambda_0)(A_0 - z)^{-1})\gamma(\lambda_0)$. Then, clearly, $\gamma(\lambda_0) \notin \operatorname{dom} A_0$. We introduce the scale (3.1) of spaces $\mathcal{H}_{\pm k}$, determined by $\mathcal{H}_0$, A_0 and the generalized element $\chi := (A - \lambda_0)^{2m+1}\gamma(\lambda_0)$. As in Section 3 we construct the Pontryagin space $\Pi_m = \mathcal{H}_0 \oplus \mathbf{C}^m \oplus \mathbf{C}^m$, equipped with the hermitian form $\langle \cdot\,, \cdot\rangle_{\Pi_m}$ defined by the Gram operator

$$\mathbf{G} = \begin{pmatrix} I_0 & 0 \\ 0 & G_m \end{pmatrix}, \quad G_m = \begin{pmatrix} 0 & I_m \\ I_m & G \end{pmatrix}. \tag{5.23}$$

Here I_0 and I_m are the identity operators in $\mathcal{H}_0$ and $\mathbf{C}^m$, respectively, and G is the operator in $\mathbf{C}^m$ given by the matrix $G = (g_{jk})_{j,k=1,\dots,m}$ with elements g_{jk} determined by $Q(z)$:

$$g_{jk} = \frac{1}{(j-1)!(k-1)!}\frac{\partial^{j-1}}{\partial(\zeta^*)^{j-1}}\frac{\partial^{k-1}}{\partial z^{k-1}}\frac{Q(z) - Q(\zeta^*)}{z - \zeta^*}\bigg|_{z=\zeta=\lambda_0}.$$

We consider in Π_m the linear relations H^∞ and S (see Section 3) and the operator function Γ_z from (3.11). The corresponding Q-function is the given function $Q(z)$, see (3.12), therefore in the operator representation of $Q(z)$ we can choose $A = H^\infty$ and $\Gamma = \Gamma_{\lambda_0}$. By the explicit form (3.5) of $R^\infty(z)(= R(z))$ we have that

$$\Gamma_{\lambda_0} 1 = \begin{pmatrix} 0 \\ 0 \\ e_1 \end{pmatrix}, \quad E_j 1 = \begin{pmatrix} 0 \\ 0 \\ e_j \end{pmatrix}, \quad \Gamma_{0z} 1 = \begin{pmatrix} R_0(z)\chi_{-1} \\ (q_j(z))_{j=1}^m \\ 0 \end{pmatrix},$$

and hence $\mathcal{M} = 0 \oplus 0 \oplus \mathbf{C}^m$ and $\gamma(z) = R_0(z)\chi_{-1}$. The m-dimensional neutral subspace $\mathcal{L}$ in the theorem is here $0 \oplus \mathbf{C}^m \oplus 0$, and

$$\mathcal{L}^\perp = \mathcal{H}_0 \oplus \mathbf{C}^m \oplus 0, \quad \mathcal{L}^\perp/\mathcal{L} = \mathcal{H}_0.$$

The formula (3.5) for the resolvent of $A = H^\infty$ implies the property $(A - z)^{-m}\mathcal{L} = \{0\}$. We have to show that for $z \in \rho(A)$, $x = x_0 + x_2$, $x_0 \in \mathcal{H}_0$, $x_2 \in \mathcal{M}$, the inclusion $R(z)x \in \mathcal{L}$ implies $x \in \mathcal{L}$. Using the matrix representation (5.20) of $R(z)$ we find $R_{00}(z)x_0 + R_{02}(z)x_2 = 0$ and $R_{22}(z)x_2 = 0$. Again by (3.5), the first equality is equivalent to $R_0(z)x_0 + P_0\Gamma_{0z}c_{m+1}(z) = 0$, $z \in \rho(A)$, with a function $c_{m+1}(z)$ which depends on the coefficients of x_2 in the canonical basis in

$\mathbf{C}^m$: $x_2 = \sum_{j=1}^m c_j e_j$; an explicit expression for $c_{m+1}(z)$ is given in (3.6). The second equality is equivalent to $c_j(z) = 0,\ j = 2, \dots, m$, with polynomials $c_j(z)$ from (3.6); it follows that $c_1 = c_2 = \cdots = c_{m-1} = 0$ and $c_{m+1}(z) = c_m$. The condition $\gamma(\lambda_0) \notin \text{dom}\, A_0$ now implies $x_0 = 0$ and $c_m = c_{m+1}(z) = 0$ which yields $x_2 = 0$, hence $x = 0$. □

The statement of the preceding theorem can be formulated in terms of the operators A in Π_m and A_0 in $\mathcal{H}_0$ from Lemma 5.6 and the function $\gamma(z) = P_0 \Gamma_{0z} 1$ as follows. Observe that for $z \in \rho(A)$ the subspace $\mathcal{L} = \ker (A - z)^{-m}$ is always contained in the root space of A at ∞.

Theorem 5.8 *Suppose that $Q(z) \in N_m$ for some $m \geq 1$ and let $A = A(Q)$ be the self-adjoint linear relation and $\Pi_m = \Pi_m(Q)$ be the Pontryagin space $\Pi_m = \Pi_m(Q)$ from the minimal operator representation* (5.10) *of $Q(z)$. Then $Q(z) \in N_m^\infty$ if and only if the following two conditions are satisfied:*

(i) *The subspace $\mathcal{L} = \ker (A-z)^{-m}$, $z \in \rho(A)$, is m-dimensional and neutral.*

(ii) *The operator A_0 from* (5.19) *is an unbounded self-adjoint operator in the Hilbert space $\mathcal{H}_0$ with the property $\gamma(\lambda_0) \notin \text{dom}\, A_0$.*

Proof: The conditions of Theorem 5.7 imply (i) and (ii). Indeed, (i) is weaker than the condition in Theorem 5.7, and (ii) was shown during the proof of the sufficiency part of Theorem 5.7.

Let us now show that (i) and (ii) imply that $\mathcal{L}$ is the full root space of A at ∞. It is sufficient to show that $R(z)x \in \mathcal{L}$ implies that $x \in \mathcal{L}$. First we observe that, by Lemma 5.6, (i) implies the decomposition (5.18) and the existence of a self-adjoint relation A_0 in $\mathcal{H}_0$ such that $(A_0 - z)^{-1}x_0 = P_0(A-z)^{-1}x_0$, $x_0 \in \mathcal{H}_0$. By (ii), A_0 is a self-adjoint operator. It remains to show that (ii) implies that for $x = x_0 + x_2$, $x_0 \in \mathcal{H}_0$, $x_2 \in \mathcal{M}$, and $z \in \rho(A)$ the relation $R(z)x \in \mathcal{L}$ yields $x \in \mathcal{L}$. To this end we use the matrix form (5.20) of $R(z)$. Then $R(z)x \in \mathcal{L}$ can be written as $R_{00}(z)x_0 + R_{02}(z)x_2 = 0$ and $R_{22}(z)x_2 = 0$. Now we can repeat the reasoning at the end of the proof of Theorem 5.7. Decompose x_2 in the basis given by $E_j 1$, $j = 1, \dots, m$: $x_2 = \sum_{j=1}^m E_j c_j$, and use the relations

$$\begin{aligned} R(z)E_j &= E_{j+1} + (z-\lambda_0)E_{j+2} + \cdots + (z-\lambda_0)^{m-j-1}E_m \\ &\quad + (z-\lambda_0)^{m-j}\Gamma_{0z}. \end{aligned}$$

Then the two conditions above can be written as a system of equations

$$R_0(z)x_0 + P_0\Gamma_{0z}c_{m+1}(z) = 0, \quad c_j(z) = 0, \quad j = 1, \dots, m,$$

where the functions $c_j(z)$ are the same as in the formula (3.6). Then the condition $\gamma(\lambda_0) \notin \text{dom}\, A_0$ implies $x_0 = 0$ and $c_j = 0$, $j = 1, \dots, m$, and hence $x_2 = 0$. □

Summing up, it follows from the considerations of this section, that the functions of the class N_m^∞ are exactly the Q-functions which arise by the constructions of Section 3. Thus the study of the models of Section 3 is equivalent to the study of the functions of the class N_m^∞.

Concluding this section we note that the analytical considerations above and also of [S1], [DT], [S3] have an intersection with the results of [JLT].

6 New Representations

6.1

We now return to the space Π_m, the operator H^∞, which is often denoted by A, and the Q-function $Q(z)$ from (3.12). With the representation (3.15) of $Q(z)$:

$$Q(z) = (z-\lambda_0)^m (z-\lambda_0{}^*)^m Q_0(z) + p_{2m-1}(z), \tag{6.1}$$

the Γ-field $\Gamma_z = (I + (z-\lambda_0)(H^\infty - z)^{-1})\Gamma_{\lambda_0}$ and its expansion

$$\Gamma_z = \sum_{j=1}^{m} (z-\lambda_0)^{j-1} E_j + (z-\lambda_0)^m \Gamma_{0z}, \tag{6.2}$$

near λ_0 and the subspaces

$$\mathcal{M} = \text{span}\,\{E_1 1, E_2 1, \dots, E_m 1\}, \quad \mathcal{L} = \ker\,(H^\infty - z)^{-m}$$

we have obtained the decomposition

$$\Pi_m = \mathcal{H}_0 \oplus (\mathcal{L} \dotplus \mathcal{M}) \tag{6.3}$$

of the space Π_m. Here $\mathcal{H}_0$ is the Hilbert space generated by the kernel

$$\frac{Q_0(z) - Q_0(\zeta)^*}{z - \zeta^*} = \Gamma_{0\zeta}^* \Gamma_{0z}.$$

If $\mathcal{L}$ and $\mathcal{M}$ are chosen as in Section 5 to be $\mathbf{C}^m$ then the inner product on $\mathcal{L} \dotplus \mathcal{M}$ is given by the Gram operator

$$\mathbf{G} = \begin{pmatrix} 0 & I_m \\ I_m & G \end{pmatrix}, \quad G = (g_{jk}). \tag{6.4}$$

In this section we show that another representation (5.3) of the function $Q(z)$ leads in a natural way to another decomposition of the space Π_m, and we describe the relation H^∞ and the operators S and H^t, $t \in \mathbf{R}$, in this decomposition.

Let Λ be a finite sequence of m not necessarily different points λ_j in $\mathbf{C}^+ \cup \mathbf{R}$ which are in the domain of holomorphy of $Q(z)$: $\lambda_j \in \rho(H^\infty) = \rho(A)$, $\operatorname{Im} \lambda_j \geq 0$, $j = 1, 2, \dots, m$. Set

$$\widehat{b}_0(z) :\equiv 1, \ \widehat{b}_j(z) := \prod_{k=1}^{j} (z - \lambda_k), \ j = 1, 2, \dots, m. \tag{6.5}$$

The construction in this section corresponds to the representation

$$Q(z) = \widehat{b}_m^*(z)\widehat{b}_m(z)\widehat{Q}_0(z) + \widehat{p}_{2m-1}(z), \tag{6.6}$$

for some function $\widehat{Q}_0(z) \in N_0^\infty$ and a self-adjoint polynomial $\widehat{p}_{2m-1}(z)$ of degree at most $2m-1$. To prove this formula and to derive formulas for $\widehat{Q}_0(z)$ and $\widehat{p}_{2m-1}(z)$, we expand the operator function Γ_z as in (6.2), but this time with respect to the functions $\widehat{b}_j(z)$ instead of $(z-\lambda_0)^j$, that is, we write

$$\Gamma_z = \sum_{j=1}^{m} \widehat{b}_{j-1}(z)\widehat{E}_j + \widehat{b}_m(z)\widehat{\Gamma}_{0z} \tag{6.7}$$

with operators $\widehat{\Gamma}_{0z}$ and $\widehat{E}_j$, $j = 1, 2, \ldots, m$, defined by this relation, $\widehat{\Gamma}_{0z}$ being holomorphic at the points λ_j, $j = 1, 2, \ldots, m$. It is easy to see that $\widehat{E}_1 = \Gamma_{\lambda_1} = (I + (\lambda_1 - \lambda_0)(A - \lambda_1)^{-1})\Gamma_{\lambda_0}$ and

$$\widehat{E}_j = \left(\prod_{l=2}^{j}(A - \lambda_l)^{-1}\right)\Gamma_{\lambda_1}, \quad \widehat{\Gamma}_{0z} = \widehat{b}_m(A)^{-1}\Gamma_z = (A-z)^{-1}\widehat{E}_m. \tag{6.8}$$

Further, introduce for $j, k = 1, 2, \ldots, m$, the numbers $\widehat{g}_{jk} := \widehat{E}_j^*\widehat{E}_k 1$. Then $\widehat{g}_{jk}^* = \widehat{g}_{kj}$ and the relation (6.8) implies

$$(\lambda_k - \lambda_j^*)\widehat{g}_{jk} = \widehat{g}_{j-1,k} - \widehat{g}_{j,k-1} \quad \text{if } \lambda_k \neq \lambda_j^*.$$

The numbers $\widehat{g}_{jk}$ can be obtained recurrently by the rule

$$\begin{aligned} \widehat{g}_{jk} \;&=\; \widehat{E}_j^*\widehat{E}_k 1 = \left(\frac{\Gamma_\zeta - \sum_{l=1}^{j-1}\widehat{b}_{l-1}(\zeta)\widehat{E}_l}{\widehat{b}_{j-1}(\zeta)}\right)^* \\ &\left.\left(\frac{\Gamma_z - \sum_{l=1}^{k-1}\widehat{b}_{l-1}(z)\widehat{E}_l}{\widehat{b}_{k-1}(z)}\right) 1\,\right|_{z=\lambda_k,\,\zeta=\lambda_j}. \end{aligned}$$

We shall show that the function $\widehat{Q}_0(z)$ in (6.6) is given by:

$$\widehat{Q}_0(z) = \widehat{\Gamma}_{0z^*}^*\widehat{E}_m 1 = \widehat{E}_m^*\widehat{\Gamma}_{0z} 1 = (z - \lambda_0^*)\widehat{\Gamma}_{0\lambda_0}^*\widehat{\Gamma}_{0z} 1 + \widehat{g}_{m+1,m}, \tag{6.9}$$

where $\widehat{g}_{m+1,m} = \widehat{g}_{m,m+1}^* = \widehat{\Gamma}_{0\lambda_0}^*\widehat{E}_m 1$. The last two equalities in (6.9) can easily be obtained by using formulas like (2.5). We need the following expressions for $\widehat{E}_j^*\widehat{\Gamma}_{0z}$, $j = 1, 2, \ldots, m$:

$$\begin{aligned} \widehat{E}_j^*\widehat{\Gamma}_{0z} \;&=\; \widehat{\Gamma}_{z^*}^*(\widehat{b}_j^*(A))^{-1}\widehat{E}_m \\ &=\; \left(\widehat{E}_{j+1}^* + \frac{\widehat{b}_{j+1}^*(z)}{\widehat{b}_j^*(z)}\widehat{E}_{j+2}^* + \cdots + \frac{\widehat{b}_{m-1}^*(z)}{\widehat{b}_j^*(z)}\widehat{E}_m^* + \frac{\widehat{b}_m^*(z)}{\widehat{b}_j^*(z)}\widehat{\Gamma}_{0z^*}^*\right)\widehat{E}_m. \end{aligned} \tag{6.10}$$

To see the second equality, we write $\widehat{b}_j^*(z)\times$ the left-hand side as

$$\begin{aligned}\widehat{b}_j^*(z)\widehat{E}_j^*\widehat{\Gamma}_{0z} &= \Gamma_{\lambda_1}^*\widehat{b}_j^*(z)\widehat{b}_{-1,j}^*(A)^{-1}\widehat{b}_m(A)^{-1}\Gamma_z\\ &= \Gamma_{\lambda_1}^*\widehat{b}_j^*(z)\widehat{b}_{-1,j}^*(A)^{-1}(A-z)^{-1}\widehat{b}_{-1,m}(A)^{-1}\Gamma_{\lambda_1},\end{aligned}$$

where $\widehat{b}_{-1,j}(A)$ is $\widehat{b}_j(A)$ without the factor $(A-\lambda_1)^{-1}$. The formula then follows from the expansion of $(A-z)^{-1}$:

$$\begin{aligned}(A-z)^{-1} &= (A-\lambda_{j+1}^*)^{-1}+(z-\lambda_{j+1}^*)(A-\lambda_{j+1}^*)^{-1}(A-z)^{-1}\\ &+\cdots+(z-\lambda_{j+1}^*)\cdots(z-\lambda_{m-1}^*)(A-\lambda_{j+1}^*)^{-1}\cdots(A-\lambda_{m-1}^*)^{-1}\\ &(A-\lambda_m^*)^{-1}+(z-\lambda_{j+1}^*)\cdots(z-\lambda_{m-1}^*)(z-\lambda_m^*)\\ &(A-\lambda_{j+1}^*)^{-1}\cdots(A-\lambda_{m-1}^*)^{-1}(A-\lambda_m^*)^{-1}(A-z)^{-1}.\end{aligned}$$

We rewrite the operator representation (5.10) of $Q(z)$ in the form

$$Q(z)=(z-\lambda_1^*)\Gamma_{\lambda_1}^*\Gamma_z 1-i\operatorname{Im}\lambda_1\widehat{g}_{11}+\operatorname{Re}Q(\lambda_1) \tag{6.11}$$

(recall that $Q(z)$ is a solution of (2.6)). Inserting the decomposition (6.7) of Γ_z into (6.11) we get

$$\begin{aligned}Q(z) &= \widehat{b}_1^*(z)\widehat{b}_m(z)\widehat{E}_1^*\widehat{\Gamma}_{0z}1\\ &+\widehat{b}_1^*(z)\sum_{j=1}^m\widehat{b}_{j-1}(z)\widehat{E}_1^*\widehat{E}_j 1-i\operatorname{Im}\lambda_1\widehat{g}_{11}+\operatorname{Re}Q(\lambda_1).\end{aligned}$$

Finally, replacing $\widehat{E}_1^*\widehat{\Gamma}_{0z}1$ on the right-hand side by the expression (6.10) with $j=1$ we obtain (6.6), where the self-adjoint polynomial $\widehat{p}_{2m-1}(z)$ has the form

$$\begin{aligned}\widehat{p}_{2m-1}(z) &= \sum_{j=2}^m\widehat{b}_m(z)\widehat{b}_{j-1}^*(z)\widehat{g}_{jm}\\ &+\sum_{j=1}^m\widehat{b}_1^*(z)\widehat{b}_{j-1}(z)\widehat{g}_{1j}-i\operatorname{Im}\lambda_1\widehat{g}_{11}+\operatorname{Re}Q(\lambda_1),\end{aligned} \tag{6.12}$$

which is similar to (3.14). That $\widehat{Q}_0(z)$ belongs to N_0^∞ follows from Theorem 5.1.

Now we decompose the space Π_m in the form (5.18) but this time corresponding to the expansion (6.7). To this end we introduce the subspace

$$\widehat{\mathcal{M}}:=\operatorname{span}\{\widehat{E}_1 1,\widehat{E}_2 1,\ldots,\widehat{E}_m 1\}.$$

This subspace is independent of the ordering of the points λ_j, in fact the following lemma holds:

Lemma 6.1 *Let $\mu_1,\mu_2,\ldots,\mu_n$ be the different values of λ_j, $j=1,2,\ldots,m$, in the sequence Λ and let l_j be the number of appearances of μ_j in Λ. Then*

$$\widehat{\mathcal{M}}=\operatorname{span}\{\Gamma_{\mu_j}^{(k)}1\mid k=0,1,\ldots,l_j-1,\ j=1,2,\ldots,n\}.$$

In particular, $\widehat{\mathcal{M}}$ is independent of the ordering of the points λ_j in Λ.

Proof: The operators $\widehat{E}_j$, $j = 1, 2, \ldots, m$, in the expansion (6.7) satisfy the relations

$$\Gamma_{\mu_i}^{(k)} = \sum_{j=1}^{m} \widehat{b}_{j-1}^{(k)}(\mu_i)\widehat{E}_j, \quad k = 0, 1, \ldots, l_i - 1, \; i = 1, 2, \ldots, n.$$

These relations are obtained from (6.7) by differentiating both sides k times, taking the limit for $z \to \mu_i$ and using that

$$\lim_{z \to \mu_i} (z - \mu_i)(A - z)^{-1}\widehat{E}_m = 0.$$

Note that $l_1 + l_2 + \cdots + l_n = m$. If we set for $l_1 + l_2 + \cdots + l_{i-1} \leq k \leq l_1 + l_2 + \cdots + l_i - 1, i = 1, 2, \ldots, n$, and $j = 1, 2, \ldots, m$,

$$\Gamma_k = \Gamma_{\mu_i}^{(k-l_1-l_2-\cdots-l_{i-1})}, \quad B_{kj} = \widehat{b}_{j-1}^{(k-l_1-l_2-\cdots-l_{i-1})}(\mu_i),$$

then these relations can be written as

$$\Gamma_k = \sum_{j=1}^{m} B_{kj}\widehat{E}_j, \quad k = 1, 2 \ldots, m.$$

The lemma follows from the fact that the matrix $(B_{kj})_{k,j=1}^{m}$ is invertible. □

Recall that $A = H^{\infty}$ and that the subspace $\mathcal{L}$ of Π_m is neutral.

Lemma 6.2 *The subspaces* $\mathcal{L} = \ker(A - z)^{-m}$ *and* $\widehat{\mathcal{M}}$ *of* Π_m *are in duality.*

Proof: Consider the original representation (6.3) of the space Π_m with $\mathcal{M} = \mathbf{C}^m$. We denote the projection onto $\mathcal{M}$, which is defined by this representation, by P_3. Then the representation (5.20) of the resolvent $(A - z)^{-1}$ and the relations (6.8) imply that

$$P_3\widehat{E}_1 1 = \begin{pmatrix} 1 \\ \vdots \\ \vdots \end{pmatrix}, \quad P_3\widehat{E}_2 1 = \begin{pmatrix} 0 \\ 1 \\ \vdots \\ \vdots \end{pmatrix}, \quad \ldots, \quad P_3\widehat{E}_m 1 = \begin{pmatrix} 0 \\ \vdots \\ 0 \\ 1 \end{pmatrix},$$

where all the entries above the indicated 1's are zeros, that is,

$$P_3\widehat{E}_j 1 \in e_j + \operatorname{span}\{e_{j+1}, \ldots, e_m\}, \quad j = 1, 2, \ldots, m.$$

Now the claim follows from the form of the Gram operator which defines the inner product on $\mathcal{L} \dot{+} \mathcal{M}$. □

Next we define the Hilbert space $\widehat{\mathcal{H}}_0$ through the decomposition

$$\Pi_m = \widehat{\mathcal{H}}_0 \oplus (\mathcal{L} \dot{+} \widehat{\mathcal{M}}), \tag{6.13}$$

where $\mathcal{L}$ is the same as in (6.3). Let $\widehat{P}_0$ be the orthogonal projection onto $\widehat{\mathcal{H}}_0$. The inner product of the space Π_m in the decomposition (6.13) is give by the

Gram operator $\widehat{\mathbf{G}}$ of the form (6.4) but with the matrix $\widehat{G} = (\widehat{g}_{jk})$ instead of G. The off-diagonal matrices in $\widehat{\mathbf{G}}$ can remain I_m by choosing a basis in $\mathcal{L}$ which is biorthonormal to the basis of $\widehat{\mathcal{M}}$; in the sequel we shall assume that this has been done.

As in Section 5 it follows now that with respect to the decomposition (6.13) the resolvent $R(z) := (A - z)^{-1}$ has the matrix representation

$$R(z) = \begin{pmatrix} \widehat{R}_{00}(z) & 0 & \widehat{R}_{02}(z) \\ \widehat{R}_{10}(z) & \widehat{R}_{11}(z) & \widehat{R}_{12}(z) \\ 0 & 0 & \widehat{R}_{22}(z) \end{pmatrix}. \tag{6.14}$$

and that $\widehat{R}_{00}(z)$ is the resolvent of a self-adjoint operator $\widehat{A}_0$ in $\widehat{\mathcal{H}}_0$, given by the formula

$$\begin{aligned} \widehat{A}_0 &= \{\{\widehat{R}_{00}(z) f_0, (I + z\widehat{R}_{00}(z)) f_0\} \mid f_0 \in \widehat{\mathcal{H}}_0\} \\ &= \{\{\widehat{P}_0 f, \widehat{P}_0 g\} \mid \{f, g\} \in A, g - zf \in \widehat{\mathcal{H}}_0\}. \end{aligned}$$

Further, we define

$$\widehat{\gamma}(z) := \widehat{P}_0 \widehat{\Gamma}_{0z} 1 = \widehat{P}_0 (\widehat{b}_m(A))^{-1} \Gamma_z 1, \qquad \widehat{\chi}_0 := \widehat{\gamma}(\lambda_0).$$

The inclusion $(\widehat{b}_m(A))^{-1}\Gamma_z 1 \in \mathcal{L}^\perp$ and the matrix representation (6.14) imply

$$\widehat{P}_0 (A - z)^{-1} \widehat{\gamma}(z) = \widehat{P}_0 (\widehat{A}_0 - z)^{-1} \widehat{P}_0 \widehat{\gamma}(z). \tag{6.15}$$

As a consequence, the elements $\widehat{\gamma}(z)$ satisfy the relation

$$\widehat{\gamma}(z) - \widehat{\gamma}(w) = (z - w)(\widehat{A}_0 - z)^{-1} \widehat{\gamma}(w).$$

Moreover,

$$\widehat{\mathcal{H}}_0 = \overline{\operatorname{span}}\, \{\widehat{\gamma}(z) \mid z \in \rho(\widehat{A}_0)\}. \tag{6.16}$$

Also $\widehat{Q}_0(z)$ in (6.9) can be expressed in terms of $\widehat{\gamma}(z)$:

$$\widehat{Q}_0(z) = (z - \lambda_0^*)\langle \widehat{\gamma}(z), \widehat{\gamma}(\lambda_0)\rangle - i \operatorname{Im} \lambda_0 \langle \widehat{\gamma}(\lambda_0), \widehat{\gamma}(\lambda_0)\rangle + \operatorname{Re} g_{m+1,m}. \tag{6.17}$$

Proposition 6.3 *In the new decomposition* (6.13) *with the basis* $\widehat{u}_j := \widehat{E}_j 1$, $j = 1, \dots, m$, *of* $\widehat{\mathcal{M}}$ *and the biorthonormal basis* $\widehat{w}_j$, $j = 1, \dots, m$, *of* $\mathcal{L}$ *the matrix representation of the self-adjoint linear relation* $A = H^\infty$ *is of the form*

$$A = \left\{ \left\{ \begin{pmatrix} h + c_{m+1}\widehat{\chi}_0 \\ \begin{pmatrix} b_1 \\ (b_j)_{j=2}^m \end{pmatrix} \\ (c_j)_{j=1}^m \end{pmatrix}, \begin{pmatrix} \widehat{A}_0 h + \lambda_0 c_{m+1}\widehat{\chi}_0 \\ \begin{pmatrix} \lambda_1^* b_1 + b_0 \\ (\lambda_j^* b_j + b_{j-1} - \widehat{g}_{jm} c_{m+1})_{j=2}^m \end{pmatrix} \\ (\lambda_j c_j + c_{j+1})_{j=1}^m \end{pmatrix} \right\} \;\middle|\; \begin{array}{l} h \in \operatorname{dom} \widehat{A}_0,\ c_1 = 0, \\ b_m = \langle (\widehat{A}_0 - \lambda_0^*)h, \widehat{\chi}_0\rangle_0 + \widehat{g}_{m,m+1} c_{m+1}, \\ c_{j+1}, b_{j-1} \in \mathbf{C}, \\ j = 1, \dots, m \end{array} \right\}.$$

Proof: Consider the linear relation B:

$$B = \left\{ \left\{ \sum_z d_z \Gamma_z 1, \sum_z z d_z \Gamma_z 1 \right\} \middle| d_z \in \mathbf{C}, \sum_z d_z = 0 \right\},$$

where $\sum_z$ stands for "finite sum over $z \in \rho(A)$." We claim that the closure $\overline{B}$ of B coincides with A. Indeed, if $\sum_z d_z = 0$,

$$\sum_z d_z \{\Gamma_z 1, z\Gamma_z 1\} = \sum_z d_z (z - \lambda_0) \{R(z)\Gamma_{\lambda_0} 1, (I + zR(z))\Gamma_{\lambda_0} 1\} \in A,$$

and hence $B \subset A$ and then also $\overline{B} \subset A$. On the other hand, for $z \neq \lambda_0$,

$$\{R(z)\Gamma_{\lambda_0} 1, (I + zR(z))\Gamma_{\lambda_0} 1\} = \frac{1}{z - \lambda_0} (\{\Gamma_z 1, z\Gamma_z 1\} - \{\Gamma_{\lambda_0} 1, \lambda_0 \Gamma_{\lambda_0} 1\}) \in B,$$

so

$$\{R(z)\Gamma_{\lambda_0} 1, (I + zR(z))\Gamma_{\lambda_0} 1\} \in \overline{B}, \quad z \in \rho(A).$$

Hence also

$$\begin{aligned} \{R(z)R(w)\Gamma_{\lambda_0} 1, (I + zR(z))R(w)\Gamma_{\lambda_0} 1\} &= \frac{1}{z - w} \\ (\{R(z)\Gamma_{\lambda_0} 1, (I + zR(z))\Gamma_{\lambda_0} 1\} &- \{R(w)\Gamma_{\lambda_0} 1, (I + wR(w))\Gamma_{\lambda_0} 1\}) \in \overline{B}, \end{aligned}$$

and by the simplicity of S, we obtain $A \subset \overline{B}$. This proves the claim.

With respect to the decomposition (6.13) an element in B can be written as (here we used the expansion (6.7) and that $\Gamma_{0z} 1 \in \mathcal{L}^\perp = \widehat{\mathcal{H}}_0 + \mathcal{L}$):

$$\sum_z d_z \{\Gamma_z 1, z\Gamma_z 1\} = \left\{ \begin{pmatrix} h + c_{m+1}\widehat{\chi}_0 \\ \sum_{j=1}^m b_j \widehat{w}_j \\ \sum_{j=1}^m c_j \widehat{u}_j \end{pmatrix}, \begin{pmatrix} h' + \lambda_0 c_{m+1}\widehat{\chi}_0 \\ \sum_{j=1}^m b'_j \widehat{w}_j \\ \sum_{j=1}^m c'_j \widehat{u}_j \end{pmatrix} \right\},$$

where

$$\begin{aligned} b_j &= \sum_z d_z \widehat{b}_m(z) \widehat{E}_j^* \widehat{\Gamma}_{0z}, \quad j = 1, \dots, m, \\ c_j &= \sum_z d_z \widehat{b}_{j-1}(z), \quad j = 1, \dots, m+1, \\ h &= \sum_z d_z \widehat{b}_m(z) (\widehat{\gamma}(z) - \widehat{\gamma}(\lambda_0)), \end{aligned}$$

and

$$\begin{aligned} b'_j &= \sum_z z d_z \widehat{b}_m(z) \widehat{E}_j^* \widehat{\Gamma}_{0z} 1, \quad j = 1, \dots, m, \\ c'_j &= \sum_z z d_z \widehat{b}_{j-1}(z), \quad j = 1, \dots, m, \\ h' &= \sum_z d_z \widehat{b}_m(z) (z\widehat{\gamma}(z) - \lambda_0 \widehat{\gamma}(\lambda_0)). \end{aligned}$$

Simple calculations show

$$\begin{aligned}
c_1 &= \sum_z d_z = 0, \\
c_j' &= \lambda_j \sum_z d_z \widehat{b}_{j-1}(z) + \sum_z (z - \lambda_j) d_z \widehat{b}_{j-1}(z) \\
&= \lambda_j c_j + c_{j+1}, \quad j = 1, \dots, m, \\
h' &= \sum_z (z - \lambda_0) d_z \, \widehat{b}_m(z) \, (\widehat{\gamma}(\lambda_0) + z(\widehat{A}_0 - z)^{-1} \widehat{\gamma}(\lambda_0)).
\end{aligned}$$

Using the relation

$$\begin{aligned}
(z - \lambda_j^*) \widehat{E}_j^* \widehat{\Gamma}_{0z} &= (z - \lambda_j^*) \widehat{E}_{j-1}^* (A - \lambda_j^*)^{-1} (A - z)^{-1} \widehat{E}_m \\
&= \widehat{E}_{j-1}^* \widehat{\Gamma}_{0z} - \widehat{E}_j^* \widehat{E}_m,
\end{aligned}$$

we see that for $j = 2, \dots, m$,

$$\begin{aligned}
b_j' &= \lambda_j^* \sum_z d_z \, \widehat{b}_m(z) \widehat{E}_j^* \widehat{\Gamma}_{0z} 1 + \sum_z d_z \, (z - \lambda_j^*) \widehat{b}_m(z) \, \widehat{E}_j^* \widehat{\Gamma}_{0z} 1 \\
&= \lambda_j^* b_j + b_{j-1} - \widehat{g}_{jm} \, c_{m+1}.
\end{aligned}$$

For $j = 1$ we find:

$$\begin{aligned}
b_1' &= \lambda_1^* \sum_z d_z \widehat{b}_m(z) \widehat{E}_1^* \Gamma_{0z} 1 + \sum_z d_z \, (z - \lambda_1^*) \widehat{b}_m(z) \, \widehat{E}_1^* \widehat{\Gamma}_{0z} 1 \\
&= \lambda_1^* b_1 + b
\end{aligned}$$

with

$$b := \sum_z d_z \, (z - \lambda_1^*) \widehat{b}_m(z) \widehat{E}_1^* \widehat{\Gamma}_{0z} 1.$$

Observing that

$$h = \sum_z d_z \, \widehat{b}_m(z) \, (z - \lambda_0)(\widehat{A}_0 - z)^{-1} \widehat{\gamma}(\lambda_0),$$

we obtain $\{h, h'\} \in \widehat{A}_0$, that is, $h' = \widehat{A}_0 h$. Finally,

$$\begin{aligned}
b_m &= \sum_z d_z \, \widehat{b}_m(z) \widehat{E}_m^* \widehat{\Gamma}_{0z} 1 \\
&= \sum_z d_z \, \widehat{b}_m(z) \widehat{E}_m^* (\widehat{\Gamma}_{0z} - \widehat{\Gamma}_{0\lambda_0}) 1 + \widehat{g}_{m,m+1} \, c_{m+1} \\
&= \sum_z d_z \, (z - \lambda_0) \widehat{b}_m(z) \widehat{E}_m^* (A - z)^{-1} \Gamma_{0\lambda_0} 1 + \widehat{g}_{m,m+1} \, c_{m+1} \\
&= \langle (\widehat{A}_0 - \lambda_0^*) h, \widehat{\chi}_0 \rangle + \widehat{g}_{m,m+1} \, c_{m+1}.
\end{aligned}$$

The last equality follows from (6.15):

$$\begin{aligned}
&(z-\lambda_0)\widehat{E}_m^*(A-z)^{-1}\widehat{\Gamma}_{0\lambda_0}1\\
&\quad=(z-\lambda_0)\langle\widehat{\Gamma}_{0\lambda_0}1,(A-z^*)^{-1}\widehat{b}_{-1,m}(A)^{-1}\Gamma_{\lambda_1}1\rangle\\
&\quad=(z-\lambda_0)\langle\widehat{\Gamma}_{0\lambda_0}1,(I+(z^*-\lambda_1)(A-z^*)^{-1})\widehat{b}_m(A)^{-1}\Gamma_{\lambda_1}1\rangle\\
&\quad=(z-\lambda_0)\langle\widehat{\gamma}(\lambda_0),(I+(z^*-\lambda_1)(\widehat{A}_0-z^*)^{-1})\widehat{\gamma}(\lambda_1)\rangle\\
&\quad=(z-\lambda_0)\langle\widehat{\gamma}(\lambda_0),(\widehat{A}_0-z^*)^{-1}(\widehat{A}_0-\lambda_1)\widehat{\gamma}(\lambda_1)\rangle\\
&\quad=(z-\lambda_0)\langle(\widehat{A}_0-z)^{-1}\widehat{\gamma}(\lambda_0),(\widehat{A}_0-\lambda_0)\widehat{\gamma}(\lambda_0)\rangle.
\end{aligned}$$

Thus the elements of B have the desired form, and since $A=\overline{B}$ the formula for A now easily follows. □

The symmetric operator S, the one-dimensional restriction of A given by (3.9), can now be written as

$$S=A\cap\{\{\Gamma_{\lambda_1}1,\lambda_1\Gamma_{\lambda_1}1\}\}^*=\{\{f,f'\}\in A\mid\widehat{E}_1^*(f'-\lambda_1 f)=0\}.$$

For $z\in\rho(A)$, analogously to (3.11), the Γ-field $\Gamma_z:\mathbf{C}\backslash\mathbf{R}\to\ker(S^*-z)$ is represented as

$$\Gamma_z1=(I+(z-\lambda_1)(A-z)^{-1})\Gamma_{\lambda_1}1=\begin{pmatrix}\widehat{b}_m(z)\widehat{\gamma}(z)\\(\widehat{b}_m(z)\widehat{q}_j(z))_{j=1}^m\\(\widehat{b}_{j-1}(z))_{j=1}^m\end{pmatrix},\tag{6.18}$$

where

$$\widehat{q}_j(z)=(\widehat{b}_j^*(z))^{-1}\left(\widehat{b}_m^*(z)\widehat{Q}_0(z)+\sum_{k=j+1}^m\widehat{b}_{k-1}^*(z)\widehat{g}_{km}\right).$$

This formula follows from (6.7)–(6.10). The adjoint linear relation S^* becomes

$$S^*=A\dotplus\operatorname{span}\{\{\widehat{E}_11,\lambda_1\widehat{E}_11\}\}=S\dotplus\{0,A(0)\}\dotplus\operatorname{span}\{\{\widehat{E}_11,\lambda_1\widehat{E}_11\}\}.$$

With the new decomposition (6.13) of Π_m the operators H^t which are the canonical self-adjoint operator extensions of S have the direct sum representation

$$H^t=S\dotplus\operatorname{span}\left\{\left\{\begin{pmatrix}0\\0\\e_1\end{pmatrix},\begin{pmatrix}0\\-(t+Q(\lambda_1))e_1\\\lambda_1e_1\end{pmatrix}\right\}\right\},\tag{6.19}$$

that is,

$$H^t=S\dotplus\operatorname{span}\{\{\widehat{u}_1,\lambda_1\widehat{u}_1-(t+Q(\lambda_1))\widehat{w}_1\}\}.$$

To see this we observe that every canonical self-adjoint extension of S is of the form

$$S\dotplus\operatorname{span}\{\{\widehat{u}_1,\lambda_1\widehat{u}_1-(s+i\widehat{g}_{11}\operatorname{Im}\lambda_1)\widehat{w}_1\}\}$$

with $s \in \mathbf{R} \cup \{\infty\}$. Hence it coincides with H^t if and only if

$$\{\widehat{u}_1, \lambda_1\widehat{u}_1 - (s + i\widehat{g}_{11}\operatorname{Im}\lambda_1)\widehat{w}_1\} \in H^t,$$

or, equivalently,

$$(H^t - \lambda_1)^{-1}\widehat{w}_1 = -\frac{\widehat{u}_1}{s + i\widehat{g}_{11}\operatorname{Im}\lambda_1}.$$

By M. G. Krein's formula 3.18 we have that

$$(H^t - \lambda_1)^{-1}\widehat{w}_1 = -\frac{\Gamma_{\lambda_1}\Gamma^*_{\lambda_1^*}\widehat{w}_1}{Q(\lambda_1) + t} = -\frac{\widehat{u}_1}{Q(\lambda_1) + t},$$

hence $s + i\widehat{g}_{11}\operatorname{Im}\lambda_1 = Q(\lambda_1) + t$, and this implies the representation (6.19) of H^t. Note $Q(\lambda_1) = \operatorname{Re} Q(\lambda_1) + i\,\widehat{g}_{11}\operatorname{Im}\lambda_1$.

We refer to the decomposition (6.13) as to the *decomposition of* Π_m *corresponding to the given sequence* Λ, and to the representations of $A = H^\infty$, S, H^t with respect to this decomposition as to the *representations corresponding to* Λ.

6.2

Now we reformulate the Theorems 4.1 and 4.2 with respect to the new space $\widehat{\mathcal{H}}_0$ and the symmetric operator

$$\widehat{S}_0 := \{\{h, \widehat{A}_0 h\} \in \widehat{A}_0 \mid \langle(\widehat{A}_0 - \lambda_0^*)h, \widehat{\chi}_0\rangle_0 = 0\} = \widehat{A}_0 \cap \{\{\widehat{\chi}_0, \lambda_0\widehat{\chi}_0\}\}^*.$$

The operator $\widehat{S}_0$ has deficiency index (1, 1), and

$$\widehat{S}_0^* = \widehat{A}_0 \dot{+} \operatorname{span}\{\{\widehat{\chi}_0, \lambda_0\widehat{\chi}_0\}\}.$$

Von Neumann's formula for $\widehat{S}_0^*$ takes the form

$$\widehat{S}_0^* = \widehat{S}_0 \dot{+} \operatorname{span}\{\{\widehat{\chi}_0, \lambda_0\widehat{\chi}_0\}\} \dot{+} \operatorname{span}\{\{\widetilde{\chi}_0, \lambda_0^*\widetilde{\chi}_0\}\},$$

where

$$\widetilde{\chi}_0 = (I + (\lambda_0{}^* - \lambda_0)\widehat{R}_0(\lambda_0^*))\widehat{\chi}_0.$$

Clearly, the Štraus relation and the generalized resolvent in $\widehat{\mathcal{H}}_0$ associated with H^t are

$$\widehat{T}_t(z) = \{\{\widehat{P}_0 f, \widehat{P}_0 g\} \mid \{f, g\} \in H^t,\ g - zf \in \widehat{\mathcal{H}}_0\}, \quad z \in \mathbf{C},$$

and

$$\widehat{R}_0^t(z) = \widehat{P}_0(H^t - z)^{-1}|_{\widehat{\mathcal{H}}_0},$$

respectively. We define the polynomial

$$\widehat{p}_{2m}(z) := \widehat{b}_m(z)\widehat{b}_m^*(z)\widehat{g}_{m,m+1} + \widehat{p}_{2m-1}(z), \tag{6.20}$$

where the polynomial $\widehat{p}_{2m-1}(z)$ is given by (6.12). Now the description of the Štraus extensions and also its proof are completely analogous to Theorem 4.1 and its proof. We just formulate this result.

Theorem 6.4 *The Štraus extension $\widehat{T}_t(z)$ of $\widehat{S}_0$ associated with H^t, $t \in \mathbf{R}$, is given by*

$$\widehat{T}_t(z) = \begin{cases} \widehat{S}_0 \dotplus \operatorname{span}\{\{\widetilde{\chi}_0 - \widehat{\theta}_t(z)\widehat{\chi}_0,\ \lambda_0^*\widetilde{\chi}_0 - \lambda_0\widehat{\theta}_t(z)\chi_0\}\}, & \operatorname{Im} z > 0, \\ \widehat{S}_0 \dotplus \operatorname{span}\{\{\widehat{\chi}_0 - \widehat{\theta}_t^*(z)\widetilde{\chi}_0,\ \lambda_0\widehat{\chi}_0 - \lambda_0{}^*\widehat{\theta}_t^*(z)\widetilde{\chi}_0\}\}, & \operatorname{Im} z < 0, \end{cases}$$

(6.21)

with

$$\widehat{\theta}_t(z) = \frac{\widehat{p}_{2m}^*(z) + t}{\widehat{p}_{2m}(z) + t}, \tag{6.22}$$

the polynomial $\widehat{p}_{2m}(z)$ being defined by (6.20).

The representation of the generalized resolvents corresponding to Λ becomes slightly different. The proof is again analogous to the proof of Theorem 4.2.

Theorem 6.5 *The generalized resolvent $\widehat{R}_0^t(z) = \widehat{P}_0(H^t - z)^{-1}|_{\widehat{\mathcal{H}}_0}$ of $\widehat{S}_0$ admits the representation*

$$\widehat{R}_0^t(z) = \widehat{R}_0(z) - \frac{\langle\,\cdot\,, \widehat{\gamma}(z^*)\rangle_0\,\widehat{\gamma}(z)}{\widehat{Q}_0(z) + \widehat{t}(z)}, \tag{6.23}$$

where $\widehat{Q}_0(z)$ is the Q-function associated with $\widehat{S}_0$ and $\widehat{A}_0$ defined by (6.9), *and*

$$\widehat{t}(z) = (\widehat{b}_m(z)\widehat{b}_m^*(z))^{-1}\,(\widehat{p}_{2m-1}(z) + t). \tag{6.24}$$

6.3

In the next theorem we show that only for points $\lambda \in \Lambda$ the root subspace $\mathcal{L}_\lambda(H^t)$ can have a nontrivial intersection with $\widehat{\mathcal{H}}_0^\perp = \mathcal{L} \dotplus \widehat{\mathcal{M}}$. To this end we enumerate the points of Λ such that the first n points $\lambda_1, \ldots, \lambda_n$ are mutually different and denote by l_j the number of appearances of λ_j, $j = 1, 2, \ldots, n$, in Λ; hence $\sum_{j=1}^n l_j = m$.

Theorem 6.6 *Let $\lambda \in \mathbf{C}^+ \cup \mathbf{R}$ be a point in $\rho(A) \cap \sigma_p(H^t)$.*

(i) *If $\lambda \notin \Lambda$, then $\mathcal{L}_\lambda(H^t) \cap \widehat{\mathcal{H}}_0^\perp = \{0\}$.*

(ii) *If $\lambda \in \Lambda$, then $\mathcal{L}_\lambda(H^t) \cap \widehat{\mathcal{H}}_0^\perp$ is spanned by the first r elements of the chain*

$$\Gamma_\lambda 1,\ \frac{1}{1!}\Gamma_\lambda^{(1)}1,\ \frac{1}{2!}\Gamma_\lambda^{(2)}1,\ \ldots,$$

of H^t at λ (and hence is part of $\widehat{\mathcal{M}}$), where $r = \min(k, l)$, $k = \dim \mathcal{L}_\lambda(H^t)$, and l is the number of appearances of λ in Λ.

Proof: (i) If $x \in \mathcal{L}_\lambda(H^t)$ then, according to the proof of Theorem 3.3, $x = \sum_{i=0}^{k-1} \xi_i \Gamma_\lambda^{(i)} 1$ with numbers $\xi_i \in \mathbf{C}$ and $k = \dim \mathcal{L}_\lambda(H^t)$. If again $\widehat{P}_0$ denotes the projection onto $\widehat{\mathcal{H}}_0$ and $\widehat{x}_0 := \widehat{P}_0 x$ it follows that $\widehat{x}_0 = \sum_{i=0}^{k-1} \xi_i \widehat{P}_0 \Gamma_\lambda^{(i)} 1$. We prove (i) by showing that if $\lambda \notin \Lambda$ and $\widehat{x}_0 = 0$, then all ξ_i are zero and hence $x = 0$. Differentiating the expression in (6.7) i-times we find

$$\Gamma_z^{(i)} = \sum_{j=1}^{m} \widehat{b}_{j-1}^{(i)}(z) \widehat{E}_j + \sum_{l=0}^{i} \binom{i}{l} \widehat{b}_m^{(i-l)}(z) R^{(l)}(z) \widehat{E}_m,$$

and, observing that $\widehat{P}_0 (A-\lambda)^{-1} \widehat{E}_m 1 = \widehat{\gamma}(\lambda)$, and $\widehat{P}_0 (A-\lambda)^{-l-1} \widehat{E}_m 1 = (\widehat{A}_0 - \lambda)^{-l} \widehat{\gamma}(\lambda)$ (see (6.15)),

$$\begin{aligned}
\widehat{x}_0 &= \sum_{i=0}^{k-1} \xi_i \sum_{l=0}^{i} \binom{i}{l} \widehat{b}_m^{(i-l)}(\lambda) \widehat{P}_0 R^{(l)}(\lambda) \widehat{E}_m 1 \\
&= \sum_{l=0}^{k-1} \sum_{i=l}^{k-1} \xi_i \binom{i}{l} \widehat{b}_m^{(i-l)}(\lambda) \widehat{P}_0 R^{(l)}(\lambda) \widehat{E}_m 1 = \sum_{l=0}^{k-1} \eta_l \widehat{P}_0 R^{(l)}(\lambda) \widehat{E}_m 1 \\
&= \sum_{l=0}^{k-1} \eta_l \widehat{P}_0 \, l! \, R(\lambda)^{l+1} \widehat{E}_m 1 = \sum_{l=0}^{k-1} \eta_l \widehat{P}_0 \, l! \, \widehat{R}_0(\lambda)^{l} \widehat{\gamma}(\lambda) \\
&= \sum_{l=0}^{k-1} \eta_l \, l! \, (\widehat{A}_0 - \lambda)^{-l} \widehat{\gamma}(\lambda)
\end{aligned} \tag{6.25}$$

with

$$\eta_l = \sum_{i=l}^{k-1} \xi_i \binom{i}{l} \widehat{b}_m^{(i-l)}(\lambda). \tag{6.26}$$

The elements $(\widehat{A}_0 - \lambda)^{-l} \widehat{\gamma}(\lambda)$, $l = 0, 1, \ldots, k-1$, are linearly independent. Indeed, if

$$\sum_{l=0}^{k-1} \alpha_l \, l! \, (\widehat{A}_0 - \lambda)^{-l} \widehat{\gamma}(\lambda) = 0$$

and for example $\alpha_0 \neq 0$, then $\widehat{\gamma}(\lambda)$ would be a linear combination of the elements $(\widehat{A}_0 - \lambda)^{-l} \widehat{\gamma}(\lambda)$ for $l = 1, 2, \ldots, k-1$, hence $\widehat{\gamma}(\lambda) \in \operatorname{dom} \widehat{A}_0$, which is impossible.

Now, assume $\widehat{x}_0 = 0$, then on account of (6.25), all of the η_i are zero. If we consider the expressions in (6.26) for the η_l as a system of equations for the ξ_i, it has a triangular form and its determinant equals $(\widehat{b}_m(\lambda))^k$, which is nonzero if $\lambda \notin \Lambda$, hence all the ξ_i are zero.

(ii) Now assume $\lambda = \lambda_j \in \sigma_p(H^t)$, and we can further assume that $j = 1$ and, just for this part of the proof we suppose that $\lambda_1 = \lambda_2 = \cdots = \lambda_{l_1}$, which we can do because of Lemma 6.1. Then, according to the proof of Theorem 3.3, $\Gamma_{\lambda_1}^{(i-1)}1 \in \mathcal{L}_{\lambda_1}(H^t)$, $i = 1, 2, \ldots, k$. Further, if $k \leq l_1$,

$$\Gamma_{\lambda_1}1 = \widehat{E}_1 1, \ \ldots, \ \frac{1}{(k-1)!}\Gamma_{\lambda_1}^{(k-1)}1 = \widehat{E}_k 1.$$

The vectors $\widehat{E}_i 1$, $i = 1, \ldots, k$, are linearly independent and belong to $\widehat{\mathcal{M}}$ by Lemma 6.1. This proves (ii) if $k \leq l_1$.

Now suppose that $k > l_1$. We show that the intersection of $\mathcal{L}_{\lambda_1}(H^t)$ and $\widehat{\mathcal{H}}_0^{\perp}$ coincides with the subspace spanned by $\Gamma_{\lambda_1}1$, $\Gamma_{\lambda_1}^{(1)}1, \ldots, \Gamma_{\lambda_1}^{(l_1-1)}1$, which is the subspace of $\widehat{\mathcal{M}}$ spanned by $\widehat{E}_1 1, \widehat{E}_2 1, \ldots, \widehat{E}_{l_1} 1$. This reasoning is close to the one in the first part of the proof. We have

$$\text{(6.27)} \quad \operatorname{span}\left\{\Gamma_{\lambda_1}1, \ \frac{1}{1!}\Gamma_{\lambda_1}^{(1)}1, \ldots, \ \frac{1}{(l_1-1)!}\Gamma_{\lambda_1}^{(l_1-1)}1\right\} \subset \mathcal{L}_{\lambda_1}(H^t) \cap \widehat{\mathcal{H}}_0^{\perp},$$

so it remains to prove the converse inclusion. Let $x \in \mathcal{L}_{\lambda_1}(H^t) \cap \widehat{\mathcal{H}}_0^{\perp}$ and assume $\widehat{x}_0 := \widehat{P}_0 x = 0$. Then $x = \sum_{i=0}^{k-1} \xi_i \Gamma_{\lambda_1}^{(i)} 1$, and we show $\xi_i = 0$, $i = l_1, \ldots, k-1$. We find analogously to (6.25) that

$$\widehat{x}_0 = \sum_{l=0}^{k-1} \eta_l l! (\widehat{A}_0 - \lambda)^{-l} \widehat{\gamma}(\lambda_1),$$

with η_l as in (6.26). Now $\widehat{x}_0 = 0$ implies as in part (i) that all the η_l are zero. If again we consider the expressions in (6.26) for the η_l as a system of equations for the ξ_i, and observe its triangular form and that $\widehat{b}_m^{(j)}(\lambda_1) = 0$ if $j = 0, 1, \ldots, l_1 - 1$ and

$$\widehat{b}_m^{(l_1)}(\lambda_1) = \prod_{i=l_1+1}^{m} (\lambda_1 - \lambda_i) \neq 0,$$

we find that $\xi_i = 0$ for $i = l_1, l_1 + 1, \ldots, k-1$, hence

$$x \in \operatorname{span}\left\{\Gamma_{\lambda_1}1, \ \frac{1}{1!}\Gamma_{\lambda_1}^{(1)}1, \ldots, \ \frac{1}{(l_1-1)!}\Gamma_{\lambda_1}^{(l_1-1)}1\right\}.$$

Thus the inclusion (6.27) is an equality. □

Next we describe the parameter sets in Theorems 6.4 and 6.5 under the assumption that all the points λ_j belong to $\rho(A)$. By k_j we denote the order of the zero λ_j of the function

$$Q(z) + t = \widehat{b}_m^*(z)\widehat{b}_m(z)\widehat{Q}_0(z) + \widehat{p}_{2m-1}(z) + t$$

(recall formulas (6.6), (6.9), and (6.12)), where $k_j = 0$ means that $Q(\lambda_j) + t \neq 0$. Here l_j denotes again the number of appearances of λ_j in Λ.

Theorem 6.7 *Assume $\Lambda \subset \rho(A)$. For the parameter functions $\widehat{\theta}_t(z)$ and $\widehat{\tau}_t(z)$ of the Štraus extension and the generalized resolvent in Theorems 6.4 and 6.5, respectively, the following statements hold:*

(i) *If $\Lambda \cap \rho(H^t) = \emptyset$, then $\widehat{\theta}_t(z) \in S_m$ and, equivalently, $\widehat{\tau}_t(z) \in N_m$.*

(ii) *If $\Lambda \cap \rho(H^t) \neq \emptyset$, then $\widehat{\theta}_t(z) \in S_{m-\kappa}$ and $\widehat{\tau}_t(z) \in N_{m-\kappa}$ with $\kappa = \sum \kappa_j$, where the sum runs over all $j \in \{1, 2, \ldots, n\}$ for which $\lambda_j \in \sigma^0(H^t)$, $\kappa_j = \min(k_j, l_j)$ if $\lambda_j \in \mathbf{C}^+$ and $\kappa_j = \min([\frac{k_j+1}{2}] - \delta_j, l_j)$ if $\lambda_j \in \mathbf{R}$, where*

$$\delta_j = \begin{cases} 1 & \text{if } k_j \text{ is odd and } Q^{(k_j)}(\lambda_j) > 0, \\ 0 & \text{otherwise.} \end{cases}$$

Proof: By Theorem 2.1 in [DLS1], $\widehat{\theta}_t$ belongs to the class $S_{m'}$, where m' equals the number of negative squares of the indefinite inner product on the set

$$\mathcal{S} := \overline{\text{span}}\,\{(H^t - z)^{-1}\widehat{\mathcal{H}}_0 \mid z \in \rho(H^t)\}.$$

Evidently, the orthogonal companion of $\mathcal{S}$ in Π_m is finite-dimensional and invariant under H^t, hence it is spanned by root vectors of H^t. These root vectors must belong to $\widehat{\mathcal{H}}_0^{\perp}$, which is in case (i) excluded by the assumptions and the first statement of Theorem 6.6.

Concerning case (ii), if some of the λ_j belong to $\sigma(H^t)$, it follows easily that the orthogonal companion of $\mathcal{S}$ coincides with the linear span of all these subspaces $\mathcal{L}_{\lambda_j}(H^t) \cap \widehat{\mathcal{H}}_0^{\perp}$. According to Theorem 6.6, $\mathcal{L}_{\lambda_j}(H^t) \cap \widehat{\mathcal{H}}_0^{\perp}$ is spanned by the first m_j elements of the chain of H^t at λ_j. The number of the negative squares of this part of the chain can now be calculated by well known formulas (see, for example, [KL3], [KL2]), and we only have to observe that for different λ_j's the root subspaces of H^t are orthogonal in Π_m. □

6.4

Let again H^t, $t \in \mathbf{R}$, be the self-adjoint extension which is given by M.G. Krein's formula (3.18) or by (3.19). Then $Q(z) + t$, the function in the denominator of the resolvent of H^t, belongs to the class N_m. In this subsection we generalize the preceeding results to the situation where the sequence $\Lambda = \{\lambda_1, \lambda_2, \ldots, \lambda_m\}$ can contain also generalized zeros of nonpositive type of the function $Q(z) + t$, that is, points where $Q(z)$ is not holomorphic. In particular, also the case that the sequence Λ consists of all points of $\sigma^0(H^t)$, each point appearing as often as the multiplicity of this point as an eigenvalue of $H^t|_{\mathcal{L}^0(H^t)}$ indicates, is allowed; it will be considered in more detail in Section 7.

We decompose Π_m analogously to (6.13) for a sequence $\Lambda = \{\lambda_1, \lambda_2, \ldots, \lambda_m\} \subset \mathbf{C}^+ \cup \mathbf{R}$ where the points λ_j are either points where $Q(z)$ is holomorphic or points which are generalized zeros of nonpositive type the function $Q(z)+t$; in the latter case, if the number λ_j appears l_j-times in Λ, then the inequality $l_j \leq \pi_j$

must hold, where π_j is the multiplicity of λ_j as a generalized zero of nonpositive type of $Q(z)+t$.

We define the polynomials $\widehat{b}_j$ according to (6.5) and expand the operator function Γ_z as in (6.7):

$$\Gamma_z = \sum_{j=1}^{m} \widehat{b}_{j-1}(z)\widehat{E}_j + \widehat{b}_m(z)(A-z)^{-1}\widehat{E}_m. \tag{6.28}$$

That this is possible can be seen as follows. Replace each λ_j in Λ by $\lambda_j + i\varepsilon$ for some $\varepsilon > 0$, then with obvious notation Γ_z has admits an expansion of the form

$$\Gamma_z = \sum_{j=1}^{m} \widehat{b}_{j-1,\varepsilon}(z)\widehat{E}_j(\varepsilon) + \widehat{b}_{m,\varepsilon}(z)(A-z)^{-1}\widehat{E}_m(\varepsilon). \tag{6.29}$$

As in proof of Lemma 6.1 the $\widehat{E}_j(\varepsilon)$ can be solved from a system of equations of the form

$$\Gamma_{k,\varepsilon} = \sum_{j=1}^{m} B_{kj}(\varepsilon)\widehat{E}_j(\varepsilon), \quad k = 1, 2\ldots, m.$$

The matrix $(B_{kj}(\varepsilon))_{j,k=1}^{n}$ is invertible and converges to the invertible matrix $(B_{kj})_{j,k=1}^{n}$ and, on account of Lemma 2.1, (the entries of) $\Gamma_{k,\varepsilon}$ converge to (the entries of) Γ_k, as $\varepsilon \downarrow 0$. It follows that all of the $\widehat{E}_j(\varepsilon)$ have nontangential limits and so the expansion (6.28) follows from (6.29) by letting $\varepsilon \downarrow 0$. It follows easily that the elements $\widehat{E}_j 1$ can be expressed as linear combinations of the elements $\Gamma_{\lambda_j}^{(k)} 1$, $k = 0, 1, \ldots, l_j - 1$. When λ_j is a generalized zero of $Q(z)+t$, these elements have to be understood in the sense of Lemma 2.1. We introduce again the subspace

$$\widehat{\mathcal{M}} = \text{span}\,\{\widehat{E}_1 1, \ldots, \widehat{E}_m 1\} = \text{span}\,\{\Gamma_{\lambda_j}^{(k)} 1 \mid k = 0, 1, \ldots, l_j - 1, j = 1, 2, \ldots n\}$$

and the decomposition (6.13) of Π_m:

$$\Pi_m = \widehat{\mathcal{H}}_0 \oplus (\mathcal{L} \dot{+} \widehat{\mathcal{M}}). \tag{6.30}$$

The spaces $\widehat{M}$ and hence also $\widehat{\mathcal{H}}_0$ are independent of the ordering of the points λ_j. With respect to this decomposition, Theorem 6.6 remains valid as it stands and the statement (ii) of Theorem 6.7 can be formulated easier by using the multiplicities of the zeros of nonpositive type of the function $Q(z)+t$ as follows.

Theorem 6.8 *For the parameter functions $\widehat{t}(z)$ and $\widehat{\theta}_t(z)$ it holds that*

$$\widehat{\theta}_t(z) \in S_{m-\kappa}, \quad \widehat{t}(z) \in N_{m-\kappa},$$

with $\kappa = \sum_{j=1}^{n} \kappa_j$, where the sum runs over all those $j \in \{1, 2, \ldots, n\}$ for which $\lambda_j \in \sigma^0(H^t)$ and $\kappa_j = \min(k_j, l_j)$, where k_j is the order of λ_j as a zero of $Q(z)+t$ and $k_j = \pi_j$ if λ_j is a generalized zero of $Q(z)+t$.

7 Spectral Factorization

7.1

In this section we consider the case that the sequence Λ consists of the points λ_j of $\sigma^0(H^t)$ each repeated as often as its algebraic multiplicity l_j as an eigenvalue of H^t (if it is in $\mathbf{C}^+$) or as an eigenvalue of $H^t|_{\mathcal{L}^0(H^t)}$ (if it is in $\mathbf{R}$) indicates. Thus, if $\lambda_j \in \mathbf{C}^+$, then l_j is the order of the zero λ_j of $Q(z)+t$, if $\lambda_j \in \mathbf{R}$ and λ_j is a generalized zero of $Q(z)+t$ of multiplicity π_j, then $l_j = \pi_j$. For this choice of Λ we write for short $\Lambda \equiv \sigma^0(H^t)$. Then the subspace $\widehat{\mathcal{M}}$ coincides with the maximal nonpositive invariant subspace $\mathcal{L}^0(H^t)$ of H^t. Whereas so far the invariance of the subspace $\mathcal{L}$ under the operator $(H^\infty - z)^{-1}$ played a central role, the following considerations are based on the fact that now the subspace $\widehat{\mathcal{M}}$ is invariant under H^t.

We denote by I the subset of all those indices $j \in \{1, 2, \dots, n\}$ for which $\lambda_j \in \mathbf{R}$ and the $(2l_j - 1)$-st nontangential derivative of $Q(z)$ is negative: $Q^{(2l_j-1)}(\lambda_j) < 0$, and we introduce the negative subspace

$$\widehat{\mathcal{M}}^- := \operatorname{span}\{\Gamma_{\lambda_j}^{(l_j-1)} 1 : j \in I\}.$$

With the isotropic subspace $\widehat{\mathcal{M}}^0$ of $\widehat{\mathcal{M}}$, which is spanned by the neutral parts of the root subspaces of H^t at the points λ_j, $j = 1, 2, \dots, n$, we have the decomposition

$$\widehat{\mathcal{M}} = \widehat{\mathcal{M}}^0 \dotplus \widehat{\mathcal{M}}^-.$$

Let $\mathcal{L}_0$ be the subspace of $\mathcal{L}$, which is orthogonal to $\widehat{\mathcal{M}}^-$ and in duality with $\widehat{\mathcal{M}}^0$. With the positive subspace

$$\mathcal{G} := (\mathcal{L} \dotplus \widehat{\mathcal{M}}) \cap ((\mathcal{L}_0 \dotplus \widehat{\mathcal{M}}^0) \oplus \widehat{\mathcal{M}}^-)^\perp$$

we form the Hilbert space $\widetilde{\mathcal{H}}_0 := \widehat{\mathcal{H}}_0 \oplus \mathcal{G}$ and obtain finally the following decomposition of the space Π_m:

$$\Pi_m = \widetilde{\mathcal{H}}_0 \oplus (\mathcal{L}_0 \dotplus \widetilde{\mathcal{M}}^0) \oplus \widetilde{\mathcal{M}}^-. \tag{7.1}$$

Evidently, $r := \dim \widehat{\mathcal{M}}^- = \dim \mathcal{G}$. Let $\widetilde{P}_0$ be the orthogonal projection in Π_m onto $\widetilde{\mathcal{H}}_0$. Define in $\widetilde{\mathcal{H}}_0$ the linear relation

$$\widetilde{A}^t := \{\{\widetilde{P}_0 f, \widetilde{P}_0 g\} \mid \{f, g\} \in H^t,\ \ g - zf \in \widetilde{\mathcal{H}}_0\},\quad z \in \rho(H^t). \tag{7.2}$$

Further, we introduce the numbers

$$\widehat{\gamma}_j := -\frac{1}{(2l_j - 1)!} Q^{(2l_j-1)}(\lambda_j) > 0,\quad j \in I,$$

and the function

$$\tau(z) = \sum_{j \in I} \frac{\widehat{\gamma}_j}{\lambda_j - z}. \tag{7.3}$$

Theorem 7.1 *Choose* $\Lambda \equiv \sigma^0(H^t)$. *Then:*

(i) *The function* $Q(z)+t$ *admits the factorization*

$$\begin{aligned} Q(z)+t &= \widehat{b}_m^*(z)\widehat{b}_m(z)\widetilde{Q}_0(z) \quad \text{with} \quad \widetilde{Q}_0(z) \\ &= \widehat{Q}_0(z)+\tau(z), \ \widehat{Q}_0(z) \in N_0^\infty. \end{aligned} \tag{7.4}$$

(ii) *The operator* $\widetilde{A}^t$ *in the subspace* $\widetilde{\mathcal{H}}_0$ *is densely defined and a self-adjoint extension of the symmetric operator* $\widehat{S}_0$ *in* $\widehat{\mathcal{H}}_0$. *The corresponding generalized resolvent* $\widehat{P}_0(\widetilde{A}^t - z)^{-1}|_{\widehat{\mathcal{H}}_0}$ *is given by the right-hand side of* (6.23) *with the parameter function* $\widehat{t}(z) = \tau(z)$ *from* (7.3).

Proof: (i) By (6.6),

$$Q(z)+t = \widehat{b}_m^*(z)\widehat{b}_m(z)(\widehat{Q}_0(z)+\widehat{t}(z))$$

with

$$\widehat{t}(z) = \frac{\widehat{p}_{2m-1}(z)+t}{\widehat{b}_m^*(z)\widehat{b}_m(z)}.$$

Theorem 6.8 with $\kappa = m$ implies that $\widehat{t}(z) \in N_0$. Hence $\widehat{t}(z)$ is a rational Nevanlinna function of the same form as $\tau(z)$ and with poles at some real points which belong to Λ. These points are generalized zeros of nonpositive type of the function $Q(z)+t$ of multiplicity l_j. The relation $\widehat{t}(z) = \tau(z)$ follows now from the integral representation of the functions of the class N_0.

(ii) The proof of the statement about the operator $\widetilde{A}^t$ is similar to the proof of the corresponding statement for the operator $\widehat{A}_0$ in the preceding section. The subspaces $\widehat{\mathcal{M}}$, $\widehat{\mathcal{M}}^\perp$ and $\widehat{\mathcal{M}}^0$ are H^t-invariant. This implies the following matrix representation of the resolvent $(H^t - z)^{-1}$ with respect to the decomposition (7.1):

$$(H^t - z)^{-1} = \begin{pmatrix} R_{00}(z) & R_{01}(z) & 0 & 0 \\ 0 & R_{11}(z) & 0 & 0 \\ R_{20}(z) & R_{21}(z) & R_{22}(z) & R_{23}(z) \\ 0 & R_{31}(z) & 0 & R_{33}(z) \end{pmatrix}. \tag{7.5}$$

It is easy to see that $R_{00}(z)$ is the resolvent of the self-adjoint operator $\widetilde{A}^t$. We show that the range of $R_{00}(z)$, which is the domain of $\widetilde{A}^t$, is dense in $\widetilde{\mathcal{H}}_0$. Otherwise an element $x_0 \in \widetilde{\mathcal{H}}_0$ would exist such that $x_0 \perp R_{00}(z)\widetilde{\mathcal{H}}_0$. Consider an element $y \in \widehat{\mathcal{M}}^\perp \oplus \widehat{\mathcal{M}}^-$ and put $y_0 := \widetilde{P}_0 y$. Then

$$\langle x_0, (H^t - z)^{-1} y\rangle_{\Pi_m} = \langle x_0, R_{00}(z) y_0\rangle_{\Pi_m} = 0,$$

therefore $R^t(z^*)x_0 \in \widehat{\mathcal{M}}^0$ for all $z \in \rho(H^t)$, consequently, $x_0 \in \widehat{\mathcal{M}}^0$ and hence $x_0 = 0$.

Further we observe that $\widehat{\mathcal{H}}_0 \subset \widetilde{\mathcal{H}}_0$ and that the compressed resolvents of H^t and of $\widetilde{A}^t$ coincide:

$$\widehat{P}_0(H^t - z)^{-1}|_{\widehat{\mathcal{H}}_0} = \widehat{P}_0(\widetilde{A}^t - z)^{-1}|_{\widehat{\mathcal{H}}_0}.$$

Now the last statement follows from the generalization of Theorem 6.5 for $\Lambda \equiv \sigma_0(H^t)$. □

Corollary 7.2 *The operator $\widetilde{A}^t$ is a canonical self-adjoint extension of $\widehat{S}_0$ if and only if the subspace $\widehat{\mathcal{M}}$ is neutral.*

Remark 7.3 The representation (7.4) of the function $Q(z)+t$ can also be written as

$$Q(z) + t = \widehat{b}_m^*(z)\widehat{b}_m(z)\widehat{Q}_0(z) + p_{2m-1}(z) + t,$$

where $\widehat{Q}_0(z) = \widetilde{Q}_0(z) - \tau(z)$. Both functions $\widehat{Q}_0(z)$ and $\widetilde{Q}_0(z)$ belong to the class N_0^∞. There are two special subspaces associated with $Q(z)$: The maximal neutral subspace $\mathcal{L} = \mathcal{L}^0(H^\infty)$, which corresponds to the generalized pole of nonpositive type of $Q(z)$ at ∞, and the maximal nonpositive subspace $\widehat{\mathcal{M}} = \mathcal{L}^0(H^t)$ which corresponds to the zeros of $Q(z)+t$ in $\mathbf{C}^+$ and to its generalized zeros of nonpositive type on $\mathbf{R}$. There are also two natural Hilbert subspaces in $\Pi_m = \Pi_m(Q)$: the space $\widehat{\mathcal{H}}_0$, which is isomorphic to the quotient space $\mathcal{L}^\perp/\mathcal{L}$ and defined through the decomposition

$$\Pi_m = \widehat{\mathcal{H}}_0 \oplus (\mathcal{L} \dotplus \widehat{\mathcal{M}}),$$

and the space $\widetilde{\mathcal{H}}_0$, which is isomorphic to $\widehat{\mathcal{M}}^\perp/\widehat{\mathcal{M}}^0$ and defined by

$$\Pi_m = \widetilde{\mathcal{H}}_0 \oplus (\mathcal{L}_0 \dotplus \widehat{\mathcal{M}}^0) \oplus \widehat{\mathcal{M}}^-.$$

The function $\widehat{Q}_0(z)$ is in correspondence with $\widehat{\mathcal{H}}_0$ (see (6.16) and (6.17)), and the function $\widetilde{Q}_0(z)$ is in correspondence with $\widetilde{\mathcal{H}}_0$ (see Theorem 7.4 (ii) and (iii) below). The spaces $\widehat{\mathcal{H}}_0$ and $\widetilde{\mathcal{H}}_0$ coincide if and only if $\widehat{\mathcal{M}}$ is neutral; in this case $\widehat{Q}_0(z) = \widetilde{Q}_0(z)$. In general $\widetilde{\mathcal{H}}_0 = \widehat{\mathcal{H}}_0 \oplus \mathcal{G}$ with an r-dimensional space $\mathcal{G}$, where r is the number of poles of $\tau(z)$.

7.2

So far we started from the Γ-field Γ_z, corresponding to $A = H^\infty$, the m-dimensional neutral invariant subspace $\mathcal{L}$ of H^∞, chose a self-adjoint extension H^t of S and, by expanding the Γ-field Γ_z at the eigenvalues of H^t in $\sigma^0(H^t)$, which are the zeros in $\mathbf{C}^+$ or the generalized zeros of nonpositive type on $\mathbf{R}$ of $Q(z)+t$, we determined the space $\widehat{\mathcal{M}}$ and thus the decomposition (7.1) of the space Π_m. Now we make a dual approach: We choose a self-adjoint extension H^t and consider the corresponding Γ-field

(7.6) $$\Gamma_z^t := (I + (z - \lambda_0)(H^t - z)^{-1})\Gamma_{\lambda_0}.$$

We show that the m-dimensional nonpositive invariant subspace of the self-adjoint relation A, which is the subspace $\mathcal{L}$ in the decomposition (6.30), can be obtained from the asymptotic expansion of the Γ-field Γ_z^t at ∞.

First we express the Γ-field Γ_z^t by the Γ-field Γ_z. To this end we replace in (7.6) the resolvent through M.G. Krein's formula and obtain

$$(7.7)\quad \begin{aligned}\Gamma_z^t &= \Gamma_{\lambda_0} + (z-\lambda_0)(A-z)^{-1}\Gamma_{\lambda_0} - (z-\lambda_0)(Q(z)+t)^{-1}\Gamma_z\Gamma_{z^*}^*\Gamma_{\lambda_0}\\ &= \Gamma_z + (Q(z)+t)^{-1}(Q(\lambda_0)-Q(z))\Gamma_z = \frac{\alpha}{Q(z)+t}\Gamma_z\end{aligned}$$

with $\alpha := Q(\lambda_0)+t$. It follows that

$$(7.8)\quad \begin{aligned}\langle \Gamma_z^t 1, \Gamma_w^t 1\rangle_{\Pi_m} &= |\alpha|^2 \frac{Q(z)-Q(w)^*}{(z-w^*)(Q(z)+t)(Q(w)^*+t)}\\ &= |\alpha|^2\left(-\frac{1}{Q(z)+t}+\frac{1}{Q(w)^*+t}\right)(z-w^*)^{-1}.\end{aligned}$$

We choose again $\Lambda \equiv \sigma^0(H^t)$ and the polynomials $\widehat{b}_0(z) \equiv 1$ and

$$\widehat{b}_k(z) = (z-\lambda_1)\cdots(z-\lambda_k),\ k = 1, 2, \ldots, m.$$

Then according to Corollary 5.5 the representation

$$(7.9)\quad Q(z)+t = b_m(z)b_m^*(z)\widehat{Q}_0^t(z)$$

holds with a function $\widehat{Q}_0^t(z) \in N_0^\infty$. We shall prove the following formula, which can be considered as an expansion of the Γ-field Γ_z^t at ∞:

$$(7.10)\quad \Gamma_z^t = \frac{1}{\widehat{b}_1^*(z)}E_1^t + \cdots + \frac{1}{\widehat{b}_m^*(z)}E_m^t + \frac{1}{\widehat{b}_m^*(z)}\Gamma_{0z}^t.$$

We start from the identity

$$(7.11)\quad \begin{aligned}&\frac{(z-\lambda_1^*)\cdots(z-\lambda_k^*)}{z-w^*} - \frac{(w^*-\lambda_1^*)\cdots(w^*-\lambda_k^*)}{z-w^*}\\ &= (z-\lambda_2^*)\cdots(z-\lambda_k^*) + (w^*-\lambda_1^*)(z-\lambda_3^*)\cdots(z-\lambda_k^*)\\ &\quad + \cdots + (w^*-\lambda_1^*)(w^*-\lambda_2^*)\cdots(w^*-\lambda_{k-1}^*),\end{aligned}$$

$k = 1, 2, \dots, m$, which gives

$$\begin{aligned}\frac{1}{z-w^*} &= \frac{1}{z-\lambda_1^*} + \frac{w^*-\lambda_1^*}{(z-\lambda_1^*)(z-\lambda_2^*)} + \frac{(w^*-\lambda_1^*)(w^*-\lambda_2^*)}{(z-\lambda_1^*)(z-\lambda_2^*)(z-\lambda_3^*)} \\ &\quad + \cdots + \frac{(w^*-\lambda_1^*)\cdots(w^*-\lambda_{k-1}^*)}{(z-\lambda_1^*)\cdots(z-\lambda_{k-1}^*)(z-\lambda_k^*)} \\ &\quad + \frac{(w^*-\lambda_1^*)\cdots(w^*-\lambda_k^*)}{(z-\lambda_1^*)\cdots(z-\lambda_k^*)(z-w^*)} \\ &= \frac{\widehat{b}_0^*(w^*)}{\widehat{b}_1^*(z)} + \frac{\widehat{b}_1^*(w^*)}{\widehat{b}_2^*(z)} + \cdots + \frac{\widehat{b}_{k-1}^*(w^*)}{\widehat{b}_k^*(z)} + \frac{\widehat{b}_k^*(w^*)}{\widehat{b}_k^*(z)(z-w^*)}.\end{aligned} \tag{7.12}$$

The existence of the operator E_1^t follows (compare the proof of Lemma 2.1) if we show that the limits

$$\lim_{z\to i\infty}(z-\lambda_1^*)\langle\Gamma_z^t 1, \Gamma_w^t 1\rangle_{\Pi_m}, \quad \lim_{z,\zeta\to i\infty}(z-\lambda_1^*)(\zeta^*-\lambda_1)\langle\Gamma_z^t 1, \Gamma_\zeta^t 1\rangle_{\Pi_m} \tag{7.13}$$

exist. Here and in the following we write for example $z \to +i\infty$ if $z = iy$ and $y \to +\infty$; then also the corresponding nontangential limits exist. If we observe the relation (7.8) the first limit in (7.13) becomes

$$\begin{aligned}\langle E_1^t 1, \Gamma_w^t 1\rangle_{\Pi_m} &= \lim_{z\to i\infty}(z-\lambda_1^*)|\alpha|^2\left(-\frac{1}{Q(z)+t}+\frac{1}{Q(w)^*+t}\right) \\ (z-w^*)^{-1} &= |\alpha|^2\frac{1}{Q(w)^*+t}.\end{aligned}$$

and the second limit

$$\begin{aligned}\langle E_1^t 1, E_1^t 1\rangle_{\Pi_m} = \lim_{z,\zeta\to i\infty}(z-\lambda_1^*)(\zeta^*-\lambda_1)|\alpha|^2 \\ \left(-\frac{1}{Q(z)+t}+\frac{1}{Q(\zeta)^*+t}\right)(z-\zeta^*)^{-1} = 0.\end{aligned}$$

Here we have used the inequalities

$$|(z-\zeta^*)^{-1}| \le \min(|z|^{-1}, |\zeta|^{-1})$$

and that for the function $\widehat{Q}_0^t(z)$ in (7.9) it holds

$$\lim_{z\to i\infty}|z Q_0^t(z)| = \infty.$$

In the same way it follows that for $j, k = 1, 2, \dots, m$:

$$\langle E_k^t 1, \Gamma_w^t 1\rangle_{\Pi_m} = \frac{|\alpha|^2}{Q(w)^*+t}\widehat{b}_{k-1}^*(w^*), \quad \langle E_k^t 1, E_j^t 1\rangle_{\Pi_m} = 0, \tag{7.14}$$

and

$$\left\langle \Gamma_z^t 1 - \frac{E_1^t 1}{\widehat{b}_1^*(z)} - \cdots - \frac{E_k^t 1}{\widehat{b}_k^*(z)}, \Gamma_\zeta^t 1 - \frac{E_1^t 1}{\widehat{b}_1^*(\zeta)} - \cdots - \frac{E_k^t 1}{\widehat{b}_k^*(\zeta)} \right\rangle_{\Pi_m}$$

$$= \langle \Gamma_z^t 1, \Gamma_\zeta^t 1 \rangle_{\Pi_m} - \sum_{j=1}^{k} \left\langle \Gamma_z^t 1, \frac{E_j^t 1}{\widehat{b}_j^*(\zeta)} \right\rangle_{\Pi_m} - \sum_{j=1}^{k} \left\langle \frac{E_j^t 1}{\widehat{b}_j^*(z)}, \Gamma_\zeta^t 1 \right\rangle_{\Pi_m}$$

$$= -\frac{|\alpha|^2}{Q(z)+t} \left(\frac{1}{z-\zeta^*} + \sum_{j=1}^{k} \frac{\widehat{b}_{j-1}(z)}{\widehat{b}_j(\zeta^*)} \right) + \frac{|\alpha|^2}{Q(\zeta)^*+t} \tag{7.15}$$

$$\left(\frac{1}{z-\zeta^*} + \sum_{j=1}^{k} \frac{\widehat{b}_{j-1}^*(\zeta^*)}{\widehat{b}_j^*(z)} \right) = |\alpha|^2 (z-\zeta^*)^{-1}$$

$$\left(-\frac{1}{Q(z)+t} \frac{\widehat{b}_k^*(z)}{\widehat{b}_k^*(\zeta^*)} + \frac{1}{Q(\zeta)^*+t} \frac{\widehat{b}_k(\zeta^*)}{\widehat{b}_k(z)} \right) = |\alpha|^2 (z-\zeta^*)^{-1}$$

$$\left(-\frac{\widehat{b}_k^*(z)}{\widehat{Q}_0^t(z) \widehat{b}_m^*(z) \widehat{b}_m(z) \widehat{b}_k^*(\zeta^*)} + \frac{\widehat{b}_k(\zeta^*)}{\widehat{Q}_0^t(\zeta)^* \widehat{b}_m(\zeta^*) \widehat{b}_m^*(\zeta^*) \widehat{b}_k(z)} \right).$$

For the third equality we have used the relation (7.12). Now it follows easily that the limit for $z \to i\infty$ of the expression in (7.15) is zero. For $k = m$ this expression equals

$$|\alpha|^2 (z-\zeta^*)^{-1} \left(-\frac{1}{\widehat{Q}_0^t(z)} + \frac{1}{\widehat{Q}_0^t(\zeta)^*} \right) \frac{1}{\widehat{b}_k(z) \widehat{b}_k^*(\zeta^*)},$$

hence with

$$\widetilde{Q}_0^t(z) := -|\alpha|^2 \widehat{Q}_0^t(z)^{-1}$$

we obtain the relation

$$\langle \Gamma_{0z} 1, \Gamma_{0\zeta} 1 \rangle_{\Pi_m} = \frac{\widetilde{Q}_0^t(z) - \widetilde{Q}_0^t(\zeta)^*}{z - \zeta^*}. \tag{7.16}$$

The first relation in (7.14) and (7.7) also imply

$$\Gamma_\zeta^* E_k^t 1 = \alpha \widehat{b}_{k-1}^*(\zeta^*), \quad k = 1, 2, \ldots, m. \tag{7.17}$$

Finally, we define the elements $\widetilde{\gamma}(z)$:

$$\widetilde{\gamma}(z) := \widetilde{P}_0 \widetilde{\Gamma}_{0z} 1, \quad z \in \rho(H^t),$$

where $\widetilde{P}_0$ is the projection onto $\widetilde{\mathcal{H}}_0$, and in the following $\widetilde{A}^t$ is the operator from (7.2).

Theorem 7.4 *Let the operators E_k^t, $k = 1, \dots, m$, and Γ_{0z}^t, $z \in \rho(H^t)$, be defined by the expansion* (7.10) *of Γ_z^t at ∞. Then:*

(i) $\mathcal{L} = \operatorname{span}\{E_1^t 1, E_2^t 1, \dots, E_m^t 1\}$, $E_1^t 1 \in \operatorname{dom}(H^t)^{m-1}$, *and*

$$E_k^t = \widehat{b}_{k-1}^*(H^t)E_1^t, \quad \Gamma_{0z}^t = \widehat{b}_m^*(H^t)(H^t - z)^{-1}E_1^t. \tag{7.18}$$

(ii) $\mathcal{M}^\perp = \overline{\operatorname{span}}\{\Gamma_{0z}^t 1, z \in \rho(H^t)\}$, $\quad \widetilde{\mathcal{H}}_0 = \overline{\operatorname{span}}\{\widetilde{\gamma}(z), z \in \rho(H^t)\}$.

(iii) $\widetilde{\gamma}(z) = (I + (z - w)(\widetilde{A}^t - z)^{-1})\widetilde{\gamma}(w)$, $\widetilde{Q}_0^t(z) \in N_0$, *and*

$$\langle \widetilde{\gamma}(z), \widetilde{\gamma}_w \rangle_{\widetilde{\mathcal{H}}_0} = \frac{\widetilde{Q}_0^t(z) - \widetilde{Q}_0^t(w)^*}{z - w^*}.$$

Proof: (i) Consider the difference of the relation (7.17) for two different points z and w:

$$(\Gamma_z^* - \Gamma_w^*)E_k^t 1 = (z^* - w^*)\Gamma_w^*(H^\infty - z^*)^{-1}E_k^t 1 = \alpha\,(\widehat{b}_{k-1}^*(z^*) - \widehat{b}_{k-1}^*(w^*)).$$

Dividing by $(z^* - w^*)$ we obtain

$$\Gamma_w^*(H^\infty - z^*)^{-1}E_k^t 1 = \alpha\,\frac{\widehat{b}_{k-1}^*(z^*) - \widehat{b}_{k-1}^*(w^*)}{z^* - w^*}. \tag{7.19}$$

The expression on the right-hand side is a polynomial in z^* and w^* of degree $k-2$. Differentiating $(k-1)$-times with respect to z^* we get

$$\Gamma_w^*(H^\infty - z^*)^{-k}E_k^t = 0.$$

Therefore $(H^\infty - z^*)^{-k}E_k^t = 0$ and hence $E_k^t 1 \in \mathcal{L}$, $k = 1, \dots, m$. The relation (7.17) also implies that these elements are linearly independent, and the first relation in (i) is proved.

From M.G. Krein's formula for the resolvent $(H^t - z)^{-1}$, (7.17), and (7.7) we get

$$\begin{aligned}(H^t - z)^{-1}E_k^t &= (H^\infty - z)^{-1}E_k^t - (Q(z) + t)^{-1}\Gamma_z\Gamma_{z^*}^*E_k^t \\ &= (H^\infty - z)^{-1}E_k^t - \widehat{b}_{k-1}^*(z)\Gamma_z^t\end{aligned}$$

For $k = 1$ the first term on the right-hand side is zero, and so

$$\Gamma_z^t = -(H^t - z)^{-1}E_1^t, \quad \Gamma_z^t 1 \in \operatorname{dom} H^t. \tag{7.20}$$

For $k \geq 2$ this term can be rewritten (via (7.19), (7.17), and (7.11)) so that

$$\begin{aligned}(H^t - z)^{-1}E_k^t &= E_{k-1}^t + (z - \lambda_{k-1}^*)E_{k-2}^t \\ &\quad + \cdots + (z - \lambda_2^*)\cdots(z - \lambda_{k-1}^*)E_1^t - \widehat{b}_{k-1}^*(z)\Gamma_z^t.\end{aligned}$$

For $k = 2$ we find $(H^t - z)^{-1} E_2^t = E_1^t - \widehat{b}_1^*(z)\Gamma_z^t$, which implies

$$E_1^t 1 \in \operatorname{dom} H^t, \quad E_2^t = (H^t - \lambda_1^*)E_1^t.$$

Indeed this follows from (7.20):

$$(H^t - z)^{-1}E_2^t = E_1^t - (z - \lambda_1^*)(H^t - z)^{-1}E_1^t = (H^t - z)^{-1}(H^t - \lambda_1^*)E_1^t.$$

For $k = 3$: $(H^t - z)^{-1}E_3^t = E_2^t + (z - \lambda_2^*)E_1^t - \widehat{b}_2^*(z)\Gamma_z^t$ which implies $E_2^t \in \operatorname{dom} H^t$, hence

$$E_1^t \in \operatorname{dom} (H^t)^2, \quad E_3^t = (H^t - \lambda_1^*)(H^t - \lambda_2^*)E_1^t.$$

Repeating this argument up to and including $k = m$, we obtain $E_1^t \in \operatorname{dom} (H^t)^{m-1}$ and the relations (7.18).

(ii) The expression for Γ_{0z}^t in (7.18) implies the inclusion

$$\Gamma_{0z}^t 1 \in \mathcal{M}^\perp = \widetilde{\mathcal{H}}_0 \oplus \mathcal{M}^0, \quad z \in \rho(H^t). \tag{7.21}$$

Since the subspaces $\mathcal{M}$ and $\mathcal{L}$ are in duality, we have $\Pi_m = \mathcal{L} \dot{+} \mathcal{M}^\perp$. On the other hand, Π_m is the closed linear span of the elements $\Gamma_z^t 1$, $z \in \rho(H^t)$ (recall the discussion in Subsection 3.6), and by (i), $\mathcal{L} = \operatorname{span}\{E_1^t 1,\ E_2^t 1, \dots, E_m^t 1\}$. So the first relation in (ii) is a consequence of the expansion (7.10), the second relation follows from $\widetilde{\mathcal{H}}_0 = \widetilde{P}_0\mathcal{M}^\perp$ and the definition of $\widetilde{\gamma}(z)$.

(iii) With the matrix representation (7.5) of the resolvent $(H^t - z)^{-1}$ corresponding to the decomposition (7.1) we obtain

$$\begin{aligned} \widetilde{\gamma}(z) - \widetilde{\gamma}(w) &= (z - w)\widetilde{P}_0(H^t - z)^{-1}\Gamma_{0w}^t 1 \\ &= (z - w)(\widetilde{A}^t - z)^{-1}\widetilde{\gamma}(w), \end{aligned} \tag{7.22}$$

what was to be shown. Since $\Gamma_{0z}^t 1 \in \mathcal{M}^\perp$ and $\mathcal{M}$ is a maximal nonpositive subspace, (7.16) implies

$$\operatorname{Im} \widetilde{Q}_0^t(z)/\operatorname{Im} z = (\Gamma_{0z}^t)^*\Gamma_{0z}^t 1 \geq 0 \ \text{ for all } z \in \mathbf{C}^+.$$

The equality sign for some $z \in \mathbf{C}^+$ is impossible since it would imply $\Gamma_{0w}^t 1 \in \mathcal{M}^0$ for all $w \in \mathbf{C}\backslash\mathbf{R}$. Then the space Π_m would be finite dimensional which is in contradiction to the unboundedness of H^t. Hence $\widetilde{Q}_0^t(z) \in N_0$. The inclusion (7.21) yields the equality

$$(\Gamma_{0w}^t)^*\Gamma_{0z}^t 1 = \langle \widetilde{\gamma}(z), \widetilde{\gamma}(w)\rangle_{\widetilde{\mathcal{H}}_0},$$

which together with (7.16) proves the last relation in (iii). □

7.3

The operator $\widetilde{A}^t$ is isomorphic to a self-adjoint Hilbert space extension of a symmetry in $\mathcal{H}_0$. In order to describe this symmetry, we first show that with the polynomial $\widehat{b}_m(z)$, corresponding to $\Lambda \equiv \sigma^0(H^t)$, there is a well defined element $\widehat{b}_m(A_0)^{-1}\chi$ in $\mathcal{H}_{-1}$. To this end we consider a representation of the originally given operator A_0 as the operator of multiplication by the independent variable in the space $L^2_{\sigma_0}$, where σ_0 is the measure in the integral representation (2.4) of the function $Q_0(z)$:

$$Q_0(z) = c + \int_{-\infty}^{+\infty} \frac{1+tz}{t-z}\, d\sigma_0.$$

If we choose here for simplicity $\lambda_0 = i$ then this representation of $Q_0(z)$ coincides with the representation (2.7) if we only observe that $\Gamma_{\lambda_0} 1 = \chi_0$ corresponds to the function $(t-\lambda_0)^{-1}$ and χ_{-1} corresponds to the function which is identically equal to 1. Now we compare the right-hand sides of the following relations (see (3.15) and (7.4)):

$$Q(z) + t = (z^2+1)^m Q_0(z) + p_{2m-1}(z) + t, \quad Q(z) + t = \widehat{b}_m(z)\widehat{b}^*_m(z)\widehat{Q}^t_0(z)$$

with $\widehat{Q}^t_0(z) \in N^\infty_0$. If we apply the Stieltjes-Livsic inversion formula we find for the measures σ_0 and $\widehat{\sigma}_0$ from the integral representations (2.4) of $Q_0(z)$ and $\widehat{Q}^t_0(z)$, the relation

$$(t^2+1)^m\, d\sigma_0(t) = |\widehat{b}_m(t)|^2\, d\widehat{\sigma}_0(t),$$

therefore $\widehat{\sigma}_0$ has no concentrated measure at the real points of $\sigma^0(H^t)$ and

$$\frac{(t^2+1)^m}{|\widehat{b}_m(t)|^2}\, d\sigma_0(t) = d\widehat{\sigma}_0(t).$$

Since in this representation $\chi = (A_0 - i)^m \chi_{-1}$ corresponds to the function $(t-i)^m$, the function $(t-i)^m\,\widehat{b}_m(t)^{-1}$ corresponds to $\widehat{b}_m(A_0)^{-1}\chi =: \chi^\Lambda_{-1}$ and, evidently, this is an element of the space $\mathcal{H}_{-1}$.

Now we consider the symmetric operator S^Λ_0 in $\mathcal{H}_0$ which is the restriction of A_0 to the domain

$$\operatorname{dom} S^\Lambda_0 = \{ f \in \operatorname{dom} A_0 \mid \langle f, \chi^\Lambda_{-1}\rangle_0 = 0 \}.$$

We introduce the corresponding Q-function

$$Q^\Lambda_0(z) := (z - \lambda^*_0)\langle R_0(z)\chi^\Lambda_{-1}, R_0(\lambda_0)\chi^\Lambda_{-1}\rangle_0 + \widehat{g}_{m+1,m} \tag{7.23}$$

and the self-adjoint extension A^Λ which is described by M.G. Krein's formula:

$$P_0(A^\Lambda - z)^{-1}\,|_{\mathcal{H}_0} = (A_0 - z)^{-1} - \frac{\langle \cdot, R_0(z^*)\chi^\Lambda_{-1}\rangle_0\, R_0(z)\chi^\Lambda_{-1}}{Q^\Lambda_0(z) + \tau(z)}, \tag{7.24}$$

with $\tau(z)$ given by (7.3).

Theorem 7.5 *Consider the operator H^t in Π_m and let $\widehat{\mathcal{M}} = \mathcal{L}^0(H^t)$ be its maximal nonpositive invariant subspace as above. Then the operator $\widetilde{A}^t$ in $\widetilde{\mathcal{H}}_0$ ($\simeq \widehat{\mathcal{M}}^\perp/\widehat{\mathcal{M}}^0$) is isomorphic to a self-adjoint extension $A^\wedge$ of the symmetric operator $S_0^\wedge$ given by the formula (7.24). If a maximal negative subspace of $\widehat{\mathcal{M}}$ is r-dimensional, then $A^\wedge$ is an extension with exit into the Hilbert space $\mathcal{H}_0 \oplus \mathbf{C}^r$, in particular, if $\widehat{\mathcal{M}}$ is neutral, then $A^\wedge$ is a canonical extension of $S_0^\wedge$.*

Proof: By Theorem 7.1 (ii), the operator $\widetilde{A}^t$ is a self-adjoint extension of the symmetric operator $\widehat{S}_0$ in $\widehat{\mathcal{H}}_0$. It is described by the generalized resolvent of $\widehat{S}_0$ with the function $\widehat{Q}_0(z)+\tau(z)$ in the denominator of the corresponding M.G. Krein formula. Further, the function $Q(z)+t$ is factored as

$$Q(z)+t = \widehat{b}_m(z)\widehat{b}_m^*(z)(\widehat{Q}_0(z)+\tau(z)).$$

But $Q(z)$ is also represented in the form (3.15):

$$Q(z) = (z-i)^m(z+i)^m Q_0(z) + p_{2m+1}(z)$$

with

$$Q_0(z) = (z+i)\langle R_0(z)\chi_{-1}, \chi_0\rangle_0 + g_{m+1,m}.$$

If we divide both these representations of $Q(z)$ by the polynomial $\widehat{b}_m(z)\widehat{b}_m^*(z)$ (see the proof of Theorem 5.4 for more details) we get the equality

$$\widehat{Q}_0(z)+\tau(z) = Q_0^\wedge(z)+\tau(z),$$

with $Q_0^\wedge(z)$ from (7.23).

On the other hand, consider the self-adjoint extension $A^\wedge$ of the symmetric operator $S_0^\wedge$ with exit into the space $\mathcal{H}_0 \oplus \mathbf{C}^r$, $r = \dim \widehat{\mathcal{M}}^-$, such that its compressed resolvent $P_0(A^\wedge - z)^{-1}|_{\mathcal{H}_0}$ is of the form (7.24). If we compare (7.24) with M.G. Krein's formula for $\widehat{P}_0(\widetilde{A}^t - z)^{-1}|_{\widehat{\mathcal{H}}_0}$ (see its description in Theorem 7.1 (ii)) we see that both formulas contain the common defining function $\widehat{Q}_0(z)+\tau(z)$ ($= Q_0^\wedge(z)+\tau(z)$). Define a mapping T:

$$T\,R_0(z)\chi_{-1}^\wedge = \widehat{\gamma}(z) \text{ for all nonreal } z.$$

Taking into account that $\widehat{Q}_0(z)$ ($= Q_0^\wedge(z)$) is the Q-function corresponding to $S_0^\wedge$ and A_0 as well as the Q-function corresponding to $\widehat{S}_0$ and $\widehat{A}_0$, we conclude (see [KL1], Satz 4.2) that T can be extended by continuity to an isometry from $\mathcal{H}_0$ onto $\widehat{\mathcal{H}}_0$ such that $T^{-1}\widehat{A}_0 T = A_0$ and $T^{-1}\widehat{S}_0 T = S_0^\wedge$. But extensions with exit of the same symmetric operator which are determined by the same parameter function $\operatorname{Re}\widehat{Q}_0(\lambda_0)+\widehat{\tau}(z)$ are isomorphic if they are minimal, see [KL]. This isomorphism is determined by an isometry which is the identity on $\mathcal{H}_0$.

By Corollary 7.2 we have $\tau(z) \equiv 0$ if and only if the subspace $\widehat{M}$ is neutral. In this case $A^\wedge$ is the canonical self-adjoint extension in $\mathcal{H}_0$ the resolvent of which is described by M.G. Krein's formula (7.24) with $\widehat{Q}_0(z)$ in the denominator. □

Summing up, the operator $\widetilde{A}^t$ (in its version A^Λ) is a self-adjoint extension of the symmetry S_0^Λ with possible exit into the Hilbert space $\mathcal{H}_0 \oplus \mathbf{C}^r$. Thus, the method presented in this paper leads to nontrivial quantum models according to the following scheme:

$$\{\mathcal{H}_0, A_0, \chi\} \longrightarrow \{\Pi_m, H^t, S\} \longrightarrow \{\widetilde{\mathcal{H}}_0, \widetilde{A}^t, \widehat{S}_0\} \simeq \{\widetilde{\mathcal{H}}_0, A^\Lambda, S_0^\Lambda\}.$$

The set of parameters $g_{j,j}$ and $\operatorname{Re} g_{j,j+1}$, $j = 1, \ldots, m$, and t define the set Λ; if Λ contains only nonreal points then just one parameter remains which fixes the canonical extension, if Λ contains also real points, then additional parameters appear which determine the Gram matrix of the Hilbert space $\mathcal{H} = \mathcal{H}_0 \oplus \mathbf{C}^r$.

Finally, the operator $\widetilde{A}^t$ (in its version A^Λ) can also be considered as a model of the singular perturbation problem

$$A_0 + \chi_{-1}^\Lambda \langle \cdot, \chi_{-1}^\Lambda \rangle, \quad \chi_{-1}^\Lambda \in \mathcal{H}_{-1},$$

which can be treated completely within the Hilbert space extension theory, see the remark in Section 1.3 and, for example, [P].

References

[AI] T.Ya. Azizov and I.S. Iokhvidov, *Foundations of the theory of linear operators in spaces with an indefinite metric*, Nauka, Moscow, 1986 (Russian); English transl.: *Linear operators in spaces with an indefinite metric*, Wiley, New York, 1989.

[AG] N.I. Achieser and I.M. Glasmann, *Theorie der linearen Operatoren im Hilbertraum*, Akademie Verlag, Berlin, 1981

[ADRS] D. Alpay, A. Dijksma, J. Rovnyak, and H.S.V. de Snoo, *Schur functions, operator colligations, and reproducing kernel Pontryagin spaces*, Operator Theory: Adv. Appl., vol. **96**, Birkhäuser Verlag, Basel, 1997.

[AGHH] S. Albeverio, F. Gesztesy, R. Høegh-Krohn and H. Holden, *Solvable models in Quantum Mechanics*, Springer-Verlag, Berlin, 1988.

[B1] Yu.M. Berezanskii, *Expansions in eigenfunctions of selfadjoint operators*, Transl. Amer. Math. Soc. 17, Providence, Rhode Island, 1968.

[B2] Yu.M. Berezanskii, *Selfadjoint operators in spaces of functions of infinitely many variables*, Transl. Amer. Math. Soc. 63, Providence, Rhode Island, 1986.

[Be] F.A. Berezin, *On the Lee model*, Matem. Sborn. **60** (1963), 425–453 (Russian).

[Bo] J. Bognar, *Indefinite inner product spaces*, Springer-Verlag, Berlin, 1974.

[C] W. Caspers, *On point interactions*, Thesis, Technical University Delft, 1992.

[DL1] K. Daho and H. Langer, *Sturm-Liouville operators with indefinite weight function*, Proc. Royal Soc. Edinburgh **78A** (1977), 161–191.

[DL2] K. Daho and H. Langer, *Matrix functions of the class N_κ*, Math. Nachr. **120** (1985), 275–294.

[DijL] A. Dijksma and H. Langer, *Operator theory and ordinary differential operators*, Lecture Series 2 in: Albrecht Böttcher et al., *Lectures on Operator theory and its applications*, Fields Institute Monographs, Amer. Math. Soc., Providence RI, 1995, 73–139.

[DS] A. Dijksma and Yu.G. Shondin, *Singular point-like perturbations of the Bessel operator in a Pontryagin space*, in preparation.

[DSn] A. Dijksma and H.S.V. de Snoo, *Symmetric and selfadjoint relations in Krein spaces I*, Operator Theory: Adv. Appl., vol. **24**, Birkhäuser Verlag, Basel, 1987, 145–166.

[DT] J.F. van Diejen and A. Tip, *Scattering from generalized point interaction using selfadjoint extensions in Pontryagin spaces*, J. Math. Phys. **32**(3) (1991), 630–641.

[DLS1] A. Dijksma, H. Langer and H.S.V. de Snoo, *Selfadjoint Π_κ-extensions of symmetric subspaces: an abstract approach to boundary problems with spectral parameter in the boundary conditions*, Integral Equations Operator Theory **7** (1984), 459–515.

[DLS2] A. Dijksma, H. Langer and H.S.V. de Snoo, *Unitary colligations in Π_κ-spaces, characteristic functions and Straus extensions*, Pacific J. Math. **125**(2) (1986), 347–362.

[DLS3] A. Dijksma, H. Langer and H.S.V. de Snoo, *Generalized coresolvents of standard isometric operators and generalized resolvents of standard symmetric relations in Krein spaces*, Operator Theory: Adv. Appl., vol. **48** (1990), 261–274.

[DLLS] A. Dijksma, H. Langer, A. Luger and Yu. Shondin, *A factorization result for generalized Nevanlinna functions of the class N_κ*, Integral Equations Operator Theory, to appear.

[HLS] S. Hassi, H. Langer and H.S.V. de Snoo, *Selfadjoint extensions for a class of symmetric operators with defect numbers* (1, 1), Topics in Operator Theory, Operator Algebras and Appl.: 15th International Conference on Operator Theory (Timisoara 1994), IMAR, Bucarest, 1995, 115–145.

[IKL] I.S. Iokhvidov, M.G. Krein and H. Langer, *Introduction to the Spectral Theory of Operators in Spaces with an Indefinite Metric*, Akademie-Verlag, Berlin, 1982.

[JLT] P. Jonas, H. Langer and B. Textorius, *Models and unitary equivalence of cyclic selfadjoint operators in Pontrjagin spaces*, Operator Theory: Adv. Appl., vol. **59**, Birkhäuser Verlag, Basel, 1992, 252–284.

[KL] M.G. Krein and H. Langer, *On defect subspaces and generalized resolvents of Hermitean operator in a space Π_κ*, Funkt. Anal. i Prilozhen. **5**(3) (1971), 54–69 (Russian); English transl.: Functional Anal. Appl. **5** (1971), 217–228.

[KL1] M.G. Krein and H. Langer, *Über die Q-Funktion eines π-hermiteschen Operators in Raume Π_κ*, Acta Sci. Math. (Szeged) **34** (1973), 191–230.

[KL2] M.G. Krein and H. Langer, *Über einige Fortzetzungsprobleme, die eng mit der Theorie hermitescher Operatoren im Raume Π_κ zusammenhängen. I. Einige Funktionenklassen und ihre Darstellungen*, Math. Nachr. **77** (1977), 187–236.

[KL3] M.G. Krein and H. Langer, *Some propositions on analytic matrix functions related to the theory of operators on the space Π_κ*, Acta Sci. Math. (Szeged) **43** (1981), 181–205.

[L] H. Langer, *A characterization of generalized zeros of negative type of functions of the class N_κ*, Operator Theory: Adv. Appl., vol. **17**, Birkhäuser Verlag, Basel, 1986, 201–212.

[LT] H. Langer and B. Textorius, *On generalized resolvents and Q-functions of symmetric linear relations (subspaces)*, Pacific J. Math. **72** (1977), 135–165.

[P] B.S. Pavlov, *Extension theory and explicitly solvable models*, Uspechi Matem. Nauk. **42**(6) (1988), 99–131 (Russian); English transl.: Russian Math. Surveys **42** (1987), 127–168.

[S1] Yu.G. Shondin, *Quantum-mechanical models in* $\mathbf{R}^n$ *associated with extensions of the energy operator in a Pontryagin space*, Teor. Mat. Fiz. **74** (1988), 331–344 (Russian); English transl.: Theor. Math. Phys. **74** (1988), 220–230.

[S2] Yu.G. Shondin, *Perturbation of differential operators on a high-codimensional manifold and the extension theory for symmetric linear relations in an indefinite metric space*, Teor. Mat. Fiz. **92**(3) (1992), 466–472 (Russian); English transl.: Theor. Math. Phys. **92** (1992), 1032–1037.

[S3] Yu.G. Shondin, *Perturbation of elliptic operators supported on subsets of high codimension, and extension theory in indefinite metric spaces*, Seminars of St. Petersburg Math. Inst., vol. **222**, Research in Linear Operators and Function Theory **23** (1995), 246–292 (Russian).

Aad Dijksma
Department of Mathematics
University of Groningen
P.O. Box 800
9700 AV Groningen
The Netherlands
dijksma@math.rug.nl

Heinz Langer
Department of Mathematics
Technical University of Vienna
Wiedner Haupsstrasse 8–10/1411
A–1040 Vienna
Austria
hlanger@email.tuwien.ac.at

Yuri Shondin
Department of Theoretical Physics
State Pedagogical University
Str. Uly'anova 1
Nizhny Novgorod 603600
Russia
shondin@shmath.nnov.ru

Chris Zeinstra
Department of Mathematics
Free University
De Boelelaan 1081a
1081 HV Amsterdam
The Netherlands
zeinstra@cs.vu.nl

AMS Classification Numbers: Primary 47B50, 47B25. Secondary 81Q10

Operator Theory:
Advances and Applications, Vol. 118

On the Uniform Convergence Theorem of Semigroups

Ya.I. Grushka

Let $\{\mathbf{T}(t) : t \geq 0\}$ be a C_0-semigroup of bounded linear operators in a Banach space $\mathbf{X}$ and $\mathbf{A}$ be the generator of this semigroup. The well known uniform convergence theorem asserts that if for the semigroup $\{\mathbf{T}(t)\}$, $\overline{\lim}_{t\to+0}\|\mathbf{T}(t) - \mathbf{I}\| < 1$, where $\mathbf{I}$ is the identity operator in $\mathbf{X}$, then the generator $\mathbf{A}$ is bounded.

In this article, a certain generalization of this result is presented.

The aim of this article is to prove the following theorem.

Theorem: *Let* $\{\mathbf{T}(t) : t \geq 0\}$ *be a* C_0*-semigroup of normal operators with generator* $\mathbf{A}$ *in a Hilbert space* $\mathbf{H}$.

If there exists a non-negative on $[0, +\infty)$ *function* $\beta(t)$ *satisfying the conditions:*

$$1)\ \lim_{t\to+0} \frac{\beta(t)}{t} = 0\,, \quad 2)\ \overline{\lim_{t\to+0}} \|T(t) - T(\beta(t))\| < 1,$$

then the generator $\mathbf{A}$ *is bounded.*

Remark: In the case of $\beta(t) \equiv 0,\ t \geq 0$, this result is known as a theorem on the uniform convergence of a semigroup in a Banach space to the identity operator ([1], [2]).

Proof of the Theorem: Let $\mathbf{E}(\cdot)$ be the spectral resolution of the generator $\mathbf{A}$. Then

$$\begin{aligned} \|\mathbf{T}(t) - \mathbf{T}(\beta(t))\| &= \left\| \int_{\sigma(\mathbf{A})} (e^{\lambda t} - e^{\lambda\beta(t)}) d\mathbf{E}(\lambda) \right\| \\ &= \sup_{\lambda\in\sigma(\mathbf{A})} |e^{\lambda t} - e^{\lambda\beta(t)}|, \quad t \geq 0, \end{aligned} \tag{1}$$

where $\sigma(\mathbf{A})$ is the spectrum of the operator $\mathbf{A}$.

By the Hille-Yosida theorem, there exists $\omega \in \mathbf{R}$ such that $\sigma(\mathbf{A}) \subseteq \{\lambda \in \mathbf{C} | \mathbf{Re}\lambda \leq \omega\}$. Taking the semigroup $e^{-(|\omega|+1)t}\mathbf{T}(t)$ instead of $\mathbf{T}(t)$ (if it is necessary), we can suppose

$$\sigma(\mathbf{A}) \subseteq \{\lambda \in \mathbf{C} | \mathbf{Re}\lambda \leq -1\}. \tag{2}$$

To prove the boundedness of the generator **A**, it suffices to show that the spectrum $\sigma(\mathbf{A})$ of the operator **A** is a bounded set. Assume the contrary. Then there exists a sequence $\{\lambda_n\}_{n=1}^{\infty} \subseteq \sigma(\mathbf{A})$ such that $|\lambda_n| \to \infty,\ n \to \infty$.

We will prove that $\lim_{n\to\infty} \frac{|\mathbf{Im}\lambda_n|}{|\mathbf{Re}\lambda_n|} = 0$. Indeed, if this is not the case, then there exist a consequence $\{\mu_k\}$ of the sequence $\{\lambda_n\}$ and a number $\eta > 0$ such that $\frac{|\mathbf{Im}\mu_k|}{|\mathbf{Re}\mu_k|} \geq \eta$ ($\forall k \in \mathbf{N}$), and we have

$$
\text{(3)}\qquad \begin{aligned} &|\mu_k| \leq \sqrt{1+\eta^{-2}}|\mathbf{Im}\mu_k| \quad (\forall k \in \mathbf{N}),\\ &\lim_{k\to\infty} |\mathbf{Im}\mu_k| = +\infty. \end{aligned}
$$

Put $t_k := \frac{\pi}{|\mathbf{Im}\mu_k|}$, $k \in \mathbf{N}$. Then $t_k \to 0$, $k \to \infty$. Using (1), we obtain

$$
\text{(4)}\qquad \begin{aligned} \|\mathbf{T}(t_k) - \mathbf{T}(\beta(t_k))\| &\geq |e^{\mu_k t_k} - e^{\mu_k\beta(t_k)}| \geq |e^{\mu_k t_k} - 1| - |e^{\mu_k\beta(t_k)} - 1|\\ &= (e^{t_k \mathbf{Re}\mu_k} + 1) - |e^{\mu_k\beta(t_k)} - 1| \geq 1 - |e^{\mu_k\beta(t_k)} - 1|. \end{aligned}
$$

Applying inequality (3) and the conditions of Theorem, we get

$$
|\mu_k\beta(t_k)| = \pi\frac{|\mu_k|}{|\mathbf{Im}\mu_k|}\frac{\beta(t_k)}{t_k} \leq \pi\sqrt{1+\eta^{-2}}\frac{\beta(t_k)}{t_k} \longrightarrow 0, \quad k \longrightarrow \infty.
$$

By virtue of (4),

$$
\underline{\lim}_{k\to\infty} \|\mathbf{T}(t_k) - \mathbf{T}(\beta(t_k))\| \geq 1,
$$

contrary to the conditions of Theorem.

So $\lim_{n\to\infty} \frac{|\mathbf{Im}\lambda_n|}{|\mathbf{Re}\lambda_n|} = 0$, whence $\lim_{n\to\infty} \frac{|\lambda_n|}{|\mathbf{Re}\lambda_n|} = 1$. Using (2), we have

$$
\text{(5)}\qquad \lim_{n\to\infty} \frac{|\lambda_n|}{\mathbf{Re}\lambda_n} = -1.
$$

Let $a \in \mathbf{R}$, $a > 0$. Set $t_n := \frac{a}{|\lambda_n|}$, $n \in \mathbf{N}$. Then $t_n \to 0$, $n \to \infty$. In view of (1),

$$
\text{(6)}\quad \|\mathbf{T}(t_n) - \mathbf{T}(\beta(t_n))\| \geq |e^{\lambda_n t_n} - e^{\lambda_n\beta(t_n)}| \geq |e^{\lambda_n\beta(t_n)}| - |e^{\lambda_n t_n}| \quad (n \in \mathbf{N}).
$$

Using (5) and the conditions of Theorem, we obtain that for any $n \in \mathbf{N}$,

$$
\text{(7)}\qquad \lambda_n t_n = \lambda_n\frac{a}{|\lambda_n|} = a\left(1 + i\frac{\mathbf{Im}\lambda_n}{\mathbf{Re}\lambda_n}\right)\frac{\mathbf{Re}\lambda_n}{|\lambda_n|} \longrightarrow -a,\ n \longrightarrow \infty
$$

$$
\text{(8)}\qquad |\lambda_n\beta(t_n)| = a\frac{\beta(t_n)}{t_n} \longrightarrow 0,\ n \longrightarrow \infty.
$$

The relations (6), (7), and (8) give $\underline{\lim}_{n\to\infty} \|\mathbf{T}(t_n) - \mathbf{T}(\beta(t_n))\| \geq 1 - e^{-a}$. Since $t_n \to 0$, $n \to \infty$, we have the inequality $\underline{\lim}_{t\to+0} \|\mathbf{T}(t) - \mathbf{T}(\beta(t))\| \geq 1 - e^{-a}$ for any $a \in \mathbf{R}, a > 0$.

Therefore,

$$\varliminf_{t\to+0} \|\mathbf{T}(t) - \mathbf{T}(\beta(t))\| \geq 1,$$

contrary to the conditions of Theorem, which is what had to be proved.

References

[1] N. Dunford and J. Shwartz, *Linear Operators*. Part 1. Intersc. Publ., London, 1958.
[2] G. Goldstein, *Semigroups Linear of Operators and their applications*. Kiev. "Vischa Shkola" (Russian).

Yaroslav Grushka
Inst of Mathematics
Tereshcenkivska str 3
Kiev 252601 MSP
Ukraine

AMS Classification Number = 47.

Operator Theory:
Advances and Applications, Vol. 118

On Stochastic Integral, Wick Calculus and Generalized Stochastic Equations on Distribution Spaces of Non-Gaussian Infinite-dimensional Analysis

N.A. Kachanovsky

We introduce an analog of the extended stochastic integral on generalized function spaces of non-Gaussian infinite dimensional analysis and discuss the properties of such integral and its relation to Wick calculus. Then we consider some examples of application of constructed objects to solving generalized stochastic equations.

0 Introduction

In 1991 Yu.L. Daletsky ([1]) proposed to consider so-called biorthogonal Appell systems, connected with non-Gaussian measures. By means of these systems a non-Gaussian infinite-dimensional analysis was constructed in [2–8] etc. A wide generalization based on the substitution of Appell polynomials by so-called Appell and Delsarte characters is constructed in [9, 10].

As is known, one of the main applications of Gaussian infinite-dimensional analysis is connected with solving stochastic equations with Wick-type nonlinearity, for more detail see e.g. [11] and references therein. In non-Gaussian analysis, it is simple to construct a Wick product and Wick versions of holomorphic functions on generalized function spaces, it was made in different cases in [5–7, 12, 8] etc. But as for construction of an analog of the extended stochastic integral—it can be made by different ways with different results (see e.g. [3] for the case of so-called smooth measure). We consider the direct generalization of extension of Skorokhod stochastic integral in Hida sense ([13]), introduced and studied in [14] (motivation and detailed references see therein). As an example of application, the Verhulst type equation was considered in that paper. In this paper, we discuss the mentioned above integral 'from Fock space point of view' and study another (more complicated) examples.

The paper is organized in the following manner. In the first section we introduce the necessary definitions and objects. In the second section we introduce and discuss the analog of extended stochastic integral. In the third section we consider the generalized stochastic equations as the examples of application of stochastic integral and Wick calculus.

Note that the most of objects introduced in Preliminaries have the classical prototypes in the works of many authors. The reader can see References of cited papers for more detailed information.

1 Preliminaries

Let us consider the chain $\mathcal{S}(\mathbf{R}) \equiv \mathcal{S} \subset L^2(\mathbf{R}, dt) \equiv L^2 \subset \mathcal{S}'(\mathbf{R}) \equiv \mathcal{S}'$, where $\mathcal{S}$ is the Schwartz space (of quickly decreasing infinitely differentiable functions), $\mathcal{S}'$ is the dual to $\mathcal{S}$ with respect to L^2 space. We denote by subindex $\mathbf{C}$ the complexifications of the spaces. It is well known (see e.g. [15]) that $\mathcal{S} = \text{pr}\lim_{p\in\mathbf{N}} \mathcal{H}_p$, $\mathcal{S}' = \text{ind}\lim_{p\in\mathbf{N}} \mathcal{H}_{-p}$, where $\mathcal{H}_p$, $p \in \mathbf{Z}$, are Hilbert spaces, $\mathcal{H}_p \subset L^2 \subset \mathcal{H}_{-p}$, and the inclusion operators $\mathcal{H}_{p+1} \hookrightarrow \mathcal{H}_p$ are of Hilbert-Schmidt type. We denote the norm in $\mathcal{H}_p$ by $|\cdot|_p$, the dual pairing between elements of $\mathcal{H}_{-p}$ and $\mathcal{H}_p$ generated by inner product in L^2 by $\langle\cdot,\cdot\rangle$ and preserve these notations for tensor powers of spaces and complexifications. Let $\text{Hol}_0(\mathcal{S}_\mathbf{C})$ and $\text{Hol}_0(\mathcal{S}_\mathbf{C}, \mathcal{S}_\mathbf{C})$ be the algebras of germs of holomorphic at zero functions $\gamma : \mathcal{S}_\mathbf{C} \to \mathbf{C}$ and $\alpha : \mathcal{S}_\mathbf{C} \to \mathcal{S}_\mathbf{C}$ respectively.

Let $\chi : \mathbf{C} \to \mathbf{C}$ be entire function, $\chi(0) = 1$ and in Taylor decomposition

$$\chi(u) = \sum_{n=0}^{\infty} \frac{1}{n!} \chi_n u^n \tag{1}$$

$\chi_n \neq 0$ for all $n \in \mathbf{Z}_+$. For $\gamma \in \text{Hol}_0(\mathcal{S}_\mathbf{C})$, $\gamma(0) \neq 0$; $\alpha \in \text{Hol}_0(\mathcal{S}_\mathbf{C}, \mathcal{S}_\mathbf{C})$ and invertible, $\alpha(0) = 0$ we consider the function $\chi^{\gamma,\alpha}(\theta; z) := \gamma(\theta)\chi(\langle z, \alpha(\theta)\rangle)$, $z \in \mathcal{S}'_\mathbf{C}$, $\theta \in \mathcal{S}_\mathbf{C}$. It is clear that $\chi^{\gamma,\alpha}(\cdot; z) \in \text{Hol}_0(\mathcal{S}_\mathbf{C})$ for all $z \in \mathcal{S}'_\mathbf{C}$ so it can be decomposed (using kernel theorem) into the series

$$\chi^{\gamma,\alpha}(\theta; z) = \sum_{n=0}^{\infty} \frac{1}{n!} \langle P_n^{\chi,\gamma,\alpha}(z), \theta^{\otimes n}\rangle,$$

$P_n^{\chi,\gamma,\alpha}(z) \in \mathcal{S}'^{\widehat{\otimes} n}_\mathbf{C}$, where $\widehat{\otimes}$ denotes the symmetric tensor product.

Definition 1 The system $\mathbf{P}^{\chi,\gamma,\alpha}$ of polynomials $\{\langle P_n^{\chi,\gamma,\alpha}(z), \varphi^{(n)}\rangle, n \in \mathbf{Z}_+ := \mathbf{N} \cup \{0\}, \varphi^{(n)} \in \mathcal{S}^{\widehat{\otimes} n}_\mathbf{C}$ is called the system of generalized Appell-like polynomials.

Let $\mathcal{P}(\mathcal{S}')$ be the set of continuous polynomials on $\mathcal{S}'$. It is clear that we can consider $\mathcal{P}(\mathcal{S}')$ as the set of polynomials of the form

$$\varphi(x) = \sum_{n=0}^{m} \langle P_n^{\chi,\gamma,\alpha}(x), \varphi^{(n)}\rangle, \ m \in \mathbf{Z}_+, \ x \in \mathcal{S}', \ \varphi^{(n)} \in \mathcal{S}^{\widehat{\otimes} n}_\mathbf{C}. \tag{2}$$

Let us introduce the norm $\|\cdot\|_{p,q}$ ($p \in \mathbf{N}_1 := \mathbf{N}\backslash\{1\}, q \in \mathbf{Z}_+$) on $\mathcal{P}(\mathcal{S}')$ by putting for φ of the form (2)

$$\|\varphi\|_{p,q}^2 := \sum_{n=0}^{m} (n!)^2 2^{qn} |\varphi^{(n)}|_p^2.$$

We denote by $(\mathcal{H}_p)_{q,\chi,\gamma,\alpha}$ the Hilbert space which is the closure of $\mathcal{P}(\mathcal{S}')$ with respect to introduced norm. Let $(\mathcal{H}_p)_\chi := \text{pr}\lim_{q\in\mathbf{N}}(\mathcal{H}_p)_{q,\chi,\gamma,\alpha}$. It was proved in [8] that $(\mathcal{H}_p)_\chi$ does not depend on γ and α as the topological space (with the projective limit topology).

Let μ be the probability measure on $\mathcal{S}'$ such that

1) There exists a nonempty open set $O \subseteq \mathcal{S}'$ such that μ is positive on any nonempty open subset of O;
2) There exist $p \in \mathbf{N}_1$, $C, K > 0$ such that $\| |P_n^{\chi,1,id}(\cdot)|_{-p} \|_{L^2(\mathcal{S}',\mu)} \leq n!C^nK$ for all $n \in \mathbf{Z}_+$.

Here $\| \cdot \|_{L^2(\mathcal{S}',\mu)}$ is the norm on the space $L^2(\mathcal{S}',\mu)$ of square integrable with respect to μ functions on $\mathcal{S}'$. Let H be the Hilbert space which is the closure of $\mathcal{P}(\mathcal{S}')$ with respect to $\| \cdot \|_{L^2(\mathcal{S}',\mu)}$.

Remark 1 Let us denote by $l_\mu(\theta) := \int_{\mathcal{S}'} e^{\langle x,\theta\rangle}\mu(dx)$ the Laplace transform of the measure μ. Note that if $l_\mu \in \text{Hol}_0(\mathcal{S}_\mathbf{C})$, then the continuous polynomials are dense in $L^2(\mathcal{S}',\mu)$ (see [16]) and therefore $H = L^2(\mathcal{S}',\mu)$.

Example: The Gaussian measure μ on $\mathcal{S}'$ has the Laplace transform $l_\mu(\theta) = e^{\frac{1}{2}\langle\theta,\theta\rangle}$; the Poisson measure π on $\mathcal{S}'$ has the Laplace transform $l_\pi(\theta) = \exp\{\langle e^\theta - 1, 1\rangle\}$. These measures satisfy the conditions 1) and 2) for $\chi = \exp$. Moreover for these measures, $H = L^2(\mathcal{S}',\mu)$ and $H = L^2(\mathcal{S}',\pi)$ respectively.

It is not difficult to prove (see [6]) that under the above assumptions there exist $p' \in \mathbf{N}_1$, $q' \in \mathbf{Z}_+$ such that for $p \geq p'$, $q \geq q'$ $(\mathcal{H}_p)_{q,\chi,\gamma,\alpha} \subset H$ densely and continuously. So we can consider the chain

$$(\mathcal{H}_p)_\chi \subset (\mathcal{H}_p)_{q,\chi,\gamma,\alpha} \subset H \subset (\mathcal{H}_{-p})'_{-q,\chi,\gamma,\alpha,\mu} \subset (\mathcal{H}_{-p})'_{\chi,\mu},$$

where $(\mathcal{H}_{-p})'_{-q,\chi,\gamma,\alpha,\mu}$ and $(\mathcal{H}_{-p})'_{\chi,\mu} = \text{ind}\lim_{q\in\mathbf{N}}(\mathcal{H}_{-p})'_{-q,\chi,\gamma,\alpha,\mu}$ are the spaces adjoint to $(\mathcal{H}_p)_{q,\chi,\gamma,\alpha}$ and $(\mathcal{H}_p)_\chi$ respectively with respect to H. Let us denote by $\langle\langle\cdot,\cdot\rangle\rangle$ the dual pairing between elements of $(\mathcal{H}_{-p})'_{-q,\chi,\gamma,\alpha,\mu}$ and $(\mathcal{H}_p)_{q,\chi,\gamma,\alpha}$, generated by inner product in H.

It was proved in [6] that there exists a system of generalized functions $Q_{\mu,m}^{\chi,\gamma,\alpha}(\Phi^{(m)};\cdot)$ ($\Phi^{(m)} \in \mathcal{S}_\mathbf{C}'^{\widehat{\otimes}m}$) which is the orthogonal basis in $(\mathcal{H}_{-p})'_{-q,\chi,\gamma,\alpha,\mu}$. To construct these generalized functions, we need the following construction.

Let us introduce on $\mathcal{P}(\mathcal{N}')$ the linear differential operator $\langle\Phi^{(m)}, D_\chi^{\otimes m}\rangle$ with constant coefficient $\Phi^{(m)} \in \mathcal{S}_\mathbf{C}'^{\widehat{\otimes}m}$ by putting on monomials

$$\langle\Phi^{(m)}, D_\chi^{\otimes m}\rangle\langle x^{\otimes n}, \varphi^{(n)}\rangle := 1_{\{n\geq m\}}\frac{n!\chi_{n-m}}{(n-m)!\chi_n}\langle x^{\otimes(n-m)}\widehat{\otimes}\Phi^{(m)}, \varphi^{(n)}\rangle, \; x \in \mathcal{S}'$$

(here $1_{\{n\geq m\}}$ is the indicator of $\{n \geq m\}$, $\{\chi_n\}_{n=0}^\infty$ are the Taylor coefficients of χ from (1)) and extending the definition by linearity. Let $\alpha^{-1} \in \text{Hol}_0(\mathcal{S}_\mathbf{C}, \mathcal{S}_\mathbf{C})$ be the inverse function of α. We define the operator $\langle \Phi^{(m)}, \alpha^{-1}(D_\chi)^{\otimes m}\rangle$ in the following way. Let $\nu(\theta) := \langle \Phi^{(m)}, \alpha^{-1}(\theta)^{\otimes m}\rangle$. Then it is obvious that $\nu \in \mathrm{Hol}_0(\mathcal{S}_{\mathbf{C}})$ and therefore the decomposition $\nu(\theta) = \sum_{n=0}^{\infty} \frac{1}{n!}\langle \nu^{(n)}, \theta^{\otimes n}\rangle$, $\nu^{(n)} \in \mathcal{S}_{\mathbf{C}}^{\prime\widehat{\otimes} n}$ holds. We put

$$\langle \Phi^{(m)}, \alpha^{-1}(D_\chi)^{\otimes m}\rangle \equiv \nu(D_\chi) := \sum_{n=0}^{\infty} \frac{1}{n!}\langle \nu^{(n)}, D_\chi^{\otimes n}\rangle.$$

Let us introduce the coordinate-wise convergence topology on the set $\mathcal{P}(\mathcal{S}')$ (see e.g. [5] for details) and consider the dual to $\mathcal{P}(\mathcal{S}')$ with respect to H space $\mathcal{P}'_\mu(\mathcal{S}')$. We denote by $\langle \Phi^{(m)}, \alpha^{-1}(D_\chi)^{\otimes m}\rangle_\mu^*$ the operator on $\mathcal{P}'_\mu(\mathcal{S}')$, adjoint to $\langle \Phi^{(m)}, \alpha^{-1}(D_\chi)^{\otimes m}\rangle$ in the sense that for all $\Phi \in \mathcal{P}'_\mu(\mathcal{S}')$ and $\varphi \in \mathcal{P}(\mathcal{S}')$ we have

$$\langle\langle \langle \Phi^{(m)}, \alpha^{-1}(D_\chi)^{\otimes m}\rangle_\mu^* \Phi, \varphi\rangle\rangle \equiv \langle\langle \Phi, \langle \Phi^{(m)}, \alpha^{-1}(D_\chi)^{\otimes m}\rangle \varphi\rangle\rangle.$$

Now let us put

$$Q_{\mu,m}^{\chi,\gamma,\alpha}(\Phi^{(m)}; \cdot) := \sum_{k=m}^{\infty} \frac{1}{(k-m)!}(\langle \Phi^{(m)} \widehat{\otimes} \widetilde{\gamma}_{k-m}, \alpha^{-1}(D_\chi)^{\otimes k}\rangle_\mu^* \delta_0)(\cdot),$$

where $\widetilde{\gamma}_k \in \mathcal{S}_{\mathbf{C}}^{\prime\widehat{\otimes} k}$ are from a decomposition $\frac{1}{\gamma(\theta)} = \sum_{k=0}^{\infty} \frac{1}{k!}\langle \widetilde{\gamma}_k, \theta^{\otimes k}\rangle$. Then $\Phi \in (\mathcal{H}_{-p})'_{-q,\chi,\gamma,\alpha,\mu}$ if and only if there exists a (unique) sequence of kernels $\Phi^{(m)} \in \mathcal{S}_{\mathbf{C}}^{\prime\widehat{\otimes} m}$ such that

$$\Phi(\cdot) = \sum_{m=0}^{\infty} Q_{\mu,m}^{\chi,\gamma,\alpha}(\Phi^{(m)}; \cdot) \tag{3}$$

and $\|\Phi\|_{-p,-q}^2 := \sum_{m=0}^{\infty} 2^{-qm} |\Phi^{(m)}|_{-p}^2 < \infty$. Moreover it is not difficult to prove that the system of generalized functions $Q_{\mu,m}^{\chi,\gamma,\alpha}(\Phi^{(m)}; \cdot)$ is *biorthogonal to* $\mathbf{P}^{\chi,\gamma,\alpha}$*-system of generalized Appell-like polynomials in the sense that*

$$\langle\langle Q_{\mu,m}^{\chi,\gamma,\alpha}(\Phi^{(m)}; \cdot), \langle P_n^{\chi,\gamma,\alpha}(\cdot), \varphi^{(n)}\rangle\rangle\rangle = \delta_{mn} n! \langle \Phi^{(n)}, \varphi^{(n)}\rangle.$$

For $\Phi \in (\mathcal{H}_{-p})'_{\chi,\mu}$ let us introduce the S-transform $(S\Phi)(\theta)$, $\theta \in \mathcal{S}_{\mathbf{C}}$ by putting

$$(S\Phi)(\theta) := \langle\langle \Phi(\cdot), \chi^{\gamma,\alpha}(\theta; \cdot)\rangle\rangle.$$

Here $2^q|\theta|_p^2 < 1$, where $q \in \mathbf{Z}_+$ is such that $\Phi \in (\mathcal{H}_{-p})'_{-q,\chi,\gamma,\alpha,\mu}$ (see [5–8]).

Now we can define the Wick product and the Wick versions of holomorphic functions on $(\mathcal{H}_{-p})'_{\chi,\mu}$ ([5–7, 12, 8, 14]). Namely, for $\Phi, \Psi \in (\mathcal{H}_{-p})'_{\chi,\mu}$ and holomorphic at $(S\Phi)(0)$ function $F : \mathbf{C} \to \mathbf{C}$ we put

$$\Phi \diamond \Psi := S^{-1}[S\Phi \cdot S\Psi]; \quad F^\diamond(\Phi) := S^{-1}F(S\Phi).$$

It was proved in [8] that these definitions are correct. Note that the Wick product is commutative and associative.

2 The Analog of Stochastic Integral

In order to recall the definition of the analog of stochastic integral, ([14]) we need the following notations. Let us introduce the family of operators $J^m_{[0,t)}$: $\mathcal{H}^{\widehat{\otimes} m}_{-p,\mathbf{C}} \otimes L^2 \to \mathcal{H}^{\widehat{\otimes}(m+1)}_{-p,\mathbf{C}}$, $m \in \mathbf{Z}_+$, acting by formula

$$(4) \qquad (J^m_{[0,t)}\Phi^{(m)})(\cdot) := (\widehat{\Phi^{(m)}(\cdot)1_{[0,t)}(\cdot)}) \equiv \widehat{\Phi}^{(m+1)}_{[0,t)},$$

where $\widehat{\cdot}$ is the symmetrization operator.

Definition 2 For $Y_{\cdot_i}(\cdot) = \sum_{m=0}^{\infty} Q^{\chi,\gamma,\alpha}_{\mu,m}(\Phi^{(m)}(\cdot_i);\cdot) \in (\mathcal{H}_{-p})'_{-q,\chi,\gamma,\alpha,\mu} \otimes L^2$ we define the analog of extended Skorokhod stochastic integral $\mathrm{J}^{\chi,\alpha,\mu}_{[0,t)}$ by putting

$$(\mathrm{J}^{\chi,\alpha,\mu}_{[0,t)} Y_s)(\cdot) := \sum_{m=0}^{\infty} Q^{\chi,\gamma,\alpha}_{\mu,m+1}(\widehat{\Phi}^{(m+1)}_{[0,t)};\cdot).$$

It was proved in [14] that the operator $\mathrm{J}^{\chi,\alpha,\mu}_{[0,t)}$ is a linear continuous operator which acts from $(\mathcal{H}_{-p})'_{-q,\chi,\gamma,\alpha,\mu} \otimes L^2$ to $(\mathcal{H}_{-p})'_{-q,\chi,\gamma,\alpha,\mu}$. Moreover for $Y_{\cdot_2}(\cdot_1) = \widetilde{Y}(\cdot_1) \otimes \Theta(\cdot_2)$, where $\widetilde{Y}(\cdot) \in (\mathcal{H}_{-p})'_{-q,\chi,\gamma,\alpha,\mu}$ and $\Theta(\cdot) \in L^2$

$$(\mathrm{J}^{\chi,\alpha,\mu}_{[0,t)} Y_s)(\cdot) = (\langle \Theta 1_{[0,t)}, \alpha^{-1}(D_\chi)\rangle^*_\mu \widetilde{Y})(\cdot).$$

Let $Z_s \in (\mathcal{H}_{-p})'_{-q,\chi,\gamma,\alpha,\mu}$, $s \geq 0$ and for all $\varphi \in (\mathcal{H}_p)_{q,\chi,\gamma,\alpha}$ the pairing $\langle\langle Z_\cdot, \varphi\rangle\rangle \in L^{1,loc}(\mathbf{R}_+, dt)$. Then the integral $\int_0^t Z_s ds$, $t > 0$ is well defined in Pettis sense, i.e. as a unique element from $(\mathcal{H}_{-p})'_{-q,\chi,\gamma,\alpha,\mu}$ such that

$$\left\langle\!\!\left\langle \int_0^t Z_s ds, \varphi \right\rangle\!\!\right\rangle \equiv \int_0^t \langle\langle Z_s, \varphi\rangle\rangle ds, \quad \varphi \in (\mathcal{H}_p)_{q,\chi,\gamma,\alpha}.$$

Using this notation, we can formulate the following result ([14]): for all $t > 0$

$$(5) \qquad (\mathrm{J}^{\chi,\alpha,\mu}_{[0,t)} Y_s)(\cdot) = \int_0^t Y_s \diamond Q^{\chi,\gamma,\alpha}_{\mu,1}(\delta_s;\cdot) ds.$$

Let us look on the introduced constructions 'from Fock space point of view'. Namely, let $\Gamma(\mathcal{H}_{-p,\mathbf{C}})$ be the Fock space generated by Hilbert space $\mathcal{H}_{-p,\mathbf{C}}$. The integrand $Y_\cdot$ may be identified by virtue of (3) with the (infinite) vector $\mathbf{Y}_\cdot := (\Phi^{(0)}(\cdot), \Phi^{(1)}(\cdot), \ldots) \in \Gamma(\mathcal{H}_{-p,\mathbf{C}}) \otimes L^2$, $\Phi^{(k)}(\cdot) \in \mathcal{H}^{\widehat{\otimes} k}_{-p,\mathbf{C}} \otimes L^2$. Further, the integral $\mathrm{J}^{\chi,\alpha,\mu}_{[0,t)} Y_s$ may be identified with the (infinite) vector $\mathbf{J}^{\chi,\alpha,\mu}_{[0,t)}\mathbf{Y} := (0, \widehat{\Phi}^{(1)}_{[0,t)}, \widehat{\Phi}^{(2)}_{[0,t)}, \ldots) \in \Gamma(\mathcal{H}_{-p,\mathbf{C}})$. So, we may interpret $\mathrm{J}^{\chi,\alpha,\mu}_{[0,t)}$ as the operator $\mathbf{J}_{[0,t)} := (0, J^1_{[0,t)}, J^2_{[0,t)}, \ldots)$ ($J^m_{[0,t)}$, $m \in \mathbf{Z}_+$ were defined in (4)) acting from $\Gamma(\mathcal{H}_{-p,\mathbf{C}}) \otimes L^2$ to $\Gamma(\mathcal{H}_{-p,\mathbf{C}})$:

$$\mathbf{J}_{[0,t)}(\Phi^{(0)}(\cdot), \Phi^{(1)}(\cdot), \ldots) = (0, \widehat{\Phi}^{(1)}_{[0,t)}, \widehat{\Phi}^{(2)}_{[0,t)}, \ldots).$$

Remark 2 The operator $\mathbf{J}_{[0,t)}$ does not depend on functions χ, α and measure μ. Therefore if we consider other functions χ', γ', α' and another measure μ' under conditions above, we obtain *another* integral $\mathrm{J}_{[0,t)}^{\chi',\alpha',\mu'}$ on *another* space $(\mathcal{H}_{-p})'_{-q,\chi',\gamma',\alpha',\mu'} \otimes L^2$, but *its action on coefficients of decomposition* (3) *for* $\Phi' \in (\mathcal{H}_{-p})'_{-q,\chi',\gamma',\alpha',\mu'} \otimes L^2$ *will be the same for all* χ', γ', α' *and* μ'. Note that there arises no paradox because the coefficients from (3), of course, depend on χ', γ', α' and μ'.

3 Generalized Stochastic Equations

As an example of application of the analog of stochastic integral we considered in [14, 8] the Verhulst type equation

$$X_t = X_0 + r\int_0^t X_s \diamond (K - X_s)ds + u\mathrm{J}_{[0,t)}^{\chi,\alpha,\mu}(X_s \diamond (K - X_s)),$$

where $X_0 \in (\mathcal{H}_{-p})'_{\chi,\mu}$, $K, r, u \in \mathbf{R}$ are constants, $K > 0, r > 0, (SX_0)(0) > 0$. It was proved in [8] that this equation has a unique solution in $(\mathcal{H}_{-p})'_{\chi,\mu}$ of the form

$$\begin{aligned} X_t(\cdot) &= K[S^{-1}1 + (KX_0^{\diamond(-1)} - S^{-1}1) \diamond \exp^{\diamond}\{-K(rtS^{-1}1 \\ &\quad + uQ_{\mu,1}^{\chi,\gamma,\alpha}(1_{[0,t)};\cdot))\}]^{\diamond(-1)}, \end{aligned}$$

where $Y^{\diamond(-1)} := S^{-1}(1/SY)$.

Following [11] we consider now more complicated examples. At first, we study the generalized stochastic equation

$$X_t = X_0 + \int_0^t g(\tau)d\tau + \int_0^t r(\tau) \diamond X_\tau d\tau + \mathrm{J}_{[0,t)}^{\chi,\alpha,\mu}(h(\tau) \diamond X_\tau), \tag{6}$$

where r, g, h are continuous functions with values in $(\mathcal{H}_{-p})'_{\chi,\mu}$, $X_0 \in (\mathcal{H}_{-p})'_{\chi,\mu}$.

Theorem 1 *The equation* (6) *has a unique solution*

$$\begin{aligned} X_t &= X_0 \diamond \exp^{\diamond}\left\{\int_0^t (r(\tau) + h(\tau) \diamond Q_{\mu,1}^{\chi,\gamma,\alpha}(\delta_\tau;\cdot))d\tau\right\} \\ &\quad + \int_0^t g(\tau) \diamond \exp^{\diamond}\left\{\int_\tau^t (r(s) + h(s) \diamond Q_{\mu,1}^{\chi,\gamma,\alpha}(\delta_s;\cdot))ds\right\} d\tau \in (\mathcal{H}_{-p})'_{\chi,\mu}. \end{aligned} \tag{7}$$

Proof: Applying S-transform to (6) and taking into account (5), we obtain

$$\widetilde{X}_t(\theta) = \widetilde{X}_0(\theta) + \int_0^t \widetilde{g}_\tau(\theta)d\tau + \int_0^t (\widetilde{r}_\tau(\theta) + \widetilde{h}_\tau(\theta)\theta(\tau))\widetilde{X}_\tau(\theta)d\tau,$$

where the notation $\widetilde{F}(\theta) := (SF)(\theta)$ is used. The solution of this ordinary differential equation is

$$\begin{aligned}\widetilde{X}_t(\theta) &= \widetilde{X}_0(\theta)\exp\left\{\int_0^t (\widetilde{r}_\tau(\theta) + \widetilde{h}_\tau(\theta)\theta(\tau))d\tau\right\} \\ &\quad + \int_0^t \widetilde{g}_\tau(\theta)\exp\left\{\int_\tau^t (\widetilde{r}_s(\theta) + \widetilde{h}_s(\theta)\theta(s))ds\right\} d\tau.\end{aligned}$$

Applying the inverse S-transform, we obtain (7). The correctness of such a procedure is based on the following Lemmas. □

Lemma 1 *For any* $\Phi \in (\mathcal{H}_{-p})'_{-q,\chi,\gamma,\alpha,\mu} \otimes L^2$ *the function* $\mathbf{R}_+ \ni t \mapsto \mathrm{J}^{\chi,\alpha,\mu}_{[0,t)}(\Phi) \in (\mathcal{H}_{-p})'_{-q,\chi,\gamma,\alpha,\mu}$ *is continuous.*

Proof: Let $\Phi = \Psi \otimes \Theta$, $\Psi(\cdot) = \sum_{m=0}^{\infty} Q^{\chi,\gamma,\alpha}_{\mu,m}(\Psi^{(m)};\cdot) \in (\mathcal{H}_{-p})'_{-q,\chi,\gamma,\alpha,\mu}$, $\Theta \in L^2$. Then

$$\begin{aligned}\|J^{\chi,\alpha,\mu}_{[0,t)}(\Phi)\|^2_{-p,-q} &= \sum_{m=1}^{\infty} 2^{-qm}|\widehat{\Psi}^{(m-1)}\widehat{\otimes}(\Theta 1_{[0,t)})|^2_{-p} \\ &\leq 2^{-q}\sum_{m=1}^{\infty} 2^{-q(m-1)}|\widehat{\Psi}^{(m-1)}|^2_{-p}\int_0^t \Theta^2(s)ds \\ &= 2^{-q}\|\Psi\|^2_{-p,-q}\int_0^t \Theta^2(s)ds\end{aligned}$$

and the continuity is obvious. Proof is completed by passing to limit. □

Lemma 2 *If the function* $\mathbf{R}_+ \ni t \mapsto \Phi(t) \in (\mathcal{H}_{-p})'_{\chi,\mu}$ *is continuous, then the function* $\mathbf{R}_+ \ni t \mapsto \exp^\diamond\{\Phi(t)\} \in (\mathcal{H}_{-p})'_{\chi,\mu}$ *is continuous.*

Proof: Let $\Phi = \Psi \otimes \Theta$, $\Psi \in (\mathcal{H}_{-p})'_{\chi,\mu}$, $\Theta : \mathbf{R}_+ \to \mathbf{C}$ be a continuous function. Then

$$\exp^\diamond\{\Phi(t)\} = \exp^\diamond\{\Psi\}\exp\{\Theta(t)\}$$

and the continuity is obvious. Proof is completed by passing to limit. □

Remark 3 The independence of result of Lemma 1 of χ, γ, α and μ is completely natural 'from Fock space point of view', see Remark 2. Moreover, it is not difficult to prove on this way that for $\Phi \in L^2(S', \mu) \otimes L^2$, the result of Lemma 1 follows from the corresponding result of [11].

Remark 4 For Wick version of arbitrary holomorphic function, the analog of Lemma 2 does not hold without additional conditions. We will study such a situation in detail in forthcoming papers.

Let us consider now the linear generalized stochastic equation

$$X_t = X_0 + \int_0^t g(\tau)d\tau + \int_0^t R(\tau) \diamond X_\tau d\tau, \tag{8}$$

where g is a continuous function with values in $(\mathcal{H}_{-p})'_{\chi,\mu}$, $X_0 \in (\mathcal{H}_{-p})'_{\chi,\mu}$, $R(t) \in (\mathcal{H}_{-p})'_{\chi,\mu}$ a.e. and $\langle\langle R(\cdot), \varphi\rangle\rangle \in L^{1,loc}(\mathbf{R}_+, dt)$ for all $\varphi \in (\mathcal{H}_p)_\chi$. Actually, we can understand this equation as a generalization of (6): if $R(\tau) = r(\tau) + h(\tau) \diamond Q_{\mu,1}^{\chi,\gamma,\alpha}(\delta_\tau; \cdot)$ then we have equation (6), see (5) and Lemma 1. The following statement is a simple corollary of Theorem 1 and Lemma 1.

Theorem 2 *Equation* (8) *has a unique solution*

$$\begin{aligned} X_t &= X_0 \diamond \exp^\diamond \left\{ \int_0^t R(\tau)d\tau \right\} \\ &+ \int_0^t g(\tau) \diamond \exp^\diamond \left\{ \int_\tau^t R(s)ds \right\} d\tau \in (\mathcal{H}_{-p})'_{\chi,\mu}. \end{aligned}$$

References

[1] Yu.L. Daletsky, *A biorthogonal analog of the Hermite polynomials and the inversion of Fourier transform with respect to a non-Gaussian measure*, Funct. Anal. Appl., v. **25**, no. 2 (1991), pp. 68–70 (Russian).

[2] S. Albeverio, Yu.G. Kondratiev and L. Streit, *How to generalize White Noise Analysis to non-Gaussian spaces*, in: 'Dynamics of Complex and Irregular Systems'. Eds. Ph. Blanchard et al., World Scientific, 1993.

[3] S. Albeverio, Yu.L. Daletsky, Yu.G. Kondratiev and L. Streit, *Non-Gaussian infinite dimensional analysis*, J. Funct. Anal., v. **138**, no. 2 (1996), pp. 311–350.

[4] G.F. Us, *Dual Appell systems in Poissonian analysis*, Methods Funct. Anal. Topol., v. **1**, no. 1 (1995), pp. 93–108.

[5] Yu.G. Kondratiev, L. Streit, W. Westerkamp and J. Yan, *Generalized functions in infinite dimensional analysis*, IIAS Reports no. 1995–002, Kyoto, 1995.

[6] N.A. Kachanovsky, *Biorthogonal Appell-like systems in a Hilbert space*, Methods Funct. Anal. Topol., v. **2**, no. 3–4 (1996), pp. 36–52.

[7] Yu.G. Kondratiev, J. Luis da Silva and L. Streit, *Generalized Appell systems*, Methods Funct. Anal. Topol., vol. **3**, no. 3 (1997), pp. 28–61.

[8] N.A. Kachanovsky, *Dual Appell-like systems and finite order spaces in non-Gaussian infinite-dimensional analysis*, Methods Funct. Anal. Topol., vol. **4**, no. 2 (1998), pp. 41–52.

[9] Yu.M. Berezansky and Yu.G. Kondratiev, *Biorthogonal systems in hypergroups: an extension of non-Gaussian analysis*, Methods Funct. Anal. Topol., v. **2**, no. 2 (1996), pp. 1–50.

[10] Yu.M. Berezansky, *Infinite dimensional analysis, connected with generalized translation operator*, Ukr. Math. J., v. **49**, no. 3 (1997), pp. 364–409 (Russian).

[11] H. Holden, B. Oksendal, J. Uboe and T.S. Zhang, *Stochastic Partial Differential Equations*, Oslo, 1996.

[12] G.F. Us, *One application of Poisson infinite-dimensional analysis*, Proc. of CROMSH–6 (1996), pp. 320–325.

[13] T. Hida, H.H. Kuo, J. Potthoff and L. Streit, *White Noise. An Infinite Dimensional Calculus*, Kluwer, Dordrecht, 1993.

[14] N.A. Kachanovsky, *On Analog of Stochastic Integral and Wick Calculus in Non-Gaussian Infinite Dimensional Analysis*, Methods Funct. Anal. Topol., v. **3**, no. 3 (1997), pp. 1–12.

[15] I.M. Gelfand and N.Ya. Vilenkin, *Generalized Function IV. Some Applications of Harmonic Analysis*, Fizmatgiz, Moscow (1961); English transl. AP, New York, 1964.

[16] A.V. Skorokhod, *Integration in a Hilbert Space*, Moscow, Nauka, 1975.

Institute of Mathematics
Tereschenkovskaya street 3
Kiev, 252601
Ukraine
nkachano@iname.com,
nkachano@sabbo.net

AMS Classification Numbers 46A11, 46F25, 60G20, 60H05, 60H20.

Operator Theory:
Advances and Applications, Vol. 118

Bistrict Plus-operators in Krein Spaces and Dichotomous Behavior of Irreversible Dynamical Systems

*Victor Khatskevich and Leonid Zelenko**

We consider the ball of angular operators determined in a Krein space, i.e. in a Hilbert space with indefinite metric, for which both positive and negative components can be infinite-dimensional in general. The weak compactness of the image and the preimage of this ball by a fractional-linear transformation is established. This transformation is generated by a continuous linear operator, which is not continuously invertible in general. We assume that this operator is a bistrict plus-operator acting from a Krein space to an another one. We apply the above result to the study of dichotomous behavior of solutions to a non-autonomous linear differential equation with an unbounded operator coefficient in a Hilbert space. The evolution operator of this equation is not continuously invertible and the corresponding unstable subspace is of infinite dimension in general. As an example we study a diffusion process on the real axis.

1 Introduction

We consider a fractional-linear transformation of the ball $\mathcal{K}^-$ of all angular operators corresponding to the set of all maximal non-positive subspaces of a Krein space, i.e, of a Hilbert space $\mathcal{H}$ with an indefinite metric, for which both positive and negative components can be infinite-dimensional in general ([Kh], [VKh], [Az-I], [MKr]). This transformation is generated by a continuous linear operator U in $\mathcal{H}$ (so called bistrict plus-operator: see for example [Kh], [VKh], [Az-I], [MKr], [Kha]). We do not suppose U to be continuously invertible. The weak compactness of the image and the preimage of $\mathcal{K}^-$ by the generated fractional-linear transformation F_U is established (Theorem 2.1). The above result is a development of the corresponding results of the papers [Kh], [VKh], [Kha], [Kh-Z]. In particular, in [Kh-Z] the positive component of the indefinite metric was assumed to be finite-dimensional. In the present paper we remove the last assumption.

We apply our result to the study of dichotomous behavior of solutions to a non-autonomous linear differential equation

$$\frac{dy}{dt} = A(t)y. \tag{1.1}$$

This behavior means that solutions with initial values belonging to some subspace of the phase space $\mathcal{H}$ stabilize themselves to zero at infinity, but all the other

*Supported in part by a grant from the German-Israeli Foundation (GIF) and in part by the Gilady Foundation.

solutions grow infinitely. This property was studied in the papers and books of O. Perron [Pr], of A.D. Maisel [Mz] and of J.L. Massera and J.J. Schaffer [Ms-Sh] for the case of a bounded operator coefficient A(t).

The study of the similar question for equation (1.1) with an unbounded operator coefficient $A(t)$ is of the particular interest because of applications to the qualitative theory of partial differential equations [Ag-Nr].

We formulate a condition of dichotomous behavior in terms of a family of indefinite metrics (Theorem 3.1) and apply the method of embedded bicones used in the paper [Z]. We obtain local and integral estimates for the velocity of dichotomy. In the case when the evolution operator of (1.1) is not continuously invertible, the application of the method of embedded bicones is connected with certain topological difficulties. To overcome these difficulties we use our Theorem 2.1. In the paper [Kh-Z] we studied the dichotomous behavior of linear non-autonomous irreversible dynamical system in the case of finite-dimensional unstable subspace. As an application of our results we have obtained in [Kh-Z] some constructive conditions of dichotomous behavior of solutions to the boundary value problem in the case of a non-autonomous linear diffusion equation in a bounded domain. The results of the present paper cover also the case of infinite-dimensional unstable subspace. This circumstance enables us to study the dichotomous behavior of diffusion processes in unbounded domains.

Let us notice that in the paper [Kh-Sh] the compactness of the above mentioned preimage of the ball $\mathcal{K}^-$ was established under the assumption that the operator U is continuously invertible. This result permits to investigate in [Kh-Sh] the dichotomy of evolution equations which describe reversible processes only.

Let us introduce some notations which will be used in the sequel.

$\mathbf{R} = (-\infty, +\infty)$;

$\mathbf{R}_+ = [0, +\infty)$;

cl(S)-the closure of a subset S of a Banach space $\mathcal{B}$;

$\mathcal{L}(\mathcal{B}_1, \mathcal{B}_2)$-the set of all linear bounded operators, acting from a Banach space $\mathcal{B}_1$ into a Banach space $\mathcal{B}_2$; if $\mathcal{B}_1 = \mathcal{B}_2 = \mathcal{B}$, we will write $\mathcal{L}(\mathcal{B})$;

$\mathcal{H}$-a complex or real Hilbert space with an inner product (x, y) and the norm $\|x\| = (x, x)^{\frac{1}{2}}$; we will use the same notation $\mathcal{H}$ with lower indexes, but in notation of norms and inner products we will omit indexes;

$\mathcal{SL}(\mathcal{H})$-the set of all selfadjoint operators from $\mathcal{L}(\mathcal{H})$;

$\sigma(A)$-the spectrum of a linear operator A acting in $\mathcal{H}$;

$(\mathcal{H}, V)$-a *Krein space*, i.e. a Hilbert space $\mathcal{H}$ with the indefinite metric $[x, y] = (Vx, y)$, where $V \in \mathcal{SL}(\mathcal{H})$ is continuously invertible;

$\mathcal{H}^+$, $\mathcal{H}^-$ (can be with lower indexes)-the invariant subspaces of V, corresponding to the positive and to the negative parts of its spectrum respectively; $d_V^+ = dim(\mathcal{H}^+)$, $d_V^- = dim(\mathcal{H}^-)$ (these numbers can be infinite in general);

$\mathcal{P}^+, \mathcal{P}^-$ (can be with lower indexes)-respectively non-negative and non-positive bicones connected with the indefinite metric $[x, y]$: $\mathcal{P}^+ = \{x \in \mathcal{H} : [x, x] \geq 0\}$, $\mathcal{P}^- = \{x \in \mathcal{H} : [x, x] \leq 0\}$.

2 Bistrict Plus-operators and Fractional-linear Transformations

2.1^0

Let $(\mathcal{H}, V)$ be a Krein space, $\mathcal{H} = \mathcal{H}^+ \oplus \mathcal{H}^-$. Denote by P^+, P^- the orthogonal projections on $\mathcal{H}+$, $\mathcal{H}^-$ respectively. It is well known that one can reduce the indefinite square form $[x, x] = (Vx, x)$ to the form:

$$[x, x] = \|P^+x\|^2 - \|P^-x\|^2,$$

passing to an equivalent norm in the space $\mathcal{H}$ [Az-I]. In the sequel we will suppose such passage to be realized. Recall that a subspace $L \subseteq \mathcal{H}$ is called the *non-positive* (*non-negative*), if it is contained in $\mathcal{P}^-$ (respectively in $\mathcal{P}^+$). A non-positive (non-negative) subspace $L \subseteq \mathcal{H}$ is called the *maximal non-positive* (*maximal non-negative*), if it is not a proper part of any non-positive (non-negative) subspace.

Consider two Krein spaces $(\mathcal{H}_i, V_i)$ $(i = 1, 2)$ with indefinite metrics in the form defined above, i.e. $V_i = P_i^+ - P_i^-$, where P_i^+, P_i^- are orthogonal projections in $\mathcal{H}_i$ on the subspaces $\mathcal{H}_i^+$, $\mathcal{H}_i^-$ respectively. This means that we have an orthogonal splitting of the spaces $\mathcal{H}_i$ $(i = 1, 2)$:

$$\mathcal{H}_i = \mathcal{H}_i^+ \oplus \mathcal{H}_i^-. \tag{2.1}$$

Assume that the signatures of the quadratic forms $[x, x]_i = (V_i x, x)$ are the same, i.e. $d_{V_1}^+ = d_{V_2}^+$ and $d_{V_1}^- = d_{V_2}^-$. Let $\mathcal{P}_i^+$, $\mathcal{P}_i^-$ $(i = 1, 2)$ be the non-negative and the non-positive bicones respectively connected with the above indefinite metrics.

Recall that a linear bounded operator U acting from $(\mathcal{H}_1, V_1)$ into $(\mathcal{H}_2, V_2)$, is called a *strict plus-operator*, if

$$[Ux, Ux]_2 \geq \mu(U)[x, x]_1 \ \ \forall x \in \mathcal{H}_1,$$

where $\mu(U) > 0$. If together with U the adjoint operator $U^\star$ is a strict plus-operator too, then U is called a *bistrict plus-operator*. There are the similar definitions for a *minus-operator (strict and bistrict)* (see [Az-I]).

Consider the block matrix representation of the operator U, corresponding to the orthogonal decompositions (2.1) of the spaces $\mathcal{H}_i$ $(i = 1, 2)$:

$$U = \{U_{i,j}\}_{i,j=1}^2. \tag{2.2}$$

Let $\mathcal{M}_i^+$ $(\mathcal{M}_i^-)$ be the set of all maximal non-negative (maximal non-positive) subspaces of $(\mathcal{H}_i, V_i)$, $i = 1, 2$. We denote by $\mathcal{K}_i^+$ $(\mathcal{K}_i^-)$ the set of all angular operators of subspaces $L_i^+ \in \mathcal{M}_i^+$ $(L_i^- \in \mathcal{M}_i^-)$, $i = 1, 2$. Recall that they are the linear operators $K_i^+ : \mathcal{H}_i^+ \to \mathcal{H}_i^-(K_i^- : \mathcal{H}_i^- \to \mathcal{H}_i^+)$, for which $L_i^+ = \text{graph}(K_i^+)$ $(L_i^- = \text{graph}(K_i^-))$ (see details in [Kha]). Define two sets:

$$\begin{aligned} \mathcal{G}_U^- &= \{K_1^- \in \mathcal{K}_1^- : U(P_1^- + K_1^-)\mathcal{H}_1^- \subset \mathcal{P}_2^-\}, \\ \mathcal{G}_U^+ &= \{K_2^+ \in \mathcal{K}_2^+ : (P_2^+ + K_2^+)\mathcal{H}_2^+ \supseteq U(P_1^+ + K_1^+)\mathcal{H}_1^+ \end{aligned}$$

for some $K_1^+ \in \mathcal{K}_1^+\}$. It is perhaps well to recall that in the case of a bistrict plus-operator U we have

$$\mathcal{G}_U^+ = F_U(\mathcal{K}_1^+), \tag{2.3}$$

and if in addition U is invertible, then $\mathcal{G}_U^- = F_{U^{-1}}(\mathcal{K}_2^-)$. Here F_U and $F_{U^{-1}}$ are the fractional-linear transformations generated by the block-matrix representations of the form (2.2) for U and U^{-1} respectively (see [Kha]).

2.2^0

We turn now to the main assertion of this section.

Theorem 2.1 *Let U be a bounded bistrict plus-operator acting from $(\mathcal{H}_1, V_1)$ into $(\mathcal{H}_2, V_2)$. Then the sets $\mathcal{G}_U^-$, $\mathcal{G}_U^+$ are compact with respect to the weak operator topology.*

Proof: In view of (2.3) the claim about $\mathcal{G}_U^+$ follows from Theorem 1.2 [Kh-Sh]. Now consider $\mathcal{G}_U^-$. First of all let us prove that it is not empty. Taking $L^+ = U^c\mathcal{H}^+$, we have for all $x_+ \in \mathcal{H}^+$, $z \in (L^+)^{[\perp]}$: $[Uz, x_+] = [z, U^c x_+] = 0$, where U^c is the conjugate operator for U with respect to the indefinite inner product $[\cdot,\cdot]$ and

$$(L^+)^{[\perp]} = \{y \in \mathcal{H} : \ [y, x] = 0 \ \forall x \in L^+\}.$$

Since (see [Kht]) $(L^+)^{[\perp]} = (P_1^- + K^-)\mathcal{H}_1^-$ for some $K^- \in \mathcal{K}_1^-$ and $Uz \in \mathcal{P}_2^-$, then $\mathcal{G}_U^- \neq \emptyset$.

Now let us prove that $\mathcal{G}_U^-$ is a closed set with respect to the weak operator topology. If $K^- \in \mathcal{G}_U^-$, then there exists $Q^- \in \mathcal{K}_2^-$ such that $U_{1,1}K^- + U_{1,2} = Q^-(U_{2,1}K^- + U_{2,2})$. Hence (see for example [Shm]):

$$(U_{2,1}K^- + U_{2,2})^*(U_{2,1}K^- + U_{2,2}) \geq (U_{1,1}K^- + U_{1,2})^*(U_{1,1}K^- + U_{1,2}).$$

It is equivalent to $(K^-)^*AK^- + (K^-)^*B + B^*K^- + C \leq 0$, where $A = U_{1,1}^*U_{1,1} - U_{2,1}^*U_{2,1} \geq 0$, $B = U_{1,1}^*U_{1,2} - U_{2,1}^*U_{2,2}$, $C = U_{1,2}^*U_{1,2} - U_{2,2}^*U_{2,2}$. Now if $K_n^- \to K^-$ as $n \to \infty$ with respect to the weak operator topology, then

$$((K^-)^*AK^-x_-, x_-) = \|A^{\frac{1}{2}}K^-x_-\|^2 \leq \liminf_{n\to\infty}((K_n^-)^*AK_n^-x_-, x_-)$$

for all $x_- \in \mathcal{H}_1^-$, and we obtain immediately the closedness of $\mathcal{G}_U^-$. Consequently, being a non-empty closed subset of $\mathcal{K}^-$ in the weak operator topology, the set $\mathcal{G}_U^-$ is compact in this topology. □

Remark 2.1 The proof of the closedness of $\mathcal{G}_U^-$ is valid also in the general case of an arbitrary bounded plus-operator U (not bistrict and even not strict). But even in the case of a strict (but not bistrict) plus-operator U the set $\mathcal{G}_U^-$ can be empty.

Example 2.1 Let $\mathcal{H} = \mathcal{H}^+ \oplus \mathcal{H}^-$,

$$\mathcal{H}^+ = \text{cl}\,(\text{span}(\{e_i\}_{i=1}^{\infty})),$$
$$(e_i, e_j) = \delta_{i,j}, \quad i, j = 1, 2, \ldots, \quad \mathcal{H}^- = \text{span}(\{e_-\}), \quad (e_-, e_-) = 1.$$

Let $U = (U_{i,j})_{i,j=1}^{2}$, where $U_{1,1}e_k = e_{k+1}$, $k = 1, 2, \ldots$, $U_{1,2}e_- = e_1$, $U_{2,1} = U_{2,2} = 0$. Then $[Ux, Ux] = |\beta|^2 + \sum_{k=1}^{\infty} |\alpha_k|^2$ for any $x = \sum_{k=1}^{\infty} \alpha_k e_k + \beta e_- \in \mathcal{H}$. Hence U is bounded and $\ker(U) = \{0\}$, $[Ux, Ux] \geq [x, x]$ for all $x \in \mathcal{H}$ and $\mathcal{G}_U^- = \emptyset$.

3 Dichotomy of Solutions to Evolution Equations

3.1[0]

Consider a linear differential equation of the form (1.1) in a Hilbert space $\mathcal{H}$, where $t \in \mathbf{R}_+$ and $A(t)$ are in general linear closed unbounded operators with a common domain $\text{Dom}(A(t)) \equiv D_0$, which is dense in $\mathcal{H}$. Further we will assume that the Cauchy problem for (1.1) is uniformly correct on the half-line $\mathbf{R}_+$ ([SKr], Chapt. II, Sect. 3, Definition 3.1). This means that there exists the continuous on D_0 evolution operator $U(t, s)$ $(0 \leq s \leq t)$ associating with each $y_0 \in D_0$ the value $y(t, s)$ of the solution to (1.1), which satisfies the initial condition: $y(s, s) = y_0$. Then for each fixed t, s the operator $U(t, s)$ may be continuously extended to the whole of $\mathcal{H}$; for this extension we conserve the same notation $U(t, s)$. For any $y_0 \in \mathcal{H}$ the function $y(t, s) = U(t, s)y_0$ is said to be *the generalized solution* to the equation (1.1), satisfying the initial condition $y(s, s) = y_0$. The following two conditions ensure the uniform correctness of the above Cauchy problem ([SKr], Chapt. II, Sect. 3, Theorem 3.11):

(A) The family $\{A(t)\}_{t\in\mathbf{R}_+}$ is strongly continuously differentiable on D_0;

(B) For each $T > 0$ the intersection of resolvent sets of the operator family $\{A(t)\}_{t\in[0,T]}$ contains a half-line (ω_T, ∞) for some $\omega_T \in \mathbf{R}$. Moreover, for their resolvents $R_{A(t)}(\lambda)$ the following estimate holds:

$$\|R_{A(t)}(\lambda)\| \leq (\lambda - \omega_T)^{-1} \quad \forall \lambda \in (\omega_T, \infty); \tag{3.1}$$

Furthermore, we shall assume that:

(C) The operators of the family $\{A^\star(t)\}_{t\in\mathbf{R}_+}$ have a common domain $D_0^\star$, which is dense in $\mathcal{H}$ and this family is strongly continuously differentiable on $D_0^\star$.

Remark 3.1 It is easy to check using the condition (B) that the family of operators $\{A^\star(t)\}_{t\in\mathbf{R}_+}$ satisfies the same condition (B). Then in view of the condition (C) for any $T > 0$ the Cauchy problem for the equation

$$\frac{dy}{dt} = A^\star(T - t)y \tag{3.2}$$

is uniformly correct on the segment [0,T].

Remark 3.2 Assume that the family $\{A(t)\}_{t\in\mathbf{R}_+}$ is uniformly bounded from the right, i.e. there exists such a number $\omega \in \mathbf{R}$, that

$$\forall t \in \mathbf{R}_+,\ \forall x \in D_0 :\ \operatorname{Re}((A(t)x, x)) \le \omega(x, x).$$

As it is well known ([SKr]), this restriction implies the condition (B) with $\omega_T = \omega$.

3.2^0

Consider an operator function $V(\cdot) : \mathbf{R}_+ \to \mathcal{SL}(\mathcal{H})$, which defines in $\mathcal{H}$ a family of indefinite metrics as follows:

$$[x, y]_t = (V(t)x, y) \quad (t \in \mathbf{R}_+). \tag{3.3}$$

We will need the following

Definition 3.1 Assume that the family $\{V(t)\}_{t\in\mathbf{R}_+}$ is strongly continuously differentiable on $\mathcal{H}$. Consider the following differential equation on an interval I

$$\frac{dy}{dt} = C(t)y \quad (t \in I), \tag{3.4}$$

where $C(t)$ are closed linear operators with a common domain D_C which is dense in $\mathcal{H}$, and the Cauchy problem for this equation is uniformly correct on I. We call a family of quadratic forms $\{Q_\tau[x]\}_{\tau\in I}$ defined on D_C *the derivative of the quadratic form* $[x, x]_t$ *with respect to the equation* (3.4), *evaluated at the point* $x \in D_c$ *and at the moment* $\tau \in I$, if the following equality holds:

$$Q_\tau[x] = \frac{d}{dt}[U_C(t, \tau)x, U_C(t, \tau)x]_t|_{t=\tau},$$

where $U_C(t, \tau)$ is the evolution operator of the equation (3.4). We denote this derivative by $\frac{d}{dt}_{(3.4)}[x, x]_t(\tau)$.

Remark 3.3 The existence and the continuous dependence of the derivative $\frac{d}{dt}_{(3.4)}$ $[x, x]_t(\tau)$ on τ are consequences of the following fact, which can be easily checked: if $\mathcal{B}_1, \mathcal{B}_2$ are Banach spaces, I is an interval, a vector function $y(\cdot) : I \to \mathcal{B}_1$ is continuously differentiable and an operator function $W(\cdot) : I \to \mathcal{L}(\mathcal{B}_1)$ is strongly continuously differentiable, then the vector function $W(\cdot)y(\cdot) : I \to \mathcal{B}_2$ is continuously differentiable.

Further we will impose the following conditions on the indefinite metrics $[x, y]_t$:

(D) The operator function $V(t)$ is strongly continuously differentiable on $\mathcal{H}$ and for any $\tau \in \mathbf{R}_+$ the derivative $\frac{d}{dt}_{(1.1)}[x, x]_t(\tau)$ is a positive definite quadratic form on D_0 with a lower bound $k(\tau)$, which is a positive locally integrable function on $\mathbf{R}_+$;

(E) For any $T > 0$ and for any $\tau \in [0, T]$ the derivative $\frac{d}{dt}_{(3.2)}[x, x]_t(\tau)$ is a nonnegative square form on $D_0^\star$;

(F) For each fixed $t \in \mathbf{R}_+$ the operator $V(t)$ is continuously invertible;

(G) The numbers $d^+_{V(t)}, d^-_{V(t)}$ do not depend on t (see subsection 2.1^0);

(H) The operator function $V(t)$ is bounded on $\mathbf{R}_+$ with respect to the operator norm and moreover:

$$\sup_{t \in \mathbf{R}_+} \|V(t)\| = 1.$$

The following statement is true:

Lemma 3.1 *Assume that the Cauchy problem for the equations* (1.1), (3.2) *is uniformly correct on* $\mathbf{R}$ *and on* $[0, T]$ *for any* $T > 0$ *and the operator function* $V(\cdot) : \mathbf{R}_+ \to \mathcal{SL}(\mathcal{H})$ *satisfies the conditions (D)–(H). Then the following relations hold for all* $0 \le \tau \le t$ *and for any* $y_0 \in \mathcal{H}$*:*

$$[U(t,\tau)y_0, U(t,\tau)y_0]_t - [y_0, y_0]_\tau \ge \int_\tau^t k(s)\|U(t,\tau)y_0\|^2 ds, \tag{3.5}$$

$$[U^\star(t,\tau)y_0, U^\star(t,\tau)y_0]_t \ge [y_0, y_0]_\tau. \tag{3.6}$$

Proof: The inequality (3.5) follows from the condition (D). It can be proved for $y_0 \in D_0$ and afterwards can be extended on the whole of $\mathcal{H}$ by using the density of D_0 in $\mathcal{H}$ (see [Kh-Z]), the proof of Lemma 2.1).

We turn to the proof of the relation (3.6). Denote by $U_T(t, \tau)$ the evolution operator of the equation (3.2) on $[0, T]$. If we will prove the equality

$$U^\star(t,\tau) = U_{(t+\tau)}(t,\tau), \tag{3.7}$$

then applying the condition (E) to the equation (3.2), we will be able to prove the relation (3.6) in a similar manner as (3.5). Now let us prove the equality (3.7). To this end consider the representation of the evolution operator $U(t, \tau)(0 \le \tau < t)$ by means of the multiplicative integral ([SKr], Chapter II, Section 3):

$$U(t,\tau) = \int_\tau^{\to t} \exp(A(s)ds),$$

where the arrow indicates the order of factors in the multiplicative integral. This integral exists in the strong sense, i.e. the integral products converge in the strong operator topology. It is clear that the integral products of the following multiplicative integral

$$\int_\tau^{\leftarrow t} \exp(A^\star(s)ds) = \int_\tau^{\to t} \exp(A^\star(t+\tau-s)ds)$$

converge to $U^\star(t, \tau)$ in the weak operator topology. On the other hand these integral products converge to $U_{(t+\tau)}(t, \tau)$ in the strong operator topology. This implies the equality (3.7). □

Corollary 3.1 *If the conditions of Lemma* 3.1 *are fulfilled and* $0 \leq \tau < t$, *then the operator* $U(t, \tau)$ *is a bistrict plus-operator from* $(\mathcal{H}, V_\tau)$ *into* $(\mathcal{H}, V_t)$.

Consider the following bicones, connected both with the metrics (3.1) and the evolution operator of (1.1):

$$C_t^- = \{y_0 \in \mathcal{H} : [U(t)y_0, U(t)y_0]_t \leq 0\} \quad (t \in \mathbf{R}_+). \tag{3.8}$$

It is easy to show using (3.5), (3.8), that for the family of these bicones the property of monotonicity holds:

$$C_t^- \subseteq C_\tau^-, \quad \text{if} \quad t > \tau. \tag{3.9}$$

We set

$$C_\infty^- = \bigcap_{t \in \mathbf{R}_+} C_t^-. \tag{3.10}$$

The following statement can be proved in the same way as Lemma 2.2 of [Kh-Z] applying Theorem 2.1 to the Krein spaces $(\mathcal{H}, V(0))$, $(\mathcal{H}, V(t))$ and to the operator $U = U(t, 0)$ and taking into account Corollary 3.1 and the property (3.9).

Lemma 3.2 *Assume that the conditions of Lemma* 3.1 *are fulfilled. Then the set* C_∞^- *defined in* (3.10) *contains a maximal non-positive subspace* L_∞^- *of the Krein space* $(\mathcal{H}, V(0))$.

The following theorem can be proved in the same manner as Theorem 2.1 of [Kh-Z], using Lemmas 3.1, 3.2 and the Bellman-Gronwall inequality [Cs].

Theorem 3.1 *Assume that the Cauchy problem for the equations* (1.1), (3.2) *is uniformly correct respectively on* $\mathbf{R}$ *and on* $[0, T]$ *for any* $T > 0$, *the operator function* $V(\cdot) : \mathbf{R}_+ \to \mathcal{SL}(\mathcal{H})$ *satisfies the conditions (D)–(H) and the function* $k(t)$ *(see (D)) satisfies the following equality:*

$$\int_0^\infty k(t)dt = +\infty. \tag{3.11}$$

Then the following statements are true:

(1) *The set* C_∞^- *coincides with a maximal non-positive subspace* L_∞^- *of the Krein space* $(\mathcal{H}, V_0)$, *where* $V_0 = V(0)$;

(2) *For any* $y_0 \in L_\infty^-$ *the generalized solution* $y(t) = U(t, 0)y_0$ *to* (1.1) *satisfies the condition:*

$$I(y) = \int_0^\infty k(s)\|y(s)\|^2 ds < \infty$$

and the following estimate holds:

$$\int_t^\infty k(s)\|y(s)\|^2 ds \leq I(y) \exp\left(-2\int_0^t k(s)ds\right);$$

(3) *For any* $y_0 \notin \mathcal{P}_0^-$ *the generalized solution* $y(t) = U(t,0)y_0$ *to* (1.1) *satisfies the inequality:*

$$\|y(t)\|^2 \geq [y_0, y_0]_0 \, exp\left(2\int_0^t k(s)ds\right),$$

where $\mathcal{P}_0^-$ *is the non-positive bicone of the Krein space* $(\mathcal{H}, V_0)$.

3.3^0

Now we intend to prove the following theorem on a non-autonomous perturbation of autonomous dynamical system.

Theorem 3.2 *Assume that the following conditions hold: in the equation* (1.1) *the operator* $A(t)$ *has the form* $A_0 + B(t)$, *where* A_0 *is a self-adjoint operator defined on* D_0 *and bounded from the right; for any* $t \in \mathbf{R}_+$ $B(t) \in \mathcal{L}(\mathcal{H})$ *and the operator functions* $B(t)$, $B^\star(t)$ *are strongly continuously differentiable on* D_0. *Furthermore, assume that the point* 0 *belongs to a gap of the spectrum* $\sigma(A_0)$ *and the following estimate holds*

$$\forall t \in \mathbf{R}_+ : \quad \|B(t)\| \leq d - k(t), \tag{3.12}$$

where $d = \operatorname{dist}(0, \sigma(A_0))$ *and* $k(t)$ *is a positive locally integrable function satisfying the condition* (3.11). *Then the Cauchy problem for equations* (1.1), (3.2) *is uniformly correct on* $\mathbf{R}$ *and on* $[0, T]$ *for any* $T > 0$ *and the conclusions of Theorem* 3.1 *are true with*

$$V_0 = P^+ - P^-, \tag{3.13}$$

where P^+, P^- *are the orthogonal projections onto the invariant subspaces of the operator* A_0, *corresponding to the positive and to the negative parts of* $\sigma(A_0)$.

Proof: As it is easy to see, the assumptions of our theorem imply conditions (A), (C) and the conditions of Remark 3.2 for the family $\{A(t)\}_{t\in\mathbf{R}}$. Then in view of Remark 3.1 the first assertion of the theorem is true. We shall show that conditions (D)–(H) are satisfied with $V(t) \equiv V_0$ (see (3.13)). Define in $\mathcal{H}$ the indefinite metric $[x, y] = (V_0 x, y)$ and calculate its derivative with respect to equation (1.1) for $x_0 \in D_0$:

$$\begin{aligned} \frac{d}{dt}_{(1.1)}[x_0, x_0](\tau) &= 2\mathrm{Re}\left(\left[\frac{d}{dt}U(t,\tau)x_0, U(t,\tau)x_0\right]\Big|_{t=\tau}\right) \\ &= 2\mathrm{Re}((V_0 A_0 x_0, x_0)) + 2\mathrm{Re}((V_0 B(\tau)x_0, x_0)). \end{aligned} \tag{3.14}$$

Since the operator V_0 is unitary, then

$$\|V_0\| = 1. \tag{3.15}$$

It is easy to see that

$$V_0 A_0 = P^+ A_0 - P^- A_0 = |A_0| = \int_{\mathbf{R}} |\lambda| E(d\lambda),$$

where $E(\Delta)$ is the spectral projection of the operator A_0, corresponding to an interval $\Delta \in \mathbf{R}$. Using the last equality, the equality (3.15) and the condition (3.12), we obtain from (3.14):

$$\begin{aligned}\frac{d}{dt}_{(1.1)}[x_0, x_0](\tau) &\geq 2((|A_0|x_0, x_0 - \|B(\tau)\|\|x_0\|^2) \\ &\geq 2(d - \|B(\tau)\|)\|x_0\|^2 \geq k(\tau)\|x_0\|^2\end{aligned}$$

for any $x_0 \in D_0$. So condition (D) is fulfilled.

In the same manner as in the case of the equality (3.14) we obtain the following equality for the derivative of the square form $[\cdot, \cdot]$ with respect to the equation (3.2):

$$\frac{d}{dt}_{(3.2)}[x_0, x_0](\tau) = 2\mathrm{Re}((V_0 A_0 x_0, x_0)) + 2\mathrm{Re}((V_0 B^\star(T - \tau)x_0, x_0))$$

for any $x_0 \in D_0$ and for any $\tau \in [0, T)$. The last equality, equality (3.15) and condition (3.12) imply condition (E). Evidently conditions (G), (H) are fulfilled for $V(t) \equiv V_0$. So all the conditions of Theorem 3.1 are fulfilled, therefore all the conclusions of this theorem hold. □

3.4^0

We will apply the last theorem to the following diffusion equation:

$$\frac{\partial u}{\partial t} = \frac{\partial^2 u}{\partial x^2} - Q(x, t)u \tag{3.16}$$

on the axis $-\infty < x < +\infty$ and $t \geq 0$. Denote $\mathcal{H} = L_2(\mathbf{R})$. We will find a solution to (3.16) satisfying the initial condition:

$$u(x, t)|_{t=0} = \phi(x), \tag{3.17}$$

where $\phi \in \mathcal{H}$, and the additional condition:

$$u(\cdot, t) \in \mathcal{H} \;\; \forall t > 0, \tag{3.18}$$

which plays the role of boundary conditions at $-\infty$ and $+\infty$. We can rewrite the boundary value problem (3.16), (3.17), (3.18) in the form of an abstract Cauchy problem:

$$\frac{dy}{dt} = A(t)y, \tag{3.19}$$

$$y(0) = \phi, \tag{3.20}$$

where $y(t)$ is the function $u(x, t)$, which is considered as a vector function with the range in the space $\mathcal{H}$: $y(t) = u(\cdot, t)$. For each $t \geq 0$ $A(t)$ is an operator defined in $\mathcal{H}$ by the differential operation $a(t) = \frac{d^2}{dx^2} - Q(x, t) \cdot (t \geq 0)$. This means that $A(t)$ is the closure of the operator, defined by $a(t)$ on the linear set $C_0^\infty(\mathbf{R})$ of smooth functions with compact supports.

Assume that the function $Q(x,t)$ satisfies the following condition:

(I) $Q(x,t) = p(x) + q(x,t)$, where $p(x)$ is a continuous T-periodic function with $T > 0$, $q(x,t)$ is continuous and bounded on the half-plane $\Pi_+ = \mathbf{R} \times \mathbf{R}_+$ and there exists the partial derivative $\frac{\partial}{\partial t} q(x,t)$ on Π_+, continuously depending on t uniformly with respect to $x \in \mathbf{R}$.

Represent $A(t)$ in the form $A(t) = A_0 + B(t)$, where A_0 is the operator, defined in $\mathcal{H}$ by the operation $a_0 = \frac{d^2}{dx^2} - p(x)\cdot$, and for each $t \geq 0$ $B(t)$ is the operator of multiplication by the function $q(x,t)$ in the space $\mathcal{H} : B(t)v(x) = q(x,t)v(x)$. It is easy to check that the operator A_0 is self-adjoint and bounded from the right, for any $t \in \mathbf{R}_+$ $B(t) \in \mathcal{SL}(\mathcal{H})$ and the operator function $B(t)$ is strongly continuously differentiable on D_0. Therefore by Theorem 3.2 the Cauchy problem (3.19), (3.20) is uniformly correct on $\mathbf{R}_+$. We will call any generalized solution to this problem *a generalized solution to the boundary value problem* (3.16), (3.17), (3.18).

Observe that since $p(x)$ is a periodic function, the spectrum $\sigma(A_0)$ consists of the closures of the zones of stability to the equation $\frac{d^2v}{dx^2} - p(x)v = \lambda v$.

Assume that the following condition is fulfilled:

(J) The point 0 belongs to a gap of the spectrum $\sigma(A_0)$.

Denote by P^+, P^- the orthogonal projections on the invariant subspaces of A_0, corresponding to the positive and to the negative parts of $\sigma(A_0)$ respectively. Set $V_0 = P^+ - P^-$ and let $\mathcal{P}_0^-$ be the non-positive bicone of the Krein space $(\mathcal{H}, V_0)$.

The following theorem is an immediate consequence of Theorem 3.2.

Theorem 3.3 *Assume that conditions (I), (J) are fulfilled and the function $q(x,t)$ satisfies the following inequality:*

$$|q(x,t)| \leq d - k(t) \quad \forall t \geq 0,$$

where $d = dist(0, \sigma(A_0))$, $k(t)$ is a positive locally integrable function satisfying the condition (3.11). *Then there exists a maximal non-negative subspace L_∞^- of the Krein space $(\mathcal{H}, V_0)$ such that the following statements are true:*

(1) *For any initial state $\phi \in L_\infty^-$ the corresponding generalized solution $u(x,t)$ to the problem* (3.16), (3.17), (3.18) *satisfies the following condition:*

$$I(u) = \int_0^\infty k(\tau)d\tau \int_{-\infty}^\infty |u(x,\tau)|^2 dx < \infty,$$

and the following estimate holds:

$$\int_t^\infty k(\tau)d\tau \int_{-\infty}^\infty |u(x,\tau)|^2 dx \leq I(u) \exp\left(-2\int_0^t k(s)ds\right);$$

(2) *For any initial state* $\phi \notin \mathcal{P}_0^-$ *the corresponding generalized solution* $u(x,t)$ *to the problem* (3.16), (3.17), (3.18) *satisfies the inequality:*

$$\int_{-\infty}^{\infty} |u(x,t)|^2 dx \geq [\phi, \phi]_0 \, exp \left(2 \int_0^t k(s) ds \right).$$

References

[Kh] V. Khatskevich, *On the symmetry of properties of plus-operator and its conjugate operator* (in Russian), Functional analysis (Ulyanovsk, Russia) **14** (1980), 177–186.

[VKh] V. Khatskevich, *On characteristic spectral properties of focused operators* (in Russian), Proceedings of Academy of Science of Armenian Republic (Doklady Acad. Nauk Arm. S.S.R.) **79** (1984), 102–105.

[Az-I] T.Ya. Azizov and I.S. Iohvidov, *Foundations of the linear operator theory in spaces with indefinite metric* (in Russian), Nauka, Moscow, 1986.

[MKr] M.G. Krein, *On one new application of fixed point principle to the operator theory in spaces with indefinite metric* (in Russian), Dokl. Acad. Nauk S.S.S.R. **154** (1964), 1023–1026.

[Kha] V. Khatskevich, *Some global properties of fractional-linear transformations*, Operator Theory: Advances and Applications **73** (1994), 355–361.

[Kh-Z] V. Khatskevich and L. Zelenko, *Indefinite metrics and dichotomy of solutions of linear differential equations in Hilbert spaces*, Chinese J. of Math. **24**, no. 2 (1996), 99–112.

[Pr] O. Perron, *Die Stabilitatsfrage bei Differentialgleichungen*, Mathematische Zeitschriften **32** (1930), 703–728.

[Mz] A.D. Maizel, *On stability of solutions to systems of differential equations* (in Russian), Proceedings of Ural Politechnic Institute (Ural. Politechn. Instit. Trudy, Russia) **51** (1954), 20–54.

[Ms-Sh] J.L. Massera and J.J. Shaffer, *Linear differential equations and function spaces*, Acad. Press, New York, London, 1966.

[Ag-Nr] S. Agmon and L. Nirenberg, *Properties of solutions of ordinary differential equations in Banach space*, Comm. Pure and Appl. Math. **16**, no. 2 (1963), 121–239.

[Z] L. Zelenko, *On manifold of square integrable solutions of nonlinear systems of differential equations* (in Russian), Differential equations and theory of functions, Saratov University, Russia **6** (1984), 19–26.

[Kh-Sh] V. Khatskevich and V. Shulman, *Operator fractional-linear transformations: convexity and compactness of image*; applications, Studia Math. **116** (2) (1995), 189–195.

[Kht] V. Khatskevich, *On some geometrical and topological properties of normed spaces with indefinite metrics* (in Russian), Izvestiya Vuzov, Matematica **6** (133) (1973), 97–101.

[Shm] Yu.L. Shmuliyan, *On divisibility in the class of plus-operators* (in Russian), Mat. Zametki **74** (1967), 516–525.

[SKr] S.G. Krein, *Linear differential equations in Banach space*, Transl. of Math. Monogr., A.M.S, vol. **29**, Providence, R.I., 1971.

[Cs] L. Cesary, *Asymptotic behavior and stability problems in ordinary differential equations*, Springer Verlag, Berlin, Gottingen, Heidelberg, 1959.

Department of Applied Mathematics
International College of Technology
Karmiel, P.O.B. 78
Karmiel 20101
Israel

Department of Mathematics
University of Haifa
Haifa 31905
Israel
zelenko math.haifa.ac.il

AMS Classification Numbers: Primary 34G10, 35B40; Secondary 47N20

Operator Theory:
Advances and Applications, Vol. 118

Singular Operator as a Parameter of Self-adjoint Extensions

Volodymyr Koshmanenko

Let $\dot{A}$ be a symmetric restriction of a self-adjoint bounded from below operator A in a Hilbert space $\mathcal{H}$ and let A_∞ denote the Friedrichs extension of $\dot{A}$. We prove that in the case, where $A_\infty \neq A$, under natural conditions, each self-adjoint extension $\tilde{A}$ of $\dot{A}$ has a unique representation in the form of a generalized sum, $\tilde{A} = A \tilde{+} V$, where V is a singular operator acting in the A-scale of Hilbert spaces, from $\mathcal{H}_1(A)$ to $\mathcal{H}_{-1}(A)$. In the particular case, where $\dot{A}$ has deficiency indices $(1, 1)$, this result has been proven by Krein and Yavrian.

1 Introduction and Preliminaries

Let $A = A^*$ be a self-adjoint bounded from below operator in a complex separable Hilbert space $\mathcal{H}$ with a norm $\|\cdot\|$ and an inner product $(\cdot, \cdot)$. We assume that A is unbounded but A^{-1} exists and is bounded. Thus $ran A \equiv \mathcal{R}(A)$ coincides with $\mathcal{H}$. Without loss of generality we can assume that $A \geq 1$.

Introduce the scale of Hilbert spaces associated with A, which usually called the A-scale:

$$\mathcal{H}_{-k} \equiv \mathcal{H}_{-k}(A) \supset \mathcal{H}_0 \equiv \mathcal{H} \supset \mathcal{H}_k \equiv \mathcal{H}_k(A), \quad k \geq 0, \tag{1}$$

where the Hilbert space $\mathcal{H}_k$ coincides with the domain of $A^{k/2}$, i.e. $dom(A^{k/2}) \equiv \mathcal{D}(A^{k/2}) = \mathcal{H}_k$ in the norm $\|\varphi\|_k := \|A^{k/2}\varphi\|$ and $\mathcal{H}_{-k}$ is the dual conjugate space to $\mathcal{H}_k$, i.e. $\mathcal{H}_{-k}$ is the completion of $\mathcal{H}$ in the norm $\|f\|_{-k} := \|A^{-k/2}f\|$. Thus each set of three spaces $\{\mathcal{H}_{-k} \supset \mathcal{H} \supset \mathcal{H}_k\}$ organizes the so-called rigged Hilbert space [8]. In what follows we use mainly only the following part of the A-scale:

$$\mathcal{H}_{-2} \supset \mathcal{H}_{-1} \supset \mathcal{H}_0 \equiv \mathcal{H} \supset \mathcal{H}_1 \supset \mathcal{H}_2, \tag{2}$$

where $\mathcal{H}_2 = \mathcal{D}(A)$ and $\mathcal{H}_1 = \mathcal{D}(A^{1/2}) = \mathcal{Q}(\gamma_A)$ is the domain of the closed quadratic form γ_A generated by A.

Obviously in the chain (2) the operator A is unitary as a map from $\mathcal{H}_2$ to $\mathcal{H}$. Moreover it acts isometrically from $\mathcal{H}_1$ to $\mathcal{H}_{-1}$ and from $\mathcal{H}$ to $\mathcal{H}_{-2}$. Thus the closure $\mathbf{A} \equiv A^{cl} : \mathcal{H} \to \mathcal{H}_{-2}$ as well as the inverse operator $\mathbf{A}^{-1}$ is unitary. The following simple relations are fulfilled in the scale (2) (see [8]):

$$\begin{aligned}(\omega, \mathbf{A}f)_{-2} &= (\mathbf{A}^{-1}\omega, f) = (A^{-1}\mathbf{A}^{-1}\omega, A^{-1}f)_2, \ \omega \in \mathcal{H}_{-2}, f \in \mathcal{H} \\ (\omega, \mathbf{A}\varphi)_{-1} &= \langle \omega, \varphi \rangle = \langle \mathbf{A}^{-1}\omega, \mathbf{A}\varphi \rangle = (\mathbf{A}^{-1}\omega, \varphi)_1, \ \omega \in \mathcal{H}_{-1}, \varphi \in \mathcal{H}_1 \\ \langle f, \varphi \rangle &= (f, \varphi), \ f \in \mathcal{H}, \varphi \in \mathcal{H}_2,\end{aligned} \tag{3}$$

where $\langle \cdot, \cdot \rangle$ denotes a dual inner product between $\mathcal{H}_{-k}$ and $\mathcal{H}_k$.

A self-adjoint in $\mathcal{H}$ operator $\tilde{A} \neq A$ is called *singularly perturbed* with respect to A whenever the domain

$$\text{(4)} \qquad \mathcal{D} := \{\varphi \in \mathcal{D}(A) \cap \mathcal{D}(\tilde{A}) \mid A\varphi = \tilde{A}\varphi\} \text{ is dense in } \mathcal{H}$$

(cf. with [2, 19]–[22]). If in addition, the range of $\tilde{A}$ coincides with the whole Hilbert space $\mathcal{H}$, i.e.

$$\mathcal{R}(\tilde{A}) = \mathcal{H},$$

then we write

$$\tilde{A} \in \mathcal{A}_s(A).$$

Due to (4) operators A and $\tilde{A}$ are different self-adjoint extensions of the symmetric operator

$$\text{(5)} \qquad \dot{A} = A \restriction \mathcal{D} = \tilde{A} \restriction \mathcal{D},$$

which is automatically closed and has non-trivial and equal deficiency indices. Moreover it is clear that A and $\tilde{A}$ are relatively prime with respect to $\dot{A}$ (see [1, 26]) in the sense that $\dot{A}$ is a maximal common part of a pair A, $\tilde{A}$.

Let A_∞ denote the Friedrichs extension of $\dot{A}$. There are only two possibilities: $A_\infty \neq A$ and $A_\infty = A$. Respectively we pick in $\mathcal{A}_s(A)$ two subsets: $\mathcal{A}_{s_1}(A)$ and $\mathcal{A}_{s_2}(A)$. In other words, for $\tilde{A} \in \mathcal{A}_s(A)$ we write

$$\tilde{A} \in \mathcal{A}_{s_1}(A) \text{ if } A_\infty \neq A, \text{ and } \tilde{A} \in \mathcal{A}_{s_2}(A) \text{ if } A_\infty = A.$$

We recall that $A_\infty \neq A$ is equivalent to the condition that the set $\mathcal{D}$ defined by (4) is not dense in $\mathcal{H}_1$.

Denote $\mathcal{M}_1 := \mathcal{D}^{cl,1}$, where $cl, 1$ stands for the closure in $\mathcal{H}_1$. In general it is possible that $\mathcal{M}_1$ contains new elements from $\mathcal{H}_2$ besides ones from $\mathcal{D}$. To exclude that in what follows we will suppose that A_∞ and A are relatively prime with respect to $\dot{A}$. This is equivalent to the condition:

$$\text{(6)} \qquad \mathcal{M}_1 \cap \mathcal{H}_2 = \mathcal{D},$$

where we used that $\mathcal{H}_2 = \mathcal{D}(A)$.

For $\tilde{A} \in \mathcal{A}_{s_1}(A)$ we will always assume that $\tilde{A}$ and A_∞ are also relatively prime with respect to $\dot{A}$.

Let V be a linear operator in the A-scale (1) with a domain $\mathcal{D}(V) \subseteq \mathcal{H}_k$ and a range $\mathcal{R}(V) \subseteq \mathcal{H}_{-k}$ for some $k > 0$. In such case we write

$$V : \mathcal{H}_k \longrightarrow \mathcal{H}_{-k}.$$

Assume V is closed and symmetric, i.e. $V \subseteq V^*$. We have to note that the adjoint operator V^* is defined with respect to a dual inner product $\langle \cdot, \cdot \rangle$. Thus

$$\langle V\varphi, \psi \rangle = \langle \varphi, V\psi \rangle, \quad \varphi, \psi \in \mathcal{D}(V) \subseteq \mathcal{D}(V^*) \subseteq \mathcal{H}_k.$$

Note else that Ker $V := \{\varphi \in \mathcal{D}(V) \mid V\varphi = 0\}$ is the closed subspace in $\mathcal{H}_k$ since V is a closed operator.

A closed symmetric operator $0 \neq V : \mathcal{H}_k \to \mathcal{H}_{-k}$ is called *singular* with respect to $\mathcal{H}$ whenever

$$\text{Ker } V \sqsubset \mathcal{H}, \tag{7}$$

where $\sqsubset$ means a dense embedding (cf. with [2, 16, 19, 20]). If V acts from $\mathcal{H}_1$ to $\mathcal{H}_{-1}$ and for Ker V the following condition is fulfilled,

$$\text{Ker } V \cap \mathcal{D}(A) \sqsubset \mathcal{H}, \tag{8}$$

then we write $V \in \mathcal{H}_{-1}(A)$-class (or more shortly $V \in \mathcal{H}_{-1}$-class).

For $V : \mathcal{H}_2 \to \mathcal{H}_{-2}$ we write $V \in \mathcal{H}_{-2}$-class whenever Ker V is dense in $\mathcal{H}_1$, i.e.

$$\text{Ker } V \sqsubset \mathcal{H}_1. \tag{9}$$

If in a more general situation for an operator $V : \mathcal{H}_k \to \mathcal{H}_{-k}$, $k \geq 2$ the set Ker$V \cap \mathcal{H}_k$ is dense in $\mathcal{H}_{k-1}(A)$, then we say that V belongs to the $\mathcal{H}_{-k}(A)$-class. Of course V is automatically singular with respect to $\mathcal{H}$.

According to Theorem A.1 from [2], a singular with respect to $\mathcal{H}$ operator V belongs to the $\mathcal{H}_{-k}(A)$-class, $k \geq 1$ whenever the range $\mathcal{R}(V)$ has a zero intersection with $\mathcal{H}_{-k+1}(A)$ and moreover

$$(\mathcal{R}(V))^{cl,-k} \cap \mathcal{H}_{-k+1} = \{0\}, \tag{10}$$

where $cl, -k$ stands for the closure in $\mathcal{H}_{-k}$. For example, suppose V is a finite-dimensional operator,

$$V\varphi = \Sigma_{i=1}^{N} \langle \varphi, \omega_i \rangle \omega_i, \ \omega_i \in \mathcal{H}_{-k}, \ \varphi \in \mathcal{H}_k.$$

If all vectors $\omega_i \in \mathcal{H}_{-k} \backslash \mathcal{H}_{-k+1}$ and are linearly independent with respect to $\mathcal{H}_{-k+1}$, then V belongs to the $\mathcal{H}_{-k}$-class. Obviously in this case (10) is fulfilled.

We interpret operators $V \in \mathcal{H}_{-k}$-class as objects carrying singular perturbations for A. One of our aim is to construct the singularly perturbed operator $\tilde{A}$ starting from V. The corresponding theory in terms of singular quadratic forms has been developed in [19].

We observe that for each $V \in \mathcal{H}_{-k}$-class, $k = 1, 2$ the set

$$\mathcal{D}_0 := \{\varphi \in \mathcal{D}(A) \cap \mathcal{D}(V) \mid V\varphi = 0\} \equiv \mathcal{D}(A) \cap \text{Ker } V \text{ is dense in } \mathcal{H}. \tag{11}$$

Therefore each singular operator $V \in \mathcal{H}_{-k}$-class, $k = 1, 2$ defines a symmetric restriction of A,

$$\dot{A}_0 = A \mid \mathcal{D}_0, \tag{12}$$

with non-trivial deficiency indices. We note that in general $\dot{A}_0$ does not coincide with $\dot{A}$. Nevertheless the Friedrichs extension of $\dot{A}_0$ we again denote by A_∞.

Obviously for $V \in \mathcal{H}_{-1}$-class the set $\mathcal{D}_0$ is dense in the closed subspace $\mathrm{Ker}V \subset \mathcal{H}_1$, i.e. $\mathcal{D}_0 \sqsubset$ Ker V in $\mathcal{H}_1$, or equivalently, $\mathcal{D}_0^{cl,1} = \mathrm{Ker}V$. Conversely, if

$$(13) \qquad (\mathcal{H}_2 \cap \text{ Ker } V)^{cl,1} = \text{ Ker } V \sqsubset \mathcal{H},$$

then a non-trivial closed symmetric operator $V : \mathcal{H}_1 \to \mathcal{H}_{-1}$ belongs to the $\mathcal{H}_{-1}$-class.

Given $\tilde{A} \in \mathcal{A}_{s_k}(A)$ and $V \in \mathcal{H}_{-k}$-class, $k = 1, 2$ we are interesting in the situation, where sets $\mathcal{D}$ and $\mathcal{D}_0$ coincide and hence operators $\dot{A}$ and $\dot{A}_0$ are identical. In this case we want to find the connection between $\tilde{A}$ and V. Note that in the case $k \geq 3$ operator $\dot{A}_0$ is essentially self-adjoint since $\mathcal{D}_0$ is dense in $\mathcal{H}_2$.

Let $V : \mathcal{H}_k \to \mathcal{H}_{-k}$, $k \geq 0$ be a closed symmetric operator. Let us introduce in accord with [7, 25, 21] a more general notion of an operator sum for A and V.

Namely let us define the *generalized sum* $A\tilde{+}V$ for operators A and V as the "restriction" of the operator sum $\mathbf{A} + V : \mathcal{H}_k \to \mathcal{H}_{-k}$ to $\mathcal{H}$. Precisely,

$$(14) \quad \mathcal{D}(A\tilde{+}V) = \{\varphi \in \mathcal{H}_k \cap \mathcal{D}(V) \mid \mathbf{A}\varphi + V\varphi \in \mathcal{H}\}, \quad (A\tilde{+}V)\varphi = \mathbf{A}\varphi + V\varphi,$$

where, recall, $\mathbf{A}$ is the closure of A as an operator from $\mathcal{H}$ to $\mathcal{H}_{-2}$.

Of course $A\tilde{+}V$ coincides with a usual operator sum $A + V$ in the case, where V is an operator in $\mathcal{H}$ since $\mathbf{A}\varphi + V\varphi \in \mathcal{H}$ only if $\varphi \in \mathcal{D}(V) \cap \mathcal{D}(A)$. We will see below that the sum in the sense of quadratic forms is also a particular case of the generalized sum.

The relevant problem arises: to find conditions when an operator $\tilde{A} = A\tilde{+}V$ is self-adjoint in $\mathcal{H}$. We assert (see below Theorem 3) that the generalized sum $\tilde{A} = A\tilde{+}V$ is a self-adjoint operator in $\mathcal{H}$ if the point -1 is regular for V. Moreover $A\tilde{+}V$ is self-adjoint if the following condition is fulfilled (for details see [18]),

$$(15) \qquad \mathcal{D}(A) \subset \mathcal{R}(\mathbf{A}^{-1}V + I),$$

where I denotes the identical transform. In fact these assertions extend the well-known KLMN-Theorem [30].

Our main result states: under natural conditions each singularly perturbed operator $\tilde{A} \in \mathcal{A}_{s_1}(A)$ has a unique representation in the form of a generalized sum, $\tilde{A} = A\tilde{+}V$ with some $V \in \mathcal{H}_{-1}$-class. It is the generalization of the well-known result by Krein and Yavrian (see Theorem 4.1 in [25]). They discussed a simple positive self-adjoint operator A and its symmetric restriction $\dot{A} = A \mid \mathcal{D}$, $\mathcal{D} \subset \mathcal{D}(A)$ with the deficiency indices $(1, 1)$. Theorem 4.1 in [25] states that each self-adjoint extension $\tilde{A} \neq A_\infty$ of $\dot{A}$ admits the representation by the formula

$$(16) \qquad \tilde{A} = A\tilde{+}\lambda \langle \cdot, \omega \rangle \omega, \quad -\infty < \lambda < \infty$$

with some $\omega \in \mathcal{H}_{-1} \setminus \mathcal{H}$. In the present paper we consider a positive restriction $\dot{A}$ of A with arbitrary deficiency indices and show that the formula (16) with a rank one singular operator $V\varphi = \lambda \langle \varphi, \omega \rangle \omega$, $\varphi \in \mathcal{H}_1$ becomes: $\tilde{A} = A\tilde{+}V$, where V belongs to the $\mathcal{H}_{-1}$-class and could have an arbitrary rank.

Theorem 1 *Let $\tilde{A} \in \mathcal{A}_{s_1}(A)$. Assume $\mathcal{D}(\tilde{A}) \subset \mathcal{H}_1$ and both pairs $\tilde{A}$, A_∞ and A, A_∞ are relatively prime with respect to the symmetric operator $\dot{A}$ defined by* (5). *Then there exists a uniquely defined symmetric operator $V : \mathcal{H}_1 \to \mathcal{H}_{-1}$, $V \in \mathcal{H}_{-1}(A)$-class such that $\tilde{A}$ admits the representation: $\tilde{A} = A\tilde{+}V$. Moreover the condition* (15) *is fulfilled, domains $\mathcal{D}$, $\mathcal{D}_0$ coincide and operators $\dot{A}$ and $\dot{A}_0$ are identical.*

Conversely, let an operator $V \in \mathcal{H}_{-1}(A)$-class satisfies the condition (15). *Then the generalized sum $\tilde{A} = A\tilde{+}V$ belongs to $\mathcal{A}_{s_1}(A)$. Moreover*

$$\mathcal{D}(\tilde{A}) \subset \mathcal{H}_1 \tag{17}$$

and both pairs $\tilde{A}$, A_∞ and A, A_∞ are relatively prime with respect to the symmetric operator $\dot{A}_0$ defined by (12).

The proof of Theorem 1 we will obtain as a consequence of Theorem 6 and Theorem 7 (see below).

In other words the above result could also be formulated as follows. The formula $\tilde{A} = A\tilde{+}V$ establishes the ono-to-one correspondence between operators $\tilde{A} \in \mathcal{A}_{s_1}(A)$ and $V \in \mathcal{H}_{-1}$-class whenever the following conditions are fulfilled: $\mathcal{D}(\tilde{A}) \subset \mathcal{H}_1$, $\mathcal{D}(A) \subset \mathcal{R}(\mathbf{A}^{-1}V + I)$, and if both pairs A, A_∞ and $\tilde{A}$, A_∞ are relatively prime with respect to $\dot{A} = \dot{A}_0$.

We will get also the explicit connections between V and the extension parameter $\tilde{B}$ the Krein resolvent formula for the operator $\tilde{A}$.

In the case $\tilde{A} \in \mathcal{A}_{s_2}(A)$ we have not the representation of the form $\tilde{A} = A\tilde{+}V$ with a symmetric operator $V : \mathcal{H}_2 \to \mathcal{H}_{-2}$ since the above defined set $\mathcal{D}$ is dense in $\mathcal{H}_1$. Nevertheless in this situation we can correspond to $\tilde{A}$ a uniquely defined singular quadratic form ν on the space $\mathcal{H}_2$. Then using ν we can construct some operator $V : \mathcal{H}_2 \to \mathcal{H}_{-2}$ (see [19]). Thus the one-to-one correspondence between $\tilde{A}$ and V exists also in the case $A = A_\infty$. This correspondence could be derived also using the Krein resolvent formula:

$$\tilde{A}^{-1} = A^{-1} + B_V^{-1} P_{\mathcal{N}_0}, \tag{18}$$

where $B_V := \mathbf{A}^{-1} V A^{-1} P_{\mathcal{N}_0}$ acts in the deficiency subspace $\mathcal{N}_0$ of the symmetric operator $\dot{A}$. For more details see [19, 20, 21].

Note that in [28] the additive representation for the one-dimensional Schrödinger operator with a point interaction has been proven using non-symmetric bilinear forms.

2 Generalized Resolvent Identity

In this section we derive the generalized resolvent identity for $\tilde{A} = A\tilde{+}V$.

We start with some discussion about self-adjointness conditions for the generalized sum $\tilde{A} = A\tilde{+}V$.

Let an operator $V \in \mathcal{H}_{-1}$-class. Suppose V is bounded from below in $\mathcal{H}_1$,

$$\langle V\varphi, \varphi\rangle \geq m_V \|\varphi\|_1^2.$$

Let ν denotes the quadratic form generated by V, i.e.

$$\nu[\varphi] = \langle V\varphi, \varphi\rangle = (\mathbf{A}^{-1}V\varphi, \varphi)_1, \ \varphi \in \mathcal{D}(V) \subset \mathcal{Q}(\nu).$$

Surely the form ν is closable in $\mathcal{H}_1$ since the operator $\mathbf{A}^{-1}V$ is closed and bounded from below. Let $\nu^{cl,1}$ denote the closure of ν in $\mathcal{H}_1$.

If ν satisfies the condition of the well-known KLMN-Theorem [30, 15],

$$|\ \nu[\varphi]\ | \leq b\gamma_A[\varphi] + a\|\varphi\|^2, \ 0 \leq b < 1, \ \ 0 \leq a, \tag{19}$$

then the quadratic form $\tilde{\gamma} = \gamma_A + \nu$ is closable and bounded from below in $\mathcal{H}$ and therefore the corresponding associated operator $A_{\tilde{\gamma}}$ is self-adjoint in $\mathcal{H}$. We assert that $A_{\tilde{\gamma}}$ coincides with the generalized sum $\tilde{A} = A\tilde{+}V$. The analogical result is true as well in a slightly more general situation (cf. with [7, 22], for more details see [6] and [18]).

Theorem 2 *Let a quadratic form ν in $\mathcal{H}_1$ is generated by a self-adjoint bounded from below operator $V : \mathcal{H}_1 \to \mathcal{H}_{-1}$. Let $\nu = \nu_- + \nu_+$ be the decomposition of ν in the positive and the negative component. Assume ν_- satisfies the condition* (19). *Then the sum $\tilde{\gamma} = \gamma_A + \nu$ is closable and bounded from below in $\mathcal{H}$ and hence the associated with its closure operator $A_{\tilde{\gamma}}$ is self-adjoint in $\mathcal{H}$. Moreover $A_{\tilde{\gamma}}$ coincides with the generalized sum $\tilde{A} = A\tilde{+}V$.*

Proof: Using the well known results (see [15], Theorem VI. 1.17, [15], Theorem VI.1.27 or [30], Theorem 2.31) one can easily check that the sum $\tilde{\gamma} = \gamma_A + \nu$ is bounded from below and closable in $\mathcal{H}$. Let $\tilde{\gamma}^{cl}$ denote the corresponding closure of $\tilde{\gamma}$. Clearly $\tilde{\gamma}^{cl} = \gamma_A + \nu^{cl,1}$ where $cl, 1$ means the closure in $\mathcal{H}_1$. By definition of the associated operator we have

$$\begin{aligned}(A_{\tilde{\gamma}}\varphi, \psi) &= \tilde{\gamma}^{cl}(\varphi, \psi) = \gamma_A(\varphi, \psi) + \nu^{cl,1}(\varphi, \psi)\\ &= \langle \mathbf{A}\varphi, \psi\rangle + \langle V\varphi, \psi\rangle, \ \varphi \in \mathcal{D}(A_{\tilde{\gamma}}), \psi \in \mathcal{H}_1,\end{aligned}$$

where we used that $\mathcal{D}(V) \subset \mathcal{D}(A_{\tilde{\gamma}})$ and $\nu^{cl,1}(\varphi, \psi) = \langle V\varphi, \psi\rangle$, $\varphi \in \mathcal{D}(A_{\tilde{\gamma}})$, $\psi \in \mathcal{H}_1$. Latter facts are true by the following arguments. For a fixed $\varphi \in \mathcal{D}(A_{\tilde{\gamma}})$ both linear functionals $l_{\tilde{\gamma}}(\psi) = (\psi, A_{\tilde{\gamma}}\varphi)$ and $l_{\mathbf{A}}(\psi) = \langle \psi, \mathbf{A}\varphi\rangle$, $\psi \in \mathcal{D}(V)$ are continuous and therefore $\varphi \in \mathcal{D}(V)$ because the operator V is self-adjoint. Since $A_{\tilde{\gamma}}\varphi \in \mathcal{H}$, we see that $\mathbf{A}\varphi + V\varphi \in \mathcal{H}$ too and therefore $A_{\tilde{\gamma}} \subset \tilde{A} = A\tilde{+}V$ (see (14)). Using $\mathbf{A} + V \geq 1$ we conclude that the range $\mathcal{R}(A\tilde{+}V) = \mathcal{H}$ and hence $A_{\tilde{\gamma}} = \tilde{A}$ because both these operators are self-adjoint. □

Now we present another sufficient condition ensuring that generalized sum $A\tilde{+}V$ is a self-adjoint operator in $\mathcal{H}$.

We say, the point -1 is regular for an operator $V : \mathcal{H}_1(A) \to \mathcal{H}_{-1}(A)$ and write

$$-1 \in \rho(V) \tag{20}$$

whenever the operator $V + \mathbf{A} : \mathcal{H}_1(A) \to \mathcal{H}_{-1}(A)$ is boundedly invertible. In other words (20) means that the inverse operator

$$(V + \mathbf{A})^{-1} : \mathcal{H}_{-1}(A) \longrightarrow \mathcal{H}_1(A)$$

exists and is defined on whole $\mathcal{H}_{-1}(A)$.

Theorem 3 *Suppose operator $V \in \mathcal{H}_{-1}$-class and satisfies the condition* (20), *then the generalized sum $\tilde{A} = A \tilde{+} V$ is self-adjoint in $\mathcal{H}$. Moreover $\tilde{A} \in \mathcal{A}_{s_1}(A)$ and $\tilde{A}$, A are relatively prime with respect to $\dot{A}_0 = A \mid \mathcal{D}_0$, where $\mathcal{D}_0 = \mathcal{D}(A) \cap$ Ker V.*

Proof: By definition (see (14) and (11)) the operator $\tilde{A} = A \tilde{+} V$ is densely defined since $\tilde{A}$ coincides with A on the set $\mathcal{D}_0$, which is dense in $\mathcal{H}$ due to (11). Using relations (3) it is easy to show that $\tilde{A}$ is symmetric in $\mathcal{H}$. In fact $\tilde{A}$ is self-adjoint thanks to the condition (20). Indeed (20) implies that the range $\mathcal{R}(V + \mathbf{A}) = \mathcal{H}_{-1}$ and therefore the range $\mathcal{R}(\tilde{A}) = \mathcal{H}$. By the same reason $\tilde{A}$ has a bounded inverse. Thus $\tilde{A} \in \mathcal{A}_s(A)$. Obviously the domain of $\tilde{A}$ is contained in $\mathcal{H}_1(A)$. Further, the set $\mathcal{D}_0$ is a maximal subset of $\mathcal{D}(\tilde{A})$, where operators $\tilde{A}$ and A coincide since always $V\varphi \neq 0$ whenever $\varphi \in \mathcal{D}(\tilde{A})$ and $\varphi \notin \mathcal{D}_0$. This means that $\mathcal{D}_0 = \mathcal{D}$, $\dot{A}_0 = A$, and therefore $\tilde{A}$ and A are relatively prime with respect to $\dot{A}_0$. Moreover setting $\mathcal{M}_1 = \mathcal{D}_0^{cl,1} = \mathrm{Ker} V$ we obtain that (6) follows from (13). So we conclude that $\tilde{A} \in \mathcal{A}_{s_1}(A)$. This finishes the proof. □

We remark that in the last theorem the operator V could be unbounded in both sides. Then the condition (20) means that $\mathbf{A}^{-1} V$ has a spectral gap containing the point -1. It is easily to see that for any symmetric operator $V : \mathcal{H}_1(A) \to \mathcal{H}_{-1}(A)$ with a gap in its spectrum there exists a real λ such that $A \tilde{+} \lambda V$ is self-adjoint in $\mathcal{H}$ (see [18]).

Theorem 4 [23]. *Let $V : \mathcal{H}_1 \to \mathcal{H}_{-1}$ be an operator belonging to the $\mathcal{H}_{-1}$-class. Assume the operator $\mathbf{A}^{-1} V + I$ in $\mathcal{H}_1$ satisfies the conditionis* (15). *Then the generalized sum $\tilde{A} = A \tilde{+} V$ is self-adjoint in $\mathcal{H}$. Moreover the operator $\tilde{A}$ belongs to the set $\mathcal{A}_{s_1}(A)$ and satisfies the condition* (17). *Besides $\tilde{A}$, A_∞ are relatively prime with respect to $\dot{A}_0 = \dot{A}$.*

Let $\rho(A)$ denote the resolvent set of A. Let us extend by closure the resolvent of A on the whole space $\mathcal{H}_{-1}$:

$$R_z^{cl} = \mathbf{R}_z := (\mathbf{A} - zI)^{-1} : \mathcal{H}_{-1} \longrightarrow \mathcal{H}_1, z \in \rho(A),$$

where I stands for the identity map.

Proposition 1 *The map*

$$\mathbf{R}_z : \mathcal{H}_{-1} \longrightarrow \mathcal{H}_1, \quad z \in \rho(A)$$

is a bounded bijection.

Proof: By definition of $\mathbf{R}_z$ we easily find that for each $\omega \in \mathcal{H}_{-1}$, $\omega = \mathbf{A}\varphi$, $\varphi \in \mathcal{H}_1$,

$$\mathbf{R}_z\omega = \varphi + zR_z\varphi, \quad R_z = (A - zI)^{-1}. \tag{21}$$

Thus the domain $\mathcal{D}(\mathbf{R}_z) = \mathcal{H}_{-1}$ and the range $\mathcal{R}(\mathbf{R}_z) = \mathcal{H}_1$. (21) implies for any $\omega \in \mathcal{H}_{-1}$:

$$\begin{aligned}\|\mathbf{R}_z\omega\|_1 &= \|A^{1/2}\mathbf{R}_z\omega\| = \|A^{1/2}(\varphi + z(A - zI)^{-1}\varphi)\| \\ &\leq (1 + |z|\|R_z\|)\|A^{1/2}\varphi\| = c\|\varphi\|_1 = c\|\omega\|_{-1}.\end{aligned}$$

□

Since for each $h \in \mathcal{H}$, $h = A\varphi$, $\varphi \in \mathcal{D}(A)$, (21) implies

$$\mathbf{R}_z h = R_z h = A^{-1}h + zR_zA^{-1}h = (I + zR_z)A^{-1}h.$$

Of course we have that the restriction of $\mathbf{R}_z$ onto $\mathcal{H}$ coincides with a usual resolvent of the operator A.

Assume $V : \mathcal{H}_1 \to \mathcal{H}_{-1}$ be a bounded operator belonging to the $\mathcal{H}_{-1}(A)$-class. Assume the generalized sum $\tilde{A} = A \tilde{+} V$ is self-adjoint and bounded from below in $\mathcal{H}$, i.e. $\tilde{A} \geq \tilde{m}$. Then by analogy with (1) one can introduce the $\tilde{A}$-scale of Hilbert spaces,

$$\tilde{\mathcal{H}}_{-k} \equiv \mathcal{H}_{-k}(\tilde{A}) \supset \mathcal{H} \supset \tilde{\mathcal{H}}_k \equiv \mathcal{H}_k(\tilde{A}), \quad k \geq 0, \tag{22}$$

where $\|\cdot\|_{\tilde{\mathcal{H}}_k} := \|(\tilde{A} + (|\tilde{m}| + 1))^{k/2}I \cdot \|$. Remark that the norm in $\mathcal{H}_k(\tilde{A})$ is not weaker than one in $\mathcal{H}_k(A)$ since V is bounded. By this reason following dense embeddings are true,

$$\tilde{\mathcal{H}}_{-1} \supseteq \mathcal{H}_{-1} \supset \mathcal{H} \supset \mathcal{H}_1 \supseteq \tilde{\mathcal{H}}_1(\tilde{A}). \tag{23}$$

By the same way as in (21) we can define the extended perturbed resolvent in the $\tilde{A}$-scale:

$$\tilde{R}_z^{cl} = \tilde{\mathbf{R}}_z : \tilde{\mathcal{H}}_{-1} \longrightarrow \tilde{\mathcal{H}}_1, \ \tilde{\mathbf{R}}_z\omega = \varphi + z\tilde{R}_z\varphi, \ \omega = (\mathbf{A} + V)\varphi, \ \varphi \in \mathcal{H}_1.$$

It is clear (see Proposition 1) that $\tilde{\mathbf{R}}_z : \tilde{\mathcal{H}}_{-1} \to \tilde{\mathcal{H}}_1$ is the bounded bijection too. Surely

$$\tilde{\mathbf{R}}_z \mid \mathcal{H} = \tilde{R}_z = (\tilde{A} - zI)^{-1}.$$

Moreover we will show now that $\tilde{\mathbf{R}}_z$ is bounded in the A-scale.

Proposition 2 *Let $V : \mathcal{H}_1 \to \mathcal{H}_{-1}$ be bounded and self-adjoint. Assume V satisfies the condition* (15) *or* (20). *Suppose the generalized sum $\tilde{A} = A \tilde{+} V$ is bounded from below in $\mathcal{H}$. Then the map*

$$\tilde{\mathbf{R}}_z : \mathcal{H}_{-1}(A) \longrightarrow \mathcal{H}_1(A), \quad z \in \rho(\tilde{A})$$

is a bounded bijection.

Proof: $\tilde{\mathbf{R}}_z$ is bijective as a map from $\mathcal{H}_{-1}$ to $\mathcal{H}_1$. Indeed the range (the domain) of $\tilde{\mathbf{R}}_z$ coincides with $\mathcal{H}_1$ (resp. with $\mathcal{H}_{-1}$) since V is bounded and norms in $\mathcal{H}_1$ and $\tilde{\mathcal{H}}_1$ (resp. in $\mathcal{H}_{-1}$ and $\tilde{\mathcal{H}}_{-1}$) are equivalent due to (15) or (20). Thus $\mathcal{H}_1 = \mathcal{R}(\tilde{\mathbf{R}}_z)$ and $\mathcal{H}_{-1} = \mathcal{D}(\tilde{\mathbf{R}}_z)$. Further, by analogy with the proof of Proposition 1, from (23) it follows that the map

$$\tilde{\mathbf{R}}_z : \mathcal{H}_{-1} \longrightarrow \mathcal{H}_1$$

is bounded. □

We are ready to derive the generalized resolvent identity for $\tilde{R}_z$.

Theorem 5 *Let $V : \mathcal{H}_1 \to \mathcal{H}_{-1}$ be a bounded operator belonging to the $\mathcal{H}_{-1}(A)$-class. Assume V satisfies one of the conditions* (15) *or* (20). *Suppose the generalized sum $\tilde{A} = A \tilde{+} V$ is the bounded from below operator in $\mathcal{H}$, $\tilde{A} \geq \tilde{m}$. Then for extended resolvents*

$$\mathbf{R}_z := (\mathbf{A} - zI)^{-1}, \ z \in \rho(A), \ \tilde{\mathbf{R}}_z = (\tilde{\mathbf{A}} - zI)^{-1}, \ z \in \rho(\tilde{A}),$$

the following identity holds:

$$\tilde{\mathbf{R}}_z - \mathbf{R}_z = -\tilde{\mathbf{R}}_z V \mathbf{R}_z = -(I + \mathbf{R}_z V)^{-1} \mathbf{R}_z V \mathbf{R}_z, \quad z \in \rho = \rho(A) \cap \rho(\tilde{A}), \tag{24}$$

where $\tilde{\mathbf{A}} = \mathbf{A} + V : \mathcal{H}_1 \to \mathcal{H}_{-1}$ and $(I + \mathbf{R}_z V)^{-1} : \mathcal{H}_1 \to \mathcal{H}_1$ are bounded operators. In particular for usual resolvents $R_z = (A - zI)^{-1}$ and $\tilde{R}_z = (\tilde{A} - zI)^{-1}$ the following generalized resolvent identity holds:

$$\tilde{R}_z = R_z - \tilde{\mathbf{R}}_z V R_z = R_z - (I + \mathbf{R}_z V)^{-1} \mathbf{R}_z V R_z, \ z \in \rho(\tilde{A}) \cap \rho(A). \tag{25}$$

Proof: Consider the difference $\tilde{\mathbf{R}}_z - \mathbf{R}_z$. By above two propositions $\tilde{\mathbf{R}}_z - \mathbf{R}_z$ is a bounded operator and its range is contained in $\mathcal{H}_1$. Therefore we can apply to this difference the bounded operator $\mathbf{A} + V - zI : \mathcal{H}_1 \to \mathcal{H}_{-1}$. By simple manipulations we obtain:

$$\begin{aligned} \tilde{\mathbf{R}}_z - \mathbf{R}_z &= \tilde{\mathbf{R}}_z (\mathbf{A} + V - zI)[\tilde{\mathbf{R}}_z - \mathbf{R}_z] \\ &= \tilde{\mathbf{R}}_z [I - (\mathbf{A} + V - zI)(\mathbf{A} - zI)^{-1}] = -\tilde{\mathbf{R}}_z V \mathbf{R}_z \end{aligned}$$

where we used that $\tilde{\mathbf{R}}_z = (\mathbf{A} + V - z)^{-1} : \mathcal{H}_{-1} \to \mathcal{H}_1$ is bounded. Thus we proved that

$$\tilde{\mathbf{R}}_z - \mathbf{R}_z = -(\mathbf{A} + V - zI)^{-1} V \mathbf{R}_z = -\tilde{\mathbf{R}}_z V \mathbf{R}_z : \mathcal{H}_{-1} \longrightarrow \mathcal{H}_1.$$

Further, again by simple manipulations we get:

$$
\begin{aligned}
\tilde{\mathbf{R}}_z - \mathbf{R}_z &= -\tilde{\mathbf{R}}_z((\mathbf{A} - zI)\mathbf{R}_z)V\mathbf{R}_z \\
&= -(\tilde{\mathbf{R}}_z(\mathbf{A} - zI))\mathbf{R}_z V\mathbf{R}_z = -(\mathbf{R}_z(\mathbf{A} + V - zI))^{-1}\mathbf{R}_z V\mathbf{R}_z \\
&= -(I + \mathbf{R}_z V)^{-1}\mathbf{R}_z V\mathbf{R}_z,
\end{aligned} \tag{26}
$$

which proves the (24). We observe that $(I + \mathbf{R}_z V)^{-1}$ is a bounded operator in $\mathcal{H}_1$ for each $z \in \rho$. In fact this operator is a product of two bounded operators: $(I + \mathbf{R}_z V)^{-1} = \tilde{\mathbf{R}}_z(\mathbf{A} - zI)$ since $\mathbf{A} : \mathcal{H}_1 \to \mathcal{H}_{-1}$ is unitary, $\|zI\varphi\|_{-1} = |z| \cdot \|\varphi\|_{-1} \leq |z| \cdot \|\varphi\|_1$, and $\tilde{\mathbf{R}}_z : \mathcal{H}_{-1} \to \mathcal{H}_1$ is a bounded bijection due to Proposition 2. Emphasize that the operator $(I + \mathbf{R}_z V)^{-1}$ is defined on the whole space $\mathcal{H}_1$. In particular it is defined on the range of the operator $\mathbf{R}_z V\mathbf{R}_z$.

The validity of (25) follows from (24) by the restriction procedure onto $\mathcal{H}$. This finishes the proof. □

In particular under the condition (20) we have $\mathcal{R}(A) = \mathcal{H} = \mathcal{R}(\tilde{A})$ since the point $0 \in \rho$. Therefore the relation (24) implies

$$
\tilde{\mathbf{A}}^{-1} - \mathbf{A}^{-1} = -\tilde{\mathbf{A}}^{-1}V\mathbf{A}^{-1} = -(I + \mathbf{A}^{-1}V)^{-1}\mathbf{A}^{-1}V\mathbf{A}^{-1}. \tag{27}
$$

The operator

$$
(I + \mathbf{A}^{-1}V)^{-1} = (\mathbf{A}^{-1}(\mathbf{A} + V))^{-1} = (\mathbf{A} + V)^{-1}\mathbf{A} : \mathcal{H}_1 \longrightarrow \mathcal{H}_1
$$

in (27) is a bounded bijection since $\mathbf{A} : \mathcal{H}_1 \to \mathcal{H}_{-1}$ is unitary and $(\mathbf{A} + V)^{-1} = \tilde{\mathbf{A}}^{-1} : \mathcal{H}_{-1} \to \mathcal{H}_1$ is a bounded bijection by Proposition 2.

Remark: The relation (27) is true also in the case, where $\tilde{\mathbf{A}}$ is not necessarily below bounded. Indeed under the condition (20) all arguments in the proof of the previous theorem remain valid since $\mathcal{R}(\tilde{\mathbf{A}}) = \mathcal{H}_{-1}$. Moreover under the condition (15) it is easily also to check that (27) is true for a not necessarily below bounded operator $\tilde{\mathbf{A}}$. In what follows we will use only the particular case of the identity (24) (namely (27)) and therefore we can consider that the operator $\tilde{\mathbf{A}}$ is not necessarily below bounded.

Now we want to present the restriction of the difference $\tilde{\mathbf{A}}^{-1} - \mathbf{A}^{-1}$ onto $\mathcal{H}$ in a more suitable form for consequent considerations. With this aim let us consider the operator $\mathbf{A}^{-1}V\mathbf{A}^{-1} : \mathcal{H}_{-1} \to \mathcal{H}_1$. Of course we assume that an operator $V : \mathcal{H}_1 \to \mathcal{H}_{-1}$ belongs to the $\mathcal{H}_{-1}$–class. Since $\mathbf{A}^{-1} : \mathcal{H}_{-1} \to \mathcal{H}_1$ is unitary and V is equal to zero on the subspace

$$
\mathcal{M}_1 := \operatorname{Ker} V
$$

in $\mathcal{H}_1$, we conclude that $\mathbf{A}^{-1}V\mathbf{A}^{-1}$ acts non-trivially only from $\mathcal{N}_{-1} := \mathbf{A}\mathcal{N}_1$ to $\mathcal{N}_1$, where

$$
\mathcal{N}_1 := \mathcal{H}_1 \ominus \mathcal{M}_1.
$$

Thus one can rewrite the restriction of $\mathbf{A}^{-1}V\mathbf{A}^{-1}$ onto $\mathcal{H}$ in the form:

$$\begin{aligned} \mathbf{A}^{-1}V\mathbf{A}^{-1} \mid \mathcal{H} &= \mathbf{A}^{-1}VA^{-1} = \mathbf{A}^{-1}VA^{-1}P_{\mathcal{N}_0} \\ &= P_{\mathcal{N}_1}\mathbf{A}^{-1}P_{\mathcal{N}_{-1}}VP_{\mathcal{N}_1}A^{-1}P_{\mathcal{N}_0}, \end{aligned} \tag{28}$$

where we used following notations: $P_{\mathcal{N}_1}$ is the orthogonal projection in $\mathcal{H}_1$ onto the subspace $\mathcal{N}_1$, $P_{\mathcal{N}_0}$ is the orthogonal projection in $\mathcal{H}$ onto the deficiency subspace $\mathcal{N}_0 = \mathrm{Ker}(\dot{A}_0)^*$, where, recall, $\dot{A}_0 = A \mid \mathcal{M}_2$ and $\mathcal{M}_2 \equiv \mathcal{D}_0 = (\mathrm{Ker}\ V) \cap \mathcal{H}_2$, and $P_{\mathcal{N}_{-1}}$ is the orthogonal projection in $\mathcal{H}_{-1}$ onto the subspace $\mathcal{N}_{-1} = \mathbf{A}\mathcal{N}_1$. Let us explain why projectors $P_{\mathcal{N}_0}$ and $P_{\mathcal{N}_1}$ stand in (28). With this aim we will show that the restriction procedure in (28) leads to a non-trivial operator in $\mathcal{N}_0$. Indeed using

$$A^{-1}P_{\mathcal{M}_0}\mathcal{H} = \mathcal{M}_2 \subset \mathcal{M}_1 = \mathrm{Ker}\ V, \quad \mathcal{M}_2 = \mathcal{M}_1^{cl,1},$$

where $P_{\mathcal{M}_0}$ is the orthogonal projection in $\mathcal{H}$ onto the subspace $\mathcal{M}_0 = \mathcal{H} \ominus \mathcal{N}_0$ and since for any $\eta_2 \in \mathcal{N}_2 = A^{-1}\mathcal{N}_0$ the equality $V\eta_2 = VP_{\mathcal{N}_1}\eta_2$ holds because $P_{\mathcal{M}_1}\eta_2 \in \mathrm{Ker}V$, we conclude that the above considered restriction acts non-trivially only on $\mathcal{N}_0$. Thus it is an operator from $\mathcal{N}_0$ onto $\mathcal{N}_1 \subset \mathcal{N}_0$, i.e. in fact it is an operator in $\mathcal{N}_0$.

In what follows we need in more details. Let again $V \in \mathcal{H}_{-1}$-class and the condition (15) or (20) is fulfilled. Recall that $\mathcal{M}_2 = \mathcal{H}_2 \cap (\mathrm{Ker}\ V)$ denotes the closed subspace in $\mathcal{H}_2 = \mathcal{D}(A)$. Therefore $\mathcal{M}_2$ coincides with $\mathcal{D}_0$. $\mathcal{M}_2$ is dense in $\mathcal{H}$ since $V \in \mathcal{H}_{-1}$-class. Let $\mathcal{M}_1$ denote the closure of $\mathcal{M}_2$ in $\mathcal{H}_1$. By (13) we have $\mathcal{M}_1 \equiv \mathrm{Ker}\ V$ and

$$\mathcal{M}_2 = \mathcal{M}_1 \cap \mathcal{H}_2. \tag{29}$$

We note that $\mathcal{M}_1 \neq \mathcal{H}_1$ as well as $\mathcal{M}_2 \neq \mathcal{H}_2$ since $V \neq 0$.

Thus in the sense of an orthogonal decomposition in a corresponding Hilbert space we have: $\mathcal{H}_2 = \mathcal{M}_2 \oplus \mathcal{N}_2$ and $\mathcal{H}_1 = \mathcal{M}_1 \oplus \mathcal{N}_1$. Let $\mathcal{M}_0 = \mathcal{R}(\dot{A})$ denote the range of the operator $\dot{A}_0 = A \mid \mathcal{D}_0$ and $\mathcal{N}_0 := \mathrm{Ker}\ (\dot{A}_0)^*$ is the deficiency subspace of $\dot{A}$. Then the space $\mathcal{H}$ admits the orthogonal decomposition: $\mathcal{H} = \mathcal{M}_0 \oplus \mathcal{N}_0$. Further, by construction $\mathcal{M}_2 \subset \mathcal{M}_1$ and $\mathcal{N}_1 \perp \mathcal{M}_1$ in $\mathcal{H}_1$. Hence for each $\eta_1 \in \mathcal{N}_1$ and $\mu_0 = A\mu_2 \in \mathcal{M}_0, \mu_2 \in \mathcal{M}_2$ we have: $(\eta_1, \mu_0) = (\eta_1, A\mu_2) = (\eta_1, \mu_2)_1 = 0$. This means that

$$\mathcal{N}_1 \subset \mathcal{N}_0. \tag{30}$$

Using (29) and (30) we can introduce the following operator B in the subspace $\mathcal{N}_0$. Namely set

$$\mathcal{D}(B) := \{\eta \in \mathcal{N}_1 | \eta = P_{\mathcal{N}_1}\varphi,\ \varphi \in \mathcal{N}_2 \subset \mathcal{D}(A)\}, \quad B\eta = A\varphi \in \mathcal{N}_0. \tag{31}$$

Due to (30) the domain $\mathcal{D}(B) \subset \mathcal{N}_0$ and due to $A\mathcal{N}_2 = \mathcal{N}_0$ the range $\mathcal{R}(B) = \mathcal{N}_0$. Thus

$$B : \mathcal{N}_0 \supset \mathcal{N}_1 \supset \mathcal{D}(B) \longrightarrow \mathcal{N}_0.$$

Let us show that the operator B is well defined. Let $P_{\mathcal{N}_1}\varphi = 0$ for some $\varphi \in \mathcal{N}_2$. This means that $\varphi \in \mathcal{M}_1$ and therefore $\varphi \in \mathcal{N}_2 \cap \mathcal{M}_1$. However from (29) it follows that

$$\mathcal{M}_1 \cap \mathcal{N}_2 = \{0\} \tag{32}$$

since $\mathcal{M}_1$ contains only such $\varphi \in \mathcal{H}_2 = \mathcal{M}_2 \oplus \mathcal{N}_2$, which simultaneously belongs to $\mathcal{M}_2$. Thus $P_{\mathcal{N}_1}\varphi = 0$, $\varphi \in \mathcal{N}_2$ imply $A\varphi = 0$.

We can describe the original operator A in terms of the operator B defined by (31) and the Friedrichs extension A_∞ of $\dot{A}$.

Lemma 1 ([4]). *The following equality holds:*

$$A^{-1} = A_\infty^{-1} + B^{-1}P_{\mathcal{N}_0}. \tag{33}$$

The equality (33) is a particular case of the Krein resolvent formula (see [1, 26, 9]). Let us recall that A is one of the self-adjoint extensions of the symmetric operator $\dot{A}_0$. Therefore for the resolvent of A the following representation is fulfilled: $(A - zI)^{-1} = (A_\infty - zI)^{-1} + B_z^{-1}P_{\mathcal{N}_z}$, where $P_{\mathcal{N}_z}$ is the orthogonal projection onto the subspace $\mathcal{N}_z = \text{Ker } (\dot{A}^* - \bar{z}I)$ and B_z provides the operator parameter fixing the extension A. The important fact is that $B_{z=0} = B$, where B is defined by (31).

Conversely from (33) we obtain the description of the Friedrichs extension A_∞:

$$\mathcal{D}(A_\infty) = \{\psi \in \mathcal{H}_1 \mid \psi = \varphi - P_{\mathcal{N}_1}\varphi, \varphi \in \mathcal{D}(A)\}, A_\infty\psi = A\varphi,$$

where we used that $P_{\mathcal{N}_1}\varphi = B^{-1}P_{\mathcal{N}_0}A\varphi$ for any $\varphi \in \mathcal{D}(A)$.

Now we want to present the relation (27) in slightly another form. With this aim let us introduce new notations. Let $\check{\mathbf{A}}$ and $\check{V}$ denote the restrictions of $\mathbf{A}$ and V onto the subspace $\mathcal{N}_1$. Then we have

$$\mathbf{A} = \check{\mathbf{A}}P_{\mathcal{N}_1} = P_{\mathcal{N}_{-1}}\mathbf{A}P_{\mathcal{N}_1} = \mathbf{A}P_{\mathcal{N}_1}, \quad V = \check{V}P_{\mathcal{N}_1} = P_{\mathcal{N}_{-1}}VP_{\mathcal{N}_1} = VP_{\mathcal{N}_1}, \tag{34}$$

where recall, $P_{\mathcal{N}_{-1}}$ denotes the orthogonal projection onto the subspace $\mathcal{N}_{-1}$ in $\mathcal{H}_{-1}$. Obviously operators $\check{\mathbf{A}}$ and $\check{V}$ are invertible since $\mathbf{A}$ is unitary and Ker $V = \mathcal{M}_1$. Noting that $V = \mathbf{A}\mathbf{A}^{-1}V = \mathbf{A}V_1$, where $V_1 = \mathbf{A}^{-1}V$ is self-adjoint in $\mathcal{H}_1$ we conclude that the range $\mathcal{R}(V) \subset \mathcal{N}_{-1} = \mathbf{A}\mathcal{N}_1$ since Ker $V = \mathcal{M}_1$. This implies that the operator $\mathbf{A}^{-1}V\mathbf{A}^{-1}$ is equal to zero on the subspace $\mathcal{M}_{-1} = \mathcal{H}_{-1} \ominus \mathcal{N}_{-1}$.

It is clear now that the restriction of $\mathbf{A}^{-1}V\mathbf{A}^{-1}$ onto $\mathcal{N}_{-1}$ have the form: $\check{\mathbf{A}}^{-1}\check{V}\check{\mathbf{A}}^{-1}$. This operator obviously is invertible and we can write:

$$(\check{\mathbf{A}}^{-1}\check{V}\check{\mathbf{A}}^{-1})^{-1} = \check{\mathbf{A}}\check{V}^{-1}\check{\mathbf{A}} : \mathcal{N}_1 \longrightarrow \mathcal{N}_{-1}.$$

Thus using above introduced notations we can rewrite (27) as follows:

$$\begin{aligned}
\tilde{\mathbf{A}}^{-1} - \mathbf{A}^{-1} &= -(\check{\mathbf{A}}\check{V}^{-1}\check{\mathbf{A}}(I + \mathbf{A}^{-1}V))^{-1}P_{\mathcal{N}_{-1}} \\
&= -(\check{\mathbf{A}}\check{V}^{-1}\check{\mathbf{A}}(I_{\mathcal{N}_1} + \check{\mathbf{A}}\check{V}))^{-1}P_{\mathcal{N}_{-1}} \\
&= -(\check{\mathbf{A}}(\check{V}^{-1}\check{\mathbf{A}} + I_{\mathcal{N}_1}))^{-1}P_{\mathcal{N}_1},
\end{aligned}$$

where $I_{\mathcal{N}_1}$ stands for the identity in $\mathcal{N}_1$. Finally we have

(35)
$$\tilde{\mathbf{A}}^{-1} - \mathbf{A}^{-1} = -(\check{V}^{-1}\check{\mathbf{A}} + I_{\mathcal{N}_1})^{-1}\check{\mathbf{A}}^{-1}P_{\mathcal{N}_{-1}}.$$

Proposition 3 *Let an operator $V \in \mathcal{H}_{-1}$-class satisfies the condition* (15). *Then the operator inverse to $\tilde{A} = A\tilde{+}V$ admits the following representation:*

(36)
$$\tilde{A}^{-1} = A^{-1} - (\check{V}^{-1}\check{\mathbf{A}} + I_{\mathcal{N}_1})^{-1}B^{-1}P_{\mathcal{N}_0},$$

where $\check{\mathbf{A}}$ and $\check{V}$ are defined by (34) *and act from $\mathcal{N}_1$ to $\mathcal{N}_{-1}$.*

Proof: At first we use Theorem 4, which ensures that $A\tilde{+}V = \tilde{A} \in \mathcal{A}_{s_1}(A)$. Further we assert that (36) follows from (35). Indeed, the restriction of the operator $\mathbf{A}^{-1}V\mathbf{A}^{-1}$ onto $\mathcal{H}$ has the following form in the subspace $\mathcal{N}_0$ (see also (28)):

(37)
$$\mathbf{A}^{-1}V\mathbf{A}^{-1} \restriction \mathcal{N}_0 = \check{\mathbf{A}}^{-1}\check{V}B^{-1},$$

where we used the equality $B^{-1} = P_{\mathcal{N}_1}A^{-1}P_{\mathcal{N}_0}$ that is valid due to (31). □

The operator $\check{\mathbf{A}}^{-1}\check{V}B^{-1}$ in (37) is invertible as it is a product of invertible operators. Therefore

(38)
$$(\check{\mathbf{A}}^{-1}\check{V}B^{-1})^{-1} = B\check{V}^{-1}\check{\mathbf{A}} = B\check{V}^{-1}\mathbf{A}P_{\mathcal{N}_1} : \mathcal{N}_0 \supset \mathcal{N}_1 \longrightarrow \mathcal{N}_0.$$

3 Connection Between V and $\tilde{B}$

Let a symmetric operator $V : \mathcal{H}_1 \to \mathcal{H}_{-1}$ belongs to the $\mathcal{H}_{-1}$-class and satisfies the condition (15). Then the generalized sum $\tilde{A} = A\tilde{+}V$ is the self-adjoint boundedly invertible operator in $\mathcal{H}$ (see (14)). Since the set $\mathcal{D}_0 = (\mathrm{Ker}\ V) \cap \mathcal{D}(A) \subset \mathcal{D}(\tilde{A})$ is dense in $\mathcal{H}$ (see (11)) we conclude that the operator $\tilde{A}$ is one of the self-adjoint extensions of the symmetric operator $\dot{A}_0 = A \restriction \mathcal{D}_0$. Hence $\tilde{A} \in \mathcal{A}_s(A)$. Further, we assert that due to (13) and since $\mathcal{D}_0$ is a proper subset in $\mathcal{H}_1$, the condition (6) is fulfilled with $\mathcal{D} = \mathcal{D}_0$. Therefore $A_\infty \neq A$ and thus $\tilde{A} \in \mathcal{A}_{s_1}(A)$. By construction the operator $\tilde{A}^{-1}$ is invertible and by Krein's resolvent formula the relation

(39)
$$\tilde{A}^{-1} = A^{-1} + \tilde{B}^{-1}P_{\mathcal{N}_0}$$

is fulfilled, where $P_{\mathcal{N}_0}$ is the orthogonal projection onto the deficiency subspace $\mathcal{N}_0 = \mathrm{Ker}\ (\dot{A}_0)^*$ of the operator $\dot{A}$. In (39) $\tilde{B}$ presents an extension parameter, i.e. $\tilde{B}$ is a self-adjoint operator in $\mathcal{N}_0$ fixing the extension $\tilde{A}$.

Now we want to establish the explicit connection between V and $\tilde{B}$.

Firstly we rewrite the result of Proposition 3 in a slightly another form. From the particular case of the generalized resolvent identity (27) (see (35) and (36)) it follows that

(40)
$$\begin{aligned} \tilde{B}^{-1}P_{\mathcal{N}_0} &= -(I + \mathbf{A}^{-1}V)^{-1}\mathbf{A}^{-1}VA^{-1}P_{\mathcal{N}_0} \\ &= -(\check{V}^{-1}\check{\mathbf{A}} + I_{\mathcal{N}_1})^{-1}B^{-1}P_{\mathcal{N}_0}. \end{aligned}$$

Due to

$$\mathbf{A}^{-1}VA^{-1}P_{\mathcal{N}_0}=\mathbf{A}^{-1}P_{\mathcal{N}_{-1}}VP_{\mathcal{N}_1}A^{-1}P_{\mathcal{N}_0}=P_{\mathcal{N}_1}\check{\mathbf{A}}^{-1}\check{V}B^{-1}P_{\mathcal{N}_0}$$

from (40) we get:

$$\tilde{B}^{-1}=-(I_{\mathcal{N}_1}+\check{\mathbf{A}}^{-1}\check{V})^{-1}\check{\mathbf{A}}^{-1}\check{V}B^{-1} \tag{41}$$

In (41) the operator $\check{\mathbf{A}}^{-1}\check{V}B^{-1}$ acts from $\mathcal{N}_0$ to $\mathcal{N}_1$ and $(I_{\mathcal{N}_1}+\check{\mathbf{A}}^{-1}\check{V})^{-1}$ is bounded in the subspace $\mathcal{N}_1\subset\mathcal{N}_0$ (see (30)). Thus the right side of (41) is the well defined operator in $\mathcal{N}_0$. Rewriting (41) in the form

$$\tilde{B}^{-1}=-((\check{\mathbf{A}}^{-1}\check{V}B^{-1})^{-1}(I_{\mathcal{N}_1}+\check{\mathbf{A}}^{-1}\check{V}))^{-1}=-(B(\check{V}^{-1}\check{\mathbf{A}}+I_{\mathcal{N}_1}))^{-1}, \tag{42}$$

and using (38) we finally obtain:

$$\tilde{B}=-B(\check{V}^{-1}\check{\mathbf{A}}+I_{\mathcal{N}_1}). \tag{43}$$

Thus looking at Proposition 3 we have proven

Theorem 6 *Let $V:\mathcal{H}_1\to\mathcal{H}_{-1}$ be an operator of the $\mathcal{H}_{-1}$-class. Assume V satisfies the condition* (15). *Then the generalized sum $\tilde{A}=A\tilde{+}V\in\mathcal{A}_{s_1}(A)$ assigns the boundedly invertible self-adjoint extension of the symmetric operator $\dot{A}_0=A\mid(\text{Ker }V\cap\mathcal{H}_2)$. Moreover the operator $\tilde{B}$ in the Krein resolvent formula* (39) *admits the representation* (43), *where B is defined by* (31).

Emphasize that $\tilde{A}$ is not necessarily bounded from below. Nevertheless $\tilde{B}^{-1}P_{\mathcal{N}_0}:=\tilde{A}^{-1}-A^{-1}$ is the well defined bounded operator. It is invertible after restriction onto $\mathcal{N}_0$. Indeed if $P_{\mathcal{N}_0}\eta\neq 0$, then $\tilde{B}^{-1}P_{\mathcal{N}_0}\eta\neq 0$ since $(\tilde{A}^{-1}-A^{-1})P_{\mathcal{N}_0}\eta=(\tilde{\mathbf{A}}^{-1}-\mathbf{A}^{-1})P_{\mathcal{N}_0}\eta\in\mathcal{N}_1\perp\text{Ker }V$.

Let us study the inverse problem. Given $\tilde{A}\in\mathcal{A}_{s_1}(A)$ we want to find an operator $V\in\mathcal{H}_{-1}$- class such that $\tilde{A}=A\tilde{+}V$.

Theorem 7 *Let $\tilde{A}\in\mathcal{A}_{s_1}(A)$. Assume $\mathcal{D}(\tilde{A})\subset\mathcal{H}_1$. Suppose both pairs $\tilde{A}$, A_∞ and A, A_∞ are relatively prime with respect to the symmetric operator $\dot{A}$ defined by* (5).

Then the operator $\tilde{A}$ admits the representation by the generalized sum:

$$\tilde{A}=A\tilde{+}V,$$

where $V:\mathcal{H}_1\to\mathcal{H}_{-1}$ is a uniquely defined symmetric operator of the $\mathcal{H}_{-1}$-class. Moreover V satisfies the condition (15) *and in turn admits the following explicit representation*

$$V=-\mathbf{A}(B^{-1}\tilde{B}+I_{\mathcal{N}_1})^{-1}P_{\mathcal{N}_1}, \tag{44}$$

where operators $\tilde{B}$, B are defined by $\tilde{B}^{-1}=(\tilde{A}^{-1}-A^{-1})P_{\mathcal{N}_0}$, $B^{-1}=(A^{-1}-A_\infty^{-1})P_{\mathcal{N}_0}$ (see also (31)) *and act from $\mathcal{N}_1\subset\mathcal{N}_0$ to $\mathcal{N}_0$.*

Proof: By the definition of $\mathcal{A}_{s_1}(A)$ the domain $\mathcal{D} := \{\varphi \in \mathcal{D}(A) \cap \mathcal{D}(\tilde{A}) \mid A\varphi = \tilde{A}\varphi\}$ is dense in $\mathcal{H}$. Therefore $\tilde{A}$ is the self-adjoint extension of the symmetric operator $\dot{A} = \tilde{A} \mid \mathcal{D} = A \mid \mathcal{D}$. So $\tilde{A}$ and A are relatively prime with respect to $\dot{A}$. Moreover both $\tilde{A}^{-1}$, A^{-1} exist and are bounded. Hence their difference defines the bounded self-adjoint operator $\tilde{B}^{-1}$ in the subspace $\mathcal{N}_0$:

$$\tilde{B}^{-1} P_{\mathcal{N}_0} = \tilde{A}^{-1} - A^{-1}.$$

Note that the difference $\tilde{A}^{-1} - A^{-1}$ equals zero on the subspace $\mathcal{M}_0 = \dot{A}\mathcal{D}$ in $\mathcal{H}$. By construction $\tilde{B}^{-1}$ is invertible since both $\tilde{A}^{-1}$, A^{-1} are self-adjoint and bounded. Besides Ker $((\tilde{A}^{-1} - A^{-1})P_{\mathcal{N}_0}) = 0$ since $\tilde{A}^{-1}$, A^{-1} are relatively prime.

Thus there exists the operator $\tilde{B}$ inverse to $\tilde{B}^{-1}$. Of course both $\tilde{B}^{-1}$ and $\tilde{B}$ are self-adjoint in $\mathcal{N}_0$ and the range $\mathcal{R}(\tilde{B}) = \mathcal{N}_0$.

Let us prove that the domain of $\tilde{B}$ belongs to the subspace $\mathcal{N}_1$ in $\mathcal{H}_1$,

$$\mathcal{D}(\tilde{B}) \subseteq \mathcal{N}_1, \tag{45}$$

where recall $\mathcal{N}_1 = (\mathcal{D}^{cl,1})^{\perp}$ in $\mathcal{H}_1$. Indeed, due to $\mathcal{D}(\tilde{A}) \subset \mathcal{H}_1$ we have $\mathcal{D}(\tilde{B}) \subset \mathcal{H}_1$. Therefore $\mathcal{D}(\tilde{B}) \subset \mathcal{H}_1 \cap \mathcal{N}_0$ since $\tilde{B}$ is an operator in $\mathcal{N}_0$. In fact $\mathcal{D}(\tilde{B}) \subset \mathcal{N}_1$ due to $(\mathcal{D}(\tilde{B}), \mathcal{M}_1)_1 = 0$, where $\mathcal{M}_1 = \mathcal{H}_1 \ominus \mathcal{N}_1$. Surely $(\mathcal{D}(\tilde{B}), \mathcal{M}_2)_1 = (\mathcal{D}(\tilde{B}), A^{-1}\mathcal{M}_2)_1 = (\mathcal{D}(\tilde{B}), \mathcal{M}_0) = 0$ since $\mathcal{D}(\tilde{B}) \subset \mathcal{N}_0 \perp \mathcal{M}_0$. Therefore $\mathcal{D}(\tilde{B}) \perp \mathcal{M}_1$ as $\mathcal{M}_2$ is dense in $\mathcal{M}_1$. This proves (45).

Let us introduce the operator

$$W := B^{-1}\tilde{B}.$$

Since $\mathcal{R}(B) = \mathcal{R}(\tilde{B}) = \mathcal{N}_0$ we conclude that the domain $\mathcal{D}(W) = \mathcal{D}(\tilde{B}) \subseteq \mathcal{N}_1$ and the range $\mathcal{R}(W) = \mathcal{D}(B) \subseteq \mathcal{N}_1$. So we have

$$W : \mathcal{N}_1 \supseteq \mathcal{D}(\tilde{B}) = \mathcal{D}(W) \longrightarrow \mathcal{R}(W) = \mathcal{D}(B) \subseteq \mathcal{N}_1. \tag{46}$$

Further since $\tilde{B}$ is self-adjoint in $\mathcal{N}_0$ the operator $W = B^{-1}\tilde{B}$ is symmetric in $\mathcal{N}_1$. Indeed for any $\eta_1, \eta_2 \in \mathcal{D}(\tilde{B})$ we can write

$$\begin{aligned}(B^{-1}\tilde{B}\eta_1, \eta_2)_1 &= (P_{\mathcal{N}_1} A^{-1}\tilde{B}\eta_1, \eta_2)_1 = (A^{-1}\tilde{B}\eta_1, \eta_2)_1 = (\tilde{B}\eta_1, \eta_2) \\ &= (\eta_1, \tilde{B}\eta_2) = \langle \mathbf{A}\eta_1, A^{-1}\tilde{B}\eta_2\rangle = (\eta_1, P_{\mathcal{N}_1} A^{-1}\tilde{B}\eta_2)_1 \\ &= (\eta_1, B^{-1}\tilde{B}\eta_2)_1.\end{aligned}$$

Remark that the inverse operator $W^{-1} = \tilde{B}^{-1}B$ exists due to $\mathcal{R}(B) = \mathcal{R}(\tilde{B}) = \mathcal{N}_0$.

Let us prove that the operator

$$W + I_{\mathcal{N}_1} = B^{-1}\tilde{B} + I_{\mathcal{N}_1}$$

also is invertible in $\mathcal{N}_1$. With this aim we will prove that

$$\mathrm{Ker}\,(B^{-1}\tilde{B}+I_{\mathcal{N}_1}) = \mathrm{Ker}\,(W+I_{\mathcal{N}_1}) = \{0\}.$$

Assume that $(W+I_{\mathcal{N}_1})\eta_1 = 0$ for some vector $0 \neq \eta_1 \in \mathcal{D}(\tilde{B}) = \mathcal{D}(W) \subset \mathcal{N}_1 \subset \mathcal{N}_0$. Then $W\eta_1 = -\eta_1$ and equivalently $\tilde{B}\eta_1 = -B\eta_1$, where we used that $W = B^{-1}\tilde{B}$ and $\mathcal{R}(W) = \mathcal{N}_1 = \mathcal{D}(B)$.

Set $\phi = \tilde{B}\eta_1 = -B\eta_1 \in \mathcal{N}_0$. Since $\mathcal{R}(\tilde{B}) = \mathcal{R}(B) = \mathcal{N}_0$ we have

$$\tilde{B}^{-1}\phi = \eta_1, \quad B^{-1}\phi = -\eta_1.$$

Using that $A^{-1} = A_\infty^{-1} + B^{-1}P_{\mathcal{N}_0}$ and $\tilde{A}^{-1} = A^{-1} + \tilde{B}^{-1}P_{\mathcal{N}_0}$ we observe:

$$\begin{aligned} \tilde{A}^{-1}\phi &= A^{-1}\phi + \tilde{B}^{-1}\phi = A_\infty^{-1}\phi + B^{-1}\phi + \tilde{B}^{-1}\phi \\ &= A_\infty^{-1}\phi - \eta_1 + \eta_1 = A_\infty^{-1}\phi. \end{aligned}$$

Thus for $0 \neq \phi \in \mathcal{N}_0$ we obtain,

$$A_\infty^{-1}\phi = \tilde{A}^{-1}\phi = \xi \in \mathcal{D}(A_\infty) \cap \mathcal{D}(\tilde{A}).$$

This gives the contradiction with the condition that A_∞ and $\tilde{A}$ are relatively prime with respect to $\dot{A}$. Indeed from the one hand side a vector ξ can not belong to the set $\mathcal{D}$ if $\phi \in \mathcal{N}_0$, (recall that $\mathcal{N}_0 = (\dot{A}\mathcal{D})^\perp$). From the other hand side if $0 \neq \xi \in \mathcal{D}(A_\infty) \cap \mathcal{D}(\tilde{A})$, $\xi \notin \mathcal{D}$ and $\tilde{A}\xi = A_\infty\xi$, then $\dot{A}$ is not a maximal common part for a pair A_∞, $\tilde{A}$ that is not true.

Thus for any $\eta_1 \in \mathcal{D}(\tilde{B}) = \mathcal{D}(W)$ we have $B^{-1}\tilde{B}\eta_1 \neq -\eta_1$. This means that $(W+I_{\mathcal{N}_1})\eta_1 = 0$ implies $\eta_1 = 0$. Therefore the operator $W + I_{\mathcal{N}_1}$ is invertible.

Consequently the operator

$$\check{V}_1 := -(B^{-1}\tilde{B}+I_{\mathcal{N}_1})^{-1} = -(W+I_{\mathcal{N}_1})^{-1}$$

is symmetric in $\mathcal{N}_1$. Moreover the operator

$$\begin{aligned} V &:= -\mathbf{A}\check{V}_1 P_{\mathcal{N}_1} = -\mathbf{A}(B^{-1}\tilde{B}+I_{\mathcal{N}_1})^{-1}P_{\mathcal{N}_1} \\ &\equiv -\mathbf{A}(W+I_{\mathcal{N}_1})^{-1}P_{\mathcal{N}_1} : \mathcal{H}_1 \longrightarrow \mathcal{H}_{-1} \end{aligned} \tag{47}$$

is symmetric in the sense of a dual pair of spaces, $V : \mathcal{H}_1 \to \mathcal{H}_{-1}$.

Evidently V belongs to the $\mathcal{H}_{-1}$-class since the set $\mathcal{D} = \mathrm{Ker}\, V \cap \mathcal{D}(A) \equiv \mathcal{M}_2$ is dense in $\mathcal{H}$ and (6) is fulfilled. Besides by construction the condition (15) is fulfilled too.

Therefore the operator V satisfies conditions of Theorem 4. Thus the generalized sum $\tilde{A}' = A\tilde{+}V$ assigns the self-adjoint operator in $\mathcal{H}$ with the range $\mathcal{R}(\tilde{A}') = \mathcal{H}$ and the domain $\mathcal{D}(\tilde{A}') \subset \mathcal{H}_1$. In other words the operator $\tilde{A}' \in \mathcal{A}_{s_1}(A)$. In fact $\tilde{A}'$ and $\tilde{A}$ coincide. The equality $\tilde{A}' = \tilde{A}$ we obtained on the following way. From (44) and (47) we see that in $\mathcal{N}_1$

$$\check{V}^{-1} = -(B^{-1}\tilde{B}+I_{\mathcal{N}_1})\check{\mathbf{A}}^{-1}. \tag{48}$$

Now after substitution (48) in (43) we obtain the desired identity. □

Corollary: *Each operator* $\tilde{A} \in \mathcal{A}_{s_1}(A)$ *with* $\mathcal{D}(\tilde{A}) \subset \mathcal{H}_1$*, under conditions that both pairs* $\tilde{A}$, A_∞ *and* A, A_∞ *are relatively prime with respect to the symmetric operator* $\dot{A}$*, admits two representations:*

$$\tilde{A}^{-1} = A^{-1} + \tilde{B}^{-1} P_{\mathcal{N}_0}$$

and

$$\tilde{A} = A\tilde{+}V.$$

Therefore formulas (43), (44), (48) *establish the one-to-one correspondence between operators* $\tilde{B}$ *and* V.

In other words formulae (43), (44), (48) establish the one-to-one correspondence between singularly perturbed operators $\tilde{A} \in \mathcal{A}_{s_1}(A)$ with $\mathcal{D}(\tilde{A}) \subset \mathcal{H}_1$ and singular perturbations $V \in \mathcal{H}_{-1}$-class under conditions of Theorem 3 and Theorem 4.

Finally we remark that an operator V defined by (47) will be bounded if the following condition is fulfilled:

$$P_{\mathcal{N}_1}\mathcal{D}(A) = \mathcal{N}_1,$$

where $\mathcal{N}_1 = (\mathcal{D}^{cl,1})^{\perp}$ in $\mathcal{H}_1$ with $\mathcal{D}$ defined by (4). Of course in this case $\mathcal{D}(B) = \mathcal{N}_1$ and $-1 \in \rho(V)$.

4 Concluding Remarks

Here we mention some papers, which influenced our investigation. First it was the paper [25] by M.G. Krein and V.A. Yavrian, where, in particular, we found the pre-conception of the generalized operator sum. To the point, in this paper a number of problems had been arisen, which are actual up to this time. Further, our considerations are based on publications [2]–[5, 10, 16, 17, 24], where different aspects of the singular perturbation theory are developed. This investigation was affected also by [11, 12, 27, 29] and by papers [13, 14].

Acknowledgment

The author is indebted to INTAS and DAAD for the support.

References

[1] N.I. Akhiezer and I.M. Glazman, *Theory of linear operators in Hilbert space*, Moscow, 1966.

[2] S. Albeverio, W. Karwowski and V. Koshmanenko, *Square Power of Singularly Perturbed Operators*, Math. Nachr. **173** (1995), 5–24.

[3] S. Albeverio and V. Koshmanenko, *Singular rank one perturbations of self-adjoint operators and Krein theory of self-adjoint extensions*, to appear in Potential Anal.

[4] S. Albeverio and V. Koshmanenko, *Form-sum approximations of singular perturbations of self-adjoint operators*, BiBoS Preprint Nr. 771/4/97, p. 23 (1997).

[5] S. Albeverio and V. Koshmanenko, *On the problem of the right Hamiltonian under singular form-sum perturbations*, SFB 237 Preprint, Nr. 375, Institute für Mathematik Ruhr-Universität-Bochum, p. 30 (1997), to appear in Rev. Math. Phys.

[6] S. Albeverio, V. Koshmanenko and K. Makarov, *Generalized eigenfunctions under singular perturbations*, Methods of Functional Analysis and Topology **5**, no. 1 (1999).

[7] Yu.M. Berezanskij, *The bilinear forms and Hilbert equipment*, In Spectral analysis of differential operators, Institute of Mathematics, Kiev (1980), 83–106.

[8] Yu.M. Berezanskij, G.F. Us and Z.G. Sheftel, *Functional Analysis*, Higher School, Kiev, 1990.

[9] M.Sh. Birman, *To self-adjoint extensions theory of positive definite operators*, Math. Sbornik **38** (80), no. 4 (1956), 431–450 .

[10] J.F. Brasche, V.D. Koshmanenko and H. Neidhardt, *New aspects of Krein's extension theory*, Ukrainian Math. J. **46** (1994), 37–54.

[11] V.A. Derkach and M.M. Malamud, *Generalized resolvent and the boundary value problems for Hermitian operators with gap*, J. Funct. Anal. **95** (1991), 1–95.

[12] F. Gesztesy and B. Simon, *Rank-One Perturbations at Infinite Coupling*, J. Funct. Anal. **128** (1995), 245–252 .

[13] F. Gesztesy, K.A. Makarov and E. Tsekanovskii, *An Addendum to Krein's Formula*, J. Math. Anal. Appl. **222** (1998), 594–606.

[14] S. Hassi and H. de Snoo, *On rank one perturbations of selfadjoint operators*, Integral equations and operator theory **29** (1997), 288–300.

[15] T. Kato, *Perturbation Theory for Linear Operators*, Springer-Verlag, Berlin-Heidelberg-New York, 1980.

[16] W. Karwowski and V.D. Koshmanenko, *On definition of the singular bilinear forms and singular linear operators*, Ukrainian Math. J. **45**, no. 8 (1993), 1084–1089.

[17] W. Karwowski, V. Koshmanenko and S. Ôta, *Schrödinger operator perturbed by operators related to null-sets*, Positivity **2**, no. 1 (1998), 77–99.

[18] T.V. Karataeva and V.D. Koshmanenko, *Generalized sum for operators* (sent to Math. Notes).

[19] V.D. Koshmanenko, *Singular bilinear forms in perturbations theory of self-adjoint operators*, Naukova Dumka, Kiev 1993, English translation: Volodymyr Koshmanenko, Singular quadratic forms in Perturbation Theory, Kluwer Academic Publishers, 1999.

[20] V.D. Koshmanenko, *Towards the rank-one singular perturbations of self-adjoint operators*, Ukrainian Math. J. **43**, no. 11 (1991), 1559–1566.

[21] V.D. Koshmanenko, *Perturbations of self-adjoint operators by singular bilinear forms*, Ukrainian Math. J. **41** (1989), 1–14.

[22] V.D. Koshmanenko, *Singularly Perturbed Operators, Operator Theory*. Advances and Applications. **70** (1994), 347–351.

[23] V.D. Koshmanenko, *Regularized approximations of singular perturbations of the $\mathcal{H}_{-2}$-class* (1999), sent to Ukrainian Math. J.

[24] V. Koshmanenko and O. Samoilenko, *Singular perturbations of finite rank*. Point spectrum, Ukrainian Math. J. **49** (1997), 1186–1212.

[25] M.G. Krein and V.A. Yavrian, *Spectral shift functions arising in perturbations of a positive operator*, J. Operator Theory **6** (1981), 155–191.
[26] M.G. Krein, *Theory of self-adjoint extensions semibounded Hermitian operators and its applications I*, Rec. Math. (Math. Sb.) **20**, no. 3 (1947), 431–495.
[27] G. Nenciu, *Removing cut-offs form singular perturbations: an abstract result*, Lett. Math. Phys. **7** (1983), 301–306.
[28] L.P. Nizhnik, *On point interaction in Quantum Mechanics*, Ukrainian Math. J. **49** (1997), 1557–1560.
[29] H. Neidhardt and V. Zagrebnov, *On the right Hamiltonian for singular perturbations: general theory*, Rev. Math. Phys. **9**, no. 5 (1997), 609–633.
[30] M. Reed and B. Simon, *Methods of Modern Mathematical Physics. II.* Fourier Analysis, Self-adjointness, Academic Press, New York-San Francisco-London, 1975.

Institute of Mathematics
252601, GSP, Kyiv 4
vul. Tereshchenkivska, 3
Kyiv
Ukraine
kosh@imath.kiev.ua

AMS Subject Classification 47A10, 47A55

Operator Theory:
Advances and Applications, Vol. 118

Few-body Krein's Formula

Pavel Kurasov and Boris Pavlov

Selfadjoint extensions of symmetric operators with infinite deficiency indices are discussed. In particular the operators describing the system of several quantum particles with generalized point interactions are investigated in detail and a few-body analog of Krein's formula for generalized resolvents is proven. New conditions for semiboundedness of M-body quantum Hamiltonian with generalized point interactions in the three-dimensional space are derived.

1 Introduction

Possibly G.R. Kirchhoff [18] was the first mathematical physicist who noticed that for electrostatic problems in composite domains consisting of two standard parts joined by several small openings the *ansatz* for the solution of the corresponding partial differential equation may be obtained in the form of linear combination of solutions of nonperturbed problems in standard domains and singular solutions- in fact Green functions - attached to the openings. At that moment this important observation was considered just as a technical device for constructing of approximate solutions ("Kirchhoff method") and the general role of it has not been noticed yet. In 1933 Fermi [14] made another important step. He suggested a solvable model for *point interaction* between neutrons and nuclei. But only few decades later the idea of Fermi was non-formally decoded by mathematicians [10] and the operator extension theory was recognized as a general and powerful tool for constructing solvable models with *zero-range potentials* for quantum systems [11]. Now we have an impressive number of solvable models of one-body quantum and acoustic systems constructed from very simple standard elements, such as matrices and differential equations with constant coefficients in simple domains [5, 6, 33, 35, 36, 37]. In general operator extension procedure the role of the Kirchhoff's Green functions in the corresponding ansatz is played by the elements of the proper the deficiency subspaces.

Generally the operator extension approach provides a large class of solvable models of wave processes which can be described roughly by the following condition: "the typical wave lengths of the process exceed the size of characteristic details of the perturbation considered", for instance, the width of the opening in acoustic problems. In quantum mechanics this geometric condition is replaced by the *low-energy* or *slow dephasing conditions*. In particular this approach gives a rigorous substitution for so-called "energy-dependent potentials" in low-energy nuclear physics [22], and for scattering problems appearing in design of nanoelectronic devices and networks similar to ones considered in [7].

The mathematical background for all mentioned *one-* and *two-body* models is given by Krein's formula for generalized resolvents of symmetric operators with finite deficiency indices. The corresponding few-body problems, see [22, 25, 34], require considering operators with *infinite deficiency indices and special algebraic structure* involving existence of several tensor decompositions - *splittings* of the *underlying* Hilbert space and underlying *nonperturbed* Hamiltonian, and a special construction of the system of corresponding integral equations in terms of *Faddeev components*. These problem are studied in detail in the recent book [6], where the most general approach to few-body problems with singular finite rank interactions is developed.

In actual paper we show that the abstract form of this construction is equivalent to some analog of Krein's formula for operators with infinite deficiency indices and several disjoint deficiency subspaces. It appears that the structure of the family of deficiency subspaces corresponds to the physical structure of *channels* which is encoded in corresponding *splittings* of the configuration space (see e.g. [8]). The deficiency elements are distributed along the channel's *interaction planes or cylinders* with densities which play the role of *Faddeev components*.

The mentioned characteristics of geometric and algebraic structures of the family of deficiency elements appeared in [6, 22, 33] as a fiber structure[1] of the deficiency subspace attached to the interaction cylinder. The relevant Krein's formula was derived in [33] as *fiber Krein's formula*. In fact even for simplest n-body problems $n > 2$ the base of the corresponding fiber space - the joining of all relevant interaction cylinders - *is not a smooth variety*, as in [6, 33], but has some conic singularities at the intersections of them, that is at the corresponding *equidomoid* (see [22]). In the simplest case of three particles with point interactions in R^3 the singular manifold contains only one point - "the triple collision point",- and the apriory estimates of operators appearing in corresponding Faddeev equations require usage of Hardy inequality [34]. In general case the singular variety may be much more sophisticated, it may contain edge points or may be unbounded, which actually creates serious obstacles in M-body scattering problems even for smooth potentials with $M \geq 4$ see [30]. In the third section of the current paper we consider few-body Krein's formula for an "elementary brick" of the whole construction which is an abstract analog of the interchange-reaction between two clusters. The semiboundedness of the corresponding few-body hamiltonian is discussed. In particular, if the corresponding deficiency subspaces form a positive angle modulo some finite dimensional subspace, hence have zero or finite dimensional intersection $\dim \mathcal{N}_1 \cap \mathcal{N}_2 < \infty$, then the semiboundedness of *every* selfadjoint extension follows from the invertibility of the corresponding Krein Q-functions for large negative values of the spectral parameter, provided the original nonperturbed operators are semibounded. But if the angle between the deficiency subspaces is equal to zero (modulo any finite dimensional subspace) or intersection of the deficiency

[1] Representing the deficiency elements as direct products, see (9) we reveal this structure in standard sense.

subspaces is infinite dimensional then the invertibility of the corresponding Krein Q-functions for large negative values of the spectral parameter implies the semiboundedness of special *Friedrichs extension* of the Hermitian operator with deficiency subspace formed of regular elements.

In the last section of the paper we consider, as an example of our approach, the Hamiltonian of the system of M distinguishable quantum particles with the generalized pairwise point interactions in Hilbert space $L_2(R^{3M-3})$. New conditions for the semiboundedness of the Hamiltonian constructed via operator extension technique are proven.

The work of the first author was partially supported by The Swedish Royal Academy of Sciences and NFR. The second author proudly recognizes partial support from Marsden Fund of the Royal Society of New Zealand, from RFFI grant of the Russian Academy of Sciences, and a support from the International Solvay Institute, Brussels.

2 Two-body Krein's Formula in Rigged Spaces

In our paper we use the approach to Operator Extensions based on analysis of the corresponding hermitian symplectic forms which is standard for differential and recently described for difference equations, see for instance [31]. This approach was developed first for common extensions of abstract and differential operators in [32], see also [6, 33] and the bibliography there. Professor S. Novikov informed the second author recently that I.M. Gelfand always underlined the importance of analysis of symplectic hermitian boundary forms in operator theory and in mathematical physics.

In what follows we use the following statement, see [6, 32] which in particular contains important Krein's formula, see [19, 20]. Let a be a selfadjoint operator[2] with the domain $\mathrm{Dom}(a)$ and acting in the Hilbert space h. Consider any finite dimensional subspace $n = n_i$ such that $\mathrm{Dom}(a) \cap n = \{0\}$. Then the restriction a_0 of the operator a to the domain $\mathrm{Dom}(a_0) = \frac{1}{a-il} h \ominus n$ is a symmetric densely defined operator. For classical description of all generalized resolvents of the operator a_0 with the finite dimensional deficiency subspaces $n_i \equiv n,\ n_{-i} = \frac{a+il}{a-il} n,\ D_0 \cap n = \{0\}$ see [1].

Theorem 2.1 *The domain* $\mathrm{Dom}(a_0^*)$ *of the adjoint operator* a_0^* *may be described using the boundary values* $\xi_\pm^u \subset n$ *as follows*

$$\mathrm{Dom}(a_0^*) = \left\{ u = u_0 + \frac{a}{a-il}\xi_+^u + \frac{1}{a-il}\xi_-^u \right\}, \tag{1}$$

[2]We use here the nonstandard notations - a small character for operator - just to be in line with our further notations involving the splittings of the total hamiltonian into channels.

where $u_0 \in \frac{1}{a-il} h \ominus n \equiv \mathrm{Dom}(a_0)$. *The action of the adjoint operator is given by the following formula*

$$(2) \qquad a^* \left(u_0 + \frac{a}{a-il}\xi_+^u + \frac{1}{a-il}\xi_-^u \right) = a u_0 - \frac{1}{a-il}\xi_+^u + \frac{a}{a-il}\xi_-^u .$$

The symplectic hermitian boundary form of the adjoint operator on elements $u, v \in \mathrm{Dom}(a_0^*)$ *is given by*

$$\mathcal{J}(u,v) \equiv \langle a_0^* u, v\rangle - \langle u, a_0^* v\rangle = \langle \xi_-^u, \xi_+^v\rangle - \langle \xi_+^u, \xi_-^v\rangle .$$

The selfadjoint extensions of the symmetric operator a_0 *are determined uniquely by the Lagrangian planes of the simplectic boundary form. Some of these planes can be determined by the boundary conditions*

$$\xi_- = \gamma \xi_+ ,$$

with any fixed selfadjoint boundary operator $\gamma : n \to n$ [3]. *The selfadjoint extensions of* a_0 *determined by the latter boundary conditions will be denoted by* a_γ *in what follows. All selfadjoint extensions of* a_0 *are defined by these Lagrangian planes and all their J-unitary transformations.*

The resolvent of any selfadjoint extension a_γ *is described by Krein's formula:*

$$(3) \qquad \frac{1}{a_\gamma - \lambda I} = \frac{1}{a-\lambda I} + \frac{a+il}{a-\lambda I} P_n \frac{I}{\gamma - P_n \frac{1+\lambda a}{a-\lambda I} P_n} P_n \frac{a-il}{a-\lambda I}$$

Proof of this "one-body version" can be found, for instance, in [6, 32]. The case of non-densely defined symmetric operators is considered in [23, 24].

Notice, that the elements $\xi_\pm^u$ play the role of boundary values for the elements from the domain of the adjoint operator, see [15]. Generally when $n \cap \mathrm{Dom}(a_0) \neq \{0\}$ Krein's formula (3) can determine resolvents of operator relations, see for instance [12, 16, 17]. But in the case when the operator a can be represented as an orthogonal sum of two operators $a = a^0 \oplus a^1$ acting in a certain orthogonal decomposition of the Hilbert space $h = h^0 \oplus h^1$ and the deficiency subspace possesses a similar decomposition $n = n^1 \oplus n^2$, $n^s \in h^s$ this formula describes the interaction between two *channels* $[h^0, a^0]$ and $[h^1, a^1]$ via the *nonadditive* perturbation γ.

If the operator a_0 is densely defined and $n_0 \cap \mathrm{Dom}(a^0) = \{0\}$, $n_1 \cap \mathrm{Dom}(a^1) \neq \{0\}$, the Krein's formula still may give a resolvent of some operator relation, see [12], but the restriction of it onto the minimal reducing subspace of the total space $h = h^0 \oplus h^1$ containing $\mathrm{Dom}(a_0^0)$ *is an operator*, see [23, 24].

[3] Of course one can consider also infinite dimensional deficiency subspaces, see below, and even sometimes unbounded operators γ. We restrict ourselves at the moment to the case of finite dimensional deficiency subspaces.

Krein's formula may be generalized in different ways. The most obvious but still useful *few-channel* generalization of it can be obtained assuming that the operator a, the corresponding Hilbert space h and the deficiency subspaces are given in the form of orthogonal decompositions:

$$h = \sum_s \oplus h_s, \quad a = \sum_s \oplus a_s, \quad n = \sum_s \oplus n_s,$$

and the boundary operator γ is defined by some operator matrix γ_{st} corresponding to the decomposition of the deficiency subspace. The calculation of components of the generalized resolvent requires elimination of some variables and solution of the matrix equations similar to ones considered in [9].

More subtle generalization of Krein's formula has been suggested in [33] as an intermediate step for the construction of a solvable model of the three-body scattering. We derive here a modified version of this *fiber Krein's formula*, see Theorem 2.3, in terms of rigged Hilbert spaces. *Using the ideas of rigged spaces one can reduce the construction of the extension a_γ to some abstract version of potential theory.*

Consider a dense linear variety $\mathrm{Dom}(a) \equiv h_a \in h$ - the domain of a positive selfadjoint operator a - and the space h^a of all bounded linear functionals defined on it:

$$h_a \subset h \subset h^a.$$

The spaces h_a, h, h^a form *Gelfand triplet* or *rigged Hilbert space*, associated with the operator a. In our case h_a and h^a are Hilbert spaces with the following dot products

$$\begin{aligned} [u, v]_a &= \langle (a + iI)u, (a + iI)v \rangle, \\ [\rho, \tau]^a &= \langle (a + iI)^{-1}\rho, (a + iI)^{-1}\tau \rangle \end{aligned}$$

for h_a and h^a respectively. In what follows we use notation $\langle , \rangle$ to denote both the dot-product in the original Hilbert space and the action of functionals. In particular we consider below linear functionals ρ which are bounded on $\mathrm{Dom}(a) \equiv h_a$ and presented as $\langle \frac{1}{a-iI} f, \rho \rangle$. For instance for a given $u = u_0 + \frac{a}{a-iI}\xi_+^u + \frac{1}{a-iI}\xi_-^u \in D_{a_0}$, we define the generalized boundary density $\rho^u \in h^a$ by the formula

$$\rho^u = (a + iI)\xi_+^u \ \in (a + iI)n_i \equiv n^i \in h^a,$$

and the *regularized boundary value* of u in h_a as follows

$$u_- \equiv \frac{1}{a - iI}\xi_-^u \ \in \frac{I}{a - iI} n_i \in h_a = \mathrm{Dom}(a),$$

then the boundary form $\mathcal{J}(u, v)$ may be rewritten as

$$\mathcal{J}(u, v) = \langle u_-, \rho^v \rangle - \langle \rho^u, v_- \rangle, \tag{4}$$

where the brackets $\langle \cdot, \cdot \rangle$ denote the action of the functionals on the functions. Note that any element $u \in \mathrm{Dom}(a)$ possesses the following representation

$$u \equiv \frac{1}{a - iI} f = \frac{1}{a - iI}(I - P_n) f + \frac{I}{a - iI} P_n f \equiv u_0 + u_-,$$

where $u_0 \in \mathrm{Dom}(a_0)$ and $u_- \in \frac{I}{a-iI} n_i$ is the regularized boundary value of u. The generalized density for any element $u \in D_0$ is trivial, $\rho^u = 0$. We call ρ^u, u_- the *rigged boundary values* of the element $u \in \mathrm{Dom}(a_0^*)$. It is important that the densities of deficiency elements $\nu_{\bar{\lambda}}$ for different values of the spectral parameter λ

$$\nu_{\bar{\lambda}} = \frac{a + iI}{a - \lambda I} \nu, \quad \nu \in n \equiv n_i,$$

may be chosen from the standard subspace $n^i \in h^a$ which does not depend on the spectral parameter:

$$\nu_{\bar{\lambda}} = \frac{1}{a - \lambda I} \rho, \quad \rho \in (a + iI) n_i = n^i \in h^a,$$

similarly to the densities of potentials which appear in potential theory. But the regularized values of these deficiency elements do depend on λ so that all spectral information on the constructed extension may be derived from this dependence via Krein's formula, see for instance [6, 33, 34].

The next statement is just a reformulation of the preceding Theorem 2.1 in terms of generalized densities and boundary values:

Theorem 2.2 *The boundary conditions defining some Lagrangian plane of the symplectic boundary form in terms of rigged variables ρ^u, u_- are parametrized by the corresponding boundary operator $\Gamma : h^a \to h_a$; acting in the relevant Gelfand triplet $h_a \subset h \subset h^a$ as follows*

$$u_- = \Gamma \rho^u.$$

This operator is connected to the operator $\gamma : n \to n$ parametrizing the boundary conditions in the previous Theorem 2.1 by the formula:

$$\Gamma = \frac{1}{a - iI} \gamma \frac{1}{a + iI}.$$

In terms of the rigged variables ρ^u, u_- the calculation of the resolvent of the extension $a^\Gamma \equiv a_\gamma$ is reduced to the solution of the equation for the density ρ^u of the element $u = \frac{1}{a^\Gamma - \lambda} f$:

$$\left[\Gamma - \frac{1}{a - iI} P_n \frac{1 + \lambda a}{a - \lambda I} P_n \frac{1}{a + iI} \right] \rho = \frac{1}{a - iI} P_n \frac{a - iI}{a - \lambda I} f,$$

which gives the Krein's formula in the rigged spaces

$$\frac{1}{a^\Gamma - \lambda} f = \frac{1}{a - \lambda} f + \frac{I}{a - \lambda} \rho.$$

Proof: Changing notations in formulas (1) and (2) we get

$$u = u_0 + u_- + \frac{a}{a^2+I}\rho^u \tag{5}$$

$$u_0 + u_- = \frac{1}{a-\lambda I}f - \left[\frac{1}{a-\lambda} - \frac{a}{a^2+1}\right]\rho^u. \tag{6}$$

Then using the boundary operator Γ acting in the rigged spaces together with $P_n(a-iI)u_0 = 0$ and (2) implies

$$\left[\Gamma - \frac{1}{a-iI}P_n\frac{1+\lambda a}{a-\lambda I}P_n\frac{1}{a+iI}\right]\rho = \frac{1}{a+iI}P_n\frac{a+iI}{a-\lambda I}f, \tag{7}$$

and then (3) gives the result announced. □

Note, that the reduction of the calculation of the resolvent of the operator $a_\gamma \equiv a^\Gamma$ to the solution of (6) results in effective lowering of the dimension of the space of the solutions due to the presence of the projection P_n onto the deficiency subspaces if $\dim n < \infty$. If for the operator a_0 we have $\dim n = \infty$, then the solution of (7) may be reduced to Riesz-Schauder problem with a compact operator, if some additional conditions are satisfied. In what follows we show, that equation (7) in the "few-body" situation plays the role of the system of Faddeev equations and the densities play the role of Faddeev components. But the consideration of the few-body situation requires a special geometrical structure of the "nonperturbed" operator a and relevant Hilbert space, see below.

Let us consider the decomposition - *splitting* - of the Hilbert space $\mathcal{H}$ which is *compatible* with the corresponding *splitting* of a certain selfadjoint operator $\mathcal{A}$:

$$\begin{aligned} \mathcal{H} &= H \times h \\ \mathcal{A} &= A \times I_h + I_H \times a \equiv A + a. \end{aligned}$$

We assume that A and a are positive selfadjoint operators acting in the Hilbert spaces H and h respectively. Then the operator $\mathcal{A}$ given by (8) can be defined on the algebraic tensor product $\mathrm{Dom}(A) \times \mathrm{Dom}(a)$ of the domains of the operators A and a. The operator defined this way is in fact essentially selfadjoint. In what follows we do not distinguish by the notations the symmetric operator $\mathcal{A}$ defined on the algebraic tensor product and the selfadjoint closure of it.

We chose a certain finite dimensional subspace $n \in h$, $n \cap \mathrm{Dom}(a) = \{0\}$ and consider the restriction $a \to a_0$ defined on $\mathrm{Dom}(a_0) \equiv \frac{1}{a-iI}h \ominus n$. Obviously this restriction is a densely defined symmetric operator with the deficiency subspace $n_i = n$. The dual deficiency subspace n_{-i} of it coincides with $\frac{a+iI}{a-iI}n$. The adjoint operator a^* is densely defined on $\mathrm{Dom}(a_0^*) = \mathrm{Dom}(a_0) + n_i + n_{-i}$ and the subspaces n_i, n_{-i} are the eigenspaces of a^* with the eigenvalues $\mp i$ respectively. The subspace

$$\frac{a+iI}{a-\bar{\mu}I}n \equiv n_\mu$$

is the deficiency subspace which corresponds to the value μ of the spectral parameter.

Consider the restriction $\mathcal{A}_0$ of $\mathcal{A}$ defined by the formula

$$\mathcal{A}_0 = A \times I_h + I_H \times a_0 \equiv A + a_0.$$

Obviously it is a symmetric operator with infinite deficiency indices if the Hilbert space H is infinite dimensional.

Lemma 2.1 *The operator $\mathcal{A}_0$ has the deficiency subspace $\mathcal{N}_i \equiv \mathcal{N}$:*

$$\mathcal{N} = \overline{\left\{ \frac{1}{\mathcal{A} + iI}(a + iI_h)\ \nu \times e \right\}}, \tag{8}$$

where $\nu \in n$, $e \in H$

Proof may be obtained by straightforward verification of the formulae

$$\langle \mathcal{N}, (\mathcal{A} - iI)\mathrm{Dom}(\mathcal{A}_0)\rangle = 0, \ \ \mathcal{H} \ominus (\mathcal{A} - iI)\mathrm{Dom}(\mathcal{A}_0) = \mathcal{N}.$$

Remark: The deficiency subspace N is equal to the closure of the set of all deficiency elements described by smooth densities e. Such a deficiency elements will be called regular. Using regular elements one can define a symmetric extension of the operator $\mathcal{A}_0$, but not always a selfadjoint operator. In what follows we are going to use the set of all regular deficiency elements assuming that extensions involving these elements are essentially selfadjoint or semibounded. The corresponding selfadjoint operators can be obtained via the closure or via the Friedrichs procedure.

One may check in a similar way that for any complex value λ of the spectral parameter we have also

$$\mathcal{N}_{\bar{\lambda}} = \overline{\frac{a + iI}{\mathcal{A} - \lambda I} n \times H}. \tag{9}$$

In what follows we describe the selfadjoint extensions of the splitted operator

$$\mathcal{A}_0 = A \times I_h + I_H \times a_0 \equiv A + a_0,$$

provided the deficiency subspaces n_i of the operator a_0 are finite dimensional. In fact we shall solve slightly more general problem concerning extensions of the operator

$$A \times I_{h \oplus E} + I_H \times [a_0 \oplus b_0] = [A + a_0] \oplus [A + b_0]$$

acting in the tensor product $\overline{H \times [h \oplus E]} = \overline{[H \times h] \oplus [H \times E]}$. The program of construction of selfadjoint extensions of splitted and reduced Laplaceans $A + a$ via some boundary conditions

$$\begin{aligned} \mathcal{A} &= A + a \longrightarrow \mathcal{A}_0 = A + a_0, \text{ in } \mathcal{H} = H \times h, \\ \mathcal{B} &= A + b \longrightarrow \mathcal{B}_0 = A + b_0, \text{ in } H \times E \end{aligned} \tag{10}$$

was realized first in [10] for trivial inner space $E = 0$ and then in [34] for zero-range potentials with nontrivial inner structure defined by $b,\ b \neq 0$. In [34] the following straightforward regularization of the boundary values of outer component of deficiency elements was used

$$\frac{1}{\mathcal{A}-\lambda I}\rho \longrightarrow \frac{1}{\mathcal{A}-\lambda I}\rho - \frac{\mathcal{A}}{\mathcal{A}^2+1}\rho, \quad \rho \in (\mathcal{A}+iI)\mathcal{N}.$$

In [34] and the semiboundedness of the total Hamiltonian constructed using operator extension technique based on this regularization has been proven.

Later investigating the solvable model for three-body scattering with "inner structure" K. Makarov noticed [26], that this straightforward regularization of the boundary values, being *translation invariant*, still is not compatible with separation of variables of the third particle on the interaction cylinder (interaction plane) for the given pair. He suggested another *pairwise* regularization of the boundary values for elements from the domain of the adjoint operator.

In the current paper we use similar regularization for general M-body problem both for inner and outer components of the boundary values and characterize it as *compatible with the splitting* or *pairwise* regularization. For instance, if just one inner channel is present and the corresponding channel Hamiltonian is defined by the operator $\mathcal{B} = A + b$ then the regularization is defined as

$$\begin{aligned} &* \frac{I}{\mathcal{A}\oplus\mathcal{B}-\lambda I}\rho \longrightarrow \frac{I}{\mathcal{A}\oplus\mathcal{B}-\lambda I}\rho - \frac{a\oplus b}{[a\oplus b]^2+I}\rho, \\ &\rho \in (a\oplus b + iI)\mathcal{N} \subset h^{a\oplus b} \times H. \end{aligned} \tag{11}$$

*, see below, Lemma 2.2. Basing on this regularization we derive in the remaining part of this section the simplest generalizations of Krein formula: the *fiber Krein formula*, see below Theorem 2.3. Then in the following section we derive the simplest two-channels Krein formula in terms of rigged boundary values which serves as an elementary block for construction of M-body Krein formula - Faddeev equations- for generalized point interactions in the last Section 4.

Note first of all that the deficiency elements of the operator $\mathcal{A}_0 \oplus \mathcal{B}_0$ in the tensor decomposition of the Hilbert Space $\mathcal{H} \oplus [E \times H] = [h \oplus E] \times H$ are represented in terms of vector-valued densities (ρ_ν, ρ_τ) in the relevant Gelfand triplet $h_{a\oplus b} \subset h \subset h^{a\oplus b}$,

$$\begin{aligned} \nu_{\bar\lambda}^{\mathcal{A}} &= \frac{a+iI}{\mathcal{A}-\lambda I}(\nu \times e) = \frac{1}{\mathcal{A}-\lambda I}(\rho_\nu \times e), \ e \in H. \\ \nu_{\bar\lambda}^{\mathcal{A}} &= \frac{b+iI}{\mathcal{A}-\lambda I}(\tau \times e) = \frac{1}{\mathcal{A}-\lambda I}(\rho_\tau \times e), \ e \in H. \end{aligned}$$

We denote the generalized vector-valued densities $\rho \in h^{(a\oplus b)} \times H$ by the same symbol and call them *two-body densities*. The dual regularized boundary values will be denoted by u_-. We denote by $P \equiv P_n$ the orthogonal projection onto

$n \times H$ in $\mathcal{H}$. The operator P is connected to the orthogonal projection onto n in h by the obvious formula:

$$P \equiv P_n = P_n^h \times I_H.$$

Similarly for any selfadjoint operator a : $\mathrm{Dom}(a) \to h$ we denote the operator $a \times I_H$ in the product space $h \times H$ just by a, having in mind that the precise meaning of the notation is clear from the context.

The following statement gives the "two-body"[4] characterization of densities and regularized boundary values for some deficiency elements. Just for the sake of brevity we formulate this statement for deficiency elements $\nu_{\bar{\lambda}}^{\mathcal{A}}$ of $\mathcal{A}$. Obviously similar statement is true also for deficiency elements for the orthogonal sum $\mathcal{A} + a \oplus b = [a+A] \oplus [b+A]$ defined by proper vector-valued densities and regularized boundary values of them.

Lemma 2.2 *The following decomposition holds for any deficiency element $n_{\bar{\lambda}}^{\mathcal{A}} = \frac{1}{\mathcal{A}-\lambda I}\rho_\nu$ of the operator $\mathcal{A}_0$:*

$$\frac{1}{\mathcal{A}-\lambda I}\rho_\nu = \frac{a}{a^2+I}\rho_\nu + \frac{1}{a-iI}P_n \frac{I+a(\lambda - A)}{\mathcal{A}-\lambda} \frac{1}{a+iI}\rho_\nu + u_0,$$

where $u_0 \in \mathrm{Dom}(a_0) \times H$.

Proof: Notice, that for any deficiency element of $\mathcal{A}$ the following decomposition holds in proper Gelfand triplet:

$$\frac{1}{\mathcal{A}-\lambda I}\rho_\nu = \frac{a}{a^2+1}\rho_\nu + \frac{I}{a-iI}\frac{I+a(\lambda - A)}{\mathcal{A}-\lambda I}\frac{I}{a+iI}\rho_\nu.$$

inserting the projection operator P we transform the latter expression as follows

$$\begin{aligned}\frac{1}{\mathcal{A}-\lambda I}\rho_\nu &= \frac{a}{a^2+1}\rho_\nu + \frac{I}{a-iI}P\frac{I+a(\lambda - A)}{\mathcal{A}-\lambda I}P\frac{I}{a+iI}\rho_\nu \\ &\quad + \frac{I}{a-iI}[I-P]\frac{I+a(\lambda - A)}{\mathcal{A}-\lambda I}P\frac{I}{a+iI}\rho_\nu.\end{aligned}$$

Note, that the second projection in front of $\frac{I}{a+iI}\rho_\nu$ does not play any role since $\frac{I}{a+iI}\rho_\nu \in \mathcal{N}$. On the other hand $\frac{I}{a-iI}[I-P]\frac{I+a(\lambda-A)}{\mathcal{A}-\lambda I}P\frac{I}{a+iI}\rho_\nu \in \mathrm{Dom}(a)$. Hence

[4] In fact we discuss here the separation of the whole system into two subsystems, one of them is a cluster and another one is a subsystem of free particles (or clusters). But each cluster is in fact a "particle" with internal degrees of freedom. Hence the example "cluster + free particles" may be considered as a representative example for few-body complex. This is our motivation for using the term "two-body" for general situation.

the density and the regularized value of the deficiency element are defined by

$$\rho_\nu = \rho_\nu,$$
$$u_- = \frac{I}{a - iI} P \frac{I + a(\lambda - A)}{\mathcal{A} - \lambda I} P \frac{I}{a + iI} \rho_\nu$$

□

To derive the *fiber* Krein formula in two-channel situation we consider the total two-channel deficiency subspace $\mathcal{N} \oplus \mathcal{M}$ of the operator $\mathcal{A}_0 \oplus \mathcal{B}_0$ and the corresponding space $\mathcal{D}$ of vector - valued densities $\mathcal{D} = (a+iI)\mathcal{N} \oplus (b+iI)\mathcal{M} \subset h^a \oplus E^b$,

$$\rho = \begin{pmatrix} \rho_h \\ \rho_E \end{pmatrix} \in \mathcal{D} = (a + iI)\mathcal{N} \oplus (b + iI)\mathcal{M} \subset h_- \oplus E_-.$$

We need also the dual space $\mathcal{C}$ of regularized boundary values

$$u_- = \begin{pmatrix} u_a \\ u_b \end{pmatrix} \in \mathcal{C} = (a + iI)^{-1}\mathcal{N} \oplus (b + iI)^{-1}\mathcal{M} \subset h_a \oplus E_b$$

The latter Lemma and the "channel-version of it" with $a \oplus b$ instead of a is essentially different from the similar "one-body" regularization of the boundary value in the Theorem 2.2 (via subtracting the term $\frac{a}{a^2+1}\rho$, which involves the total operator a). Really, using for the total Hamiltonian the splitted operator $\mathcal{A} = A \times I_h + I_H \times a \equiv A + a$ we achieve the regularization by subtracting the term which involves the component a only. Thus the local values of the corresponding densities and regularized values *are compatible with the separation of variables related to the component* A. We say that the regularization $\frac{1}{\mathcal{A}-\lambda I}\rho_\nu \to \frac{1}{\mathcal{A}-\lambda I}\rho_\nu - \frac{a}{a^2+I}\rho_\nu$ *is compatible with the splitting* $\mathcal{A} \to (A + a)$ or call it *two-body regularization*, still having in mind, that each of the objects a, A may have a sophisticated inner structure, see the footnote on the previous page.

Consider a bounded operator Γ acting from $\mathcal{D}$ to C and associated with the 2×2 operator matrix within the decompositions above:

$$\Gamma = \begin{pmatrix} G_{hh} & G_{hE} \\ G_{Eh} & G_{EE} \end{pmatrix}.$$

We assume , that Γ is defined via some finite Hermitian block matrix γ acting from $n \oplus m$ into $n \oplus m$:

$$\gamma = \begin{pmatrix} \gamma_{hh} & \gamma_{hE} \\ \gamma_{Eh} & \gamma_{EE} \end{pmatrix}, \Gamma = \begin{pmatrix} \frac{I}{a-iI} & 0 \\ 0 & \frac{I}{b-iI} \end{pmatrix} \gamma \begin{pmatrix} \frac{I}{a+iI} & 0 \\ 0 & \frac{I}{b+iI} \end{pmatrix}.$$

The next statement describes the construction of the selfadjoint extension $\mathcal{A}_\Gamma$ of the operator $\mathcal{A}_0 \oplus \mathcal{B}_0 = A + [a_0 \oplus b_0]$ in $H \times [h \oplus E]$ with the abstract boundary conditions defined by the operator Γ.

Theorem 2.3 *The boundary form of the adjoint operator $\mathcal{A}_0 \oplus \mathcal{B}*_0$ is represented in terms of densities and regularized boundary values as*

$$J(u, v) = \langle u_-, \rho_v \rangle - \langle \rho_u, v_- \rangle.$$

The boundary condition

$$u_- = \Gamma \rho$$

defines a Lagrangian plane of this form and the corresponding selfadjoint operator $\mathcal{A}_\Gamma$. The resolvent of $\mathcal{A}_\Gamma$ is given by the formula

$$\frac{1}{\mathcal{A}_\Gamma - \lambda I} f = \frac{1}{\mathcal{A} - \lambda I} f + \frac{1}{\mathcal{A} - \lambda I} \rho,$$

where the vector-valued density ρ satisfies the equation

$$(12) \quad \begin{pmatrix} \Gamma_{hh} - \frac{I}{a-iI} P_n \frac{I+(\lambda I - A)a}{A+a-\lambda I} P_n \frac{I}{a+iI} & \Gamma_{hE} \\ \Gamma_{Eh} & \Gamma_{EE} - \frac{I}{b-iI} P_m \frac{I+(\lambda I - A)b}{A+b-\lambda I} P_m \frac{I}{b+iI} \end{pmatrix}$$

$$(13) \quad \begin{pmatrix} \rho_h \\ \rho_E \end{pmatrix} = \begin{pmatrix} \frac{I}{a-iI} P_n \frac{a-iI}{a-\lambda I} f_h \\ \frac{I}{b-iI} P_m \frac{b-iI}{a-\lambda I} f_E, \end{pmatrix}$$

and (f_h, f_E) are the components of $f \in \mathcal{H}$ with respect to the orthogonal decomposition $\mathcal{H} = (H \times h) \oplus (H \times E)$.

Proof: Let us consider the equation involving the adjoint operator

$$\begin{pmatrix} \mathcal{A}_0 & 0 \\ 0 & \mathcal{B}_0 \end{pmatrix}^* u - \lambda u = f \in (H \times h) \oplus (H \times E).$$

Using the two-body regularization we represent the element u as

$$\begin{aligned} u &= \frac{a \oplus b}{[a \oplus b]^2 + I} \rho \\ &\quad + \frac{I}{a \oplus b - iI} \left[\frac{I + (\lambda I - A) a \oplus b}{A + a \oplus b - \lambda I} \right] \frac{I}{a \oplus b + iI} \rho + \frac{I}{\mathcal{A} - \lambda I} f \\ &= \frac{a \oplus b}{[a \oplus b]^2 + I} \rho + \frac{I}{a \oplus b - iI} P \left[\frac{I + (\lambda I - A) a \oplus b}{A + a \oplus b - \lambda I} \right] \frac{I}{a \oplus b + iI} \rho \\ &\quad + \frac{I}{a \oplus b - iI} P \frac{a \oplus b - iI}{\mathcal{A} - \lambda I} f \\ &\quad + \frac{I}{a \oplus b - iI} [I - P] \left[\frac{I + (\lambda I - A) a \oplus b}{A + a \oplus b - \lambda I} \right] \frac{I}{a \oplus b + iI} \rho \\ &\quad + \frac{I}{a \oplus b - iI} [I - P] \frac{a \oplus b - iI}{\mathcal{A} - \lambda I} f \\ &\equiv \frac{a \oplus b}{[a \oplus b]^2 + I} \rho + u_- + u_0, \end{aligned}$$

where $u_0 \in D_A \times D_{a_0 \oplus b_0}$ and u_- is the regularized boundary value of the element u. Then using the boundary conditions we get for ρ the inhomogeneous equation

$$\left\{\gamma - P\left[\frac{I + (\lambda I - A)\, a \oplus b}{A + a \oplus b - \lambda I}\right] P\right\} \frac{I}{a \oplus b + iI}\, \rho = P \frac{a \oplus b - iI}{A - \lambda I} f. \tag{14}$$

If we can construct the solution of it then inserting the result into the expression for u we get *the fiber Krein's formula* derived first in [33]

$$\begin{aligned} u &= \frac{1}{A + a \oplus b - \lambda I} f \\ &\quad + \frac{a \oplus b + iI}{A + a \oplus b - \lambda I} P \left[\frac{I}{\gamma - P \frac{I + (\lambda - A)(a \oplus b)}{A + a \oplus b - \lambda I} P}\right] P \frac{a \oplus b - iI}{A + a \oplus b - \lambda I} f. \end{aligned}$$

Using rigged variables $\rho = (\rho_h \oplus \rho_E) = (a + iI)\xi_h \oplus (b + iI)\xi_E$ one can reduce the latter formula to the announced form. □

Remark: In applications one often needs to calculate the restriction of the resolvent calculated above to only one of the orthogonal components of the Hilbert space $\mathcal{H} = (H \times h) \oplus (H \times E)$. Let vector f have only one nonzero component $f = (f_h \oplus 0)$. Then the variable ρ_E may be excluded using the standard techniques of the operator matrices, see [9]:

$$\rho_E = \frac{I}{\Gamma_{EE} - \frac{I}{b - iI} P_m \frac{I + (\lambda - A) b}{A + b - \lambda I} P_m \frac{I}{b + iI}} \Gamma_{Eh} \rho_h.$$

Then the problem is reduced to the calculation of the density ρ_h from the equation

$$\begin{aligned} &\left[\Gamma_{hh} - \frac{I}{a - iI} P_n \frac{I + (\lambda - A) a}{A + a - \lambda I} P_n \frac{I}{a + iI}\right. \\ &\qquad \left. - \Gamma_{hE} \frac{I}{\Gamma_{EE} - \frac{I}{b - iI} P_m \frac{I + (\lambda - A) b}{A + [-\lambda \mathcal{I}} P_m \frac{I}{b + iI}} \Gamma_{Eh}\right] \rho_h \\ &\quad = \frac{I}{a - iI} P_n \frac{a - iI}{A + a - \lambda I} f_h. \end{aligned}$$

The interaction constructed above, though formed by operator extensions tools, is quite strong since the deficiency indices of operators are infinite. Still the discrete spectrum of the perturbed operator may be characterized in terms of solutions of the homogeneous equation

$$\left[\gamma_{hh} - P_n \frac{I + (\lambda - A) a}{A + a - \lambda I} P_n - \gamma_{hE} \frac{I}{\gamma_{EE} - P_m \frac{I + (\lambda - A) b}{A + b - \lambda I} P_m} \gamma_{Eh}\right] \frac{I}{a + iI} \rho_h = 0 \tag{15}$$

The operator $\mathcal{A}_\Gamma$ just defined is semibounded from below just because it admits separation of variables. Really, it possesses the tensor decomposition

$$\mathcal{A}_\Gamma = A \times I_{h \oplus E} + I_H \times (a \oplus b)_\gamma,$$

where $(a \oplus b)_\gamma$ is a certain selfadjoint extension of the symmetric operator $a_0 \oplus b_0$. The operator $a_0 \oplus b_0$ has finite deficiency and the operator A is positive, therefore the operator $\mathcal{A}_\Gamma$ is semibounded.

3 The Simplest Few-body Krein's Formula

In the previous section we discussed "two-body" Krein's formula which gives the description of the simplest two-body Hamiltonians corresponding to two channels only [8]. We may interpret the channel $(A+a,\ H \times h)$ as a channel of free particles and the channel $(A+b,\ H \times E)$ as a channel where one cluster described by the Hamiltonian b coexists with a group of free particles described by the Hamiltonian A. Neither equidomoids, nor other channels are present in this case.

In this section we consider the simplest three-channels situation when the channel of free particles admits two splittings:

$$A_1 + a_1,\ H_1 \times h_1$$

$$A_2 + a_2,\ H_2 \times h_2,$$

and there are two more channels

$$(A_1 + b_1,\ H_1 \times E_1)$$

$$(A_2 + b_2,\ H_2 \times E_2)$$

where the clusters with inner Hamiltonians b_1, b_2 coexist with the groups of free particles with Hamiltonians $A_1,\ A_2$. We assume that recombination of one cluster into another or just the interchange between them passes through the free channel. Generally this is not true, but for point interactions we prove that this is the only possible way, since intermediate clusters containing total number of particles more than two, give no contribution to the scattering process, see Section 4. Hence three channels process of recombination of two clusters mentioned above may be considered as a representative example of few-body scattering.

Consider a Hilbert space $\mathcal{H}$ and a selfadjoint operator $\mathcal{A}$ in $\mathcal{H}$ admitting two splittings

$$\mathcal{H} = H_1 \times h_1,\ \mathcal{A} = A_1 \times I_{h_1} + I_{H_1} \times a_1,$$
$$\mathcal{H} = H_2 \times h_2,\ \mathcal{A} = A_2 \times I_{h_2} + I_{H_2} \times a_2,$$

where $A_s,\ a_t,\ s, t \in (1, 2)$ are selfadjoint positive operators acting in the Hilbert spaces $H_s,\ h_t$ respectively.

Let b_s, $s \in (1,2)$ be selfadjoint operators acting in the Hilbert spaces E_s. We consider the spaces

$$\mathcal{H}_1 = H_1 \times E_1, \text{ and } \mathcal{H}_2 = H_2 \times E_2,$$

and the selfadjoint operators

$$\mathcal{B}_1 = A_1 \times I_{E_1} + I_{H_1} \times b_1,$$
$$\mathcal{B}_2 = A_2 \times I_{E_2} + I_{H_2} \times b_2,$$

acting in these spaces. Restricting the operators a_s, $s = 1, 2$ to a_{s0}, $\mathrm{Dom}(a_{s0}) = \frac{I}{a_s - iI} h_s \ominus \{n_s\}$, $s = 1, 2$ and the operators b_s, $s = 1, 2$ to b_{s0}, $\mathrm{Dom}(b_{s0}) = \frac{I}{b_s - iI} E_s \ominus \{m_s\}$, $s = 1, 2$ we get the symmetric operator$\mathcal{A}_0$ in $\mathcal{H}$ with the total deficiency subspace $\overline{\mathcal{N}_1 + \mathcal{N}_2} \equiv \mathcal{N}_i$ generated by the densities ρ_s

$$\mathcal{N}_s = \left\{ \frac{I}{\mathcal{A} + iI} \rho_s \right\}, \quad \rho_s \in n^{a_s} \times H_s \; s = 1, 2.$$

and two symmetric operators $\mathcal{B}_{s0}$ with the deficiency subspaces $\mathcal{M}_s$ generated by the corresponding densities ν_s:

$$\mathcal{M}_s = \left\{ \frac{I}{\mathcal{B} + iI} \nu_s \right\} \quad \nu_s \in m^{b_s} \times H_s \; s = 1, 2.$$

In what follows we denote by P_{ns} the orthogonal projections in $\mathcal{H}$ onto the deficiency subspaces of operators $a_s \times I_{H_s}$, and by P_{m1}, P_{m2} the orthogonal projections onto the deficiency subspaces $b_{0s} \times I_{H_s}$.

In this section we discuss an analog of Krein's formula describing selfadjoint extensions $\mathcal{A}_\Gamma$ of the orthogonal sum generated by the fixed splitting of the operator $\mathcal{A}$

$$\mathbf{A} = \mathcal{A}_0 \oplus \mathcal{B}_{10} \oplus \mathcal{B}_{20}$$

in the Hilbert space

$$\mathbf{H} = \mathcal{H} \oplus [H_1 \times E_1] \oplus [H_2 \times E_2].$$

Our aim is to derive the "non-collaps condition" which guarantees the semiboundedness of the constructed extension. This condition is relevant to the condition of *physical stability* of this model quantum system when interacting with electromagnetic field. We plan to discuss this interesting connection in forthcoming publications.

We consider first the case when the deficiency subspaces $\mathcal{N}_1$, $\mathcal{N}_2$ have trivial or finite dimensional intersection, $\dim \mathcal{N}_1 \cap \mathcal{N}_2 < \infty$, but generally we may admit the zero angle between them and even infinite dimensional intersections. In the first case when the intersection is finite-dimensional the semiboundedness of *all* selfadjoint extensions may be easily derived. For the infinite dimensional intersection and zero angle we consider first reduced deficiency subspaces constructed of regular elements and forming positive angle and construct some symmetric extension. If this extension proves to be semibounded then using closure or Friedrichs procedure we may construct a special semibounded extension as an operator corresponding to the strong limit of corresponding quadratic forms. The formal construction will be described in forthcoming publications.

The densities corresponding to the reduced deficiency subspaces will be characterized as regular densities. In the last section these densities appear as densities vanishing near equidomoid.

Our nearest aim is to construct the Hamiltonian describing the exchange between two clusters via extension of the restricted operators

$$\mathcal{A}_0 \oplus \mathcal{B}_{10} \oplus \mathcal{B}_{20}$$

to some selfadjoint operator in the total Hilbert space

$$\mathbf{H} = \mathcal{H} \oplus [H_1 \times E_1] \oplus [H_2 \times E_2].$$

Practically we need to construct the resolvent of this extension. We reduce the construction of the resolvent to the solution of some intermediate equation. In the classical case when the deficiency subspace is finite dimensional this intermediate equation is just a matrix equation involving Krein's Q-functions, see the previous section. In our case this equation is obviously infinite dimensional. We shall see that it is similar to the system of Faddeev equations in three-body scattering problem.

We derive the equation for corresponding rigged variables and then investigate the solvability of this equation both for general situation in this section and for some basic example in the Section 4.

The selfadjoint extension of $\mathcal{A}_0 \oplus \mathcal{B}_{10} \oplus \mathcal{B}_{20}$ will be constructed via imposing on the generalized boundary values some *pairwise* boundary conditions attached to the interaction cylinders (planes) L^s. These conditions connect the rigged variables ρ_s, u_s of the adjoint operator $\mathcal{A}_0$ with rigged variables ν_s, U_s for the operators $\mathcal{B}_s$:

$$\begin{pmatrix} u_s \\ U_s \end{pmatrix} = \Gamma_s \begin{pmatrix} \rho_s \\ \nu_s \end{pmatrix}, \tag{16}$$

here Γ_s is a *constant* hermitian matrix. The dimension of it is equal to (dim n_s + dim m_s) $\times$ (dim n_s + dim m_s), where n_s, m_s are the deficiency indices of the operators a_{s0}, b_{s0}:

$$\Gamma_s = \begin{pmatrix} \Gamma_{h_s,h_s} & \Gamma_{h_s,E_s} \\ \Gamma_{E_s,h_s} & \Gamma_{E_s,E_s} \end{pmatrix}.$$

The described boundary conditions incorporate the data of a_s, b_s which correspond to the particles interacting at given cylinder L^s. The other data, which correspond to free particles, described by A^s may be separated as in [26, 27, 28].

One may check directly, that this boundary conditions (14) define a symmetric operator on the dense domain in the total Hilbert space corresponding to the regular densities ρ^δ. If we prove, that this operator is semibounded, then we can accomplish the construction of the corresponding selfadjoint operator by the Friedrich's procedure. It means, that we need to derive the operator equation corresponding to the equation for densities similar to the equation (12) in the theorem and then prove, that it has a unique solution on a dense domain in the total Hilbert space for large negative values of the spectral parameter.

We fulfill the described program deriving in this section the relevant system of equations for densities in abstract model. The resulting statement summarizing our analysis will be formulated at the end of this section. Basing on this statement we investigate in the following section the conditions of semiboundedness of M body quantum hamiltonian with a generalized point interaction (with inner structure).

Consider the solutions u, U_1, U_2 of the adjoint equations:

$$(17)\qquad \begin{aligned}(\mathcal{A}_0^* - \lambda I)u &= f,\\ (\mathcal{B}_{10}^* - \lambda I)U_1 &= f_1 \in E_1 \times H_1,\\ (\mathcal{B}_{20}^* - \lambda I)U_2 &= f_2 \in E_2 \times H_2.\end{aligned}$$

Note that on a dense domain of $f \in \frac{I}{a_1 - iI}\mathcal{H}$ the element $\frac{I}{\mathcal{A}-\lambda I} f$ may be represented as

$$\frac{I}{a_1 - iI} P_{1n} \frac{a_1 - iI}{\mathcal{A} - \lambda I} f + \frac{I}{a_1 - iI}\left[I - P_{1n}\right] \frac{a_1 - iI}{\mathcal{A} - \lambda I} f \equiv F_{11} + F_{10},$$

where F_{11} is a regularized boundary value and F_{10} belongs to the domain of a_{10}.

Solutions of the latter equations can be represented by the following ansatzes

$$\begin{aligned}
u &= \frac{I}{\mathcal{A}-\lambda I}\rho_1 + \frac{I}{\mathcal{A}-\lambda I}\rho_2 + \frac{I}{\mathcal{A}-\lambda I} f\\
&= \frac{a_1}{a_1^2 + I}\rho_1 + \frac{1}{a_1 - iI}\,\frac{I + (\lambda - A_1)a_1}{\mathcal{A} - \lambda I}\,\frac{1}{a_1 + iI}\rho_1\\
&\quad + \frac{I}{\mathcal{A}-\lambda I}\rho_2 + \frac{1}{\mathcal{A}-\lambda I} f\\
&= \frac{a_2}{a_2^2 + I}\rho_2 + \frac{1}{a_2 - iI}\,\frac{I + (\lambda - A_2)a_2}{\mathcal{A} - \lambda I}\,\frac{1}{a_2 + iI}\rho_2\\
&\quad + \frac{I}{\mathcal{A}-\lambda I}\rho_1 + \frac{1}{\mathcal{A}-\lambda I} f,\\
U_1 &= \frac{b_1}{b_1^2 + I}\nu_1 + \frac{1}{b_1 - iI}\,\frac{I + (\lambda - A_1)b_1}{\mathcal{B}_1 - \lambda I}\,\frac{1}{b_1 + iI}\nu_1,\\
U_2 &= \frac{b_1}{b_2^2 + I}\nu_2 + \frac{1}{b_2 - iI}\,\frac{I + (\lambda - A_1)b_2}{\mathcal{B}_2 - \lambda I}\,\frac{1}{b_2 + iI}\nu_2.
\end{aligned}$$

To calculate the regularized boundary values of the solution u (on the proper interaction cylinders L^1, L^2) we multiply u by $(a_1 - iI)$ and $(a_2 - iI)$ respectively and apply the projections P_{1n}, P_{2n} onto the deficiency subspaces $\mathcal{N}_1$, $\mathcal{N}_2$. If taking into account, that the domain of $\mathcal{A}_0$ is

$$\frac{I}{a_1 - iI}(h_1 \ominus n_1) \times \mathrm{Dom}(A_1) \cap \frac{I}{a_2 - iI}(h_2 \ominus n_2) \times \mathrm{Dom}(A_2),$$

we get the following expressions for the regularized values $u_1,\ u_2$ of the solution u

$$\begin{aligned} u_1 &= \frac{I}{a_1 - iI} P_{1n} \frac{I + a_1(\lambda - A_1)}{\mathcal{A} - \lambda I} \frac{I}{a_1 - iI}\rho_1 \\ &+ \frac{I}{a_1 - iI} P_{1n} \frac{a_1 - iI}{\mathcal{A} - \lambda I}\rho_2 - \frac{I}{a_1 - iI} P_{1n} \frac{a_1 - iI}{\mathcal{A} - \lambda I} f, \\ u_2 &= \frac{I}{a_2 - iI} P_{2n} \frac{I + a_2(\lambda I - A_2)}{\mathcal{A} - \lambda I} \frac{I}{a_2 - iI}\rho_2 \\ &+ \frac{I}{a_2 - iI} P_{2n} \frac{a_2 - iI}{\mathcal{A} - \lambda I}\rho_1 - \frac{I}{a_2 - iI} P_{2n} \frac{a_2 - iI}{\mathcal{A} - \lambda I} f. \end{aligned}$$

The regularized values of the solutions of the adjoint equations exist if $\rho_s = \rho_s^\delta,\ s = 1, 2$ and belong to the domain of A_s.

The regularized values of the solutions $U_1,\ U_2$ may be obtained similarly to the regularized values of the solutions appeared in the Theorem 2.3:

$$\begin{aligned} U_1 &= \frac{1}{b_1 - iI} P_{1m} \frac{I + b_1(\lambda I - A_1)}{A_1 + b_1 - \lambda I}\nu_1 \\ &+ \frac{I}{b_1 - iI} P_{1m} \frac{b_1 - iI}{\mathcal{B}_1 - \lambda I} f_1, \\ U_2 &= \frac{I}{b_2 - iI} P_{2m} \frac{I + b_2(\lambda I - A_2)}{A_2 + b_2 - \lambda I}\nu_2 \\ &+ \frac{I}{b_2 - iI} P_{2m} \frac{b_2 - iI}{\mathcal{B}_2 - \lambda I} f_2, \end{aligned}$$

Now we apply the boundary conditions (14) and receive the following system of equations for rigged variables $\rho_s,\ \nu_s,\ s = 1, 2$

$$\begin{aligned} & -\frac{I}{a_1 - iI} P_{1n} \frac{a_1 - iI}{\mathcal{A} - \lambda I}\rho_2 + [\Gamma^1 - Q(a_1, b_1)] \\ & \begin{pmatrix} \rho_1 \\ \nu_1 \end{pmatrix} = \begin{pmatrix} \frac{1}{a_1 - iI} P_{1n} \frac{a_1 - iI}{\mathcal{A} - \lambda I} f \\ \frac{1}{b_1 - iI} P_{1m} \frac{b_1 - iI}{\mathcal{B} - \lambda I} f_1 \end{pmatrix} \\ (18) \qquad & \\ & -\frac{I}{a_2 - iI} P_{2n} \frac{a_2 - iI}{\mathcal{A} - \lambda I}\rho_1 + [\Gamma^2 - Q(a_2, b_2)] \\ & \begin{pmatrix} \rho_2 \\ \nu_2 \end{pmatrix} = \begin{pmatrix} \frac{1}{a_2 - iI} P_{2n} \frac{a_2 - iI}{\mathcal{A} - \lambda I} f \\ \frac{1}{b_2 - iI} P_{2m} \frac{b_2 - iI}{\mathcal{B} - \lambda I} f_2 \end{pmatrix} \end{aligned}$$

where

$$Q(a_s, b_s) =$$

$$\begin{pmatrix} \frac{1}{a_s - iI} P_{sn} \frac{1 + a_s(\lambda I - A_s)}{A_s + a_s - \lambda I} P_{sn} \frac{1}{a_s + iI} & 0 \\ 0 & \frac{1}{b_s - iI} P_{sn} \frac{1 + b_s(\lambda I - A_s)}{A_s + b_s - \lambda I} P_{sn} \frac{1}{a_s + iI} \end{pmatrix}.$$

Together with the latter system it is convenient to use the equivalent system with respect to non-rigged variables

$$\begin{pmatrix} \frac{1}{a_s - iI} \rho_s \\ \frac{1}{b_s - iI} \nu_s \end{pmatrix} \equiv \begin{pmatrix} e_{h_s} \\ e_{E_s} \end{pmatrix},$$

$$(19) \qquad -P_s n \frac{(a_s - iI)(a_t - iI)}{\mathcal{A} - \lambda I} e_{h_t} + [\gamma^s - q^s] \begin{pmatrix} e_{h_s} \\ e_{E_s} \end{pmatrix}$$

$$= \begin{pmatrix} P_n^s \frac{a_s - iI}{\mathcal{A} - \lambda I} f \\ P_m^s \frac{b_s - iI}{\mathcal{B} - \lambda I} f_s \end{pmatrix} = \mathcal{F}_s, s = 1, 2, \ t \neq s.$$

Here the rigged Q-functions $Q(a_s, b_s)$ are connected to the non-rigged Q-functions q_s via multiplication by factors $\frac{I}{a_s - iI}, \frac{I}{b_s - iI}$ from the left side and adjoint factors from the right side.

The last system can be simplified by eliminating variables of the inner parts of both channels ν_s from each second equation.

Practically we need to solve the equations [15, 16] and then to restore the Krein formula basing on [14]. At the same time these equations play the role of Faddeev equations for the components represented by the channel's densitied ρ_1, ρ_2.

We analyse the solvability of these equations assuming dim m_s = dim $n_s = 1$ and considering separately three cases:

1) $n_s \cap \mathrm{Dom}(\sqrt{a_s}) = \{0\}, \ \ \gamma_{EE}^s + P_{ms} b_s P_{ms} \neq 0, \ \ s = 1, 2,$
2) $\gamma_{EE}^s + P_{ms} b_s P_{ms} = 0, \ s = 1, 2,$

and

3) $n_1 \cap \mathrm{Dom}(\sqrt{a_1}) = \{0\}, \ \ \gamma_{EE}^1 + P_{m1} b_1 P_{m1} \neq 0, \ \gamma_{EE}^2 + P_{m2} b_2$
$P_{m2} = 0.$

Denoting by $\tilde{O}(\frac{I}{\lambda})$ the vectors and operators which admit the estimate by $|\frac{1}{\lambda}|$ for large negative λ and denoting by $\mathcal{F}_s$ the expression which appears in the right hand side after the elimination the inner variables e_{E_s} we get the following systems in these cases.

In the first case we have:

$$
(20)\quad
\begin{aligned}
&\left[\gamma^s_{hh} + \int_0^\infty \frac{\alpha\,(A_s - \lambda I)}{A_s + \alpha - \lambda I} P_n d\mathcal{E}^s_\alpha P_n \right.\\
&\left. -\gamma^s_{hE}\left[\gamma^s_{EE} + P_{sm} b_{sm} P_{sm} + \tilde{O}\left(\frac{I}{\lambda}\right)\right]^{-1} \gamma^s_{Eh}\right] e_h\\
&+ P_{ns}\frac{a_s - iI}{\mathcal{A} - \lambda I}(a_t + iI)e_t = P_{ns}\frac{a_s - iI}{\mathcal{A} - \lambda I} f\\
&- \gamma^s_{hE}\left[\gamma^s_{EE} + P_{ms} b_s P_{ms} + \tilde{O}\left(\frac{I}{\lambda}\right)\right]^{-1} P_{ms}\frac{b_s - iI}{\mathcal{B}_s - \lambda I} f_{E_s} \equiv \mathcal{F}_s.
\end{aligned}
$$

In the second case we have:

$$
(21)\quad
\begin{aligned}
&P_{sn}\frac{a_s - iI}{\mathcal{A} - \lambda I}(a_t + iI)e_{ht} + \left[\int_0^\infty \frac{\alpha\,(A_s - \lambda I)}{A + \alpha - \lambda I} P_{sn} d\mathcal{E}^s_\alpha P_{sn} - \gamma^s_{hh}\right.\\
&- \gamma^s_{hE}\left([\lambda I - A_s][P_{sm}[I + b_s^2]P_{sm}]^{-1} - \frac{P_{sm}[I + b_s^2]b_s P_{sm}}{P_{sm}[I + b_s^2]P_{sm}}\right.\\
&\left.\left.+\tilde{O}\left(\frac{1}{\lambda}\right)\right)\gamma^s_{Eh}\right] e_{sh} = P_{sn}\frac{a_s - iI}{\mathcal{A} - \lambda I} f_h - \left([\lambda I - A][P_{sm}[I + b^2]P_{sm}]^{-1}\right.\\
&\left.- \frac{P_{sm}[I + b_s^2]b P_{sm}}{P_{sm}[I + b_s^2]P_{sm}} + \tilde{O}\left(\frac{1}{\lambda}\right)\right) P_{sm}\frac{b_s - iI}{\mathcal{B}_s - \lambda I} f_{Es} \equiv \mathcal{F}_s,
\end{aligned}
$$

and for the last one we have

$$
\begin{aligned}
&P_{2n}\frac{a_2 - iI}{\mathcal{A} - \lambda I}(a_1 + iI)e_{h1} + \left[\int_0^\infty \frac{\alpha(A_2 - \lambda I)}{A_2 + \alpha - \lambda I} P_{2n} d\mathcal{E}^2_\alpha P_{2n} - \gamma^2_{hh} - \gamma^2_{hE}\right.\\
&\left.\left([\lambda I - A_2][P_{2m}[I + b_2^2]P_{2m}]^{-1} - \frac{P_{2m}[I + b_2^2]b_2 P_{2m}}{P_{2m}[I + b_2^2]P_{2m}} + \tilde{O}\left(\frac{1}{\lambda}\right)\right)\gamma^2_{Eh}\right]
\end{aligned}
$$

$$
(22)\quad
\begin{aligned}
&e_{2h} = P_{2n}\frac{a_2 - iI}{\mathcal{A} - \lambda I} f_h - \gamma_h E\left([\lambda I - A_2](P_{2m}[I + b_2^2]P_{2m})^{-1}\right.\\
&\left.-\frac{P_{2m}[I + b_2^2]b_2 P_{2m}}{P_{2m}[I + b_2^2]P_{2m}} + \tilde{O}\left(\frac{1}{\lambda}\right)\right) P_{2m}\frac{b_2 - iI}{\mathcal{B}_2 - \lambda I} f_{E2} \equiv \mathcal{F}_2,
\end{aligned}
$$

We have used here the notation $\mathcal{E}^s_\alpha$ for the family of spectral projections of the operator a_s.

The only essential complication of each of the above systems is the presence of *crossing terms* in them, for instance:

$$
P_{1n}\frac{a_1 - iI}{\mathcal{A} - \lambda I}(a_2 - \lambda I)e_{h2},\ P_{2n}\frac{a_2 - iI}{\mathcal{A} - \lambda I}(a_1 - \lambda I)e_{h1}.
$$

If the whole construction is reduced to the domain of regular densities ρ_s^δ, $s = 1, 2$, $\delta > 0$ then these crossing terms are not essential, since the angles between the deficiency subspaces $\mathcal{N}_1$, $\mathcal{N}_2$ is positive and the crossing terms are bounded. Then the constructed extensions are semibounded from below.

We consider now the class of regular densities $\hat{\rho}_s$. We assume the smoothness of densities with respect to the corresponding channel operators A_s. This class of densities is a dense linear set which will be used to derive the apriory estimates and develop the Friedrich's procedure. For this class of densities the bilinear forms of crossing terms obviously exist. Our aim is to find the conditions which guarantees the uniform (in δ) domination the forms of crossing terms by the quadratic forms of the diagonal terms of the matrix $\Gamma - Q$.

Retaining only leading terms and crossing terms in equations corresponding to the cases considered we get the following results:

$$\begin{aligned} &1) \left[\int_0^\infty \frac{\alpha(A_s - \lambda I)}{A_s + \alpha - \lambda I} P_{sn} d\mathcal{E}_\alpha^s P_{sn} + \tilde{O}(1)\right] e_s \\ &\quad + P_{ns} \frac{a_s - il}{\mathcal{A} - \lambda I}(a_t + il)e_t = \mathcal{F}_s, \ s, t = 1, 2. \end{aligned} \tag{23}$$

$$\begin{aligned} &2) \int_0^\infty \frac{\alpha\,(A_s - \lambda I)}{A_s + \alpha - \lambda I} P_{sn} d\mathcal{E}_\alpha^s P_{sn} e_{sh} + \gamma_{hE}^s \frac{(A_s - \lambda I)}{P_{sm}[I + b_s^2]P_{sm}} \gamma_{Eh}^s e_{sh} \\ &\quad + \tilde{O}(1)e_{sh} + P_{sn} \frac{a_s - il}{\mathcal{A} - \lambda I}(a_t + il)e_{th} = \mathcal{F}_s, \ s, t = 1, 2. \end{aligned} \tag{24}$$

$$\begin{aligned} &3) \int_0^\infty \frac{\alpha(A_1 - \lambda I)}{A_1 + \alpha - \lambda I} P_{1n} d\mathcal{E}_\alpha^1 P_{1n} e_{1h} + \tilde{O}(1)e_{1h} \\ &+P_{n1} \frac{a_1 - il}{\mathcal{A} - \lambda I}(a_2 + il)e_{2h} = \mathcal{F}_1. \end{aligned}$$

$$\begin{aligned} &P_{2n} \frac{a_2 - il}{\mathcal{A} - \lambda I}(a_1 + il)e_{1h} + \int_0^\infty \frac{\alpha(A_2 - \lambda I)}{A_2 + \alpha - \lambda I} P_{2n} d\mathcal{E}_\alpha^2 P_{2n} e_{2h} \\ &\quad - \gamma_{hE}^2 \frac{\lambda I - A_2}{P_{2m}[I + b_2^2]P_{2m}} \gamma_{Eh}^2 e_{2h} + \tilde{O}(1)e_{2h} = \mathcal{F}_2. \end{aligned} \tag{25}$$

Here only the leading diagonal terms are presented in explicit form. The dominated diagonal terms are denoted by $\tilde{O}(1)$.

Theorem 3.1 *Provided some of equations* (21), (22), (23) *have unique solutions in proper rigged spaces for large negative* λ, *the corresponding operators defined by the corresponding boundary conditions are semibounded and we can accomplish the extension procedure for these operators by Friedrichs method. Then the corresponding resulting few-body Krein's formulae for resolvents of these extensions may be obtained from the solution of the system of adjoint equations* (15) *submitted to the boundary conditions* (14).

We continue the discussion of solvability of the equations (21), (22), (23) in the following section.

It is clearly seen now that the relevant densities in rigged-space setting of the problem play the role of Faddeev components, see [13] and each of systems (21), (22), (23) is a system of Faddeev equations for the resolvent of the corresponding few-body Hamiltonian generated by switching on the interaction via the boundary conditions between two groups of *quantum particles* arranged in two different clusters and the channel of free particles. This means that during the collision both clusters are virtually destroyed and then new clusters recreated. Of course we may consider more subtle setting of the problem when only partial destruction takes place. This requires more delicate treating of intersections of deficiency subspaces. This will be done elsewhere.

The discussed problem for three channels (two scattering channels and a free channel) may be considered, to some extent, as an elementary brick of the real problem involving several channels.

4 Semiboundedness of the Few-body Schrödinger Operator with Generalized Point Interaction

In fact even description of the simplest three-body quantum processes requires investigation of several channels, for instance (for distinguishable particles)

$$[(1),(2),(3)],\ [(3,1)(2)],\ [(1,2)(3)],\ [(2,3)(1)],\ [(1,2,3)].$$

The corresponding *equidomoid* - the intersection of the interaction planes of these channels $L_{31} \equiv (x : x_3 = x_1)$, L_{12}, L_{23}, contains only one common point 0, so called *triple collision point*. Construction of the interaction via operator-extensions techniques in this situation has been described in [2, 3, 4, 6, 10, 22, 34]. The description of the similar procedure for M-body Hamiltonians requires considering channels with non smooth and unbounded equidomoids. For instance, to switch the interaction in the 4-body quantum problem we should split the 4-body free Hamiltonian into several channels $[(1)(2)(3)(4)]$, $[(1,2)(3)(4)]$, $[(1,3)(2)(4)]$, $[(1,4)(2)(3)]$, $[(2,3)(4)(1)]$, $[(2,3)(4)(1)]$, $[(3,4)(1)(2)]$ $[(1,2)(3,4)]$, $[(1,3)(2,4)]$, $[(1,4)(2,3)]$, $[(1,2,3)(4)]$, $[(2,3,4)(1)]$, $[(3,4,1)(2)]$, $[(4,1,2)(3)]$, $[(1,2,3,4)]$. The intersections of the interaction planes of some channels, for instance $[(1,2,3)(4)]$, and $[(2,3,4)(1)]$, are unbounded. A representative example of clusters with a bounded equidomoid is given by the triple of channels

$$[(1)(2)(3)(4)],\ [(1,2)(3)(4)],\ [(3,4)(1)(2)],$$

an example of the triple of channels with an unbounded equidomoid is given by the triple of channels

$$[(1)(2)(3)(4)],\ [(1,2)(3)(4)],\ [(1,4)(3)(2)].$$

In this section we discuss the following *basic example*. Consider M *distinguishable* quantum particles in Euclidean Space R_3 with a free Hamiltonian represented by Laplace operator with respect to cartesian coordinates in $R^{(3M-3)}$ associated with the center of masses of the system, and the corresponding Hilbert space $\mathcal{H} = L_2(R^{(3M-3k)})$. Selecting a plane $l \in R^{(3M-3)}$, $\dim l = 3k-3$ we assign the Hilbert space $h = L_2(l)$ to it. This space may play the role of inner Hilbert space of some cluster of k particles and the complementary space $H = L_2(L)$, supported by the complementary plane L, $\dim L = 3(M-k)$, (such that $l \times L = R_{(3M-3)}$) will play the role of the Hilbert space of the complementary group of free particles[5]. The direct product of these spaces gives the total space: $\mathcal{H} = h \times H$. The space L plays the role of the interaction plane, so that the elements from the deficiency subspace $\mathcal{N}$ of the total Hamiltonian $\mathcal{A}_0 = -\Delta_0 = [-\Delta_l]_0 - \Delta_L \equiv a_0 + A$ reduced by the condition of vanishing near L

$$a = -\Delta_\perp \longrightarrow -\Delta_{\perp 0} \equiv a_0, \ \mathcal{A} \longrightarrow a_0 + A$$

are represented via densities supported by this plane:

$$e_{\bar{\lambda}} = \frac{I}{-\Delta - \lambda I}\rho$$

In terms of Fourier transform $\mathcal{F}$ the condition $\operatorname{supp}\rho \in L$ means that $\mathcal{F}\rho \equiv \tilde{\rho}$ depends only on the longitudinal component of the momentum in the orthogonal decomposition $p = p_\perp + p_\parallel$ with respect to L, $p_\perp \perp L$ or $p_\perp \in l$, $\dim L = 3(M-k)$, $\dim l = 3(k-1)$:

$$\tilde{\rho}(p) \equiv \tilde{\rho}(p_\parallel),$$

hence

$$\tilde{e_{\bar{\lambda}}} = \frac{\tilde{\rho}(p_\parallel)}{p_\perp^2 + p_\parallel^2 - \lambda}.$$

Generally the splitting of Laplace operator with respect to the decomposition $\mathcal{H} = L_2(l) \times L_2(L)$ is given in the Fourier representation by the formula

$$\mathcal{A} = -\Delta \xrightarrow{\mathcal{F}} p_\perp^2 + p_\parallel^2 \equiv a + A.$$

In the case of finite deficiency index of a_0 the dot products in representation of the boundary forms with respect to the rigged variables are reduced to the integration over the plane L, for instance

$$\langle u, \rho_L^v \rangle - \langle \rho_L^u, v \rangle = \int_L [\langle u, \rho_L^v \rangle_h - \langle \rho_L^u, v \rangle_h] dL.$$

[5] The latter condition means that further splitting of these particles is not observed during the process considered below. But we do not exclude the possibility, that some of them can be splitted in other processes.

We consider now all possible splittings of the configuration space R_{3M-3}, $R_{3M-3} = l_s \times L_s$ and assign the proper index s to all relevant quantities. For instance we consider two pairs of mutually complementary subspaces of the Hilbert space: $h_s, H_s; \mathcal{H} = h_s \times H_s$. We associate these pairs with two clusters, $h_s = L_2(l^2)$, $H_s = L_2(L^s)$. In what follows, if not mentioned specially, we consider the restriction of the Laplacian by the conditions of vanishing near the planes L^s

$$\mathcal{A} \longrightarrow \mathcal{A}_0.$$

The domain of the restricted operator is defined as intersection of domains of partial restrictions $\mathcal{A}_s = a_{0s} + A_s$. The total deficiency subspace is a sum of deficiency subspaces $\mathcal{N}_s$

$$\overline{\sum_s \mathcal{N}_s}$$

of the splitted operators $\mathcal{A}_s$. The angles between the deficiency subspaces are obviously equal to zero. The following basical statement permits to consider in our model only the recombination of two-body clusters.

Lemma 4.1 *Consider the symmetric operator defined as a Laplacian $-\Delta$ on all compactely supported smooth functions vanishing near the plane L_k, $dim L_k = 3(M-k)$. If $k > 2$ then this operator is essentially selfadjoint.*

Proof: Really, the deficiency elements of this operator are Fourier-represented as

$$\frac{\tilde{\rho}(p_{\|})}{p_{\perp}^2 + p_{\|}^2 - \lambda}$$

with proper generalised densities $\tilde{\rho}(p_{\|})$. These elements are square integrable if

$$\int \frac{|\tilde{\rho}(p_{\|})|^2}{|p_{\perp}^2 + p_{\|}^2 - \lambda|^2} dp_{\|} dp_{\perp} < \infty.$$

The last condition is fulfilled for proper densities if

$$\dim p_{\perp} = 3(M-1) - 3(M-k) = 3(k-1) < 4$$

since it implies the convergence of the integral

$$\int \frac{1}{|p_{\perp}^2 + p_{\|}^2 - \lambda|^2} dp_{\perp} \approx |p_{\|}^2 - \lambda|^{\frac{3k-3}{2} - 2}.$$

Hence only $k = 1, 2$ are possible and the densities for the deficiency elements corresponding to two-body clusters should satisfy the condition

$$\int \frac{|\tilde{\rho}(p_{\|})|^2}{\sqrt{p_{\|}^2 + 1}} dp_{\|}.$$

□

Hence remaining in the Hilbert space $L_2(R_{3M-3})$ we can construct solvable models with point interactions only for processes of scattering and recombination of several two-body clusters. For larger clusters, $k > 2$, the interaction planes are too slim so that the restriction of the Laplacian onto smooth functions vanishing near them is essentially selfadjoint. One may derive from it that the deficiency subspaces in our case have finite dimensional intersection, but still have zero angle between them.

Consider the linear space of smooth regular densities ρ_s[6]. Since the complementary planes l^s are three-dimensional, the deficiency subspaces n_s of operators $a_{s0} = -\Delta$ in $L_2(l^s)$ and the corresponding projections P_{sn} onto them are one-dimensional. The relevant projections onto $n_s \times H_s = n_s \times L_2(L^s)$ are represented as $P_{sn} \times I_{H_s}$. We use for them the former notations $P_{sn} \times I_{H_s} \equiv P_{sn}$.

We investigate now the solvability of analogs of equations (21), (22), (23) in our model.

Note first that the quadratic and bilinear forms of the diagonal terms and crossing terms on regular deficiency elements

$$e_s = \frac{\tilde{\rho}(p_{\|}^s)}{p^2 - \lambda I}$$

may be represented now in terms of Fourier transforms of the corresponding densities $\tilde{\rho}^s(p_{\|}^s)$ the following way:

$$\left\langle \int_0^\infty \frac{\alpha(A_s - \lambda I)}{A_s + \alpha - \lambda I} P_{sn} d\mathcal{E}_\alpha^s P_{sn} e_s, e_s \right\rangle = \int \frac{([p_{\|}^s]^2 - \lambda)|\tilde{\rho}(p_{\|}^s)|^2}{|[p_{\perp}^s]^2 + i|^2} dp.$$

$$\left\langle P_{ns} \frac{a_s - iI}{A_f - \lambda I}(a_t + iI)e_t, e_s \right\rangle = \int \frac{\tilde{\rho}(p_{\|}^t)\overline{\tilde{\rho}(p_{\|}^s)}}{p^2 - \lambda} dp. \tag{26}$$

$$\left\langle \gamma_{hE}^t \frac{(A_t - \lambda I)}{P_{tm}[I + b_t^2]P_{tm}} \gamma_{Eh}^t e_t, e_s \right\rangle = \frac{1}{P_{tm}[I + b_t^2]P_{tm}} \int \frac{[p_{\perp}^t]^2[\tilde{\rho}(p_{\|}^t)]^2}{(p^2 - \lambda)|[p_{\perp}^t]^2 + i|^2}$$

Note that the first spectral integral

$$\left\langle \int_0^\infty \frac{\alpha(A_s - \lambda I)}{A_s + \alpha - \lambda I} P_{sn} d\mathcal{E}_\alpha^s P_{sn} e_s, e_s \right\rangle.$$

may be estimated for large negative λ from both sides by a simpler expression with proper coefficients:

$$\int \frac{([p_{\|}^s]^2 - \lambda)|\tilde{\rho}(p_{\|}^s)|^2}{|[p_{\perp}^s]^2 + i|^2} dp \approx \int ([p_{\|}^s]^2 - \lambda)^{\frac{1}{2}} |\tilde{\rho}(p_{\|}^s)|^2 dp_{\|}^s$$

[6]Vanishing near the equidomoid and at infinity on interaction planes L^s.

Having in mind that the operator-extension construction can be developed for two-body clusters only we see that all interaction planes are $3M-6$-dimensional. This means that two interaction planes $L_s,\ L_t$ may have either infinite intersection codim 9 like L_{12} and L_{34}, or infinite intersection codimension 6 like R_{3M-3} as L_{12} and L_{23}. We estimate below the crossing term by the product of the diagonal terms assuming that one of them, with the index t, contains the differential factor $A_t - \lambda I$ and the other one is may have the same form or may be represented just by the spectral integral

$$\left\langle \int_0^\infty \frac{\alpha(A_s - \lambda I)}{A_s + \alpha - \lambda I} P_{sn} d\mathcal{E}_\alpha^s P_{sn} e_s, e_s \right\rangle.$$

We consider here most interesting case when the component of the equidomoid has codimension 6. The case of codimension 9 may be studied the same way.

Consider the intersection L_{st} of interaction planes $L_s,\ L_t$ in the momentum space. We denote by p_{st} the component of the Fourier variable p along L_{st} and by $q_s,\ q_t$ the complementary components of the momentum in $l_s,\ l_t$. Since the angle between $q_s,\ q_t$ is positive, we have generally the direct decomposition of the total momentum space R_{3M-3}, $\dim L_{st} = 3M-6$, $\dim l_s = 3 \dim l_t = 3$:

$$p = p_{st} + q_s + q_t;\ R^{3M-3} \equiv L_{st} + [l_s + l_t],\ q_s \in l_s,\ s \neq t.$$

Note, that $p_\parallel^s$ admits a unique decomposition in form $p_{st} + q_s$, and $p_\parallel^t$ admits a unique decomposition in form $p_{st} + q_t$.

Lemma 4.2 *The following estimates for the bilinear forms of crossing terms via diagonal leading terms are valid for large negative λ:*

$$\left| \int \frac{\rho^s(p_\parallel^s)\overline{\rho^t(p_\parallel^t)}}{p^2-\lambda} dp \right| \leq \frac{C_1}{\sqrt{-\lambda}} \left[\int (|p_\parallel^s|^2 - \lambda) \| \rho^s(p_\parallel^s)|^2 dp_\parallel^s + \int (|p_\parallel^t|^2 - \lambda) |\rho^t(p_\parallel^t)|^2 dp_\parallel^t \right] \tag{27}$$

$$\left| \int \frac{\rho^s(p_\parallel^s)\overline{\rho^t(p_\parallel^t)}}{p^2-\lambda} dp \right| \leq \frac{C_2}{|-\lambda|^{1/4}} \left[\int (|p_\parallel^s|^2 - \lambda) |\rho^s(p_\parallel^s)|^2 dp_\parallel^s + \int (|p_\parallel^t|^2 - \lambda)|^{1/2} |\rho^t(p_\parallel^t)|^2 dp_\parallel^t \right], \tag{28}$$

with certain positive constants $C_1,\ C_2$

Proof: Introducing new functions $f_s,\ f_t$ as follows $f_s(q_s + p_{st}) \equiv \sqrt{p_{st}^2 + q_s^2 - \lambda}$ $\rho^s(p_\parallel^s)$ and denoting by J Jacobians corresponding to the nonorthogonal pais of coordinates q_s, q_t we get the first estimate:

$$\int \frac{|\rho^s(p_{\|}^s)\overline{\rho^t(p_{\|}^t)}|}{|p^2-\lambda|}dp \leq \frac{J}{2}\int \frac{|f_s(q_s+p_{st})|^2 dp_{st}dq_s dq_t}{|p_{st}^2+q_t^2-\lambda|\;|p_{st}^2+(q_s+q_t)^2-\lambda|}$$
$$+\frac{J}{2}\int \frac{|f_t(q_t+p_{st})|^2 dp_{st}dq_s dq_t}{|p_{st}^2+q_s^2-\lambda|\;|p_{st}^2+(q_s+q_t)^2-\lambda|}$$
$$\leq \frac{J}{2}\int |f_s(q_s+p_{st}|^2)dp_{st}dq_s \;\sup_{p_{st},q_s}$$
$$\int \frac{dq_t}{|p_{st}^2+q_t^2-\lambda|\;|p_{st}^2+(q_s+q_t)^2-\lambda|}$$
$$+\frac{J}{2}\int |f_t(q_t+p_{st})|^2 dp_{st}dq_t \;\sup_{p_{st},q_t}$$
$$\int \frac{dq_s}{|p_{st}^2+q_s^2-\lambda|\;|p_{st}^2+(q_s+q_t)^2-\lambda|}$$
$$\leq \frac{C_1}{\sqrt{|\lambda|}}\left[\int |f_s(q_s+p_{st})|^2 dp_{st}dq_s + \int |f_t(q_t+p_{st})|^2 dp_{st}dq_t\right].$$

Returning to the original functions ρ we get the first estimate.

To prove the second estimate we need proper distribution of powers in the brackets of the denominator. Now we are using the following notations

$$f_s(q_s+p_{st}) \equiv \sqrt{q_s^2+p_{st}^2-\lambda}\;\rho^s,\quad f_t(q_t+p_{st}) \equiv (q_t^2+p_{st}^2-\lambda)^{1/4}\;\rho^t.$$

Then we may estimate the second integral as

$$\int \frac{|\rho^s(p_{\|}^s)\overline{\rho^t(p_{\|}^t)}|}{p^2-\lambda}dp$$
$$=\int \frac{|f(q_s+p_{st})\overline{f(q_2+p_{st})}|dq_s dq_t dp_{st}}{|p_{st}^2+q_s^2-\lambda|^{\frac{1}{2}}\;|p_{st}^2+(q_s+q_t)^2-\lambda||p_{st}^2+q_t^2-\lambda|^{\frac{1}{4}}}$$
$$\leq \frac{J}{2}\int \frac{|f(q_s+p_{st})|^2 dq_s dq_t dp_{st}}{|p_{st}^2+q_s^2-\lambda|^{\frac{1}{2}}\;|p_{st}^2+(q_s+q_t)^2-\lambda|^{5/4}}$$
$$+\frac{J}{2}\int \frac{|f(q_t+p_{st})|^2 dq_s dq_t dp_{st}}{|p_{st}^2+q_s^2-\lambda||p_{st}^2+(q_s+q_t)^2-\lambda|^{3/4}}$$
$$\leq \frac{C_2}{|\lambda|^{1/4}}\left[\int |f_s(q_s+p_{st})|^2 dq_s dp_{st} + \int |f_t(q_t+p_{st})|^2 dq_t dp_{st}\right]$$

□

Remark: The estimates (24), (25) permit to investigate the solubility of equations (22), (23) for large negative λ, but still not the equation (31). Really, the attempt to estimate the crossing terms for the equation (21) we get the following integral for "renormalized densities" $f_s(q_s + p_{st}) \equiv (q_s^2 + p_{st}^2 - \lambda)^{1/4}\rho^s(q_s + p_{st})$, $f_t(q_t + p_{st}) \equiv (q_t^2 + p_{st}^2 - \lambda)^{1/4}\rho^t(q_t + p_{st})$:

$$\int \frac{|f_s(q_s + p_{st}) f_t(q_t + p_{st})| dq_s \, dq_t \, dP_{st}}{|q_t^2 + p_{st}^2 - \lambda|^{1/4} |p_{st} + (q_s + q_t)^2 - \lambda| |q_s^2 + p_{st}^2 - \lambda|^{1/4}}. \tag{29}$$

The straightforward estimate as above gives just

$$\begin{aligned}
&\frac{J}{2} \int |q_s^2 + p_{st}^2 - \lambda|^{\epsilon} |f_s(q_s + p_{st})|^2 dq_s \, dp_{st} \\
&\sup_{q_s,\, p_{st}} \int \frac{dq_t}{|p_{st} + (q_s + q_t)^2 - \lambda| \; |q_t^2 + p_{st}^2 - \lambda|^{1/2+\epsilon}} \\
&\quad + \frac{J}{2} \int |q_t^2 + p_{st}^2 - \lambda|^{\epsilon} |f_t(q_t + p_{st})|^2 dq_t \, dp_{st} \\
&\sup_{q_t,\, p_{st}} \int \frac{dq_s}{|p_{st} + (q_s + q_t)^2 - \lambda| \; |q_s^2 + p_{st} \bar{-} \lambda|^{1/2+\epsilon}} \\
&\le \frac{C_3}{|\lambda|^{\epsilon}} \left[\int |(q_t^2 + p_{st}^2 - \lambda|^{\epsilon} |f_t(q_t + p_{st})|^2 dq_t \, dp_{st} \right. \\
&\quad \left. + \int |q_s^2 + p_{st}^2 - \lambda|^{\epsilon} |f_s(q_s + p_{st})|^2 dq_s \, dp_{st} \right]
\end{aligned}$$

for each positive ϵ. This is not sufficient for our aims.

Summarizing results of the last Lemma in general case of few two-body clusters and previous analysis of equations (21), (22), (23) we obtain the following statement for M-body scattering problem with generalized point interaction switched on between pairs (in two-body clusters):

Theorem 4.1 *The Hamiltonian of M distinguishable quantum particles with point interaction is semibounded if at least for* $\frac{M(M-1)}{2} - 1$ *interaction planes of 2-body clusters the condition* $\gamma_s + P_m^s b_s P_m^s = 0$ *is satisfied.*

This statement is a natural generalization of the result [27, 28] for M particles.

References

[1] N.I. Akhiezer and I.M. Glazman, *Theory of linear operators in Hilbert space*. vol. II, Pitman, Boston, 1981.

[2] S. Albeverio, W. Hunziker, W. Schneider and R. Schrader, *A note on L.D. Faddeev's three-particle theory*, Helv. Phys. Acta **40** (1967), 745–748.

[3] S. Albeverio, R. Høegh-Krohn and L. Streit, *Energy forms, Hamiltonians, and distorted Brownian paths*, J. Math. Phys. **18** (1977), 907–917.

[4] S. Albeverio, R. Høegh-Krohn and T.T. Wu, *A class of exactly solvable three-body quantum mechanical problems and the universal low energy behavior*, Phys. Lett. A **83** (1981), 105–109.

[5] S. Albeverio, F. Gesztesy, R. Høegh-Krohn and H. Holden, *Solvable models in quantum mechanics*, Springer-Verlag, New York, 1988.

[6] S. Albeverio and P. Kurasov, *Singular perturbations of differential operators*, Cambridge Univ. Press, to appear.

[7] B.L. Altshuler, P.A. Lee and R.A. Webb, eds., *Mesoscopic Phenomena in Solids*, North Holland, New York, 1991.

[8] W. Amrein, J. Jauch and K. Sinha, *Scattering theory in quantum mechanics*, Benjamin, Reading, Mass, 1977.

[9] F.V. Atkinson, H. Langer, R. Mennicken and A. Shkalikov, *The essential spectrum of some matrix operators*, Math. Nachr. **167** (1994), 5–20.

[10] F.A. Berezin and L.D. Faddeev, *Remark on the Schrödinger equation with singular potential*, Dokl. Akad. Nauk SSSR **137** (1961), 1011–1014.

[11] Yu.N. Demkov and V.N. Ostrovsky, *Zero-range potentials and their applications in atomic physics*, Plenum Press, New York and London, 1988.

[12] V.A. Derkach and M.M. Malamud, *Generalized resolvents and the boundary value problems for Hermitian operators with gaps*, J. Funct. Anal. **95** (1991), 1–95.

[13] L.D. Faddeev and S.P. Merkuriev, *Quantum scattering theory for several particle systems*, Kluwer, Dordrecht, 1993.

[14] E. Fermi, *Sul moto dei neutroni nelle sostanze idrogenate*, Ricerce Scientifica, 7: 13–52, 1936 (in Italian), English translation in E. Fermi, Collected papers, vol. I, Italy 1921–1938, Univ. of Chicago Press, Chicago-London, 1962, pp. 980–1016.

[15] V.I. Gorbachuk and M.L. Gorbachuk, *Boundary value problems for operator differential equations*, Kluwer, Dordrecht, 1991.

[16] S. Hassi and H. de Snoo, *On rank one perturbations of selfadjoint operators*, Integr. Eq. Oper. Theory **29** (1997), 288–300.

[17] S. Hassi, H. Langer and H. de Snoo, *Selfadjoint extensions for a class of symmetric operators with defect numbers* $(1, 1)$. In Topics in operator theory, operator algebras and applications (Timisoara, 1994), pp. 115–145. Rom. Acad., Bucharest, 1995.

[18] G.R. Kirchoff, *Gesammelte Abhandlungen* Publ. Leipzig: Barth, 1882, p. 641.

[19] M. Krein, *Concerning the resolvents of an Hermitian operator with the deficiency-index* (m, m), Comptes Rendue (Doklady) Acad. Sci. URSS (N.S.) **52** (1946), 651–654.

[20] M. Naimark. *Self-adjoint extensions of the second kind of a symmetric operatots* Bull. Acad. Sci. URSS., Ser. Math. **4** (1940), 53–104.

[21] V. Kostrykin and R. Schrader *Kirchhoff's rule for quantum wires*, J. Phys. A: Math. Gen. **32** (1999), 595–630.

[22] Yu.A. Kuperin and S.P. Merkurev, *Selfadjoint extensions and scattering theory for several-body systems*, in: Spectral theory of operators (Novgorod, 1989), Amer. Math. Soc. Transl. Ser. 2 **150**, Amer. Math. Soc., Providence, RI, 1992, 141–176.

[23] P.B. Kurasov and B.S. Pavlov, *Scattering problem with physical behavior of scattering matrix and operator relations*, to be published in Vinogradov's memorial volume.

[24] P.B. Kurasov and B.S. Pavlov, *Generalized perturbations and operator relations*, to be published in Proc. Conf. "Hyperfunctions, Operator and Dynamical Systems", Brussel, 1997.

[25] P.B. Kurasov, *Energy dependent boundary conditions and few-body scattering problem*, Rev. Math. Phys. **9** (1997), 853–906.

[26] K.A. Makarov, *Semiboundedness of the energy operator of a system of three particles with paired interactions of δ-function type*, Algebra i Analiz **4** (1992), 155–171.

[27] K.A. Makarov, V.V. Melezhik and A.K. Motovilov, *Point interactions in the problem of three quantum particles with internal structure*, Teoret. Mat. Fiz. **102** (1995), 258–282.

[28] K.A. Makarov and V.V. Melezhik, *Two sides of a coin: the Efimov effect and collapse in a three-body system with point interactions. I*, Teoret. Mat. Fiz. **107** (1996), 415–432.

[29] R. Mennicken and A. Shkalikov, *Spectral Decomposition of Symmetric Operator Matrices*. Mathematische Nachrichten. **179** (1996), 259–273.

[30] S.P. Merkuriev and S.L. Yakovlev, *On the quantum scattering problem for four identical particles interacting in the S state*, Soviet J. Nuclear Phys. **39** (1984), 1002–1006.

[31] S.P. Novikov, *Schr"odinger operators on graphs and symplectic geometry*, in: The Arnol'dfest (Proceedings of the Fields Institute Conference in Honour of the 60th Birthday of Vladimir I. Arnol'd), eds. E. Bierstone, B. Khesin, A. Khovanskii, and J. Marsden, to appear in the Fields Institute Communications Series.

[32] B. Pavlov, *A model of zero-radius potential with internal structure*, Teor. Mat. Fiz. vol. **59**, 3 (1984), 345–354 (English translation: Theoret. and Math. Phys **59** (1984), no. 3, 544–550).

[33] B.S. Pavlov, *Operator Extension Theory and explicitly solvable models*, Uspekhi Mat. Nauk **42** (1987), 99–131, 247.

[34] B.S. Pavlov, *Boundary conditions on thin manifolds and the semiboundedness of the three-body Schrödinger operator with point potential*, Mat. Sb. (N.S.) **136**(178) (1988), 163–177, 301.

[35] I.Yu. Popov, *Theory of extensions and the localization of resonances for domains of trap type*, Mat. Sb. **181** (1990), 1366–1390.

[36] I.Yu. Popov, *The Helmholtz resonator and operator extension theory in a space with indefinite metric*, Mat. Sb. **183** (1992), 3–27.

[37] Yu.G. Shondin, *On the three-particle problem with δ-potentials*, Teoret. Mat. Fiz. **51** (1982), 181–191.

Dept. of Mathematics
Stockholm University
106 91 Stockholm
Sweden
pak@matematik.su.se

Dept. of Mathematics
Auckland University
Private Bag 92019 Auckland
New Zealand
pavlov@math.auckland.ac.nz

31B35, 35Q40, 47N50, 81Q10

Operator Theory:
Advances and Applications, Vol. 118

Linearization and Compact Perturbation of Self-adjoint Analytic Operator Functions

H. Langer, A. Markus and V. Matsaev*

This paper is dedicated to M.G. Krein whose work and personality were of great influence to each of us

Spectral points of positive and of negative type of a self-adjoint analytic operator function A are introduced and their behavior under bounded and compact perturbations is studied. An essential tool is a linearization of the function A, which is a self-adjoint operator in some Krein space.

1 Introduction

Given an analytic operator function A with values in the algebra of bounded linear operators in some Hilbert space, the investigation of the spectral properties of A is often reduced to the study of the spectral properties of an associated linear operator M, a *linearization* of A, in a larger Hilbert space. In the simplest case of a monic operator polynomial A, the linearization M can be chosen to be the companion operator of the polynomial, for general analytic operator functions the construction of a linearization is more complicated, see [GKL], [KMR]. In the papers [KMR] the method of linearization was included into the more general theory of matrix nodes. In these papers no special attention was paid to the case of a *self-adjoint* analytic operator function. However, for monic operator polynomials it was observed and used long ago (see, e.g., [L1], [L2]), that in this case the linearization M can be chosen to be a self-adjoint operator in some Krein space.

In the present paper, for a given analytic self-adjoint operator function A in Sections 2 and 3 we define a linearization Λ which is self-adjoint in some Krein space $\mathcal{F}$. This gives the possibility to introduce and to study spectral points of definite type of the analytic operator function A, see Section 5. In fact, e.g. the definition of a spectral point of positive type of A is such that this point is a spectral point of positive type of the self-adjoint linearization Λ in the Krein space $\mathcal{F}$ in the sense of [LancMM1], see also [LMM]. In the proof of the final result of Section 5 (Theorem 5.5) the local spectral function of the self-adjoint operator Λ, which exists on the set $\sigma_+(A) \cup (\rho(A) \cap \mathbb{R})$ according to [LMM], plays an essential role.

*This author acknowledges support of the Fonds zur Förderung der wissenschaftlichen Forschung of Austria, Project P 12176 MAT.

It is shown in Section 4, that the linearization Λ is equivalent to the linearization of [GKL], see also [KMR]. However, we introduce additionally a Krein space structure, and the self-adjointness of the analytic operator function is reflected by the self-adjointness of the linearization in this Krein space inner product. This is used in Section 6, which contains the main result of this paper. It concerns the stability of the set $\sigma_+(A)$ of spectral points of positive type of A under a compact perturbation. In fact, it is shown in Theorem 6.1, that if a compact valued analytic self-adjoint operator function B is added to A : $A_1 = A + B$, then, roughly speaking, the points of $\sigma_+(A) \cup (\rho(A) \cap \mathbb{R})$ belong also to $\sigma_+(A_1) \cup (\rho(A_1) \cap \mathbb{R})$ with the possible exception of isolated points, which can become eigenvalues of A_1 of "finite negative index" (for the definition see Section 6). A simpler result about bounded perturbations of A, which are sufficiently small, is contained in Theorem 5.2. For operator polynomials and in more particular situations similar results for compact and bounded perturbations were proved in [LancMM2] and [FMM].

In this paper, we restrict ourselves to the consideration of spectral points of positive type. It is clear that, mutatis mutandis, these results also hold for spectral points of negative type. Finally we mention that the linearization Λ, which is introduced in Sections 2 and 3, can be considered also for operator functions which are not self-adjoint. Also in this case Theorems 3.1 and 5.1 remain valid.

After an earlier version of this paper was completed M.A. Kaashoek read it carefully and made a number of important remarks. For us it was particularly useful that he pointed out various connections between the general theory of spectral nodes from [KMR] and the linearization Λ from Section 3. Not only the Gohberg-Kaashoek-Lay linearization M (see Section 4) can be embedded in a spectral node $\Theta = (M, B, C; \mathcal{K}, \mathcal{H})$[1] but also the quintet $(\Lambda, P, P^*; \mathcal{F}, \mathcal{H})$ is a spectral node. The fact that the self-adjointness of the operator function A implies that the inner operator M is self-adjoint in a Krein space follows from the uniqueness statement in [KMR], Part I. Indeed, if an operator function is self-adjoint then with the corresponding node $\Theta = (M, B, C; \mathcal{K}, \mathcal{H})$ also $\Theta_* = (M^*, C^*, B^*; \mathcal{K}, \mathcal{H})$ is a spectral node. But then according to Theorem 1.2 of [KMR], Part I, there exists a unique invertible operator S on $\mathcal{K}$ such that

$$M^* = SMS^{-1}, \; C^* = SB, \; B^* = CS^{-1}, \tag{1.1}$$

and the uniqueness of S implies $S = S^*$ (cf. [R], p. 178). Therefore $\mathcal{K}$ endowed with the indefinite inner product $[x, y] = (Sx, y)$, $x, y \in \mathcal{K}$, is a Krein space. The first relation in (1.1) shows that M is self-adjoint in this Krein space. However, in our proof of the perturbation results in Sections 6 and 7 a particular form of the Krein space inner product is important.

The above mentioned uniqueness theorem for spectral nodes in [KMR] implies also that the Krein spaces $(\mathcal{F}, \langle \cdot, \cdot \rangle)$ and $(\widehat{\mathcal{F}}, [\cdot, \cdot]_{L^2})$, see Lemma 4.2, and also the corresponding linearizations Λ and M are isomorphic. An explicit form of this isomorphism was obtained in Theorem 4.1.

[1] In [KMR] the inner operator is denoted by A.

2 The Krein Space

Let $\mathcal{H}$ be a Hilbert space and denote by $\mathcal{L}(\mathcal{H})$ the set of bounded linear operators in $\mathcal{H}$. We consider a bounded and simply connected domain $\mathcal{D}$ in the complex plain $\mathbb{C}$ which is symmetric with respect to the real axis $\mathbb{R}$, that is $\mathcal{D} = \mathcal{D}^*$[2], and an $\mathcal{L}(\mathcal{H})$-valued function A on $\mathcal{D}$ with the following properties:

(i) A is analytic in $\mathcal{D}$;

(ii) A is self-adjoint in $\mathcal{D}$, i.e. $A(\lambda^*) = A(\lambda)^*$ $(\lambda \in \mathcal{D})$;

(iii) $\sigma(A)$ is a compact subset of $\mathcal{D}$, i.e. $A(\lambda)^{-1}$ exists and is an analytic function of λ outside of a compact subset of $\mathcal{D}$.

Recall that $\sigma(A) := \{\lambda \in \mathcal{D} : 0 \in \sigma(A(\lambda))\}$, the resolvent set $\varrho(A)$ and the point spectrum (set of eigenvalues) $\sigma_p(A)$ are defined correspondingly.

In order to define the Krein space where the linearization will act, we choose a simply connected domain $\mathcal{D}_0$ such that $\sigma(A) \subset \mathcal{D}_0$, $\mathcal{D}_0 = \mathcal{D}_0^*$, its closure $\overline{\mathcal{D}_0}$ is contained in $\mathcal{D}$ and its boundary γ_0 is sufficiently smooth. By $L^2 := L^2(\gamma_0, \mathcal{H})$ we denote the Hilbert space of all strongly measurable $\mathcal{H}$-valued functions f on γ_0 such that

$$\oint_{\gamma_0} (f(t), f(t))_{\mathcal{H}} |dt| < \infty.$$

The Hilbert inner product $(\cdot\,,\,\cdot)_{L^2}$ on L^2 is defined as follows:

$$(f, g)_{L^2} = \oint_{\gamma_0} (f(t), g(t))_{\mathcal{H}} |dt|.$$

Let L^2_+ (L^2_-, *respectively*) be the subspace of L^2 which consists of those functions $f \in L^2$ which can be analytically continued into $\mathcal{D}_0$ (into $\mathbb{C} \setminus \overline{\mathcal{D}_0}$ and which vanish at ∞, respectively). Recall that

$$L^2 = L^2_+ \dotplus L^2_-, \tag{2.1}$$

where $\dotplus$ denotes the direct sum. A proof of (2.1) can be found in [GL]. We mention that it can easily be reduced to the scalar case if one considers an orthonormal basis in $\mathcal{H}$.

The set

$$\mathcal{M} := A\,L^2_+ \tag{2.2}$$

is a closed subspace of L^2_+. We define an in general indefinite and degenerate inner product $\langle \cdot\,,\,\cdot \rangle$ on L^2_+:

$$\langle f, g \rangle := \frac{1}{2\pi i} \oint_{\gamma_0} (A(t)^{-1} f(t), g(t^*))_{\mathcal{H}}\, dt$$

---

[2] If $\lambda = \alpha + i\beta$ with $\alpha, \beta \in \mathbb{R}$, then $\lambda^* = \alpha - i\beta$, and for a set $\mathcal{D} \subset \mathbb{C}$ we denote $\mathcal{D}^* = \{\lambda^* : \lambda \in \mathcal{D}\}$.

It has, evidently, the following properties:

(i) $\langle f, g\rangle^* = \langle g, f\rangle$ $(f, g \in L^2)$;

(ii) If $f \in \mathcal{M}$ then $\langle f, g\rangle = 0$ for all $g \in L^2_+$.

Finally we introduce the quotient space

$$\mathcal{F} := L^2_+/\mathcal{M},$$

which is a Hilbert space with respect to the corresponding quotient norm. According to (ii) the inner product $\langle\cdot\,,\,\cdot\rangle$ induces an inner product on $\mathcal{F}$ which we denote also by $\langle\cdot\,,\,\cdot\rangle$.

Theorem 2.1 *The space $(\mathcal{F}, \langle\cdot\,,\,\cdot\rangle)$ is a Krein space.*

The following two lemmas will be used in the proof of Theorem 2.1 and also later.

Lemma 2.1 *Each $f \in \mathcal{F}$ (considered as an equivalence class of elements of L^2_+) contains one and only one function u_f of the form*

$$u_f(t) = A(t)\widehat{f}_-(t) \qquad (t \in \gamma_0)$$

with $\widehat{f}_- \in L^2_-$. According to the last equality the function $\widehat{f}_-$ has a natural analytic extension onto $\mathbb{C}\backslash\sigma(A)$.

Proof: If $w \in f$, then according to (2.1) the function $A^{-1}w \in L^2$ can be written as

$$A(t)^{-1}w(t) = \widetilde{v}_+(t) + \widetilde{v}_-(t) \quad (\widetilde{v}_\pm \in L^2_\pm,\ t \in \gamma_0). \tag{2.3}$$

Define

$$\widehat{f}_-(\lambda) := \begin{cases} \widetilde{v}_-(\lambda) & \lambda \in \mathbb{C}\backslash\overline{\mathcal{D}_0}, \\ A(\lambda)^{-1}w(\lambda) - \widetilde{v}_+(\lambda) & \lambda \in \mathcal{D}_0\backslash\sigma(A). \end{cases} \tag{2.4}$$

Because of (2.3) on γ_0 the boundary values of $\widetilde{v}_-(\lambda)$ and of $A(\lambda)^{-1}w(\lambda) - \widetilde{v}_+(\lambda)$ coincide, hence $\widehat{f}_-(\lambda)$ is analytic on $\mathbb{C}\backslash\sigma(A)$. Since

$$w(\lambda) = A(\lambda)\widetilde{v}_+(\lambda) + A(\lambda)\widetilde{v}_-(\lambda)$$

the function $u_f = A\widehat{f}_-$ has all the stated properties. □

Lemma 2.2 *The transformation $f \mapsto \widehat{f}_-$ from $\mathcal{F}$ into L^2_- as defined in Lemma 2.1 is a linear continuous bijection from the Hilbert space $\mathcal{F}$ onto a closed subspace of L^2_-. For each $\lambda \in \rho(A)$ there exists a number C_λ such that*

$$\|\widehat{f}_-(\lambda)\|_{\mathcal{H}} \le C_\lambda \|\widehat{f}_-\|_{L^2}. \tag{2.5}$$

Proof: Let $f \in \mathcal{F}$. For any $w \in f$ we consider the decomposition (2.3) and the elements

$$\widetilde{v}_{\pm} = P_{\pm}[A^{-1}(t)w(t)] \tag{2.6}$$

where P_+ is the projection onto L^2_+ which vanishes on L^2_-, and $P_- := I - P_+$. From (2.3) and (2.4) it follows that $\widehat{f}_- = \widetilde{v}_-$, and by (2.6)

$$\|\widehat{f}_-\|_{L^2} = \|\widetilde{v}_-\|_{L^2} \leq \|P_-\| \max_{t\in\gamma_0} \|A^{-1}(t)\| \; \|w\|_{L^2}.$$

Here w is an arbitrary function from f, hence

$$\|\widehat{f}_-\|_{L^2} \leq C\|f\|_{\mathcal{F}} \quad \left(C = \|P_-\| \max_{t\in\gamma_0} \|A^{-1}(t)\|\right).$$

For the inverse estimate we use (2.4):

$$A(t)\widehat{f}_-(t) = w(t) - A(t)\widehat{v}_+(\lambda).$$

Since $w - A\widetilde{v}_+ \in f$, we have

$$\|f\|_{\mathcal{F}} \leq \|w - A\widetilde{v}_+\|_{L^2} = \|A\widehat{f}_-\|_{L^2} \leq \max_{t\in\gamma_0} \|A(t)\| \; \|\widehat{f}_-\|_{L^2},$$

and the first assertion is proved.

We turn to the proof of the second assertion. If $\lambda \in \mathbb{C}\backslash\overline{D_0}$ then the inequality (2.5) follows from Cauchy's formula applied to the function $\widehat{f}_- \in L^2_-$:

$$\widehat{f}_-(\lambda) = -\frac{1}{2\pi i}\oint_{\gamma_0} \frac{f(t)\,dt}{t-\lambda}.$$

If $\lambda \in D_0\backslash\sigma(A)$ we use the equality (see (2.4))

$$\widehat{f}_-(\lambda) = A^{-1}(\lambda)w(\lambda) - \widetilde{v}_+(\lambda). \tag{2.7}$$

Applying Cauchy's formula to the functions $w,\ \widetilde{v}_+ \in L^2_+$ we obtain

$$\|w(\lambda)\|_{\mathcal{H}} \leq C'_\lambda \|w\|_{L^2}, \quad \|\widetilde{v}_+(\lambda)\|_{\mathcal{H}} \leq C''_\lambda \|\widetilde{v}_+\|_{L^2}. \tag{2.8}$$

From (2.7), (2.8) and (2.6) we see that

$$\begin{aligned}\|\widehat{f}_-(\lambda)\|_{\mathcal{H}} &\leq \|A^{-1}(\lambda)\|\, C'_\lambda \|w\|_{L^2} + C''_\lambda \|\widetilde{v}_+\|_{L^2}\\ &\leq \left(C'_\lambda \|A^{-1}(\lambda)\| + C''_\lambda \|P_+\| \max_{t\in\gamma_0} \|A^{-1}(t)\|\right) \|w\|_{L^2}.\end{aligned}$$

Here w is an arbitrary function from f, and hence we can replace $\|w\|_{L^2}$ by $\|f\|_{\mathcal{F}}$. By the first part of the proof the last norm is equivalent to $\|\widehat{f}_-\|_{L^2}$, and we obtain (2.5) for $\lambda \in D_0\backslash\sigma(A)$ as well. It remains only to consider the case $\lambda \in \gamma_0$.

Choose an additional simply connected domain D_1 such that $\sigma(A) \subset D_1$, $D_1 = D_1^*$ and $\overline{D_1} \subset D_0$. Denote the (sufficiently smooth) boundary of D_1 by γ_1. We shall consider the Hilbert space $L^2(\gamma_1) = L^2(\gamma_1, \mathcal{H})$. With respect to the domain D_1 points $\lambda \in \gamma_0$ are exterior points, and by the proof of the inequality (2.5) for $\lambda \in \mathbb{C}\backslash\overline{D_0}$ we have

$$\|\widehat{f}_-(\lambda)\|_{\mathcal{H}} \leq C_\lambda \, \|\widehat{f}_-\|_{L^2(\gamma_1)} \qquad (\lambda \in \gamma_0).$$

Thus, for the proof of (2.5) for $\lambda \in \gamma_0$ it is sufficient to obtain an estimate

$$\|\widehat{f}_-\|_{L^2(\gamma_1)} \leq C_1 \, \|\widehat{f}_-\|_{L^2}. \tag{2.9}$$

Put $u(t) = A(t)\, \widehat{f}_-(t)$. Then

$$\|\widehat{f}_-\|_{L^2(\gamma_1)} = \|A^{-1}u\|_{L^2(\gamma_1)} \leq \max_{t \in \gamma_1} \|A^{-1}(t)\| \, \|u\|_{L^2(\gamma_1)}.$$

But $\|u\|_{L^2(\gamma_1)} \leq C \, \|u\|_{L^2}$ since $u \in L^2_+$. On the other hand

$$\|u\|_{L^2} = \|A\widehat{f}_-\|_{L^2} \leq \max_{t \in \gamma_0} \|A(t)\| \|\widehat{f}_-\|_{L^2},$$

and we obtain (2.9). □

Lemma 2.3 *Each continuous linear functional Φ on the space L^2_+ can be represented by means of a function $\varphi \in L^2_-$ in the form*

$$\Phi(f) = \oint_{\gamma_0} (f(t), \varphi(t^*))_{\mathcal{H}} \, dt \qquad (f \in L^2_+). \tag{2.10}$$

This correspondence between $(L^2_+)^*$ and L^2_- is an isomorphism of these Hilbert spaces.

Proof: We consider on L^2 also the following bilinear form:

$$[f, g] := \oint_{\gamma_0} (f(\zeta), g(\zeta))_{\mathcal{H}} \, d\zeta = \oint_{\gamma_0} \left(f(\zeta), \frac{d\zeta^*}{|d\zeta|} g(\zeta) \right)_{\mathcal{H}}$$
$$|d\zeta| = \left(f, \frac{d\zeta^*}{|d\zeta|} g \right)_{L_2}.$$

The operator of multiplication by the function $\frac{d\zeta^*}{|d\zeta|}$ is unitary in L_2. Therefore for each continuous linear functional ψ on L_2 there exists a vector function h in L_2 such that

$$\psi(f) = [f, h], \quad \|h\|_{L^2} = \|\psi\|.$$

The set of elements in L_2 which are orthogonal on L_2^+ with respect to the form $[\cdot, \cdot]$ is the set of all functions $h \in L_2$ such that the function $k(\zeta) := h(\zeta^*)$ is in L_2^+. Indeed, if $f \in L_2^+$ and $h \in L_2^+$, then the scalar function $(f(\zeta), h(\zeta^*))_{\mathcal{H}}$ belongs to the Smirnov (Hardy) class $E_1(\gamma_0)$, and hence

$$[f, k] = \oint_{\gamma_0} (f(\zeta), h(\zeta^*))_{\mathcal{H}} d\zeta = 0.$$

Conversely, if

$$\oint_{\gamma_0} (f(\zeta), h(\zeta^*))_{\mathcal{H}}\, d\zeta = 0$$

for all $f \in L_2^+$, we can choose $f(\zeta) = \frac{u}{\zeta - z}$, where z is an arbitrary point of $\mathbb{C}\backslash\overline{\mathcal{D}_0}$ and $u \in \mathcal{H}$. It follows that

$$\begin{aligned} 0 &= \int_{\gamma_0} (f(\zeta), h(\zeta^*))_{\mathcal{H}} d\zeta = \oint_{\gamma_0} \left(u, \frac{h(\zeta^*)}{\zeta^* - z^*} \right)_{\mathcal{H}} \\ d\zeta &= \left(u, \oint_{\gamma_0} \frac{h(\zeta^*)}{\zeta^* - z^*} d\zeta^* \right)_{\mathcal{H}}. \end{aligned}$$

Since $u \in \mathcal{H}$ is arbitrary, this yields $\oint_{\gamma_0} \frac{h(\zeta^*)}{\zeta^* - z^*}\, d\zeta^* = 0$, or $\oint_{\gamma_0} \frac{h(t)dt}{t-w} = 0$ for all $w \in \mathbb{C}\backslash\overline{\mathcal{D}_0}$, which means $h \in L_2^+$.

Now we recall (see [Ru], Theorem 4.9) that the adjoint space of a subspace $\mathcal{M}$ of a Banach space E is the quotient space $E^*/\mathcal{M}^\perp$, where $\mathcal{M}^\perp$ is the set of all elements of E^* which vanish on $\mathcal{M}$. Therefore this adjoint space is isomorphic to an arbitrary complement of $\mathcal{M}^\perp$ in E^*. Since $L_2^+ \dot{+} L_2^- = L_2$, it follows from the above that the adjoint space of L_2^+ with respect to $[\cdot\,, \cdot]$ is of the form (2.10) with $\varphi \in L_2^-$ (up to an isomorphism). □

Proof of Theorem 2.1: According to [L3], Lemma I.3, we have to show that the space $(\mathcal{F}, \langle \cdot\,, \cdot \rangle)$ is weakly sequentially complete, that is, $(f^n) \subset \mathcal{F}$ and

$$\langle f^n - f^m, g \rangle \longrightarrow 0 \quad (n, m \longrightarrow \infty) \tag{2.11}$$

for all $g \in \mathcal{F}$ imply the existence of an $f \in \mathcal{F}$ such that $\langle f^n - f, g \rangle \to 0$ for all $g \in \mathcal{F}$ if $n \to \infty$. Given such a sequence (f^n), we choose as a representative for f^n the function $A(\lambda)\widehat{f_-^n}(\lambda)$ with $\widehat{f_-^n} \in L_-^2$. Then the relation (2.11) implies s

$$\oint_{\gamma_0} (\widehat{f_-^n}(t) - \widehat{f_-^m}(t), g(t^*))_{\mathcal{H}}\, dt \longrightarrow 0 \quad (n, m \longrightarrow \infty)$$

for all $g \in L_+^2$. According to Lemma 2.3, the sequence $\widehat{f_-^n}$ is a weak Cauchy sequence in L_-^2. As the Hilbert space L_-^2 is weakly complete there exists a function $\widehat{f}_- \in L_-^2$ such that $\widehat{f_-^n} \to \widehat{f}_-$ if $n \to \infty$, weakly in L_-^2. This relation and

$$(\widehat{f_-^n}, \varphi)_{L^2} = (\widehat{f}_-, \varphi)_{L^2} = 0 \qquad (\varphi \perp L_-^2)$$

imply $\widehat{f_-^n} \to \widehat{f}_-$, weakly in L^2, hence $A\widehat{f_-^n} \to A\widehat{f}_-$ if $n \to \infty$, weakly in L^2, and $A\widehat{f}_- \in L_+^2$. The equivalence class f which contains the function $A\widehat{f}_-$ has all the desired properties. □

3 The Linearization Λ

In the space L_+^2 we consider the (bounded linear) operator Λ_0 of multiplication by the independent variable: $(\Lambda_0 u)(t) := tu(t)$ $(t \in \gamma_0, u \in L_+^2)$. As the subspace $\mathcal{M}$ in (2.2) is invariant under Λ_0, this operator generates a bounded linear operator Λ in $\mathcal{F} = L_+^2/\mathcal{M}$. It is easy to see that Λ is self-adjoint in the Krein space $\mathcal{F}$. In this section connections between the given self-adjoint operator function A in $\mathcal{H}$ and the self-adjoint operator Λ in $\mathcal{F}$ are established.

Theorem 3.1 *The following relations hold true:*

$$\sigma(A) = \sigma(\Lambda), \quad \sigma_p(A) = \sigma_p(\Lambda).$$

Proof: Consider the equation

$$(\Lambda - z)x = f \tag{3.1}$$

for $f \in \mathcal{F}$ and $z \notin \sigma(A)$. If $f = 0$ this equation has only the trivial solution. Indeed, a function $x(\lambda)$ of the equivalence class $x \in \mathcal{F}$ satisfies then the relation

$$(\lambda - z)x(\lambda) = A(\lambda)g(\lambda) \qquad (\lambda \in \mathcal{D}_0) \tag{3.2}$$

with some $g \in L_+^2$, and if $z \in \mathcal{D}_0$ this also implies $A(z)g(z) = 0$, hence $g(z) = 0$. Therefore the function $x(\lambda) = A(\lambda)\frac{g(\lambda)-g(z)}{\lambda-z}$ belongs to the equivalence class $0 \in \mathcal{F}$. If $z \notin \mathcal{D}_0$ we use for x the representing function $u_x(\lambda) = A(\lambda)\widehat{x}_-(\lambda)$ with some $\widehat{x}_- \in L_-^2$, and (3.2) implies

$$(\lambda - z)A(\lambda)\widehat{x}_-(\lambda) = A(\lambda)g(\lambda).$$

It follows that $g(\lambda)$ can be analytically continued onto γ_0, and that $g(z) = 0$ and hence

$$A(t)\left(\widehat{x}_-(t) - \frac{g(t)}{t-z}\right) = 0 \qquad (t \in \gamma_0).$$

Since A is boundedly invertible on γ_0, we find that

$$\widehat{x}_-(t) = \frac{g(t)}{t-z}, \qquad (t \in \gamma_0),$$

hence $\widehat{x}_-$ is analytic on $\mathcal{D}_0$. Since it belongs also to L_-^2 this implies $\widehat{x}_-(t) \equiv 0$ and $x = 0$.

Now consider the inhomogenous equation (3.1), first for $z \in \mathcal{D}_0 \backslash \sigma(A)$. We choose for f the representative u_f of the form $u_f(\lambda) = A(\lambda)\widehat{f}_-(\lambda)$ with some $\widehat{f}_- \in L_-^2$, which is analytic on $\mathbb{C}\backslash\sigma(A)$, see Lemma 2.1. Consider the function

$$u(\lambda) = \frac{\widehat{f}_-(\lambda) - \widehat{f}_-(z)}{\lambda - z}.$$

Since it is analytic on $\mathbb{C}\backslash\sigma(A)$ and vanishes at ∞, it belongs to L_-^2. Further,

$$A(\lambda)u(\lambda) = \frac{A(\lambda)\widehat{f_-}(\lambda) - A(z)\widehat{f_-}(z)}{\lambda - z} - \frac{A(\lambda) - A(z)}{\lambda - z}\widehat{f_-}(z),$$

hence $Au \in L_+^2$. If x denotes the element of $\mathcal{F}$ which contains Au, it follows that the function $A(\lambda)\widehat{f_-}(\lambda) - A(\lambda)\widehat{f_-}(z)$ belongs to the equivalence class $(\Lambda - z)x$ and that this function is a representative of f. Observing the definition of $\widehat{f_-}$ from (2.4), with a representative w of f we obtain that the function

$$x(\lambda) = \frac{w(\lambda) - A(\lambda)\widetilde{v}_+(\lambda) - A(\lambda)\widehat{f_-}(z)}{\lambda - z} \tag{3.3}$$

is a representative of x. If $z \in \mathbb{C}\backslash\overline{\mathcal{D}_0}$, then the function $x(\lambda) = \frac{f(\lambda)}{\lambda - z}$ is a representative of x, it belongs to L_+^2 and satisfies (3.1); if $z \in \gamma_0$ then with $u_f = Av_-$, $v_- \in L_-^2$, we have $A(z)v_-(z) = 0$ and the function $x(\lambda) = \frac{A(\lambda)v_-(\lambda)}{\lambda - z}$ has again an analytic continuation into $\overline{\mathcal{D}_0}$, hence it belongs to L_+^2 and satisfies (3.1). Thus $\sigma(\Lambda) \subset \sigma(A)$.

In order to prove the inclusion $\sigma(A) \subset \sigma(\Lambda)$ we show that $\lambda_0 \in \varrho(\Lambda) \cap \mathcal{D}_0$ yields $\lambda_0 \in \varrho(A)$. The operator $A(\lambda_0)$ is injective. Indeed, $A(\lambda_0)h = 0$ implies that the function

$$u(\lambda) = \frac{A(\lambda)h}{\lambda - \lambda_0} = \frac{A(\lambda) - A(\lambda_0)}{\lambda - \lambda_0}h \tag{3.4}$$

belongs to L_+^2 and $(\lambda - \lambda_0)u(\lambda) \in \mathcal{M}$, hence for $f \in L_+^2/\mathcal{M}$, which contains u, we have $(\Lambda - \lambda_0)f = 0$ and, since $\lambda_0 \in \varrho(\Lambda)$, $f = 0$ (in $\mathcal{F}$) follows. Thus $u(\lambda) = A(\lambda)v(\lambda)$ with $v \in L_+^2$, and therefore by (3.4)

$$A(\lambda)\Big(\frac{h}{\lambda - \lambda_0} - v(\lambda)\Big) = 0.$$

Since $A(\lambda)$ is boundedly invertible on γ_0 we find $\frac{h}{\lambda - \lambda_0} = v(\lambda) \in L_+^2$, which is possible only if $h = 0$.

Now let $u_0 \in \mathcal{H}$. If u is the element of $\mathcal{F}$ which contains the constant function $u_0(\lambda) = u_0$ ($\lambda \in \overline{\mathcal{D}_0}$), then, since $\lambda_0 \in \varrho(\Lambda)$, there exists a solution $h \in \mathcal{F}$ of the equation $(\Lambda - \lambda_0)h = u$. Thus for any function v of the equivalence class h it holds

$$(\lambda - \lambda_0)v(\lambda) = u_0 + A(\lambda)w(\lambda)$$

where $w \in L_+^2$. Choosing $\lambda = \lambda_0$ we find $A(\lambda_0)(-w(\lambda_0)) = u_0$, hence $u_0 \in A(\lambda_0)\mathcal{H}$, and the equality $\sigma(A) = \sigma(\Lambda)$ is proved.

If $\lambda_0 \in \sigma_p(A)$ and u_0 is a corresponding eigenvector, that is $A(\lambda_0)u_0 = 0$, then the element $\widetilde{u}_0$ of $\mathcal{F}$, given by the function $\widetilde{u}_0(\lambda) = \frac{A(\lambda)u_0}{\lambda - \lambda_0}$, is an eigenvector of Λ at λ_0, since $(\lambda - \lambda_0)\widetilde{u}_0(\lambda) = A(\lambda)u_0$ is a representative of the zero element of $\mathcal{F}$ and the function on the right hand side is a representative of $(\Lambda - \lambda_0)\widetilde{u}_0$. Conversely, if $\widetilde{u}_0$ is an eigenvector of Λ at λ_0 with representative $\widetilde{u}_0(\lambda)$, then

$$(3.5)\qquad (\lambda - \lambda_0)\widetilde{u}_0(\lambda) = A(\lambda) f_0(\lambda)$$

with some $f_0 \in L^2_+$, or $\widetilde{u}_0(\lambda) = \frac{A(\lambda) f_0(\lambda)}{\lambda - \lambda_0}$. It follows from (3.5) that $A(\lambda_0) f_0(\lambda_0) = 0$. Here $f_0(\lambda_0) \neq 0$, since $f_0(\lambda_0) = 0$ would imply that the function

$$\widetilde{u}_0(\lambda) = \frac{A(\lambda)(f_0(\lambda) - f_0(\lambda_0))}{\lambda - \lambda_0}$$

belongs to the zero class of $\mathcal{F}$, a contradiction. Observe that together with $\frac{A(\lambda) f_0(\lambda)}{\lambda - \lambda_0}$ also the function $\frac{A(\lambda) f_0(\lambda_0)}{\lambda - \lambda_0}$ is a representative of $\widetilde{u}_0$ in $\mathcal{F}$. □

We mention that there is a bijective correspondence between the Jordan chains of the operator function A and of the operator Λ at the same eigenvalue. It will be shown in Section 3 that also the approximate point spectra of A and of Λ coincide. A more direct relation between the given self-adjoint operator function A in $\mathcal{H}$ and the self-adjoint operator Λ in $\mathcal{F}$ is established in Theorem 3.2 below. In order to formulate it, denote by P the mapping from $\mathcal{H}$ to $\mathcal{F}$ which associates with an element $\varphi \in \mathcal{H}$ the equivalence class in $L_2^+/\mathcal{M}$, which contains the function $u(t) = \varphi,\ t \in \gamma_0$. We consider here $\mathcal{F}$ as a Krein space with the inner product $\langle \cdot\,, \cdot \rangle$. Then the adjoint operator P^* is a mapping from the Krein space $(\mathcal{F}, \langle \cdot\,, \cdot \rangle)$ into the Hilbert space $(\mathcal{H}, (\cdot\,, \cdot)_{\mathcal{H}})$.

Theorem 3.2 *For $z \in \mathcal{D}_0 \backslash \sigma(A)$ the following relation holds true:*

$$(3.6)\qquad P^*(\Lambda - z)^{-1} P = -A(z)^{-1} + B(z)$$

where B is given by

$$B(z) = \frac{1}{2\pi i} \oint_{\gamma_0} \frac{A(t)^{-1}}{t - z} dt$$

and hence analytic in $\mathcal{D}_0$. Moreover, it holds

$$(3.7)\qquad \mathcal{F} = c.l.s.\{(\Lambda - z)^{-1} P\mathcal{H} : z \in \mathbb{C}\backslash\overline{\mathcal{D}_0}\}.$$

Proof: If $\varphi, \psi \in \mathcal{H}$, $f := P\psi$ and we apply the expression (3.3) for $(\Lambda - z)^{-1} f$, it follows that

$$\begin{aligned}
&\langle (\Lambda - z)^{-1} P\varphi, P\psi \rangle \\
&= \frac{1}{2\pi i} \oint_{\gamma_0} \left(A(t)^{-1} \frac{\varphi - A(t)\widetilde{v}_+(t) - A(t)\widetilde{v}_-(z)}{t - z}, \psi \right)_{\mathcal{H}} dt \\
&= \frac{1}{2\pi i} \oint_{\gamma_0} \frac{(A(t)^{-1}\varphi, \psi)_{\mathcal{H}}}{t - z} dt - (\widetilde{v}_+(z) + \widetilde{v}_-(z), \psi)_{\mathcal{H}}.
\end{aligned}$$

Observing (2.3), the last expression equals

$$\frac{1}{2\pi i}\left(\oint_{\gamma_0}\frac{A(t)^{-1}}{t-z}dt\,\varphi,\psi\right)_{\mathcal{H}} - (A(z)^{-1}\varphi,\psi)_{\mathcal{H}},$$

and the relation (3.6) is proved.

In order to prove the relation (3.7) we choose as a representative of $(\Lambda - z)^{-1}P\varphi$ the function $\frac{\varphi}{t-z}$, $t \in \gamma_0$, see the proof of Theorem 3.1. If $f^0 \in \mathcal{F}$ is orthogonal to all elements $(\Lambda - z)^{-1}P\varphi$, $\varphi \in \mathcal{H}$, $z \in \mathbb{C}\backslash\overline{\mathcal{D}_0}$, then with $f^0 = A\widehat{f}_-^0$ where $\widehat{f}_-^0 \in L^2_-$ we find

$$\begin{aligned} 0 &= \langle(\Lambda - z)^{-1}P\varphi, f^0\rangle = \frac{1}{2\pi i}\oint_{\gamma_0}\left(\frac{\varphi}{t-z}, \widehat{f}_-^0(t^*)\right)_{\mathcal{H}} dt \\ &= -(\varphi, \widehat{f}_-^0(z^*))_{\mathcal{H}}, \end{aligned}$$

hence $\widehat{f}_-^0 = 0$ and $f^0 = 0$. □

Remark 1 If $A(z)^{-1}$ is analytic in $\mathbb{C}\backslash\overline{\mathcal{D}_0}$ and vanishes at ∞, then in relation (3.6) we have $B(z) \equiv 0$ and it reduces to

$$P^*(\Lambda - z)^{-1}P = -A(z)^{-1} \quad (z \in \varrho(A)).$$

This holds, in particular, if A is an operator polynomial with invertible leading coefficient.

Remark 2 In the relation (3.7) the set $\mathbb{C}\backslash\overline{\mathcal{D}_0}$ can be replaced by any nonempty open subset of $\mathbb{C}\backslash\overline{\mathcal{D}_0}$ and also by $\varrho(A) \cap \mathcal{D}_0$. Moreover, (3.7) is equivalent to the relation

$$\mathcal{F} = \text{c.l.s.}\{\Lambda^k P\mathcal{H} : k = 0, 1, \ldots\}.$$

Recall that two bounded linear operators T, T' in Krein spaces $\mathcal{K}$, $\mathcal{K}'$ are called *weakly isomorphic* if there exist dense linear subsets $\mathcal{W}$, $\mathcal{W}'$ of $\mathcal{K}$, $\mathcal{K}'$, respectively, and a linear mapping U from $\mathcal{W}$ onto $\mathcal{W}'$ such that

$$[T\varphi, T\psi] = [T'U\varphi, T'U\psi] \quad (\varphi, \psi \in \mathcal{W}).$$

The relations (3.6) and (3.7) determine the operator Λ up to a weak isomorphism. This follows easily from the relation

$$\begin{aligned} \langle\Lambda^k P\varphi, \Lambda^l P\psi\rangle &= \langle\Lambda^{k+l}P\varphi, P\psi\rangle \\ &= \left(-\frac{1}{2\pi i}\oint_{\gamma_0} z^{k+l}P^*(\Lambda - z)^{-1}dzP\varphi, \psi\right)_{\mathcal{H}} \\ &= -\frac{1}{2\pi i}\left(\oint_{\gamma_0} z^{k+l}A(z)^{-1}dz\,\varphi, \psi\right)_{\mathcal{H}}. \end{aligned}$$

In the following, the self-adjoint operator Λ in the Krein space $\mathcal{F}$ is called a *linearization* of the self-adjoint operator function A in the Hilbert space $\mathcal{H}$. We do not give a formal definition of a linearization here, but we only mention that such a definition of a (minimal) linearization of A (or of A^{-1}) by the relations (3.6) and (3.7) would correspond to the linearization, which was given in [DLS1], [DL] for Sturm-Liouville problems with eigenvalue depending boundary conditions or for the corresponding Titchmarsh-Weyl coefficients.

4 The Linearization of Gohberg, Kaashoek and Lay

In this section we show that the operator Λ of Section 3 is isomorphic to the linearization M of the operator function A as defined by Gohberg, Kaashoek and Lay [GKL], see also [KMR], [GGK], Section III.2 and (for operator polynomials) [R], Theorem 1.3.1. Moreover, we shall equip the space $L^2 = L^2(\gamma_0, \mathcal{H})$, which is the basic space for the construction of M, with a G-inner product which coincides under this isomorphism with the Krein space inner product of $\mathcal{F}$.

The linearization M is defined as follows. Consider the space L^2 as above. It is supposed without loss of generality that $0 \in \mathcal{D}_0$. The operator T in L^2 is defined by the relation

$$(4.1) \qquad (Tf)(t) = tf(t) - \frac{1}{2\pi i}\oint_{\gamma_0} f(s)ds + \frac{1}{2\pi i}\oint_{\gamma_0} A(s)f(s)ds.$$

It is a bounded linear operator in L^2.

Lemma 4.1 *The resolvent $(T-z)^{-1}$ exists for $z \in \mathcal{D}_0 \backslash \sigma(A)$ and is given by the relation*

$$(4.2) \quad ((T-z)^{-1}f)(t) = \frac{f(t)}{t-z} + \frac{A(z)^{-1}}{t-z}\frac{1}{2\pi i}\oint_{\gamma_0} \frac{(I-A(s))f(s)}{s-z}ds.$$

Proof: We have

$$\begin{aligned}
&(T-z)\left(\frac{f(t)}{t-z} + \frac{A(z)^{-1}}{t-z}\frac{1}{2\pi i}\oint_{\gamma_0}\frac{(I-A(s))f(s)}{s-z}ds\right) \\
&= f(t) + A(z)^{-1}\frac{1}{2\pi i}\oint_{\gamma_0}\frac{(I-A(s))f(s)}{s-z}ds \\
&\quad - \frac{1}{2\pi i}\oint_{\gamma_0}\left(\frac{f(s)}{s-z} + \frac{A(z)^{-1}}{s-z}\frac{1}{2\pi i}\oint_{\gamma_0}\frac{(I-A(s'))f(s')}{s'-z}ds'\right)ds \\
&\quad + \frac{1}{2\pi i}\oint_{\gamma_0}\left(\frac{A(s)f(s)}{s-z} + \frac{A(s)A(z)^{-1}}{s-z}\frac{1}{2\pi i}\oint_{\gamma_0}\frac{(I-A(s'))f(s')}{s'-z}ds'\right)ds \\
&= f(t).
\end{aligned}$$

Here the last equality follows easily if we observe that

$$\begin{aligned}
&\frac{1}{2\pi i}\oint_{\gamma_0}\frac{A(s)A(z)^{-1}}{s-z}\frac{1}{2\pi i}\oint_{\gamma_0}\frac{(I-A(s'))f(s')ds'}{s'-z}ds\\
&\quad=\frac{1}{2\pi i}\oint_{\gamma_0}\frac{A(s)-A(z)}{s-z}A(z)^{-1}\frac{1}{2\pi i}\oint_{\gamma_0}\frac{(I-A(s'))f(s')ds'}{s'-z}ds\\
&\qquad+\frac{1}{2\pi i}\oint_{\gamma_0}\frac{1}{s-z}\frac{1}{2\pi i}\oint_{\gamma_0}\frac{(I-A(s'))f(s')ds'}{s'-z}ds.
\end{aligned}$$

The first integral on the right hand side vanishes, and the second integral equals

$$\frac{1}{2\pi i}\oint_{\gamma_0}\frac{(I-A(s))f(s)ds}{s-z}.$$

In a similar way it can be shown that for the expression of the operator $(T-z)^{-1}$ from (4.2) the relation $(T-z)^{-1}(T-z)=I$ holds. □

Next we choose a contour γ_1 in $\mathcal{D}_0$ which surrounds $\sigma(A)$ and introduce the Riesz projection Q onto the spectral subspace of T corresponding to $\sigma(A)$. The linearization M of [GKL] is the restriction $M=T|_{\mathcal{R}(Q)}$. We have for $f\in L^2$:

$$\begin{aligned}
(Qf)(t) &= \left(-\frac{1}{2\pi i}\oint_{\gamma_1}(T-z)^{-1}f dz\right)(t)\\
&= -\frac{1}{2\pi i}\oint_{\gamma_1}\frac{A(z)^{-1}}{t-z}\oint_{\gamma_0}\frac{(I-A(s))f(s)ds}{s-z}dz.
\end{aligned}\tag{4.3}$$

Lemma 4.2 *The projection Q maps L^2 onto the set*

$$\widehat{\mathcal{F}}:=\{\widehat{f}_-\in L^2_-:A\widehat{f}_-\in L^2_+\},$$

hence the mapping

$$\widehat{f}_-\longmapsto A\widehat{f}_-=f\tag{4.4}$$

from $\widehat{\mathcal{F}}$ onto $\mathcal{F}$ (where we consider for the elements of $\mathcal{F}$ the representatives $A\widehat{f}_-$, $\widehat{f}_-\in\widehat{\mathcal{F}}$) is a bijection.

Proof: As $(Qf)(t)$ can be analytically continued outside γ_1 and vanishes at ∞ we have $Qf\in L^2_-$. Further,

$$\begin{aligned}
A(t)(Qf)(t) &= -\frac{1}{2\pi i}\oint_{\gamma_1}\frac{A(t)-A(z)}{t-z}A(z)^{-1}\oint_{\gamma_0}\frac{(I-A(s))f(s)ds}{s-z}dz\\
&\quad-\frac{1}{2\pi i}\oint_{\gamma_1}\frac{1}{t-z}\oint_{\gamma_0}\frac{(I-A(s))f(s)ds}{s-z}dz.
\end{aligned}$$

The first integral can be continued analytically as a function of t to all of $\mathcal{D}$, the second integral vanishes. Hence $QL^2 \subset \widehat{\mathcal{F}}$. Now let $\widehat{f}_- \in \widehat{\mathcal{F}}$. Then

$$(Q\widehat{f}_-)(t) = -\frac{1}{2\pi i}\oint_{\gamma_1} \frac{A(z)^{-1}}{t-z}\frac{1}{2\pi i}\oint_{\gamma_0} \frac{(I-A(s))\widehat{f}_-(s)ds}{s-z}dz,$$

and $\widehat{f}_- \in L^2_-$ implies for $z \in \mathcal{D}_0$

$$\frac{1}{2\pi i}\oint_{\gamma_0} \frac{\widehat{f}_-(s)}{s-z}ds = 0,$$

$A\widehat{f}_- \in L^2_+$ implies for $z \in \mathcal{D}_0$

$$\frac{1}{2\pi i}\oint_{\gamma_0} \frac{A(s)\widehat{f}_-(s)}{s-z}ds = (A\widehat{f}_-)(z)$$

and we get

$$(Q\widehat{f}_-)(t) = \frac{1}{2\pi i}\oint_{\gamma_1} \frac{\widehat{f}_-(z)}{t-z}dz = \widehat{f}_-(t).$$

□

Consider the action of the operator Λ in $\mathcal{F}$. The representative $A\widehat{f}_-$ of $f \in \mathcal{F}$ is mapped onto $(\Lambda A\widehat{f}_-)(t) = tA(t)\widehat{f}_-(t)$, which can be written as $A\widehat{g}_- + A\widehat{g}_+$ with $\widehat{g}_\pm \in L^2_\pm$,

$$\widehat{g}_-(t) = t\widehat{f}_-(t) - \frac{1}{2\pi i}\oint_{\gamma_0} \widehat{f}_-(s)ds.$$

Indeed, in order to decompose $t\widehat{f}_-(t)$ into its L^2_+– and L^2_--components we consider for $\lambda \in \mathbb{C}\backslash\overline{\mathcal{D}_0}$ the relation

$$\begin{aligned}(\lambda\widehat{f}_-(\lambda))_- &= -\frac{1}{2\pi i}\oint_{\gamma_0} \frac{t\widehat{f}_-(t)dt}{t-\lambda} \\ &= -\frac{1}{2\pi i}\oint_{\gamma_0} \widehat{f}_-(t)dt - \lambda\frac{1}{2\pi i}\oint_{\gamma_0} \frac{\widehat{f}_-(t)dt}{t-\lambda} \\ &= -\frac{1}{2\pi i}\oint_{\gamma_0} \widehat{f}_-(t)dt + \lambda\widehat{f}_-(\lambda).\end{aligned}$$

Thus, the action of Λ in $\mathcal{F}$ with these representatives is given by

$$\Lambda\widehat{f}_- = \widehat{g}_-, \quad \widehat{g}_-(t) = t\widehat{f}_-(t) - \frac{1}{2\pi i}\oint_{\gamma_0} \widehat{f}_-(s)ds.$$

On the other hand, the action of T in $\widehat{\mathcal{F}} = \mathcal{R}(Q)$ is

$$(T\widehat{f}_-)(t) = t\widehat{f}_-(t) - \frac{1}{2\pi i}\oint_{\gamma_0} \widehat{f}_-(s)ds + \frac{1}{2\pi i}\oint_{\gamma_0} A(s)\widehat{f}_-(s)ds$$

and here the last integral vanishes since $A\widehat{f}_- \in L^2_+$. Thus we have shown that the operators Λ and $M = T|_{\mathcal{R}(Q)}$ are isomorphic.

Now we consider the inner products. We equip L^2 with the possibly indefinite inner product

$$[f, g]_{L^2} := \frac{1}{2\pi i}\oint_{\gamma_0} (A(t)f(t), g(t^*))_{\mathcal{H}}dt,$$

which we can write also as

$$[f, g]_{L^2} = (Gf, g)_{L^2}$$

with some bounded operator G in L^2. If we define the following bounded operators in L^2 :

$$\begin{aligned} (\mathcal{A}f)(t) &:= A(t)f(t), \quad (\mathcal{U}f)(t) := f(t^*), \\ S &:= \frac{1}{2\pi i}\frac{dt^*}{|dt|}\mathcal{U} = \frac{1}{2\pi i}\mathcal{U}\frac{dt}{|dt|}, \end{aligned}$$

then we obtain

(4.5) $$G = S\mathcal{A}.$$

It is easy to see, that the operator $\mathcal{U}$ is unitary and self-adjoint in L^2. The operators S and G are self-adjoint:

$$\begin{aligned} S^* &= -\frac{1}{2\pi i}\frac{(dt)^*}{|dt|}\mathcal{U} = \frac{1}{2\pi i}\frac{dt^*}{|dt|}\mathcal{U} = S, \\ (Gf, f)^*_{L^2} &= \left(\frac{1}{2\pi i}\oint_{\gamma_0} (A(t)f(t), f(t^*))_{\mathcal{H}}\, dt\right)^* \\ &= -\frac{1}{2\pi i}\oint_{\gamma_0} (A(t)f(t), f(t^*))^*_{\mathcal{H}}\, dt^* \\ &= -\frac{1}{2\pi i}\oint_{\gamma_0} (f(t^*), A(t)f(t))_{\mathcal{H}}\, dt^* \\ &= \frac{1}{2\pi i}\oint_{\gamma_0} (f(t), A(t^*)f(t^*))_{\mathcal{H}}\, dt \\ &= \frac{1}{2\pi i}\oint_{\gamma_0} (A(t)f(t), f(t^*))_{\mathcal{H}}\, dt = (Gf, f)_{L^2}. \end{aligned}$$

The operator G is also boundedly invertible: $G^{-1} = \mathcal{A}^{-1}S^{-1}$.

Now consider $f \in \mathcal{F}$. Under the above isomorphism between $\mathcal{F}$ and $\widehat{\mathcal{F}} \subset L^2$ it corresponds to the element $\widehat{f}_- \in \widehat{\mathcal{F}} \subset L^2$, where $A(t)\widehat{f}_-(t)$ is a representative of $f \in \mathcal{F}$, and we have for two such elements $f, g \in \mathcal{F}$ with evident notation

$$\begin{aligned}\langle f, g\rangle &= \frac{1}{2\pi i}\oint_{\gamma_0}(A(t)^{-1}A(t)\widehat{f}_-(t), A(t^*)\widehat{g}_-(t^*))_{\mathcal{H}}dt\\ &= \frac{1}{2\pi i}\oint_{\gamma_0}(A(t)\widehat{f}_-(t), \widehat{g}_-(t^*))_{\mathcal{H}}dt\\ &= [\widehat{f}_-, \widehat{g}_-]_{L^2}.\end{aligned}$$

Thus the spaces $\mathcal{F}$ and $\widehat{\mathcal{F}}$ are even isomorphic as Krein spaces. Summing up we have proved the following theorem.

Theorem 4.1 *The mapping* (4.4) *is a Krein space isomorphism between the Krein spaces* $(\mathcal{F}, \langle\,.\,,\,.\,\rangle)$ *and* $(\widehat{\mathcal{F}}, [\,.\,,\,.\,]_{L^2})$. *Under this isomorphism the operator* Λ *in* $\mathcal{F}$ *becomes the operator* M *in* $\widehat{\mathcal{F}}$.

We mention that the operator T is not self-adjoint in $(L^2, [\,.\,,\,.\,]_{L^2})$. Indeed,

$$\begin{aligned}[Tf, g]_{L^2} &= \frac{1}{2\pi i}\oint_{\gamma_0}\left(A(t)\left(tf(t) - \frac{1}{2\pi i}\oint_{\gamma_0} f(s)ds\right.\right.\\ &\quad \left.\left.+ \frac{1}{2\pi i}\oint_{\gamma_0} A(s)f(s)ds\right), g(t^*)\right)_{\mathcal{H}} dt\\ &= \frac{1}{2\pi i}\oint_{\gamma_0} t(A(t)f(t), g(t^*))_{\mathcal{H}}dt\\ &\quad -\left(\frac{1}{2\pi i}\oint_{\gamma_0} f(t)dt, \frac{1}{2\pi i}\oint_{\gamma_0} A(t^*)g(t^*)dt^*\right)_{\mathcal{H}}\\ &\quad +\left(\frac{1}{2\pi i}\oint_{\gamma_0} A(t)f(t)dt, \frac{1}{2\pi i}\oint_{\gamma_0} A(t)g(t)dt\right)_{\mathcal{H}}.\end{aligned}$$

Here, if $f = g$, the first and the last term on the right hand side are real, whereas this is in general not the case for the middle term.

5 Spectral Points of Positive Type of *A*

Recall that the approximate point spectrum $\sigma_{ap}(B)$ of an operator B is the set of all $\lambda \in \mathbb{C}$ for which there exists a sequence (x_n), $\|x_n\| = 1, n = 1, 2, \ldots$, such that $\|(B - \lambda)x_n\| \to 0$ if $n \to \infty$. In the sequel, let A be an analytic operator function as in Section 2. The point $\lambda \in \mathcal{D}$ belongs to the approximate point spectrum $\sigma_{ap}(A)$ of this operator function if $0 \in \sigma_{ap}(A(\lambda))$.

Theorem 5.1 *If* Λ *denotes the linearization of the operator function* A *from Section* 3, *then*

$$\sigma_{ap}(A) = \sigma_{ap}(\Lambda).$$

Proof: Let $h_n \in \mathcal{H}$, $\|h_n\|_{\mathcal{H}} = 1$, $\|A(\lambda_0)h_n\|_{\mathcal{H}} \to 0$ if $n \to \infty$ for some $\lambda_0 \in \mathcal{D}$. We consider the functions $u_n \in L^2_+$:

$$u_n(t) = \frac{(A(t) - A(\lambda_0))h_n}{t - \lambda_0} \tag{5.1}$$

and

$$\widehat{u}_n(t) := A(t)^{-1}u_n(t) = \widehat{u}_n^-(t) + \widehat{u}_n^+(t).$$

Then for z outside of γ_0 we have

$$\begin{aligned} \widehat{u}_n^-(z) &= -\frac{1}{2\pi i}\oint_{\gamma_0} \frac{\widehat{u}_n^-(t)}{t-z}\,dt = -\frac{1}{2\pi i}\oint_{\gamma_0} \frac{A(t)^{-1}u_n(t)}{t-z}\,dt \\ &= -\frac{1}{2\pi i}\oint_{\gamma_0} \frac{h_n - A(t)^{-1}A(\lambda_0)h_n}{(t-\lambda_0)(t-z)}\,dt. \end{aligned} \tag{5.2}$$

Since $(\Lambda - \lambda_0)u_n = A(t)h_n - A(\lambda_0)h_n$, it follows that

$$\|(\Lambda - \lambda_0)u_n\|_{\mathcal{F}} \le \|A(\lambda_0)h_n\|_{L^2} \le C\|A(\lambda_0)h_n\|_{\mathcal{H}} \longrightarrow 0 \text{ if } n \longrightarrow \infty.$$

Further, according to the isomorphism of $\mathcal{F}$ and $\widehat{\mathcal{F}}$ and $\|u\|_{\mathcal{F}} \ge C_1\|\widehat{u}^-\|_{L^2}$ ($u \in \mathcal{F}$), the norms $\|\cdot\|_{\mathcal{F}}$ on $\mathcal{F}$ and $\|\cdot\|_{L^2}$ on $\widehat{\mathcal{F}}$ are equivalent, in particular, $\|u_n\|_{\mathcal{F}} \ge C\|\widehat{u}_n^-\|_{L^2}$. Here and in the sequel all the constants C, C_1 etc. are positive. If $z \in \mathbb{C}\backslash\overline{\mathcal{D}_0}$, then

$$\|\widehat{u}_n^-(z)\|_{\mathcal{H}} \le C_z\|\widehat{u}_n^-\|_{L^2}.$$

It follows (observe (5.2)) that

$$\begin{aligned} \|u_n\|_{\mathcal{F}} &\ge C\|\widehat{u}_n^-\|_{L^2} \ge \frac{C}{C_z}\|\widehat{u}_n^-(z)\|_{\mathcal{H}} \\ &= \frac{C}{C_z}\left\|\frac{1}{2\pi i}\oint_{\gamma_0}\frac{h_n - A(t)^{-1}A(\lambda_0)h_n}{(t-\lambda_0)(t-z)}\,dt\right\|_{\mathcal{H}} \\ &\ge C_z'\|h_n\|_{\mathcal{H}} + o(1), \end{aligned}$$

hence $\underline{\lim}\|u_n\|_{\mathcal{F}} > 0$.

Conversely, given a sequence $(f_n) \subset \mathcal{F}$, $\|f_n\|_{\mathcal{F}} = 1$, $\|(\Lambda - \lambda_0)f_n\|_{\mathcal{F}} \to 0$ if $n \to \infty$. With representatives $u_n \in f_n$ we can write

$$(t - \lambda_0)u_n(t) = A(t)v_n(t) + w_n(t) \tag{5.3}$$

where $v_n, w_n \in L^2_+$ and $\|w_n\|_{L^2} \to 0$ if $n \to \infty$. The relation (5.3) implies

$$\|A(\lambda_0)v_n(\lambda_0)\|_{\mathcal{H}} = \|w_n(\lambda_0)\|_{\mathcal{H}} \le C\|w_n\|_{L^2} \longrightarrow 0.$$

On the other hand,

$$u_n(t) = A(t)\frac{v_n(t) - v_n(\lambda_0)}{t - \lambda_0} + \frac{A(t)v_n(\lambda_0)}{t - \lambda_0} + \frac{w_n(t)}{t - \lambda_0},$$

and

$$\begin{aligned} 1 = \|f_n\|_{\mathcal{F}} &\le \left\| \frac{A(\cdot)v_n(\lambda_0)}{\cdot - \lambda_0} \right\|_{L^2} + \left\| \frac{w_n}{\cdot - \lambda_0} \right\|_{L^2} \\ &\le \left\| \frac{A(\cdot)v_n(\lambda_0)}{\cdot - \lambda_0} \right\|_{L^2} + C_{\lambda_0}\|w_n\|_{L^2}. \end{aligned} \tag{5.4}$$

This yields $\underline{\lim}\|v_n(\lambda_0)\|_{\mathcal{H}} > 0$, since otherwise for some subsequence the expression on the right hand side in (5.4) would tend to zero, and $\lambda_0 \in \sigma_{ap}(A)$. □

The real point $\lambda_0 \in \sigma_{ap}(A) \cap \mathcal{D} \cap \mathbb{R}$ is said to be a *spectral point of positive type* for the self-adjoint analytic operator function A, if for each sequence (x_n) such that $\|x_n\|_{\mathcal{H}} = 1$, $\|A(\lambda_0)x_n\|_{\mathcal{H}} \to 0$ we have

$$\underline{\lim}\,(A'(\lambda_0)x_n, x_n)_{\mathcal{H}} > 0.$$

The set of points of positive type for A is denoted by $\sigma_+(A)$. The following statements can easily be proved (comp. [LMM]):

1. The real interval $[a, b] \subset \mathcal{D}$ is contained in $\varrho(A) \cup \sigma_+(A)$ if and only if there exist numbers $\varepsilon, \delta > 0$ such that

$$\lambda \in [a, b], \quad \|x\|_{\mathcal{H}} = 1, \quad \|A(\lambda)x\|_{\mathcal{H}} < \varepsilon \Longrightarrow (A'(\lambda)x, x)_{\mathcal{H}} \ge \delta.$$

   Here the case $a = b$ is not excluded.

2. $(\varrho(A) \cap \mathbb{R}) \cup \sigma_+(A)$ is an open subset of $\overline{\mathcal{D}} \cap \mathbb{R}$.

From statement 1 a result about bounded perturbations follows immediately.

Theorem 5.2 *Suppose that* $[a, b] \subset \rho(A) \cup \sigma_+(A)$, *and let* γ *be a contour in* $\mathcal{D}$ *which surrounds the interval* $[a, b]$. *Then there exists a number* $\delta > 0$, *such that for each operator function* $B(\lambda)$ *which is analytic and self-adjoint in* $\mathcal{D}$ *and such that*

$$\max_{\lambda \in \gamma} \|B(\lambda) - A(\lambda)\|_{\mathcal{H}} < \delta$$

the inclusion $[a, b] \subset \rho(B) \cup \sigma_+(B)$ *holds.*

Recall that for a self-adjoint operator T in a Krein space $\mathcal{K}$ with indefinite scalar product $\langle \cdot, \cdot \rangle$ the point $\lambda_0 \in \sigma(T)$ is called a spectral point of positive type, if for each sequence $(f_n) \subset \mathcal{K}$ it holds

$$\|f_n\| = 1, \quad \|(T - \lambda_0) f_n\| \longrightarrow 0 \quad \Longrightarrow \quad \underline{\lim} \langle f_n, f_n \rangle > 0.$$

The set of all spectral points of positive type of T is denoted by $\sigma_+(T)$. It is easy to see (see [LMM]), that $\sigma_+(T)$ is real. Evidently, $\lambda \in \sigma_+(T)$ if and only if $\lambda \in \sigma_+(A)$ for the analytic function $A(\lambda) := \lambda J - JA$; where J is a fundamental symmetry in $\mathcal{K}$ and $A(\lambda)$ is considered in the Hilbert space, which is $\mathcal{K}$ equipped with the inner product $(\cdot, \cdot) = \langle J\cdot, \cdot \rangle$.

The main result of this section is the following theorem.

Theorem 5.3 *Let the operator function A be given as in Section 2, and let Λ be its linearization in the Krein space $\mathcal{F}$, see Section 3. Then $\sigma_+(A) = \sigma_+(\Lambda)$.*

Proof: Let (h_n) be as in the first part of the proof of Theorem 5.1 and suppose further that $\underline{\lim}\, (A'(\lambda_0)h_n, h_n)_{\mathcal{H}} > 0$, $\lambda_0 \in \mathbb{R}$. Since $\|A(\lambda_0)h_n\|_{\mathcal{H}} \to 0$, we get for the element u_n from the relation (5.1)

$$\begin{aligned} \langle u_n, u_n \rangle &= \frac{1}{2\pi i} \oint_{\gamma_0} \left(A(t)^{-1} \left(\frac{A(t)h_n}{t - \lambda_0} - \frac{A(\lambda_0)h_n}{t - \lambda_0} \right), \right. \\ &\quad \left. \frac{A(t^*)h_n}{t^* - \lambda_0} - \frac{A(\lambda_0)h_n}{t^* - \lambda_0} \right)_{\mathcal{H}} dt = (A'(\lambda_0)h_n, h_n)_{\mathcal{H}} + o(1), \end{aligned}$$

hence $\underline{\lim}\, \langle u_n, u_n \rangle > 0$.

Now let (f_n) be as in the second part of the proof of Theorem 5.1 and suppose that $\underline{\lim}\, \langle u_n, u_n \rangle > 0$. Then, since $\|w_n\|_{L^2} \to 0$, in the same way,

$$\begin{aligned} \langle f_n, f_n \rangle &= \frac{1}{2\pi i} \oint_{\gamma_0} (A(t)^{-1} u_n(t), u_n(t^*))_{\mathcal{H}} dt = \frac{1}{2\pi i} \\ &\quad \oint_{\gamma_0} \left(A(t)^{-1} \frac{A(t)v_n(t)}{t - \lambda_0} + \frac{w_n(t)}{t - \lambda_0}, \frac{A(t^*)v_n(t^*)}{t^* - \lambda_0} + \frac{w_n(t^*)}{t^* - \lambda_0} \right)_{\mathcal{H}} dt \\ &= (A'(\lambda_0)v_n(\lambda_0), v_n(\lambda_0))_{\mathcal{H}} + o(1) \end{aligned}$$

and hence $\underline{\lim}\, (A'(\lambda_0)v_n(\lambda_0), v_n(\lambda_0))_{\mathcal{H}} > 0$. Theorem 5.3 is proved. □

As we have mentioned already, there is a bijective correspondence between the Jordan chains of the operator function A and of the operator Λ at the same eigenvalue. The reasoning in the proof of Theorem 5.3 shows that also the signatures of the eigenspaces of A and Λ at the same eigenvalue coincide.

The following two theorems are now immediate consequences of the results of [LMM].

Theorem 5.4 *Suppose that*

$$[a,b] \subset \rho(A) \cup \sigma_+(A). \tag{5.5}$$

Then there exist numbers $\delta > 0$ *and* $\eta > 0$, *such that*

$$\{\lambda : a-\delta \le \operatorname{Re}\lambda \le b+\delta,\ 0 < |\operatorname{Im}\lambda| \le \eta\} \subset \rho(A). \tag{5.6}$$

Proof: According to Theorem 5.3 the assumption (5.5) implies $[a,b] \subset \rho(\Lambda) \cup \sigma_+(\Lambda)$. If we apply [LMM], Corollary 1.4, the two strips in (5.6) belong to $\rho(\Lambda)$, and Theorem 3.1 implies that they belong to $\rho(A)$. □

Theorem 5.5 *Suppose that for the closed interval* $[a,b]$ *the inclusion* (5.5) *holds. Then* $A(z)^{-1}$ *admits the representation*

$$A(z)^{-1} = -\int_{[a,b]} \frac{dF(t)}{t-z} + B_1(z) \tag{5.7}$$

where F *is a nonnegative* $\mathcal{L}(\mathcal{H})$*-valued measure on* $[a,b]$ *and* $B_1(z)$ *is analytic on the open interval* (a,b).

Proof: According to Theorem 5.3, for the operator Λ we have $[a,b] \subset \sigma_+(\Lambda) \cup \varrho(\Lambda)$, hence, according to Theorem 3.1 of [LMM], there exists a local spectral function E of positive type for the operator Λ on some open interval containing $[a,b]$. It follows that

$$(\Lambda - z)^{-1} = \int_{[a,b]} \frac{dE(t)}{t-z} + (I - E([a,b]))(\Lambda - z)^{-1}, \qquad (z \in \varrho(\Lambda)),$$

where the last term can be analytically continued into (a,b). Inserting this representation into (3.6) we get (5.7) with

$$F(dt) = P^* E(dt) P, \quad B_1(z) = -P^*(I - E([a,b]))(\Lambda - z)^{-1} P + B(z).$$

The relation

$$(F(dt)x, x)_{\mathcal{H}} = \langle E(dt)Px, Px \rangle \quad (t \in [a,b],\ x \in \mathcal{H})$$

implies that F is a nonnegative measure. □

According to (5.7), the singularities of the function A^{-1} at points of definite type of the operator function A are of the same (simple) structure as the singularities of an operator valued Nevanlinna function.

6 Compact Perturbations

The following lemma seems to be known; for the case of a circle see [P], Lemma 2. We give its proof for the sake of completeness. Here $\mathcal{H}$, $A(\lambda)$, γ_0 are the same as in Section 1 and P_+ (P_-) is again the projection onto L^2_+ (L^2_-) parallel to L^2_- (L^2_+, respectively).

Lemma 6.1 *Let T be an operator valued function on γ_0 which is continuous in the operator norm and the values of which are compact operators in $\mathcal{H}$. Then the Hankel operator*

$$\mathcal{T} := P_- T(\cdot) P_+$$

is compact as an operator from L^2_+ into L^2_-.

Proof: Let $\varepsilon > 0$ be given. Choose a partition $\{t_k\}_1^n$ of γ_0 such that on the arc γ_k of γ_0 with endpoints t_k, t_{k+1}, $k = 1, 2, \ldots, n$, $t_{n+1} = t_1$, for the operator function T we have $\|T(\lambda) - T(\mu)\|_{\mathcal{H}} \le \varepsilon$ if $\lambda, \mu \in \gamma_k$. Further, choose finite dimensional operators S_k such that

$$\|T(t_k) - S_k\|_{\mathcal{H}} < \varepsilon \qquad (k = 1, 2, \ldots, n).$$

We define an operator function S on γ_0 such that $S(t_k) = S_k$ and that on each arc γ_k it is a linear function of the length of γ_0. Then for $t \in \gamma_k$ we have

$$\|T(t) - S(t_k)\|_{\mathcal{H}} \le \|T(t) - T(t_k)\|_{\mathcal{H}} + \|T(t_k) - S_k\|_{\mathcal{H}} < 2\varepsilon,$$

and, in the same way, $\|T(t) - S(t_{k+1})\|_{\mathcal{H}} \le 2\varepsilon$, hence with some $\alpha(t) \in [0, 1]$,

$$\begin{aligned}\|T(t) - S(t)\|_{\mathcal{H}} &\le \alpha(t)\|T(t) - S(t_k)\|_{\mathcal{H}} \\ &\quad + (1 - \alpha(t))\|T(t) - S(t_{k+1})\|_{\mathcal{H}} < 2\varepsilon,\end{aligned}$$

and it follows that

$$\|T(t) - S(t)\|_{\mathcal{H}} < 2\varepsilon \qquad (t \in \gamma_0). \tag{6.1}$$

Denote by $\mathcal{L}$ the finite dimensional subspace of $\mathcal{H}$ which contains all the ranges of the operators S_k and S_k^*, $k = 1, 2, \ldots, n$. Then $S_k\mathcal{L} \subset \mathcal{L}$ and $S_k\mathcal{L}^\perp = 0$. We choose an orthonormal basis $\{e_j\}_1^l$ in $\mathcal{L}$ and consider the matrix representation $(s_{jk}(t))_1^l$ of the operator $S(t)|_{\mathcal{L}}$ with respect to this basis. The functions s_{jk} are continuous and piecewise linear on γ_0 and the operator $S(t)$ has the form

$$S(t) = \sum_{j,k=1}^{l} s_{jk}(t)(\,\cdot\,, e_k)_{\mathcal{H}} e_j.$$

We set

$$q_{jk}^{\pm}(t) := P_{\pm} s_{jk}(t) \quad (t \in \gamma_0).$$

The functions q_{jk}^{+} (q_{jk}^{-}) are analytic in the interior (exterior, respectively) domain of γ_0 and continuous in the closed domains, and $q_{j,k}^{-}(\infty) = 0$. According to a well known theorem (see [W], Section 2.4) there exist rational functions $r_{jk}(t)$ which vanish at ∞ and have their singularities in the interior of γ_0 such that

$$\sup_{t\in\gamma_0} |q_{jk}^{-}(t) - r_{jk}(t)| < \varepsilon/l \qquad (j,k = 1,2,\ldots,l). \tag{6.2}$$

After a small perturbation of the functions $r_{jk}(t)$ we can suppose that they have only simple poles, that is, that they are of the form

$$r_{jk}(t) = \sum_{i=1}^{n_{jk}} \frac{c_{jk,i}}{t - \lambda_{jk,i}} \qquad (\lambda_{jk,i} \in \mathcal{D}_0). \tag{6.3}$$

We define

$$R(t) := \sum_{j,k=1}^{l} (q_{jk}^{+}(t) + r_{jk}(t))(\,\cdot\,, e_k)_{\mathcal{H}}\, e_j \tag{6.4}$$

and denote by $\mathcal{R}$ the Hankel operator from L_+^2 into L_-^2, given by

$$\mathcal{R}f := P_- Rf \quad (f \in L_+^2). \tag{6.5}$$

From the relations (6.3)–(6.5) it follows that

$$\begin{aligned} \mathcal{R}f &= P_- \sum_{j,k=1}^{l} \sum_{i=1}^{n_{jk}} \frac{c_{jk,i}}{\cdot - \lambda_{jk,i}} (f(\cdot), e_k)_{\mathcal{H}} e_j \\ &= \sum_{j,k=1}^{l} \sum_{i=1}^{n_{jk}} c_{jk,i} e_j P_- \frac{(f(\cdot), e_k)_{\mathcal{H}}}{\cdot - \lambda_{jk,i}}. \end{aligned}$$

Since $\frac{f(\cdot)-f(\lambda_{jk,i})}{\cdot-\lambda_{jk,i}} \in L_+^2$ we find

$$(\mathcal{R}f)(t) = \sum_{j,k=1}^{l} \sum_{i=1}^{n_{jk}} c_{jk,i} e_j \frac{(f(\lambda_{jk,i}), e_k)_{\mathcal{H}}}{t - \lambda_{jk,i}},$$

therefore $\mathcal{R}$ is a finite dimensional operator from L_+^2 into L_-^2. We estimate the norm of $\mathcal{T} - \mathcal{R}$:

$$\begin{aligned} \|(\mathcal{T} - \mathcal{R})f\|_{L^2} &= \|P_-(\mathcal{T} - \mathcal{R})f\|_{L^2} \\ &\le \|P_-\|_{L^2} \sup_{t\in\gamma_0} \|T(t) - R(t)\|_{\mathcal{H}} \|f\|_{L_+^2}, \end{aligned}$$

hence

$$\|\mathcal{T}-\mathcal{R}\|_{L^2_+\to L^2_-} \le \|P_-\|_{L^2} \sup_{t\in\gamma_0} \|T(t)-R(t)\|_{\mathcal{H}}. \tag{6.6}$$

The relations

$$S(t)-R(t) = \sum_{j,k=1}^{l} (q^-_{jk}(t) - r_{jk}(t))(\,\cdot\,, e_k)_{\mathcal{H}}\, e_j$$

and (6.2) yield for $t \in \gamma_0$

$$\|S(t)-R(t)\|_{\mathcal{H}} \le \left(\sum_{j,k=1}^{l} |q^-_{jk}(t) - r_{jk}(t)|^2 \right)^{1/2} < \varepsilon. \tag{6.7}$$

If we now observe (6.6), (6.1) and (6.7) it follows that

$$\begin{aligned} \|\mathcal{T}-\mathcal{R}\|_{L^2} \le \|P_-\|_{L^2} &\sup_{t\in\gamma_0}(\|T(t)-S(t)\|_{\mathcal{H}} \\ &+ \|S(t)-R(t)\|_{\mathcal{H}}) < 3\|P_-\|_{L^2}\varepsilon. \end{aligned}$$

This means that $\mathcal{T}$ can be approximated in operator norm by finite dimensional operators as closely as we want, hence $\mathcal{T}$ is compact. □

The main result of this paper is the following theorem. In order to formulate it, we need some more definitions. An $\mathcal{L}(\mathcal{H})$-valued function C, which is meromorphic on $\mathbb{C}\backslash\mathbb{R}$ and such that $C(z^*) = C(z)^*$ for all $z \in \mathcal{D}(C)$, the domain of analyticity of C, belongs to the class $\mathbf{N}_\kappa$, if for arbitrary $n \in \mathbb{N}$, elements $x_1, x_2, \dots, x_n \in \mathcal{H}$ and points $z_1, z_2, \dots, z_n \in \mathcal{D}(C)$ the $n \times n$-matrix

$$\left(\left(\frac{C(z_i) - C(z_j^*)}{z_i - z_j^*} x_i, x_j \right)_{\mathcal{H}} \right)_{i,j=1}^{n}$$

has at most κ negative eigenvalues and for at least one choice of $n, x_1, x_2, \dots, x_n$ and $z_1, z_2, \dots, z_n$ it has exactly κ negative eigenvalues. For properties of this class of functions we refer to [DLS2], [HSW] and [ADRS]. If A is an $\mathcal{L}(\mathcal{H})$-valued function as above, the set $\sigma_{-,f}(A)$ consists of all real points α with the following property: There exists an open interval Δ containing α, such that $\Delta\backslash\{\alpha\} \subset \sigma_+(A) \cup \rho(A)$ and that the operator function A^{-1} admits a representation

$$A(\lambda)^{-1} = A^1_\Delta(\lambda) + A^2_\Delta(\lambda), \tag{6.8}$$

where A^1_Δ is analytic in $\mathcal{D}\backslash\overline{\Delta}$ and belongs to some class N_κ, $\kappa \in \mathbb{N} \cup \{0\}$, and A^2_Δ is analytic in Δ.

For a self-adjoint operator function A as above a decomposition (6.8) is obtained as follows. Consider the corresponding G-self-adjoint operator Λ in $\mathcal{F}$, and suppose that the point α belongs to $\sigma_{-,f}(\Lambda)$. We recall that this means (see [LMM]) that for some open real interval Δ containing α it holds

$$\{\lambda : 0 < |\mathrm{Im}\lambda| < \eta, \quad \mathrm{Re}\lambda \in \Delta\} \subset \rho(\Lambda),$$

that

$$\Delta\backslash\{\alpha\} \subset \sigma_+(\Lambda) \cup \rho(\Lambda),$$

and that on the maximal spectral subspace $\mathcal{L}_\Delta(\Lambda)$ (see [LMM], [LM]) the indefinite inner product has a finite number of negative squares. Then there exists an orthogonal projection $E(\Delta)$ onto $\mathcal{L}_\Delta(\Lambda)$ in $\mathcal{F}$. The decomposition (6.8) is now obtained from (3.6) as follows (here we use P also for the canonical embedding of $\mathcal{H}$ into L^2):

$$\begin{aligned} A(z)^{-1} &= -P^*(\Lambda - z)^{-1}P + B(z) \\ &= -P^*E(\Delta)(\Lambda - z)^{-1}P - P^*(I - E(\Delta))(\Lambda - z)^{-1}P + B(z), \end{aligned}$$

and we choose

$$\begin{aligned} A_\Delta^1(z) &= -P^*E(\Delta)(\Lambda - z)^{-1}P, \\ A_\Delta^2(z) &= -P^*(I - E(\Delta))(\Lambda - z)^{-1}P + B(z). \end{aligned}$$

It follows from the relation

$$\left(\frac{A_\Delta^1(z_i) - A_\Delta^1(z_j^*)}{z_i - z_j^*}x_i, x_j\right)_{\mathcal{H}} = \langle E(\Delta)(\Lambda - z_i)^{-1}x_i, (\Lambda - z_j)^{-1}x_j\rangle$$

that the function $A_\Delta^1(z)$ belongs to some class N_κ since on the range of $E(\Delta)$ the G-inner product has a finite number of negative squares.

Since a point $\alpha \in \sigma_{-,f}(\Lambda)$ is an eigenvalue of Λ with a nonpositive eigenvector (see [LMM], p. 418), also the points of $\sigma_{-,f}(A)$ are eigenvalues of A with a nonpositive eigenvector.

Theorem 6.1 *Let A and $A_1 := A + B$ be two $\mathcal{L}(\mathcal{H})$-valued functions which are defined and analytic on the domain $\mathcal{D}$ $(= \mathcal{D}^*)$ of the complex plane and such that*

$$A(\lambda^*) = A(\lambda)^*, \quad A_1(\lambda^*) = A_1(\lambda)^*$$

and that $\sigma(A)$, $\sigma(A_1)$ are compact subsets of $\mathcal{D}$. Suppose that the values $B(\lambda) = A_1(\lambda) - A(\lambda)$ are compact operators. Consider an open interval $I \subset \sigma_+(A) \cup (\varrho(A) \cap \mathbb{R})$ and suppose that at least one point of this interval does not belong to $\mathrm{int}\,\sigma(A_1)$. Then the following inclusion holds:

$$I \subset \sigma_+(A_1) \cup \varrho(A_1) \cup \sigma_{-,f}(A_1).$$

In particular, the points of I cannot be accumulation points of the nonreal spectrum of the operator function A_1 or of real points from $\sigma_{-,f}(A_1)$.

The **proof** of this theorem is divided into three steps.

Step 1: With the operator functions A and A_1 in the Hilbert space $L^2 = L^2(\gamma_0, \mathcal{H})$ we consider the operators T and T_1, see (4.1), the Riesz projections Q and Q_1 corresponding to the spectral subspaces of T and T_1 at $\sigma(A)$ and $\sigma(A_1)$, respectively, see (4.3), and the operators G and G_1, see (4.5). Then the operators

$$Q - Q_1,\quad TQ - T_1Q_1,\quad Q^*GQ - Q_1^*G_1Q_1 \tag{6.9}$$

are compact. Indeed, if $\mathcal{A}$, $\mathcal{A}^{-1}$ etc. denote the operators of multiplication by $A(\lambda)$, $A(\lambda)^{-1}$ etc. in L^2, then from (4.3) and from

$$P_+(I - \mathcal{A}) = (I - \mathcal{A})P_+ - P_+\mathcal{A}P_-,$$

we have

$$Q = P_-\mathcal{A}^{-1}P_+(I - \mathcal{A}) = P_-\mathcal{A}^{-1}(I - \mathcal{A})P_+ - P_-\mathcal{A}^{-1}P_+\mathcal{A}P_-$$

and a corresponding relation for Q_1. Hence

$$\begin{aligned} Q - Q_1 &= P_-(\mathcal{A}^{-1} - \mathcal{A}_1^{-1})P_+ - P_-(\mathcal{A}^{-1} - \mathcal{A}_1^{-1})P_+\mathcal{A}P_- \\ &\quad - P_-\mathcal{A}_1^{-1}P_+(\mathcal{A} - \mathcal{A}_1)P_-. \end{aligned}$$

According to Lemma 6.1 the operators $P_-(\mathcal{A}^{-1} - \mathcal{A}_1^{-1})P_+$ and $P_+(\mathcal{A} - \mathcal{A}_1)P_-$ are compact, therefore $Q - Q_1$ is compact. The operator TQ can be written as

$$(TQf)(t) = -\frac{1}{2\pi i}\oint_{\gamma_0} \frac{zA(z)^{-1}}{t - z} \oint_{\gamma_0} \frac{(I - A(s))f(s)ds}{s - z}\, dz,$$

therefore the compactness of $TQ - T_1Q_1$ follows in the same way as the compactness of $Q - Q_1$ if we only replace the operator functions $A(z)^{-1}$, $A_1(z)^{-1}$ by $zA(z)^{-1}$, $zA_1(z)^{-1}$, respectively. In order to prove the compactness of the last difference in (6.9) we observe that

$$Q^*GQ - Q_1^*G_1Q_1 = (Q^* - Q_1)GQ + Q_1^*(G - G_1)Q + Q_1^*G_1(Q - Q_1).$$

This equality implies that we only need to prove the compactness of the second term on the right-hand side. Using the definition of Q in (4.3) and of G in (4.5) we obtain

$$Q_1^*(G - G_1)Q = (I - \mathcal{A}_1^*)P_+^*\mathcal{A}_1^{-*}P_-^*S(\mathcal{A} - \mathcal{A}_1)P_-\mathcal{A}^{-1}P_+(I - \mathcal{A}),$$

and it is enough to prove the compactness of the operator $P_-^* S(\mathcal{A} - \mathcal{A}_1)P_-$. We shall prove that the relation

$$P_-^* S = S P_+ \tag{6.10}$$

holds, then

$$P_-^* S(\mathcal{A} - \mathcal{A}_1)P_- = SP_+(\mathcal{A} - \mathcal{A}_1)P_-,$$

and the last operator is compact by Lemma 6.1. It remains to prove the equality (6.10):

$$(SP_+ f, g)_{L^2} = \frac{1}{2\pi i} \oint_{Gg_0} ((SP_+ f)(t), g(t))_{\mathcal{H}} |dt|$$

$$= \frac{1}{2\pi i} \oint_{\gamma_0} \left(\mathcal{U} \left(\frac{dt}{|dt|} P_+ f \right), g \right)_{\mathcal{H}} |dt| = \frac{1}{2\pi i} \oint_{\gamma_0} \left(\frac{dt}{|dt|} P_+ f, \mathcal{U} g \right)_{\mathcal{H}} |dt|$$

$$= \frac{1}{2\pi i} \oint_{\gamma_0} ((P_+ f)(t), g(t^*))_{\mathcal{H}} dt = \frac{1}{2\pi i} \oint_{\gamma_0} (f(t), (P_- g(t))_{\mathcal{H}} dt$$

$$= \frac{1}{2\pi i} \oint_{\gamma_0} \left(f(t), \frac{(dt)^*}{|dt|} (\mathcal{U} P_- g)(t) \right)_{\mathcal{H}} |dt|$$

$$= \frac{1}{2\pi i} \oint_{\gamma_0} (f, SP_- g)_{\mathcal{H}} |dt| = (f, SP_- g)_{L^2};$$

here we have used that the operator P_- is the adjoint of P_+ with respect to the bilinear form (2.10). This means that $SP_+ = (SP_-)^*$. But $(SP_-)^* = P_-^* S^* = P_-^* S$, and we obtain (6.10).

Step 2: In this part we prove slight generalizations of the statements (i) and (ii) of [LMM], Section 5. Let $\mathcal{K}$ be a Hilbert space and consider two bounded and boundedly invertible self-adjoint operators G, G_1 in $\mathcal{K}$. Further, let T, $T_1 \in \mathcal{L}(\mathcal{K})$ be such that GT and $G_1 T_1$ are self-adjoint. Consider components σ and σ_1 of $\sigma(T)$ and $\sigma(T_1)$ with corresponding Riesz projections Q and Q_1, respectively. Further, suppose that $Q - Q_1$, $TQ - T_1 Q_1$ and $Q^* G Q - Q_1^* G_1 Q_1$ are compact operators. Then the following statement holds:

If $\lambda_0 \in \sigma \cap \sigma_+(T)$ and $\lambda_0 \notin \operatorname{int} \sigma(T_1)$, then:

(i) $\lambda_0 \in \sigma_+(T_1) \cup \rho(T_1)$, *or* $\lambda_0 \in \sigma_p(Q_1 T_1)$ *with a nonpositive eigenvector;*

(ii) *there exists a* $\rho > 0$ *such that*

$$\{\lambda \in \mathbb{C} : 0 < |\lambda - \lambda_0| < \rho\} \subset \sigma_+(T_1) \cup \rho(T_1).$$

Proof: (i) Suppose that $\lambda_0 \notin \sigma_+(T_1) \cup \varrho(T_1)$. As λ_0 is a boundary point of $\sigma(T_1)$, there exists a sequence (x_n), $\|x_n\| = 1$, such that

$$\|(T_1 - \lambda_0)x_n\| \longrightarrow 0, \tag{6.11}$$

$$\underline{\lim}\, (G_1 x_n, x_n) \leq 0. \tag{6.12}$$

Without loss of generality we can assume that $(x_n) \subset \mathcal{R}(Q_1)$ and that $x_n \xrightarrow{w} x_0$ if $n \to \infty$ for some $x_0 \in \mathcal{R}(Q_1)$ where $\xrightarrow{w}$ denotes weak convergence. Let at first be $x_0 = 0$. Then the compactness of $Q - Q_1$ and the conditions $\|Q_1 x_n\| = \|x_n\| = 1$ imply $\|(Q - Q_1)x_n\| \to 0$ and $\|Qx_n\| \to 1$, the compactness of $TQ - T_1Q_1$ and (6.11) imply $\|(T - \lambda_0)Qx_n\| \to 0$, the compactness of $Q^*GQ - Q_1^*G_1Q_1$ and (6.12) imply $\underline{\lim}\,(GQx_n, Qx_n) \leq 0$. This is impossible since $\lambda_0 \in \sigma_+(T)$. If $x_0 \neq 0$, we find $(T_1 - \lambda_0)x_0 = 0$ and it remains to show that

$$(G_1 x_0, x_0) \leq 0. \tag{6.13}$$

Let $y_n = x_n - x_0$. If $\|y_n\| \to 0$ $(n \to \infty)$, then (6.13) follows from (6.12). Hence we can suppose that

$$\inf \|y_n\| > 0.$$

Now $y_n \xrightarrow{w} 0$ implies $\|(TQ - T_1Q_1)y_n\| \to 0$, and from (6.11) we get

$$\|(T - \lambda_0)Qy_n\| \longrightarrow 0,$$

hence $\underline{\lim}\,(GQy_n, Qy_n) > 0$ as $\lambda_0 \in \sigma_+(T)$, and since $Q^*GQ - Q_1^*G_1Q_1$ is compact this yields

$$\underline{\lim}\,(G_1 y_n, y_n) > 0. \tag{6.14}$$

Further

$$(G_1y_n, y_n) = (G_1x_n, x_n) - (G_1x_n, x_0) - (x_0, G_1x_n) + (G_1x_0, x_0)$$

and

$$\underline{\lim}\,(G_1y_n, y_n) = \underline{\lim}\,(G_1x_n, x_n) - (G_1x_0, x_0).$$

It follows from (6.12) and (6.14) that

$$(G_1x_0, x_0) = \underline{\lim}\,(G_1x_n, x_n) - \underline{\lim}\,(G_1y_n, y_n) < 0.$$

(ii) Assume the contrary. Then we find a sequence (λ_n), such that $\lambda_n \to \lambda_0$ $(n \to \infty)$ and $\lambda_n \notin \sigma_+(T_1Q_1) \cup \rho(T_1)$. Moreover, $\lambda_n \notin \operatorname{int}\sigma(T_1)$ if $n \geq n_0$: Since T_1Q_1 is a compact perturbation of TQ and $\lambda_0 \notin \operatorname{int}\sigma(T_1)$, for some $\varepsilon > 0$, $\eta > 0$ both strips

$$\{\lambda \in \mathbb{C} : 0 < |\mathrm{Re}\lambda - \lambda_0|,\ 0 < |\mathrm{Im}\lambda| < \eta\}$$

can contain only isolated points of $\sigma(T_1)$. Here we use the fact that the spectrum of a self-adjoint operator in a Krein space is symmetric with respect to the real axis. Thus, for sufficiently large n, $\lambda_n \notin \operatorname{int}\sigma(T_1)$. If $\lambda_n \neq \lambda_n^*$ then λ_n is an eigenvalue of T_1 with a neutral eigenvector ϕ_n. If $\lambda_n \in \mathbb{R}$, then $\lambda_n \in \sigma(T_1)\backslash\sigma_+(T_1)$. As for n sufficiently large $(n \geq n_1)$ we have $\lambda_n \in \sigma_+(TQ) \cup \rho(T)$ and $\lambda_n \notin \sigma_+(T_1Q_1) \cup \rho(T_1)$, it follows from (i) that λ_n is an eigenvalue of T_1Q_1 with a corresponding nonpositive eigenvector ϕ_n. If we suppose $\lambda_n \neq \lambda_m^*$ if $n \neq m$, the eigenvectors ϕ_n can be chosen mutually G_1-orthogonal. Let $\mathcal{L}_1$ be the closed linear span of the

$\phi_n, n \geq \max(n_0, n_1)$. It is a nonpositive invariant subspace of $T_1 Q_1$. We consider the restriction $S_1 := (T_1 Q_1 - \lambda_0)|_{\mathcal{L}_1}$. There are two possibilities:

1. S_1 is a Φ_+-operator, that is the range $\mathcal{R}(S_1)$ is closed and $\dim \ker S_1 < \infty$. As all $\lambda_n - \lambda_0$ are eigenvalues of S_1 there exists a neighbourhood of zero which consists only of eigenvalues of S_1 (see [GK]). But this is impossible since λ_0 is not an inner point of $\sigma(T_1)$.
2. S_1 is not a Φ_+-operator. Then for any $\varepsilon > 0$ and an arbitrary subspace $\mathcal{M}_1 \subset \mathcal{L}_1$ of finite codimension ($\dim \mathcal{L}_1/\mathcal{M}_1 < \infty$) there exists an $f \in \mathcal{M}_1$ such that $\|f\| = 1$ and $\|S_1 f\| < \varepsilon$. Hence there exists an orthonormal sequence $(f_n) \in \mathcal{M}_1$ such that $\|S_1 f_n\| \to 0$, that is $\|(T_1 Q_1 - \lambda_0) f_n\| \to 0$ $(n \to \infty)$. Since $TQ - T_1 Q_1$ is compact $\|(TQ - \lambda_0) f_n\| \to 0$ follows, which implies $\underline{\lim}(GQf_n, Qf_n) > 0$. Since $Q^*GQ - Q_1^* G_1 Q_1$ is compact, it follows that $\underline{\lim}(G_1 Q_1 f_n, Q_1 f_n) > 0$. On the other hand, $f_n \in \mathcal{L}_1$, hence $(G_1 Q_1 f_n, Q_1 f_n) \leq 0$, a contradiction. □

Step 3: In this step we set $\sigma := \sigma(A)$, $\sigma_1 =: \sigma(A_1)$, that is σ is the spectrum of T in $\mathcal{D}$, and σ_1 is the spectrum of T_1 in $\mathcal{D}$. Suppose that $\lambda_0 \in I$. We choose intervals $[a, b]$, $[a_1, b_1]$ and $[a_2, b_2]$ such that

$$a_2 < a < a_1 < \lambda_0 < b_1 < b < b_2$$

and that $[a_2, b_2] \in I$. Let p and $\widehat{p}$ be polynomials with zeros of order at least 3 in a, b and in a_2, b_2, respectively, and such that p is positive on the interval (a, b) and $\widehat{p}$ is positive on (a_2, b_2). We consider two Jordan contours Γ and $\widehat{\Gamma}$:

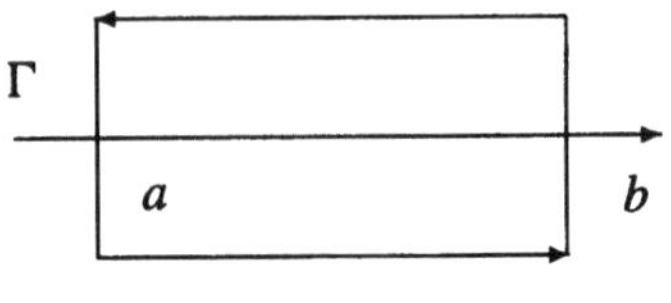

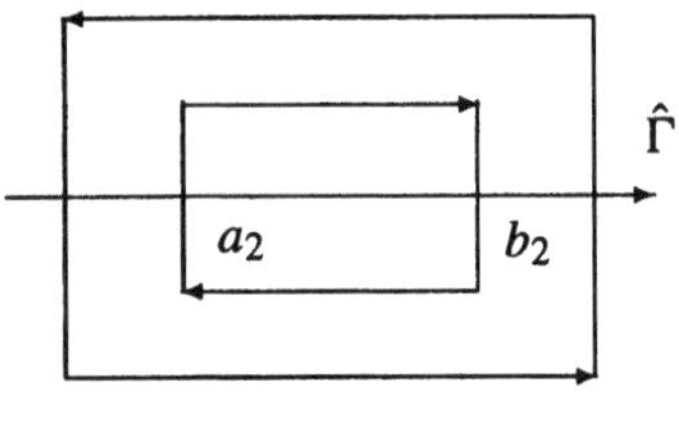

which are symmetric with respect to the real axis and such that Γ and the inner part of $\widehat{\Gamma}$ do not surround nonreal parts of $\sigma(A)$, and the outer part of $\widehat{\Gamma}$ surrounds all of $\sigma(A)$ and is contained in Γ_0, and define the following operators:

$$D: = -\frac{1}{2\pi i}\oint_{\Gamma} p(\lambda)(T-\lambda)^{-1}d\lambda,\ \widehat{D} := -\frac{1}{2\pi i}\oint_{\widehat{\Gamma}} \widehat{p}(\lambda)(T-\lambda)^{-1}d\lambda,$$
$$D_1: = -\frac{1}{2\pi i}\oint_{\Gamma} p(\lambda)(T_1-\lambda)^{-1}d\lambda,\ \widehat{D}_1 := -\frac{1}{2\pi i}\oint_{\widehat{\Gamma}} \widehat{p}(\lambda)(T_1-\lambda)^{-1}d\lambda.$$

The operators $D, \widehat{D}$ map $\widehat{\mathcal{F}} = QL^2$ into itself and vanish on $(I - Q)L^2$, and $D_1, \widehat{D}_1$ map $\widehat{\mathcal{F}}_1 = Q_1 L^2$ into itself and vanish on $(I - Q_1)L^2$. Therefore in the definitions of D etc. T can be replaced by TQ and T_1 by T_1Q_1. The operators TQ, D and $\widehat{D}$ are self-adjoint with respect to the G-inner product and the same holds for T_1Q_1, D_1 and $\widehat{D}_1$ with respect to the G_1-inner product on L^2 or on $\widehat{\mathcal{F}}, \widehat{\mathcal{F}}_1$, respectively. Moreover, the operator

$$D - D_1 = -\frac{1}{2\pi i}\oint_\Gamma p(\lambda)(TQ-\lambda)^{-1}(TQ - T_1Q_1)(T_1Q_1-\lambda)^{-1}d\lambda$$

is compact since $TQ - T_1Q_1$ is compact.

For a closed interval Γ and a linear operator C we denote by $\mathcal{L}_\Gamma(C)$ the maximal spectral subspace of C corresponding to Γ ([LMM], [LM]). Evidently, the restrictions $T|_{\mathcal{L}_{[a_1,b_1]}(T)}$ and $T_1|_{\mathcal{L}_{[a_1,b_1]}(T_1)}$ have spectrum in $\sigma(T)\cap[a_1,b_1]$ and $\sigma(T_1)\cap[a_1,b_1]$, respectively. The restrictions $D|_{\mathcal{L}_{[a_1,b_1]}(T)}$ and $D_1|_{\mathcal{L}_{[a_1,b_1]}(T_1)}$ are functions of the restrictions $T|_{\mathcal{L}_{[a_1,b_1]}(T)}$ and $T_1|_{\mathcal{L}_{[a_1,b_1]}(T_1)}$, respectively. By the spectral mapping theorem, $\sigma(D|_{\mathcal{L}_{[a_1,b_1]}(T)}) \subset \Delta$, where $\Delta := p([a_1,b_1])$, and hence $\mathcal{L}_{[a_1,b_1]}(T)\subset\mathcal{L}_\Delta(D)$. Similarly it follows that

$$\mathcal{L}_{[a_1,b_1]}(T_1)\subset\mathcal{L}_\Delta(D_1). \tag{6.15}$$

Further, $\widehat{D}D = 0$ and $\ker\widehat{D} = \mathcal{L}_{[a_2,b_2]}(T)$. Since $[a_2,b_2]\subset\sigma_+(T)\cup\rho(T)$, $\ker\widehat{D}$ is uniformly G-positive, and from the inclusion $\mathcal{R}(D)\subset\mathcal{L}_{[a_2,b_2]}(T)$ it follows that $\mathcal{R}(D)$ is uniformly positive. This means that there exists an $\alpha > 0$ such that

$$[Df, Df]_{L^2} \ge \alpha\|Df\|^2 \qquad (f\in\widehat{\mathcal{F}}),$$

or

$$Q^*D^*GDQ \ge \alpha Q^*D^*DQ. \tag{6.16}$$

Further, $D - D_1$ and $Q^*GQ - Q_1^*G_1Q_1$ are compact, hence also

$$Q^*D^*GDQ - Q_1^*D_1^*G_1D_1Q_1 = D^*Q^*GQD - D_1^*Q_1^*G_1Q_1D_1$$

is compact. Therefore (6.16) implies

$$Q_1^*D_1^*G_1D_1Q_1 - \alpha Q_1^*D_1^*D_1Q_1 = Q^*D^*GDQ - \alpha Q^*D^*DQ + K \ge K$$

for some compact operator K. It follows that

$$D_1^*G_1D_1 - \alpha D_1^*D_1 \ge Q_1^*KQ_1.$$

This inequality holds, in particular, in $\mathcal{L}_\Delta(D_1)$ and from the spectral mapping theorem we find $0\notin\sigma(D_1|\mathcal{L}_\Delta(D_1))$. Therefore the inner product $(G_1\,\cdot,\,\cdot)_{L^2}$ has at most finitely many negative squares on $\mathcal{L}_\Delta(D_1)$, and by (6.15) also on $\mathcal{L}_{[a_1,b_1]}(T_1)$. It follows that

$$\lambda_0\in\sigma_+(A_1)\cup\rho(A_1)\cup\sigma_{-,f}(A_1).$$

If $\lambda_0\in\rho(A)\cap\mathbb{R}$ then, since T_1Q_1 is a compact perturbation of TQ, λ_0 belongs either to $\rho(A_1)$, or it is an isolated eigenvalue of A_1 such that the corresponding Riesz projection of T_1 is finite dimensional and G_1-self-adjoint. Theorem 6.1 is proved.

Corollary 6.1 *If $\sigma(A) = \sigma_+(A)$ then the spectrum of A_1 outside of $\sigma_+(A)$ consists of an at most finite number of eigenvalues and the operator function A^{-1} is finitely meromorphic there.*

References

[ADRS] D. Alpay, A. Dijksma, J. Rovnyak and H. de Snoo, *Schur Functions, Operator Colligations, and Reproducing Kernel Pontryagin Spaces*. Operator Theory: Advances and Applications, vol. **96** (1997).

[DL] A. Dijksma and H. Langer, *Operator theory and ordinary differential operators*, in: Lectures on Operator Theory and its Applications, Fields Institute Monographs, vol. **3**, pp. 73–139. American Mathematical Society, 1995.

[DLS1] A. Dijksma, H. Langer and H.S.V. de Snoo, *Symmetric Sturm-Liouville operators with eigenvalue depending boundary conditions*, Canadian Math. Soc., Conference Proceedings **8** (1987), 87–116.

[DLS2] A. Dijksma, H. Langer and H.S.V. de Snoo, *Representations of holomorphic operator functions by means of resolvents of unitary or self-adjoint operators in Krein spaces*, Operator Theory: Advances and Applications, vol. **24** (1987), 123–143.

[FMM] I. Feldman, A. Markus and V. Matsaev, *Perturbations of isolated part of definite spectrum for G-self-adjoint operators and self-adjoint operator polynomials*, Integral Equations Operator Theory **23** (1995), 435–440.

[GGK] I. Gohberg, S. Goldberg and M.A. Kaashoek, *Classes of Linear Operators*. vol. I, Operator Theory: Advances and Applications, vol. **49** (1990).

[GK] I.C. Gohberg and M.G. Krein, *Introduction to the Theory of Linear Nonselfadjoint Operators*, Amer. Math. Soc., Transl. Math. Monographs, vol. **18** (1969).

[GKL] I. Gohberg, M.A. Kaashoek and D.C. Lay, *Equivalence, linearization and decomposition of holomorphic operator functions*, J. Funct. Anal. **28** (1978), 102–144.

[GL] I. Gohberg and J. Leiterer, *General theorems on the canonical factorization of operator functions with respect to a contour*, Matem. Issled. Kishinev **7** (1972), 87–134 [Russian].

[HSW] S. Hassi, H. de Snoo and H. Woracek, *Some interpolation problems of Nevanlinna-Pick type. The Krein-Langer method*, Operator Theory: Advances and Applications (to appear).

[KMR] M.A. Kaashoek, C.V.M. van der Mee and L. Rodman, *Analytic operator functions with compact spectrum*. I: Spectral nodes, linearization and equivalence, Integral Equations Operator Theory **4** (1981), 504–547; II: Spectral pairs and factorization, Integral Equations Operator Theory **5** (1982), 791–827; III: Hilbert space case: Inverse problems and applications, J. Oper. Theory **10** (1983), 219–250.

[L1] H. Langer, *Über stark gedämpfte Scharen im Hilbertraum*, J. Math. Mech. **17** (1968), 685–706.

[L2] H. Langer, *Über eine Klasse polynomialer Scharen selbstadjungierter Operatoren im Hilbertraum*, II, J. Funct. Anal. **16** (1974), 221–234.

[L3] H. Langer, *Zur Spektraltheorie J-selbstadjungierter Operatoren*, Math. Ann. **146** (1962), 60–85.

[LancMM1] P. Lancaster, A. Markus and V. Matsaev, *Definitizable operators and quasi-hyperbolic operator polynomials*, J. Funct. Anal. **131** (1995), 1–28.

[LancMM2] P. Lancaster, A. Markus and V. Matsaev, *Perturbations of G-self-adjoint operators and operator polynomials with real spectrum*, Operator Theory: Advances and Applications, vol. **87** (1996), 207–221.

[LM] Yu.I. Lyubich and V.I. Matsaev, *On operators with decomposable spectrum*, Mat. Sbornik **56** (98) (1962), 433–468 (Russian). Engl. transl.: AMS Transl. (2) **47** (1965), 89–129.

[LMM] H. Langer, A. Markus and V. Matsaev, *Locally definite operators in indefinite inner product spaces*, Math. Ann. **308** (1997), 405–424.

[P] L.B. Page, *Bounded and compact vectorial Hankel operators*, Trans. Amer. Math. Soc. **150** (1970), 529–534.

[R] L. Rodman, *An Introduction to Operator Polynomials*, Operator Theory: Advances and Applications, vol. **38** (1989).

[Ru] W. Rudin, *Functional Analysis*, Mc Graw-Hill, New York, 1973.

[W] J.L. Walsh, *Interpolation and Approximation by Rational Functions in the Complex Domain*, Amer. Math. Soc., Providence, R.I., 1969.

H. Langer
Institut für Analysis
und Technische Mathematik
Technische Universität Wien
A-1040 Wien
Austria

A. Markus
Department of Mathematics
and Computer Sciences
Ben-Gurion University of the Negev
Beer-Sheva 84105
Israel

V. Matsaev
School of Mathematical Sciences
Tel-Aviv University
Ramat-Aviv 69978
Israel

Mathematics Subject Classification (1991): 47B50, 47A56

Operator Theory:
Advances and Applications, Vol. 118

Operator Interpretation of Resonances Generated by Some Operator Matrices

R. Mennicken and A.K. Motovilov

We consider the analytic continuation of the transfer function for a 2×2 operator matrix into the unphysical sheets of the energy Riemann surface. We construct a family of non-selfadjoint operators which reproduce certain parts of the transfer-function spectrum including resonances situated on the unphysical sheets neighboring the physical sheet. On this basis, completeness and basis properties for the root vectors of the transfer function (including those for the resonances) are proved.

1 Introduction

In this paper we deal with 2×2 operator matrices

$$\mathbf{H} = \begin{pmatrix} A_0 & B_{01} \\ B_{10} & A_1 \end{pmatrix} \tag{1}$$

acting in the orthogonal sum $\mathcal{H} = \mathcal{H}_0 \oplus \mathcal{H}_1$ of separable Hilbert spaces $\mathcal{H}_0$ and $\mathcal{H}_1$. The entries $A_0 : \mathcal{H}_0 \to \mathcal{H}_0$, and $A_1 : \mathcal{H}_1 \to \mathcal{H}_1$, are assumed to be self-adjoint operators with domains $\mathcal{D}(A_0)$ and $\mathcal{D}(A_1)$, respectively. It is assumed that the couplings $B_{ij} : \mathcal{H}_j \to \mathcal{H}_i$, $i, j = 0, 1$, $i \neq j$, are bounded operators (i.e., $B_{ij} \in \mathbf{B}(\mathcal{H}_j, \mathcal{H}_i)$) and $B_{01} = B_{10}^*$. Under these assumptions the matrix $\mathbf{H}$ is a self-adjoint operator in $\mathcal{H}$ with domain $\mathcal{D}(\mathbf{H}) = \mathcal{D}(A_0) \oplus \mathcal{D}(A_1)$. Note that operators of the form (1) arise in many quantum-physical problems (see e.g., [1]–[7]).

In the spectral theory of operator matrices (1) an important role is played by the *transfer functions*

$$\begin{aligned} M_i(z) &= A_i - z + V_i(z) \text{ where} \\ V_i(z) &= -B_{ij}(A_j - z)^{-1} B_{ji}, \quad i = 0, 1, \quad j \neq i. \end{aligned} \tag{2}$$

A particular role of the functions $M_i(z)$ can be understood already from the fact that the resolvent of the operator (1) can be expressed explicitly in terms of the inverse transfer functions $M_0^{-1}(z)$ or $M_1^{-1}(z)$. Therefore, in studying the spectral properties of the transfer functions one studies at the same time the spectral properties of the operator matrix $\mathbf{H}$.

In the papers [8], [9] the following question was raised: Is it possible to introduce the operators H_i, $i = 0, 1$, independent of the spectral parameter z, such that the equality $H_i \psi^{(i)} = z \psi^{(i)}$, $\psi^{(i)} \in \mathcal{H}_i$, implies $M_i(z) \psi^{(i)} = 0$? Obviously, having found such an operator one would reduce the spectral problem for the transfer-function $M_i(z)$ to the standard spectral problem for the operator H_i and, thus, the completeness and basis properties for the eigenvectors of M_i could be studied in terms of the operator H_i referring to well known facts from operator theory. A rigorous answer to the above question was found in [10], [11] in the case where the spectra $\sigma(A_0)$ and $\sigma(A_1)$ of the entries A_0 and A_1 are separated from each other,

$$\operatorname{dist}\{\sigma(A_0), \sigma(A_1)\} > 0. \tag{3}$$

To this end an operator-valued function $V_i(Y_i)$ on the space of linear operators in $\mathcal{H}_i$ was constructed in [10], [11] such that $V_i(Y_i)\psi^{(i)} = V_i(z)\psi^{(i)}$ for any eigenvector $\psi^{(i)}$ corresponding to an eigenvalue z of the operator Y_i. The desired operator H_i was constructed as a solution of the operator equation

$$H_i = A_i + V_i(H_i), \qquad i = 0, 1. \tag{4}$$

Notice that an equation of the form (4) first appeared explicitly in the paper [12] by M.A. Braun.

The solvability of the equation (4) was announced in [10] and proved in [11] under the assumption $\|B_{ij}\|_2 < \frac{1}{2}\operatorname{dist}\{\sigma(A_0), \sigma(A_1)\}$ where $\|B_{ij}\|_2$ stands for the Hilbert-Schmidt norm of the couplings B_{ij}. It was found in [11] that the problem of constructing the operators H_i is closely related to the problem of characterizing the invariant subspaces of the matrix $\mathbf{H}$ which admit representations in the form of the graphs for some bounded operators $Q_{ji} : \mathcal{H}_i \to \mathcal{H}_j$. The point is that under the conditions of [10], [11] the solutions H_i, $i = 0, 1$, of Eqs. (4) read as $H_i = A_i + B_{ij}Q_{ji}$ where Q_{ji} are contractions. The operators Q_{ji} determine a similarity transform reducing the matrix $\mathbf{H}$ to the block-diagonal form $\mathbf{H}' = \operatorname{diag}\{H_0, H_1\}$ (see [11]).

The idea of the block diagonalization of 2×2 operator matrices in terms of the invariant subspaces allowing a graph representation already goes back to the paper [3] by S. Okubo (regarding applications of Okubo's approach to particle physics see, e.g., Refs. [3]–[5]). In a mathematically rigorous way this idea was applied to a class of selfadjoint operators in statistical physics by V.A. Malyshev and R.A. Minlos [6]. The techniques of Ref. [6] are restricted to the case where the norms of the entries B_{ij} are sufficiently small and the separation condition (3) holds, too. V.M. Adamyan and H. Langer [13] proved the existence of invariant subspaces allowing a graph representation for arbitrary bounded entries B_{ij} however assuming, instead of the condition (3), the essentially different assumption that the spectrum of one of the entries A_i, $i = 0, 1$ is situated strictly below the spectrum of the other one, say $\max \sigma(A_1) < \min \sigma(A_0)$. Recently, the result of [13] was extended by V.M. Adamyan, H. Langer, R. Mennicken and J. Saurer [14] to the case where $\max \sigma(A_1) \le \min \sigma(A_0)$ and where the couplings B_{ij} were allowed to be unbounded operators such that, for $\alpha_0 < \min \sigma(A_0)$, the product $(A_0 - \alpha_0)^{-1/2} B_{01}$ makes sense as a bounded operator. The mentioned conditions were then somewhat weakened by R. Mennicken and A.A. Shkalikov [15] in the case of a bounded entry A_1 and the same type of entries B_{ij} as in [14]. Instead of the explicit conditions on the spectra of A_i, the paper [15] uses a condition on the spectrum of the transfer function $M_1(z)$ itself. One can check that the spectral component H_i of the matrix $\mathbf{H}$ constructed in [13]–[15] satisfies the equation (4), at least in the case where for $j \neq i$ the entry A_j is bounded.

In the present work we study the equation (4) in a case which is totally different from the spectral situations considered in [6], [10], [11], [13]–[15]. From the beginning, we suppose that $\sigma(A_0) \cap \sigma(A_1) \neq \emptyset$. Moreover, we are especially interested just in the case where the spectrum of, say A_1, is partly or totally embedded into the continuous spectrum of A_0. We work under the assumption that the coupling operators B_{ij} are such that the transfer function $M_1(z)$ admits an analytic continuation, as an operator-valued function, through the cuts along the branches of the absolutely continuous spectrum $\sigma_{ac}(A_0)$ of the entry A_0. In Sect. 2 we describe the conditions on B_{ij} making such a continuation of $M_1(z)$ possible.

The problem considered is closely related to the investigation of the resonances generated by the matrix $\mathbf{H}$. Regarding a definition of resonances and the history of the subject, see e.g. the book [16]. A recent survey of the literature on resonances can be found in [17]. Throughout the present paper we treat resonances as the discrete spectrum of the transfer function $M_1(z)$ situated on the unphysical sheets of its Riemann surface. We assume that the absolutely continuous spectrum of the entry A_0 consists of m $(1 \leq m < \infty)$ distinct intervals. As a result in Sect. 3 we get 2^m variants of the function $V_1(Y)$ and, consequently, 2^m different variants of the equation (4) which read now as Eq. (12). The solutions of (12) represent non-selfadjoint operators the spectrum of which includes the resonances on the unphysical sheets connected directly with the physical sheet.

In Sect. 4 we first prove a factorization theorem for the transfer function $M_1(z)$. It follows from this theorem that there exist certain domains surrounding the set $\sigma(A_1)$ and lying partly in the unphysical sheet(s) where the spectrum of M_1 is represented only by the spectrum of the corresponding solutions of the basic equation (12). Since the root vectors of these solutions are also root vectors for M_1, this fact allows us to consider further, in Sect. 5 and 6, completeness and basis properties of the root vectors of the transfer function M_1 corresponding to its spectrum in the above domains, including the resonance spectrum. To prove these properties we rely mainly on the certain results from the books by I.C. Gohberg and M.G. Krein [18] and by T. Kato [19].

2 Analytic Continuation of the Transfer Function

The transfer functions $M_i(z)$, $i = 0, 1$, considered on the resolvent set $\varrho(A_j)$ of the entry A_j, $j \neq i$, represent a particular case of holomorphic operator-valued functions. In the present work we use the standard definition of holomorphy of an operator-valued function with respect to the operator norm topology (see, e.g., [14]). One can extend the usual definitions of the spectrum and its components to operator-valued functions. Each transfer function $M_i(z)$ is holomorphic at least on the resolvent set $\varrho(A_j)$ of the entry A_j, $j \neq i$. Since the inverse transfer functions $M_i^{-1}(z)$ coincide with the respective block components $\mathbf{R}_{ii}(z)$ of the resolvent $\mathbf{R}(z) = (\mathbf{H} - z)^{-1}$, they are both holomorphic at least in the set $\varrho(\mathbf{H})$.

Let E_j be the spectral measure for the entry A_j, $A_j = \int_{\sigma(A_j)} \lambda dE_j(\lambda)$, $j = 0, 1$, $\sigma(A_j) \subset \mathbb{R}$. Then the functions $V_i(z)$ can be written

$$V_i(z) = B_{ij} \int_{\sigma(A_j)} dE_j(\mu) \frac{1}{z-\mu} B_{ji}.$$

Thus, it is convenient to introduce the quantities $\mathcal{V}_j(B) = \sup_{\{\delta_k\}} \sum_k \|B_{ij} E_j(\delta_k) B_{ji}\|$, with $\{\delta_k\}$ being a finite or countable complete system of Borel subsets of $\sigma(A_j)$ such that $\delta_k \cap \delta_l = \emptyset$, if $k \neq l$ and $\cup_k \delta_k = \sigma(A_j)$. The number $\mathcal{V}_j(B)$ is called the variation of the operators B_{ij} with respect to the spectral measure E_j. Notice that the value of $\mathcal{V}_j(B)$ satisfies the estimates $\|B_{ij}\|^2 \leq \mathcal{V}_j(B) \leq \|B_{ij}\|_2^2$. This value can be finite even if the Hilbert-Schmidt norm $\|B_{ij}\|_2$ is infinite.

Along with the "total" variation $\mathcal{V}_j(B)$ we shall use the "truncated" variations $\mathcal{V}_j(B)|_\Delta = \sup_{\{\delta_k\}} \sum_k \|B_{ij} E_j(\delta_k \cap \Delta) B_{ji}\|$ where Δ is a certain Borel subset of $\sigma(A_j)$. Obviously $\mathcal{V}_j(B)|_\Delta \leq \mathcal{V}_j(B)$.

We assume that the spectrum of the operator A_1 only intersects the absolutely continuous spectrum $\sigma_{ac}(A_0)$ of the entry A_0 and this intersection only takes place on the pairwise nonintersecting open intervals $\Delta_k^0 = (\mu_k^{(1)}, \mu_k^{(2)}) \subset \sigma_{ac}(A_0)$, where $\mu_k^{(1)} < \mu_k^{(2)}$, $k = 1, 2, \ldots, m$, $m < \infty$, and $-\infty \leq \mu_1^{(1)}$, $\mu_m^{(2)} \leq +\infty$. Hence, we assume that $\Delta_k^0 \cap \sigma(A_1) \neq \emptyset$ for all $k = 1, 2, \ldots, m$ and $\sigma(A_1) \cap \sigma'(A_0) = \emptyset$ where $\sigma'(A_0) = \sigma(A_0) \setminus \cup_{k=1}^m \Delta_k^0$.

We shall suppose that the product $K_B(\mu) = B_{10} E^0(\mu) B_{01}$ is differentiable in $\mu \in \Delta_k^0$, $k = 1, 2, \ldots, m$, in the operator norm topology, where $E^0(\mu)$ stands for the spectral function of A_0, $E^0(\mu) = E_0((-\infty, \mu))$. The derivative $K_B'(\mu)$ is non-negative, $K_B'(\mu) \geq 0$, since $K_B(\mu)$ is a non-decreasing function. Obviously, $\mathcal{V}_0(B)|_{\Delta_k^0} = \int_{\Delta_k^0} d\mu \|K_B'(\mu)\|$.

Further, we suppose that the function $K_B'(\mu)$ is continuous within the closed intervals $\overline{\Delta_k^0}$ and, moreover, that it admits analytic continuation from each of these intervals to a simply connected domain situated, say, in $\mathbb{C}^-$. Let this domain be called D_k^- for the interval Δ_k^0. We assume that the boundary of each domain D_k^- includes the entire spectral interval Δ_k^0 and the domains D_k^- and D_j^- for different k and j do not intersect each other. Since $K_B'(\mu)$ represents a selfadjoint operator for $\mu \in \Delta_k^0$ and $\Delta_k^0 \subset \mathbb{R}$, the function $K_B'(\mu)$ also automatically admits analytic continuation from Δ_k^0 into the domain $D_k^+ = \{z : \overline{z} \in D_k^-\}$ which is symmetric to D_k^- with respect to the real axis. For the continuation into D_k^+ we retain the same notation $K_B'(\mu)$. The selfadjoitness of $K_B'(\mu)$ for $\mu \in \Delta_k^0$ implies

$[K'_B(\mu)]^* = K'_B(\bar{\mu})$, $\mu \in D_k^{\pm}$. Also, we shall always suppose the $K'_B(\mu)$ satisfies the Hölder condition at the (finite) end points $\mu_k^{(1)}$, $\mu_k^{(2)}$ of the spectral intervals Δ_k^0, $\|K'_B(\mu) - K'_B(\mu_k^{(i)})\| \leq C|\mu - \mu_k^{(i)}|^{\gamma}$, $i = 1, 2$, $\mu \in D_k^{\pm}$, with some positive C and γ.

Let $l = (l_1, l_2, \ldots, l_m)$ be a multi-index having the components $l_k = +1$ or $l_k = -1$, $k = 1, 2, \ldots, m$. In what follows we consider the domains $D_l = \bigcup_{k=1}^{m} D_k^{l_k}$. Let $\Gamma_k^{l_k}$ be a rectifiable Jordan curve in $D_k^{l_k}$ resulting from continuous deformation of the interval Δ_k^0, the (finite) end points of this interval being fixed. With the exception of the end points, the closure $\overline{\Gamma}_k^{l_k}$ of the contour $\Gamma_k^{l_k}$ should have no other common points with the set $\sigma_c(A_0)$. By Γ_l, $l = (l_1, l_2, \ldots, l_m)$, we shall denote the union of the contours $\Gamma_k^{l_k}$, $\Gamma_l = \bigcup_{k=1}^{m} \Gamma_k^{l_k}$. Also, we extend the definition of the variation $\mathcal{V}_0(B)$ to the set $\sigma'(A_0) \cup \Gamma_l$ by introducing the modified variation

$$\mathcal{V}_0(B, \Gamma_l) = \mathcal{V}_0(B)|_{\sigma'(A_0)} + \int_{\Gamma_l} |d\mu| \|K'_B(\mu)\| \tag{5}$$

where $|d\mu|$ is the Lebesgue measure on Γ_l. We suppose that the operators B_{ij} are such that there exist contours Γ_l where the value $\mathcal{V}_0(B, \Gamma_l)$ is finite, $\mathcal{V}_0(B, \Gamma_l) < \infty$, including also the case of an unbounded set $\bigcup_{k=1}^{m} \Delta_k^0$. The contours Γ_l satisfying the condition $\mathcal{V}_0(B, \Gamma_l) < \infty$ are said to be *K_B-bounded* contours.

Lemma 1 *The analytic continuation of the transfer function $M_1(z)$, $z \in \mathbb{C}\backslash\sigma(A_0)$, through the spectral intervals Δ_k^0 into the subdomain $D(\Gamma_l) \subset D_l$ bounded by the set $\bigcup_{k=1}^{m} \Delta_k^0$ and a K_B-bounded contour Γ_l is given by*

$$M_1(z, \Gamma_l) = A_1 - z + V_1(z, \Gamma_l) \tag{6}$$

where

$$\begin{aligned} V_1(z, \Gamma_l) &= \int_{\sigma'(A_0)\cup\Gamma_l} \frac{K_B(d\mu)}{z - \mu} \\ &:= \int_{\sigma'(A_0)} \frac{B_{10}E_0(d\mu)B_{01}}{z - \mu} + \int_{\Gamma_l} d\mu \frac{K'_B(\mu)}{z - \mu}. \end{aligned} \tag{7}$$

For $z \in D_k^{l_k} \cap D(\Gamma_l)$ the function $M_1(z, \Gamma_l)$ may be written as

$$M_1(z, \Gamma_l) = M_1(z) + 2\pi i l_k K'_B(z). \tag{8}$$

Proof: The proof is reduced to the observation that the function $M_1(z, \Gamma_l)$ is holomorphic for $z \in \mathbb{C}\backslash[\sigma'(A_0) \cup \Gamma_l]$ and coincides with $M_1(z)$ for $z \in \mathbb{C}\backslash[\sigma'(A_0) \cup \overline{D(\Gamma_l)}]$. Eq. (8) is obtained from (7) using the Residue Theorem. □

The formula (8) shows that in general the transfer function M_1 has a multi-sheeted Riemann surface. The sheet of the complex plane where the transfer function $M_1(z)$ is considered together with the resolvent $\mathbf{R}(z)$ initially is said to be the *physical sheet*. The remaining sheets of the Riemann surface of M_1 are said to be *unphysical sheets*. In the present work we only deal with the unphysical sheets closest to the physical sheet, i.e., with the sheets which are connected directly with the physical sheet through the intervals Δ_k^0 for some $k \in \{1, 2, \dots, m\}$.

3 The Basic Equation and its Solutions

If an operator-valued function $T : \sigma'(A_0) \cup \Gamma \to \mathbf{B}(\mathcal{H}_1, \mathcal{H}_1)$ is continuous and bounded on a K_B-bounded contour Γ,

$$\|T\|_{\infty,\Gamma} = \sup_{\mu \in \sigma'(A_0) \cup \Gamma} \|T(\mu)\| < \infty,$$

and satisfies the Lipschitz condition on $\sigma'(A_0)$, then the integral

$$\begin{aligned}\int_{\sigma'(A_0)\cup\Gamma} K_B(d\mu)T(\mu) \;:=\; &\int_{\sigma'(A_0)} B_{10}E_0(d\mu)B_{01}T(\mu) \\ &+ \int_\Gamma d\mu K_B'(\mu)T(\mu)\end{aligned} \tag{9}$$

exists in the sense of the operator norm topology (see Lemma 7.2 in [14]) and

$$\left\| \int_{\sigma'(A_0)\cup\Gamma} K_B(d\mu)T(\mu) \right\| \le \mathcal{V}_0(B, \Gamma)\|T\|_{\infty,\Gamma}. \tag{10}$$

In particular, if $T(z)$ is the resolvent of an operator Y, $T(z) = (Y - zI_1)^{-1}$, the spectrum of which is separated from $\sigma'(A_0) \cup \Gamma$, then one can define the operator

$$V_1(Y, \Gamma) = \int_{\sigma'(A_0)\cup\Gamma} K_B(d\mu)(Y - \mu)^{-1}.$$

This operator is bounded, $V_1(Y, \Gamma) \in \mathbf{B}(\mathcal{H}_1, \mathcal{H}_1)$, and, because of (10), its norm admits the estimate

$$\|V_1(Y, \Gamma)\| \le \mathcal{V}_0(B, \Gamma_l) \sup_{\mu \in \sigma'(A_0)\cup\Gamma} \|(Y - \mu)^{-1}\|. \tag{11}$$

In what follows we consider the equation

$$Y = A_1 + V_1(Y, \Gamma). \tag{12}$$

This equation possesses the following property: If an operator H_1 is a solution of (12) and u_1 is an eigenvector of H_1, $H_1u_1 = zu_1$, then automatically $zu_1 = A_1u_1 + V_1(H_1, \Gamma)u_1 = A_1u_1 + V_1(z, \Gamma)u_1$. This implies that any eigenvalue z of such an operator H_1 is automatically an eigenvalue for the continued transfer function $M_1(z, \Gamma_l)$ and u_1 is its eigenvector. Thus, having found the solution(s) of the equation (12) one obtains an effective tool for studying the spectral properties of the transfer function $M_1(z, \Gamma)$ itself. Often it is convenient to rewrite Eq. (12) in the form

$$X = V_1(A_1 + X, \Gamma) \tag{13}$$

where $X = Y - A_1$.

Theorem 1 *Let the contour* Γ *be* K_B*-bounded and*

$$\mathcal{V}_0(B, \Gamma) < \frac{1}{4} d_0^2(\Gamma) \tag{14}$$

where $d_0(\Gamma) = \text{dist}\{\sigma(A_1), \sigma'(A_0) \cup \Gamma\}$. *Then Eq.* (13) *is uniquely solvable in any ball* $\mathcal{S}_1(r) \subset \mathbf{B}(\mathcal{H}_1, \mathcal{H}_1)$ *including operators* $X : \mathcal{H}_1 \to \mathcal{H}_1$ *the norms of which are bounded as* $\|X\| \le r$ *with* r *such that*

$$r_{\min}(\Gamma) \le r < r_{\max}(\Gamma). \tag{15}$$

Here

$$\begin{aligned} r_{\min}(\Gamma) &= d_0(\Gamma)/2 - \sqrt{d_0^2(\Gamma)/4 - \mathcal{V}_0(B, \Gamma)}, \\ r_{\max}(\Gamma) &= d_0(\Gamma) - \sqrt{\mathcal{V}_0(B, \Gamma)}. \end{aligned} \tag{16}$$

The solution X *of Eq.* (13) *is the same for any* r *satisfying* (15) *and in fact it belongs to the smallest ball* $\mathcal{S}_1(r_{\min})$, $\|X\| \le r_{\min}(\Gamma)$.

Proof: Let $F(X) = V_1(A_1 + X, \Gamma)$ with $X \in \mathcal{S}_1(r)$. To begin with we search for a condition under which the function F is a contracting mapping of the ball $\mathcal{S}_1(r)$ into itself. Since, in view of (15) and (16), the condition $0 < r < d_0$, $d_0 = d_0(\Gamma)$ automatically holds, the spectrum of the operator $A_1 + X$ does not intersect the set $\sigma'(A_0) \cup \Gamma$. This means that for all $\mu \in \sigma'(A_0) \cup \Gamma$ the resolvent $(A_1 + X - \mu I_1)^{-1}$ exists as a bounded operator in $\mathcal{H}_1$. It follows from the estimate (11) that

$$\|F(X)\| \le \mathcal{V}_0(B, \Gamma) \sup_{\mu \in \sigma'(A_0) \cup \Gamma} \|(A_1 + X - \mu)^{-1}\| \le \mathcal{V}_0(B, \Gamma) \frac{1}{d_0 - r}$$

while

$$\|F(X) - F(Y)\| \le \mathcal{V}_0(B, \Gamma) \frac{1}{(d_0 - r)^2} \|Y - X\|.$$

Thus, the ball $\mathcal{S}_1(r)$ is mapped into itself by the function F and this mapping is a contraction if the radius r and the value $\mathcal{V}_0(B, \Gamma)$ are such that

$$\mathcal{V}_0(B, \Gamma) \le r(d_0 - r), \qquad \mathcal{V}_0(B, \Gamma) < (d_0 - r)^2. \tag{17}$$

Under the condition (14) the inequalities in (17) are equivalent to the condition (15). Hence, if this condition is valid then Eq. (13) has a solution in any ball $\mathcal{S}_1(r)$ with r satisfying (15) and this solution is unique. This means that the solution is the same for all the radii satisfying (15). Moreover, it belongs to the ball $\mathcal{S}_1(r_{\min})$ with the radius $r_{\min}$ defined in (16). □

Theorem 2 *Let the conditions of* Theorem 1 *be valid for a K_B-bounded contour $\Gamma \subset D_l$ and let X be the solution of Eq.* (13) *mentioned in that theorem. Then X coincides with the analogous solution $\tilde{X}$ for any other K_B-bounded contour $\tilde{\Gamma} \subset D_l$ satisfying the estimate $\mathcal{V}_0(B, \tilde{\Gamma}) < \tilde{d}_0^2/4$ with $0 < \tilde{d}_0 = \operatorname{dist}\{\sigma(A_1), \sigma'(A_0) \cup \tilde{\Gamma}\} \le d_0(\Gamma)$. Moreover, this solution satisfies the inequality $\|X\| \le r_0(B)$ where $r_0(B) = \inf\limits_{\Gamma:\omega(B,\Gamma)>0} r_{\min}(\Gamma)$ with $r_{\min}(\Gamma)$ given by* (16) *and $\omega(B, \Gamma) = d_0^2(\Gamma) - 4\mathcal{V}_0(B, \Gamma)$.*

Proof of this theorem is reduced to an appropriate continuous deformation of the integration paths. The essential point is the checking of independence of the radius $r_0(B)$ of the multi-index l. To this end we consider an arbitrary K_B-bounded contour $\Gamma_l \subset D_l$, $\Gamma_l = \cup_{k=1}^m \Gamma_k^{l_k}$. Denote by $\Gamma_{l'}$ the contour resulting from Γ_l after replacing a part of the curves $\Gamma_k^{l_k}$ with the curves $\Gamma_k^{(-l_k)} = \{\mu : \overline{\mu} \in \Gamma_k^{l_k}\}$, symmetric to corresponding $\Gamma_k^{l_k}$ with respect to the real axis. Obviously, such replacements generate, additionally to Γ_l, $2^m - 1$ different contours $\Gamma_{l'}$ where $l' = (l_1', l_2', \ldots, l_m')$ with $l_k' = \pm l_k$, $k = 1, 2, \ldots, m$. For every of such contours the value of $\mathcal{V}_0(B, \Gamma_{l'})$ is the same, namely $\mathcal{V}_0(B, \Gamma_{l'}) = \mathcal{V}_0(B, \Gamma_l)$, since the replacement of Γ_l with $\Gamma_{l'}$ does not change $\int_{\Gamma_l} |d\mu| \|K_B'(\mu)\|$. But this means that $r_0(B)$ does not depend on l.

So, for a given holomorphy domain D_l the solutions X and H_1, $H_1 = A_1 + X$, do not depend on the K_B-bounded contours $\Gamma_l \subset D_l$ satisfying the condition (14). But when the index l changes, X and H_1 can also change. For this reason we shall supply them in the following, when it is necessary, with the index l writing, respectively, $X^{(l)}$ and $H_1^{(l)}$, $H_1^{(l)} = A_1 + X^{(l)}$. Therefore, Theorem 1 guarantees us, in general, the existence of the 2^m solutions $X^{(l)}$ to the basic equation (12) and, hence, 2^m corresponding solutions $H_1^{(l)}$ to the basic equation (13). Surely, Eqs. (12) and (13) are non-linear equations and, outside the balls $\|X\| < r_{\max}(\Gamma)$, they may, in principle, have other solutions, different from the $X^{(l)}$ or $H_1^{(l)}$ the existence of which is guaranteed by Theorem 1. In the following we shall deal only with the solutions $X^{(l)}$ or $H_1^{(l)}$.

4 Factorization of the Transfer Function

Theorem 3 *Let Γ_l be a K_B-bounded contour satisfying the condition* (14) *and $H_1^{(l)} = A_1 + X^{(l)}$ where $X^{(l)}$ is the above solution of the basic equation* (12). *Then,* *for $z \in \mathbb{C}\backslash(\sigma'(A_0) \cup \Gamma_l)$, the transfer function $M_1(z, \Gamma_l)$ admits the factorization*

$$M_1(z, \Gamma_l) = W_1(z, \Gamma_l)(H_1^{(l)} - z) \tag{18}$$

where $W_1(z, \Gamma_l)$ is the following bounded operator in $\mathcal{H}_1$:

$$W_1(z, \Gamma_l) = I_1 - \int_{\sigma'(A_0)\cup\Gamma_l} K_B(d\mu)\frac{1}{\mu - z}(H_1^{(l)} - \mu)^{-1}. \tag{19}$$

Here, I_1 denotes the identity operator in $\mathcal{H}_1$.

If $\operatorname{dist}\{z, \sigma(A_1)\} \le d_0(\Gamma_l)/2$, *the operator $W_1(z, \Gamma_l)$ is boundedly invertible and*

$$\|[W_1(z, \Gamma_l)]^{-1}\| \le \left(1 - \frac{\mathcal{V}_0(B, \Gamma_l)}{d_0^2(\Gamma_l)/4}\right)^{-1} < \infty. \tag{20}$$

Note that the above statement recalls the known factorization theorem by A.I. Virozub and V.I. Matsaev [20] being valid for a class of selfadjoint operator-valued functions (see also [21]). However, in the case we deal with in the present paper, the function $M_1(z, \Gamma_l)$ is not a selfadjoint operator-valued function in the sense of [20].

Proof: For $z \in \mathbb{C}\backslash(\sigma'(A_0) \cup \Gamma_l)$ the boundedness of the operator $W_1(z, \Gamma_l)$ given by (19) is evident. To prove the factorization (18) we note that for any $z \notin \sigma'(A_0) \cup \Gamma_l$

$$\begin{aligned} & W_1(z, \Gamma_l)(H_1^{(l)} - z) \\ & = H_1^{(l)} - z - \int_{\sigma'(A_0)\cup\Gamma_l} K_B(d\mu)\frac{(H_1^{(l)} - \mu)^{-1}(H_1^{(l)} - z)}{\mu - z}. \end{aligned} \tag{21}$$

Since $(H_1^{(l)} - \mu)^{-1}(H_1^{(l)} - z) = I_1 + (\mu - z)(H_1^{(l)} - \mu)^{-1}$ one finds

$$\begin{aligned} & \int_{\sigma'(A_0)\cup\Gamma_l} K_B(d\mu)\frac{(H_1^{(l)} - \mu)^{-1}(H_1^{(l)} - z)}{\mu - z} \\ & = \int_{\sigma'(A_0)\cup\Gamma_l} \frac{K_B(d\mu)}{\mu - z} + \int_{\sigma'(A_0)\cup\Gamma_l} K_B(d\mu)(H_1^{(l)} - \mu)^{-1}. \end{aligned}$$

But according to (6) and (7) $\int_{\sigma'(A_0)\cup\Gamma_l} K_B(d\mu)(\mu - z)^{-1} = A_1 - z - M_1(z, \Gamma_l)$, while according to (12) $\int_{\sigma'(A_0)\cup\Gamma_l} K_B(d\mu)(H_1^{(l)} - \mu)^{-1} = H_1^{(l)} - A_1$. Making use of these expressions one immediately obtains Eq. (18).

Further, we prove that the factor $W_1(z, \Gamma_l)$ is a boundedly invertible operator if the condition $\operatorname{dist}\{z, \sigma(A_1)\} \le d_0(\Gamma_l)/2$ is valid. Indeed, under this condition $|\mu - z| \ge \operatorname{dist}\{z, \sigma'(A_0) \cup \Gamma_l\} \ge d_0(\Gamma_l)/2$ since $\operatorname{dist}\{\sigma(A_1), \sigma'(A_0) \cup \Gamma_l\} =$

$d_0(\Gamma_l)$. On the other hand $H_1^{(l)} = A_1 + X^{(l)}$ and $\|X^{(l)}\| < d_0(\Gamma_l)/2$. Thus for $\mu \in \sigma'(A_0) \cup \Gamma_l$ we have $\|(H_1^{(l)} - \mu)^{-1}\| < 2/d_0(\Gamma_l)$. Consequently

$$(22) \qquad \left\| \int_{\sigma'(A_0)\cup\Gamma_l} K_B(d\mu) \frac{1}{\mu - z} (H_1^{(l)} - \mu)^{-1} \right\| < \frac{\mathcal{V}_0(B, \Gamma_l)}{(d_0(\Gamma_l)/2)^2} < 1$$

and, thus, the estimate (20) is true. □

It is easy to write some simple but useful relations between some operators $H_1^{(l)}$. In particular, we derive such relations between $H_1^{(l)}$ and $H_1^{(-l)}$, $(-l) = (-l_1, -l_2, \ldots, -l_m)$ where l_k, $k = 1, 2, \ldots, m$, stand for the components of the multi-index $l = (l_1, l_2, \ldots, l_m)$. According to our convention, $\Gamma_{(-l)} \subset D_{(-l)}$ is a contour which is obtained from the contour Γ_l by replacing all the components $\Gamma_k^{l_k}$ by the conjugate ones $\Gamma_k^{(-l_k)}$.

Lemma 2 *Let $\Gamma_l \subset D_l$ be a K_B-bounded contour for which the conditions of Theorem 1 are valid. Then for any $z \in \mathbb{C}\backslash(\sigma'(A_0) \cup \Gamma_l)$ the following equality holds:*

$$(23) \qquad W_1(z, \Gamma_l)(H_1^{(l)} - z) = (H_1^{(-l)*} - z)[W_1(\overline{z}, \Gamma_{(-l)})]^*.$$

Therefore the spectrum of $H_1^{(-l)*}$ coincides with the spectrum of $H_1^{(l)}$.

Theorem 4 *The spectrum $\sigma(H_1^{(l)})$ of the operator $H_1^{(l)} = A_1 + X^{(l)}$ belongs to the closed $r_0(B)$-neighborhood $\mathcal{O}_{r_0}(A_1)$ of the spectrum of A_1, $\mathcal{O}_{r_0}(A_1) = \{z \in \mathbb{C} : \operatorname{dist}\{z, \sigma(A_1)\} \le r_0(B)\}$. If a contour $\Gamma_l \subset D_l$ satisfies (14), then the complex spectrum of $H_1^{(l)}$ belongs to $D_l \cap \mathcal{O}_{r_0}(A_1)$ while outside D_l the spectrum of $H_1^{(l)}$ is real. Moreover, the spectrum $\sigma(H_1^{(l)})$ coincides with a subset of the spectrum of the transfer function $M_1(\cdot, \Gamma_l)$. More precisely, the spectrum of $M_1(\cdot, \Gamma_l)$ in $\mathcal{O}_{d_0/2}(A_1) = \{z : z \in \mathbb{C}, \operatorname{dist}\{z, \sigma(A_1)\} \le d_0(\Gamma_l)/2\}$ is equal to the spectrum of $H_1^{(l)}$, i.e. $\sigma(M_1(\cdot, \Gamma_l)) \cap \mathcal{O}_{d_0/2}(A_1) = \sigma(H_1^{(l)})$. In fact such a statement is true separately for the point and continuous spectra.*

Theorem 5 *Suppose that two different domains $D_{l'}$ and $D_{l''}$ include the same subdomain $D_k^{l_k}$ for some $k = 1, 2, \ldots, m$, i.e., $l'_k = l''_k = l_k$. Then the spectra of the operators $H_1^{(l')}$ and $H_1^{(l'')}$ in $D_k^{l_k}$ coincide.*

Consider the operator

$$\Omega^{(l)} = \int_{\sigma'(A_0)\cup\Gamma_l} (H_1^{(-l)*} - \mu)^{-1} K_B(d\mu)(H_1^{(l)} - \mu)^{-1}$$

where as previously Γ_l stands for a K_B-bounded contour satisfying the condition (14). This operator does not depend (for a fixed l) on the choice of such a Γ_l. Moreover, $\|\Omega^{(l)}\| < 1$ and $\Omega^{(-l)} = \Omega^{(l)*}$.

Theorem 6 *The operators $\Omega^{(l)}$ possess the following properties (cf.* [15], [20], [21]):

$$-\frac{1}{2\pi i}\int_{\gamma} dz[M_1(z,\Gamma_l)]^{-1} = (I_1+\Omega^{(l)})^{-1}, \tag{24}$$

$$\begin{aligned} -\frac{1}{2\pi i}\int_{\gamma} dz z[M_1(z,\Gamma_l)]^{-1} &= (I_1+\Omega^{(l)})^{-1}H_1^{(-l)*} \\ &= H_1^{(l)}(I_1+\Omega^{(l)})^{-1}, \end{aligned} \tag{25}$$

where γ stands for an arbitrary rectifiable closed contour surrounding the spectrum of $H_1^{(l)}$ inside the set $\mathcal{O}_{d_0(\Gamma)/2}(A_1)$ in the positive direction and including the point infinity if the entry A_1 is unbounded. The integration over γ is understood in the strong sense.

The formulas (24) and (25) allow us, in principle, to construct the operators $H_1^{(l)}$ and, thus, to resolve the equation (13) by a contour integration of the inverse transfer function $[M_1(z,\Gamma^l)]^{-1}$.

5 Properties of the Real Eigenvalues

If λ is a real isolated eigenvalue of the operator $H_1^{(l')} = A_1 + X^{(l')}$, $l' = (l'_1, l'_2, \ldots, l'_m)$, then it is such an eigenvalue also for the remaining 2^{m-1} operators $H_1^{(l)} = A_1 + X^{(l)}$ where $l = (l_1, l_2, \ldots, l_m)$ with arbitrary $l_k = \pm 1$, $k = 1, 2, \ldots, m$. The resolvents of each of the 2^m operators $H_1^{(l)}$ at $z = \lambda$ have a first order pole. Simultaneously such an eigenvalue λ belongs to the point spectrum of the block operator $\mathbf{H}$.

An isolated real eigenvalue λ of the operator $H_1^{(l)}$ can not belong to the spectrum $\sigma'(A_0)$ of the entry A_0 lying outside $\cup_{k=1}^{m}\Delta_k^0$. Indeed, according to Theorem 4, the spectrum of $H_1^{(l)}$ for arbitrary l is situated in the $r_0(B)$-neighborhood $\mathcal{O}_{r_0}(A_1)$ of the set $\sigma(A_1)$ and in any case $r_0(B) < \frac{1}{2}\operatorname{dist}\{\sigma'(A_0), \sigma(A_1)\}$ so that automatically $\sigma'(A_0) \cap \sigma(H_1^{(l)}) = \emptyset$. Hence, such a λ belongs either to the resolvent set $\varrho(A_0)$ of the entry A_0 or it is embedded into the continuous spectrum of A_0 in $\cup_{k=1}^{m}\Delta_k^0$.

Lemma 3 *If a vector $\psi^{(1)} \in \mathcal{D}(A_1)$ is an eigenvector of $H_1^{(l)}$ corresponding to a real eigenvalue $\lambda \in \varrho(A_0)$ then the vector $\Psi = (\psi^{(0)}, \psi^{(1)}) \in \mathcal{H}$ with*

$$\psi^{(0)} = -(A_0-\lambda)^{-1}B_{01}\psi^{(1)} \tag{26}$$

is an eigenvector of $\mathbf{H}$, $\mathbf{H}\Psi = \lambda\Psi$. The converse statement is also true: if $\lambda \in \varrho(A_0)$ is a real eigenvalue of $H_1^{(l)}$ and $\mathbf{H}\Psi = \lambda\Psi$ for some $\Psi = (\psi^{(0)}, \psi^{(1)})$ with $\psi^{(0)} \in \mathcal{D}(A_0)$ and $\psi^{(1)} \in \mathcal{D}(A_1)$, then $\psi^{(0)}$ is related to $\psi^{(1)}$ as in (26) and $H_1^{(l)}\psi^{(1)} = \lambda\psi^{(1)}$.

If an eigenvalue λ of $H_1^{(l)}$ belongs to $\Delta_k^0 = (\mu_k^{(1)}, \mu_k^{(2)})$ for some $k = 1, 2, \ldots, m$, then $|\lambda - \mu_k^{(i)}| \geq \operatorname{dist}\{\mu_k^{(i)}, \sigma(A_1)\} - r_0(B), i = 1, 2$. Recall that according to our assumption the entry A_0 has no point spectrum inside Δ_k^0. Since Δ_k^0 is a part of the continuous spectrum of A_0, the resolvent $(A_0 - z)^{-1}$ for $z = \lambda \pm \mathrm{i}0$ exists being however an unbounded operator. Nevertheless a statement analogous to Lemma 3 is valid in this case, too.

Lemma 4 *If a vector $\psi^{(1)} \in \mathcal{D}(A_1)$ is an eigenvector of $H_1^{(l)}$ corresponding to a real eigenvalue $\lambda \in \Delta_k^0 = (\mu_k^{(1)}, \mu_k^{(2)})$, $k = 1, 2, \ldots, m$, $H_1^{(l)}\psi^{(1)} = \lambda\psi^{(1)}$, then either*

a) $E^0(\mu)B_{01}\psi^{(1)} = 0$ *for all* $\mu \leq \mu_k^{(2)}$

or

b) $E^0(\mu)B_{01}\psi^{(1)} \neq 0$ *for any* $\mu \in \Delta_k^0$,

c) *the function* $\|E^0(\mu)B_{01}\psi^{(1)}\|$ *is differentiable in* μ *on* Δ_k^0

and

d) $\frac{d}{d\mu}\|E^0(\mu)B_{01}\psi^{(1)}\||_{\mu=\lambda} = 0.$

In both cases the vector $\psi^{(0)}$ given by (26) *belongs to $\mathcal{D}(A_0)$ and $\Psi = (\psi^{(0)}, \psi^{(1)})$ is an eigenvector of* $\mathbf{H}$, $\mathbf{H}\Psi = \lambda\Psi$.

The converse statement is also true. Namely, if a vector $\Psi = (\psi^{(0)}, \psi^{(1)})$ with $\psi^{(0)} \in \mathcal{D}(A_0)$ and $\psi^{(1)} \in \mathcal{D}(A_1)$ is an eigenvector of $\mathbf{H}$, $\mathbf{H}\Psi = \lambda\Psi$, *corresponding to an eigenvalue λ of $H_1^{(l)}$, $\lambda \in \Delta_k^0$, then either the condition* (a) *holds or the conditions* (b–d) *are valid. In both cases the relation* (26) *is meaningful, in particular, $\psi^{(1)} \neq 0$ and $\psi^{(1)}$ is an eigenvector of $H_1^{(l)}$ corresponding to the eigenvalue λ.*

Let $\sigma_{pri}(H_1^{(l)})$ be the set of all real isolated eigenvalues of the operator $H_1^{(l)}$. As we already established, this set coincides with the part $\sigma_{pri}(M_1(\cdot, \Gamma_l))$ of the set of the real isolated eigenvalues of the transfer function $M_1(\cdot, \Gamma_l)$ inside the set $\mathcal{O}_{d_0/2}(A_1)$ for any K_B-bounded contour Γ_l satisfying the condition (14).

Since in the remainder of the section we will consider different eigenvalues $\lambda \in \sigma_{pri}(H_1^{(l)})$, we will use a more specific notation, $\psi_{\lambda,j}^{(1)}$, $j = 1, 2, \ldots, m_\lambda$, for the corresponding eigenvectors of $H_1^{(l)}$. The notation m_λ, $m_\lambda \leq \infty$, stands for the multiplicity of the eigenvalue λ. Recall that every $\psi_{\lambda,j}^{(1)}$ is an eigenvector simultaneously for all the $H_1^{(l)}$ and $M_1(\lambda \pm \mathrm{i}0, \Gamma_l)$, $l = (l_1, l_2, \ldots, l_m)$ with $l_k = \pm 1$, $k = 1, 2, \ldots, m$. In the considered case the multiplicity m_λ is both the geometric and algebraic multiplicity of λ. The corresponding eigenvectors of the block matrix $\mathbf{H}$ will be denoted by $\Psi_{\lambda,j}$, $\Psi_{\lambda,j} = (\psi_{\lambda,j}^{(0)}, \psi_{\lambda,j}^{(1)})$. It will be supposed that the $\psi_{\lambda,j}^{(1)}$ are chosen in such a way that the vectors $\Psi_{\lambda,j}$ are orthonormal,

$\langle \Psi_{\lambda,j}, \Psi_{\lambda',j'} \rangle = \delta_{\lambda\lambda'}\delta_{jj'}$. Obviously, the statements of Lemmas 3 and 4 imply that the eigenvectors $\Psi_{\lambda,j}$, $\lambda \in \sigma_{pri}(H_1^{(l)})$, $j = 1, 2, \ldots, m_\lambda$, form an orthonormal basis in the invariant subspace of the operator $\mathbf{H}$ corresponding to the subset $\sigma_{pri}(H_1^{(l)})$ of the point spectrum $\sigma_p(\mathbf{H})$ of $\mathbf{H}$.

Lemma 5 *Let $\mathcal{H}_1^{(pri)}$, $\mathcal{H}_1^{(pri)} \subset \mathcal{H}_1$, be the closed linear span of the eigenvectors $\psi_{\lambda,j}^{(1)}$ for $\lambda \in \sigma_{pri}(H_1^{(l)})$, $j = 1, 2, \ldots, m_\lambda$. For any $l = (l_1, l_2, \ldots, l_m)$, $l_k = \pm 1$, $k = 1, 2, \ldots, m$, the operator $\Omega^{(l)}$ is non-negative on $\mathcal{H}_1^{(pri)}$.*

This statement implies that one can introduce a new inner product in $\mathcal{H}_1^{(pri)}$, $[u_1, v_1]_{\mathcal{H}_1^{(pri)}} = \langle (I_1 + \Omega^{(l)})u_1, v_1 \rangle$. Then, with the help of a theorem of N.K. Bari (Theorem VI.2.1 of [18]) one can prove the following statement.

Theorem 7 *The system of vectors $\psi_{\lambda,j}^{(1)}$, $\lambda \in \sigma_{pri}(H_1^{(l)})$, $j = 1, 2, \ldots, m_\lambda$, forms a Riesz basis of the subspace $\mathcal{H}_1^{(pri)}$.*

6 Completeness and Basis Properties

In this section we restrict ourselves to the case where the entry A_1 has a pure discrete spectrum, i.e., the resolvent $(A_1 - z)^{-1}$ is a compact operator in $\mathcal{H}_1$ for any $z \in \varrho(A_1)$. In this case the operators $H_1^{(l)}$ have compact resolvents, too. This is a consequence of Theorem V.3.17 of [19] since the difference $H_1^{(l)} - A_1 = X^{(l)}$ is a bounded operator (see Theorem 1). Further, the operators $X^{(l)}$ are compact, too.

Denote by $\mathcal{H}_{1,\lambda}^{(l)}$ the algebraic eigenspace of $H_1^{(l)}$ corresponding to an eigenvalue λ. Let m_λ be the algebraic multiplicity, $m_\lambda = \dim \mathcal{H}_{1,\lambda}^{(l)}$, $m_\lambda < \infty$, and $\psi_{\lambda,i}^{(l)}$, $i = 1, 2, \ldots, m_\lambda$, be the root vectors of $H_1^{(l)}$ forming a basis of the subspace $\mathcal{H}_{1,\lambda}^{(l)}$. In the following we will try to give an answer to the question when the union of such bases in λ forms a basis of the total space $\mathcal{H}_1$.

Theorem 8 *The system $\{\psi_{\lambda,i}^{(l)}, \lambda \in \sigma(H_1^{(l)}), i = 1, 2, \ldots, m_\lambda\}$ is complete in $\mathcal{H}_1$, i.e., the closure of the linear span of this system coincides with $\mathcal{H}_1$.*

This assertion is a particular case of Theorem V.10.1 from [18].

We shall consider the case where the intersection $(\cup_{k=1}^m \Delta_k^0) \cap \sigma(A_1)$ includes infinitely many points and the entry A_1 is semibounded from below. Such an assumption means that at least the interval Δ_m^0 is infinite, $\Delta_m^0 = (\mu_m^{(1)}, +\infty)$. The eigenvalues $\lambda_i^{(A_1)}$, $i = 1, 2, \ldots,$ of the operator A_1 will be enumerated in increasing order, $\lambda_1^{(A_1)} < \cdots < \lambda_i^{(A_1)} < \lambda_{i+1}^{(A_1)} < \cdots$ with $\lim_{i\to\infty} \lambda_i^{(A_1)} = +\infty$.

Suppose further that there is a number i_0 such that for any $i \geq i_0$ and for some fixed $r > r_0(B)$

$$\lambda_i^{(A_1)} - \lambda_{i-1}^{(A_1)} > 2r. \tag{27}$$

Let γ_0 be a circle centered at $z = (\lambda_1^{(A_1)} + \lambda_{i_0-1}^{(A_1)})/2$ and having radius $(\lambda_{i_0-1}^{(A_1)} - \lambda_1^{(A_1)})/2 + r$ while γ_i for $i \geq i_0$ are the circles with centers $\lambda_i^{(A_1)}$ and radius r. Let us introduce the projections $\mathsf{Q}_i^{(l)} = -\frac{1}{2\pi \mathrm{i}} \int_{\gamma_i} dz (H_1^{(l)} - z)^{-1}$, $i = 0, i_0, i_0+1, \ldots$. Every projection $\mathsf{Q}_i^{(l)}$ represents a sum of the eigenprojections corresponding to the eigenvalues $\lambda^{(l)}$ of $H_1^{(l)}$ lying inside γ_i and $\mathsf{Q}_i^{(l)}\mathsf{Q}_j^{(l)} = \delta_{ij}\mathsf{Q}_i^{(l)}$. The subspaces $\mathcal{N}_i^{(l)} = \mathsf{Q}_i^{(l)}\mathcal{H}_1$ are invariant under $H_1^{(l)}$. We introduce also the (orthogonal) projections $\mathsf{P}_i^{(A_1)} = -\frac{1}{2\pi \mathrm{i}} \int_{\gamma_i} dz (A_1 - z)^{-1}$, $i = 0, i_0, i_0 + 1, \ldots$.

Lemma 6 *Under the condition* (27) *the sequence* $\mathcal{N}_i^{(l)}$, $i = 0, i_0, i_0 + 1, \ldots$, *is* ω*-linearly independent and complete in* $\mathcal{H}_1$. *If instead of* (27) *the condition*

$$\lambda_i^{(A_1)} - \lambda_{i-1}^{(A_1)} > 2r > 4r_0(B) \qquad \forall i \geq i_0, \tag{28}$$

is satisfied then $\dim \mathcal{N}_i^{(l)} = \dim \mathsf{P}_i^{(A_1)}\mathcal{H}_1$, $i = 0, i_0, i_0 + 1, \ldots$.

Theorem 9 *Assume* $\lambda_{i+1}^{(A_1)} - \lambda_i^{(A_1)} \to \infty$ *as* $i \to \infty$. *Let* i_0 *be the number from which the inequality* (28) *starts to hold. Then the following limit exists*

$$s - \lim_{n\to\infty} \sum_{i=0, i\geq i_0}^{n} \mathsf{Q}_i^{(l)} = I_1. \tag{29}$$

Additionally, assume that

$$\sum_{i=1}^{\infty} (\lambda_{i+1}^{(A_1)} - \lambda_i^{(A_1)})^{-2} < \infty. \tag{30}$$

Then (29) *is true for any renumbering of* $\mathsf{Q}_i^{(l)}$. *Moreover, there exists a constant* C *such that* $\| \sum_{i\in\mathcal{I}} \mathsf{Q}_i^{(l)} \| \leq C$ *for any finite set* $\mathcal{I}$ *of integers* $i = 0$, $i \geq i_0$.

This theorem represents a slightly extended statement of Theorems V.4.15 and V.4.16 of [19] (the extension only concerns a possible degeneracy of the eigenvalues of A_1).

Remark 1 Eq. (29) implies that

$$s - \lim_{n\to\infty} \sum_{i=0, i\geq i_0}^{n} \sum_{\lambda \in \mathrm{Int}\,\gamma_i} \mathsf{P}_\lambda^{(l)} = I_1, \tag{31}$$

where λ stand for the eigenvalues of the operator $H_1^{(l)}$ and $\mathsf{P}_\lambda^{(l)}$ for the corresponding eigenprojections. If, additionally, the inequality (30) holds and all the eigenvalues $\lambda_i^{(A_1)}$ are simple, then one can renumber the eigenprojections $\mathsf{P}_\lambda^{(l)}$ in Eq. (31) in any way (see Theorem V.4.16 of [19]).

Theorem 10 *As before, assume* $\Delta_m = (\mu_m^{(1)}, +\infty)$. *Also, suppose that there is a* K_B*-bounded contour* $\Gamma_l \subset D_l$ *satisfying* (14) *and such that a part of its component* $\Gamma_m^{l_m}$ *coincides with the ray* $\tilde{\Delta}_m^0 = [\mu_0, ib_0 + \infty)$ *where* $\mu_0 \in D_m^{l_m}$, $\mu_0 = a_0 + ib_0$ *with* $a_0, b_0 \in \mathbb{R}$. *Additionally, suppose that the remaining part* $\tilde{\Gamma}_l = \Gamma_l \backslash \tilde{\Delta}_m^0$ *of the contour* Γ_l *belongs to the half-plane* $\operatorname{Re}\mu < a_0$, *and for* $\mu \in \tilde{\Delta}_m^0$ $\|K_B'(\mu)\| \leq \tilde{C}(1 + |\operatorname{Re}\mu|)^{-\theta}$, *with* $\tilde{C} > 0$ *and* $\theta > 1$. *Also, let the condition* (30) *hold. The sequence of the subspaces* $\mathcal{N}_i^{(l)} = \mathsf{Q}_i^{(l)}\mathcal{H}_1$, $i = 0, i_0, i_0 + 1, \ldots,$ *forms a basis of the space* $\mathcal{H}_1$, *quadratically close to an orthogonal one. If, additionally,* $\dim \mathsf{P}_i^{(A_1)}\mathcal{H}_1 \leq n$, *for some* $n \in \mathbb{N}$ *and for all* $i = 0, i_0, i_0 + 1, \ldots,$ *then the union of orthonormal vector bases of the subspaces* $\mathcal{N}_i^{(l)}$, $i = 0, i_0, i_0 + 1, \ldots,$ *forms a Bari basis of the space* $\mathcal{H}_1$.

References

[1] R.L. Jaffe and F.E. Low, *Connection between Quark-Model Eigenstates and Low-Energy Scattering*, Phys. Rev. **D19** (1979), 2105–2118.

[2] Yu.A. Simonov, *Hadron-Hadron Interaction in the Compound-Bag Model*, Yadernaya Fizika **36** (1982), 722–731 [Russian].

[3] S. Okubo, *Diagonalization of the Hamiltonian and Tamm-Dancoff Equation*, Progr. Theor. Phys. **12** (1954), 603–622.

[4] W. Glöckle and L. Müller, *Relativistic Theory of Interacting Particles*, Phys. Rev. C **23** (1981), 1183–1195.

[5] A.Yu. Korchin and A.V. Shebeko, *The Method of Okubo's Effective Operators and the Relativistic Model of Nuclear Structure*, Phys. At. Nucl. **56** (1993), 1663–1671.

[6] V.A. Malyshev and R.A. Minlos, *Invariant Subspaces of Clustering Operators. I.*, J. Stat. Phys. **21** (1979), 231–242; *Invariant Subspaces of Clustering Operators. II.*, Comm. Math. Phys. **82** (1981), 211–226.

[7] B.S. Pavlov and A.A. Shushkov, *The Theory of Extensions, and Null-Range Potentials with Internal Structure*, Math. USSR Sb. **65** (1990), 147–184.

[8] B.H.J. McKellar and C.M. McKay, *Formal Scattering Theory for Energy-Dependent Potentials*, Aust. J. Phys. **36** (1983), 607–616.

[9] E.W. Schmid, *The Problem of Using Energy-Dependent Nucleon-Nucleon Potentials in Nuclear Physics*, Helv. Phys. Acta **60** (1987), 394–397.

[10] A.K. Motovilov, *Potentials Appearing after the Removal of an Energy-Dependence and Scattering by Them*, In: Proc. of the Intern. Workshop "Mathematical Aspects of the Scattering Theory and Applications", St. Petersburg University, St. Petersburg, 1991, 101–108.

[11] A.K. Motovilov, *Elimination of Energy from Interactions Depending on the Energy as a Resolvent*, Theor. Math. Phys. **104** (1995), 989–1007.

[12] M.A. Braun, *On a Relation between the Quasipotential Equation and the Schrödinger Equation*, Teor. Mat. Fiz. **72** (1987), 394–402 [Russian].

[13] V.M. Adamjan and H. Langer, *Spectral Properties of a Class of Operator-Valued Functions*, J. Operator Theory **33** (1995), 259–277.

[14] V. Adamyan, H. Langer, R. Mennicken and J. Saurer, *Spectral Components of Selfadjoint Block Operator Matrices with Unbounded Entries*, Math. Nachr. **178** (1996), 43–80.

[15] R. Mennicken and A.A. Shkalikov, *Spectral Decomposition of Symmetric Operator Matrices*, Math. Nachr. **179** (1996), 259–273.

[16] M. Reed and B. Simon, *Methods of Modern Mathematical Physics*, III: Scattering Theory, Academic Press, N.Y., 1979.

[17] A.K. Motovilov, *Representations for the Three-Body T-Matrix, Scattering Matrices and Resolvent on Unphysical Energy Sheets*, Math. Nachr. **187** (1997), 147–210.

[18] I.C. Gohberg and M.G. Krein, *Introduction to the Theory of Linear Non-selfadjoint Operators*, American Mathematical Society, Providence, 1988.

[19] T. Kato, *Perturbation Theory for Linear Operators*, New York: Springer-Verlag, 1966.

[20] A.I. Virozub and V.I. Matsaev, *The Spectral Properties of a Certain Class of Selfadjoint Operator Functions*, Funct. Anal. Appl. **8** (1974), 1–9.

[21] A.S. Markus and V.I. Matsaev, *On the Basis Property for a Certain Part of the Eigenvectors and Associated Vectors of a Selfadjoint Operator Pencil*, Math. USSR Sb. **61** (1988), 289–307.

Department of Mathematics
University of Regensburg
D-93040 Regensburg
Germany

Laboratory of Theoretical Physics
Joint Institute for Nuclear Research
141980 Dubna, Moscow Region
Russia

AMS Classification Numbers: Primary 47A56, 47Nxx; Secondary 47N50, 47A40.

Operator Theory:
Advances and Applications, Vol. 118

Generalized Resolvents of Isometric Linear Relations in Pontryagin Spaces, I: Foundations

Oleg Nitz

This is the first of three papers, devoted to a generalization of the Kreĭn-Langer formula for the generalized resolvents of an isometric relation in a Pontryagin space and to its application to the Nehari-Takagi problem. This paper contains some useful results about isometric and unitary relations in Pontryagin spaces. We define open and closed linear conservative scattering dynamical systems with a Pontryagin state space and describe the embedding of an open system into a closed system.

Introduction

The parametric description of the generalized resolvents of a symmetric and an isometric operator in a Pontryagin space by Kreĭn and Langer [KL]) have been further developed, also for symmetric linear relations, by many authors (see, for example, Dijksma, Langer and de Snoo [DLS], Bruinsma [B], Constantinescu and Geondea [CG], Derkach [D], Kaltenbäck and Woracek [KW]).

The problem of describing the generalized resolvents of an isometric linear relation in a Pontryagin space arose in a natural way during our investigation of the Nehari-Takagi problem. We used an operator approach, which was developed by Adamyan, Arov and Kreĭn for the Nehari problem in [AAK], [A]. That the operator approach might also be effective in the Nehari-Takagi problem was indicated by Adamyan, Arov, Kreĭn in [AAK2]. An isometric operator in a Pontryagin space can be build from the data of the problem, such that the unitary extensions of this operator correspond to the solutions of the Nehari-Takagi problem. But there arise some difficulties. A solution of the Nehari-Takagi problem may have a pole at 0, and in this case it does not correspond to any operator, it naturally corresponds to a linear relation. One cannot a priori choose a transformation of the data of the problem, which guarantees the holomorphy of all the solutions at 0. That is why the investigation of isometric and unitary linear relation play an essential role in this series of papers. Sections 1 and 2 of this paper contain general definitions and results related to Kreĭn spaces and to linear relations. An important class of completely closed linear relations in Kreĭn spaces is introduced in Section 3. Section 4 begins with properties of a unitary relations in a Kreĭn space and ends with a description of the spectrum of unitary relations in a Pontryagin space (Theorem 4.6). In Section 5 we consider open and closed linear conservative scattering dynamical systems with a Pontryagin state space and describe the embedding of an open system into a closed system. In the second paper of the series we shall use these notions to reformulate and then to solve the problem of the description of the generalized resolvents of an isometric relation in a Pontryagin space. In the third paper we shall apply the results of the previous papers and obtain a description of the set of all the solutions of the Nehari-Takagi problem.

1 Definitions and Preliminaries

We start with some basic definitions and results related to Kreĭn spaces. The proofs of all the statements in this section can be found in [AI]. All topological linear spaces are assumed separable.

In this article by a scalar product we mean a Hermitian sesquilinear form on a complex linear space. A **Kreĭn space** $\mathfrak{K}$ is a topological complex linear space $\mathfrak{K}$ endowed with a scalar product $[\cdot,\cdot]$, such that for some continuous linear operator J in the space $\mathfrak{K}$ with the property $J^2 = I$, the new scalar product $(x, y)_J := [Jx, y]$ turns $\mathfrak{K}$ into a Hilbert space. It follows, that the topology of the Hilbert space is equivalent to the topology of the Kreĭn space. The operator J is called a **fundamental symmetry**. In general, a fundamental symmetry is not unique. It is easy to see, that J is selfadjoint with respect to both scalar products $(\cdot,\cdot)_J$ and $[\cdot,\cdot]$. A **direct sum** $\mathfrak{K}_1 \oplus \mathfrak{K}_2$ of Kreĭn spaces $\mathfrak{K}_1$ and $\mathfrak{K}_2$ is the direct sum of the topological linear spaces endowed with the scalar product $\left[\begin{pmatrix} x_1 \\ x_2 \end{pmatrix}, \begin{pmatrix} y_1 \\ y_2 \end{pmatrix}\right] = [x_1, y_1]_{\mathfrak{K}_1} + [x_2, y_2]_{\mathfrak{K}_2}$.

A vector $h \in \mathfrak{K}$ is called **positive** (**neutral**, **negative**) if $[h, h] > 0$ $(= 0, < 0)$. A **subspace** of a Kreĭn space $\mathfrak{K}$ is a nonempty linear set $\mathfrak{L} \subset \mathfrak{K}$ (not necessarily closed). A subspace $\mathfrak{L}$ is called positive (neutral, negative) if every nonzero vector $h \in \mathfrak{L}$ is positive (neutral, negative). Below some standard notation is given:

$$\begin{aligned}
x\,[\perp]\,y \quad &: \iff [x, y] = 0 \\
\mathfrak{L}_1 \dot{+} \mathfrak{L}_2 &:= \mathfrak{L}_1 + \mathfrak{L}_2 \text{ if } \mathfrak{L}_1 \cap \mathfrak{L}_2 = \{0\} \\
\mathfrak{L}^{[\perp]} &:= \{x \in \mathfrak{K} : \forall y \in \mathfrak{L}\; x\,[\perp]\,y\} \\
\mathfrak{L}_1\,[+]\,\mathfrak{L}_2 &:= \mathfrak{L}_1 + \mathfrak{L}_2 \text{ if } \mathfrak{L}_1\,[\perp]\,\mathfrak{L}_2 \\
\mathfrak{L}_2\,[-]\,\mathfrak{L}_1 &:= \mathfrak{L}_2 \cap \mathfrak{L}_1^{[\perp]} \text{ if } \mathfrak{L}_1 \subset \mathfrak{L}_2 \\
\mathfrak{L}_1\,[\dot{+}]\,\mathfrak{L}_2 &:= \mathfrak{L}_1 + \mathfrak{L}_2 \text{ if } \mathfrak{L}_1 \cap \mathfrak{L}_2 = \{0\},\ \mathfrak{L}_1\,[\perp]\,\mathfrak{L}_2 \\
\mathfrak{L}_1 \vee \mathfrak{L}_2 &:= \overline{\mathfrak{L}_1 + \mathfrak{L}_2}, \text{ i.e. the closure of } \mathfrak{L}_1 + \mathfrak{L}_2
\end{aligned}$$

A subspace $\mathfrak{L}$ is called **hypermaximal neutral** if $\mathfrak{L} = \mathfrak{L}^{[\perp]}$. By a **regular** subspace of a Kreĭn space we mean a closed subspace $\mathfrak{L} \subset \mathfrak{K}$ which is a Kreĭn space in the scalar product of $\mathfrak{K}$. A subspace $\mathfrak{L} \subset \mathfrak{K}$ is regular if and only if $\mathfrak{L}\,[+]\,\mathfrak{L}^{[\perp]} = \mathfrak{K}$.

Every closed subspace $\mathfrak{L}$ of a Kreĭn space $\mathfrak{K}$ admits a decomposition of the form $\mathfrak{L} = \mathfrak{L}_+\,[\dot{+}]\,\mathfrak{L}_-\,[\dot{+}]\,\mathfrak{L}_0$, where $\mathfrak{L}_+, \mathfrak{L}_-, \mathfrak{L}_0$ are positive, negative and neutral closed subspaces respectively. The subspace $\mathfrak{L}_0$ is uniquely defined and can be found by the formula $\mathfrak{L}_0 = \mathfrak{L} \cap \mathfrak{L}^{[\perp]}$. It is called the **isotropic part** of $\mathfrak{L}$. In general, the subspaces $\mathfrak{L}_\pm$ are not unique but their dimensions do not depend on the choice and are called the **signatures** of $\mathfrak{L}$. We set $\kappa^0[\mathfrak{L}] = \dim \mathfrak{L}_0$, $\kappa^\pm[\mathfrak{L}] = \dim \mathfrak{L}_\pm$. The whole Kreĭn space $\mathfrak{K}$ has no isotropic part, i.e. $\kappa^0[\mathfrak{K}] = 0$, so it has a decomposition $\mathfrak{K} = \mathfrak{K}_+ [\dot{+}] \mathfrak{K}_-$, which is called a **fundamental decomposition**. The number $\kappa^-[\mathfrak{K}]$ is called the **number of negative squares** of a Kreĭn space $\mathfrak{K}$. A Kreĭn space $\mathfrak{K}$ is called a **Pontryagin space** if $\kappa^-[\mathfrak{K}] < \infty$. A closed subspace $\mathfrak{L}$ of a Pontryagin space is regular if and only if $\kappa^0[\mathfrak{L}] = 0$.

Two neutral finite-dimensional subspaces $\mathfrak{L}_1, \mathfrak{L}_2$ are called **skewly linked** if $\mathfrak{L}_1 + \mathfrak{L}_2$ is a regular subspace. In this case $\mathfrak{L}_1 \cap \mathfrak{L}_2 = 0$, $\dim \mathfrak{L}_1 = \dim \mathfrak{L}_2 = \kappa^-[\mathfrak{L}_1 \dot{+} \mathfrak{L}_2] = \kappa^+[\mathfrak{L}_1 \dot{+} \mathfrak{L}_2]$ and $\forall x \in \mathfrak{L}_1, x \neq 0 \, \exists y \in \mathfrak{L}_2 : [x, y] \neq 0$. For any nontrivial finite-dimensional neutral subspace $\mathfrak{L}_1$ one can find infinitely many subspaces $\mathfrak{L}_2$ such that $\mathfrak{L}_1, \mathfrak{L}_2$ are skewly linked.

Lemma 1.1 *Let $\mathfrak{K}$ be a Pontryagin space and let $\mathfrak{L} \subset \mathfrak{K}$ be a closed subspace. Then there is a finite-dimensional neutral subspace Z such that $\mathfrak{L} \dot{+} Z$ is regular.*

Proof: It is sufficient to find a neutral subspace Z such that Z and the isotropic part of $\mathfrak{L}$ are skewly linked subspaces. □

2 Operators and Linear Relations in Kreĭn Spaces

We denote by $\mathcal{L}(\mathfrak{K}_1, \mathfrak{K}_2)$ the set of all continuous and everywhere defined linear operators from the Kreĭn space $\mathfrak{K}_1$ to the Kreĭn space $\mathfrak{K}_2$; we write $\mathcal{L}(\mathfrak{K})$ instead of $\mathcal{L}(\mathfrak{K}, \mathfrak{K})$. We define the **graph-sum** $\mathfrak{K}_1 \boxplus \mathfrak{K}_2$ of Kreĭn spaces $\mathfrak{K}_1$ and $\mathfrak{K}_2$ as the direct sum of the topological linear spaces endowed with the scalar product $\left[\begin{bmatrix} x_1 \\ x_2 \end{bmatrix}, \begin{bmatrix} y_1 \\ y_2 \end{bmatrix}\right] = [x_1, y_1]_{\mathfrak{K}_1} - [x_2, y_2]_{\mathfrak{K}_2}$. We use square brackets for vectors from $\mathfrak{K}_1 \boxplus \mathfrak{K}_2$ to emphasize the disparity of the two components of the vector with respect to the scalar product. When writing block-matrices for the graph-sums of spaces we shall also use square brackets. The rules for the multiplication of a block-matrix by another one or by a block-vector are as usual. However, note that the rules for taking the adjoint of such block-matrices differ from the usual ones. Let $\mathfrak{K}, \mathfrak{K}_1, \mathfrak{K}_2, \mathfrak{K}_1', \mathfrak{K}_2'$ be Kreĭn spaces, $A \in \mathcal{L}(\mathfrak{K}, \mathfrak{K}_1)$, $B \in \mathcal{L}(\mathfrak{K}, \mathfrak{K}_2)$ and $A_{ij} \in \mathcal{L}(\mathfrak{K}_i, \mathfrak{K}_j')$ for $i, j = 1, 2$. Then

$$\begin{bmatrix} A \\ B \end{bmatrix}^\star = [\, A^\star \; - B^\star \,] \quad \text{and} \quad \begin{bmatrix} A_{11} & A_{12} \\ A_{21} & A_{22} \end{bmatrix}^\star = \begin{bmatrix} A_{11}^\star & -A_{21}^\star \\ -A_{12}^\star & A_{22}^\star \end{bmatrix}$$

The **graph** of a linear operator $A \in \mathcal{L}(\mathfrak{K}_1, \mathfrak{K}_2)$ is a closed subspace of $\mathfrak{K}_1 \boxplus \mathfrak{K}_2$, defined by $\operatorname{gr} A = \left\{\begin{bmatrix} x \\ Ax \end{bmatrix} : x \in \mathfrak{K}_1\right\}$. Let $A \in \mathcal{L}(\mathfrak{K}_1, \mathfrak{K}_2)$ and let $\mathfrak{L} \subset \mathfrak{K}_1$ be a regular subspace, then the restriction of A to $\mathfrak{L}$ is denoted by $A|_{\mathfrak{L}}$ and is considered as an element of $\mathcal{L}(\mathfrak{L}, \mathfrak{K}_2)$. Let $\mathfrak{L}$ be a regular subspace of a Kreĭn space $\mathfrak{K}$, then the orthogonal projection onto $\mathfrak{L}$ (with respect to the indefinite scalar product) is denoted by $P_{\mathfrak{L}}$ and $P_{\mathfrak{L}} \in \mathcal{L}(\mathfrak{K})$.

By a **multi-valued mapping** from a set X to a set Y we mean a mapping from X to the set of all subsets of Y. If $\mathcal{A}$ is a multi-valued mapping, then we denote by $\mathcal{A}x$ the set, which is the value of $\mathcal{A}$ on the element $x \in X$. We also define the action of $\mathcal{A}$ on the set $L \subset X$ by formula $\mathcal{A}L := \cup_{x \in L} \mathcal{A}x$. We introduce notation dom $\mathcal{A} = \{x \in X : \mathcal{A}x \neq \emptyset\}$, ran $\mathcal{A} = \mathcal{A}X$.

A **linear relation** $\mathcal{A}$ is a multi-valued mapping from a Kreĭn space $\mathfrak{K}_1$ to a Kreĭn space $\mathfrak{K}_2$ such that the set gr $\mathcal{A} = \left\{ \begin{bmatrix} x \\ y \end{bmatrix} \in \mathfrak{K}_1 \boxplus \mathfrak{K}_2 : y \in \mathcal{A}x \right\}$ is a subspace of $\mathfrak{K}_1 \boxplus \mathfrak{K}_2$. It is easy to see that $\mathcal{A}$ maps subspaces to subspaces, i.e. for any subspace $\mathfrak{L} \subset \mathfrak{K}_1$ the set $\mathcal{A}\mathfrak{L}$ is a subspace.

For subsets X and Y of a linear space we write λX instead of $\{\lambda x : x \in X\}$ and $X + Y$ instead of $\{x + y : x \in X, y \in Y\}$. Let $\mathcal{A}, \mathcal{A}_1, \mathcal{A}_2$ be linear relations from $\mathfrak{K}_1$ to $\mathfrak{K}_2$, and let $\mathcal{B}$ be a linear relation from $\mathfrak{K}_2$ to $\mathfrak{K}_3$. Then we define the linear relations $\lambda\mathcal{A}$, $\mathcal{A}_1 + \mathcal{A}_2$, $\mathcal{A}^{-1}$, $\mathcal{B} \circ \mathcal{A}$ and $\mathcal{A}^\star$ by

$$\begin{aligned}
\forall x \in \mathfrak{K}_1 \quad (\lambda\mathcal{A})x &= \lambda(\mathcal{A}x), \quad \lambda \in \mathbb{C} \\
\forall x \in \mathfrak{K}_1 \quad (\mathcal{A}_1 + \mathcal{A}_2)x &= \mathcal{A}_1 x + \mathcal{A}_2 x \\
\forall y \in \mathfrak{K}_2 \quad \mathcal{A}^{-1}y &= \{x \in \mathfrak{K}_1 : y \in \mathcal{A}x\} \\
\forall x \in \mathfrak{K}_1 \quad (\mathcal{B} \circ \mathcal{A})x &= \mathcal{B}(\mathcal{A}x) \\
\text{gr } \mathcal{A}^\star &= (\text{gr } \mathcal{A}^{-1})^{[\perp]}
\end{aligned}$$

Sometimes we write $\mathcal{B}\mathcal{A}$ instead of $\mathcal{B} \circ \mathcal{A}$. Let $\mathcal{A}$ and $\mathcal{B}$ be as above. The following assertions are straightforward consequences of the above definitions:

$$(\mathcal{A}^{-1})^{-1} = \mathcal{A} \tag{2.1}$$

$$(\mathcal{B} \circ \mathcal{A})^{-1} = \mathcal{A}^{-1} \circ \mathcal{B}^{-1} \tag{2.2}$$

$$(\mathcal{A}^\star)^{-1} = (\mathcal{A}^{-1})^\star \tag{2.3}$$

For a subspace $\mathfrak{L} \subset \mathfrak{K}_1$ it is easy to prove that

$$\mathcal{A}^\star \mathfrak{L}^{[\perp]} \subset (\mathcal{A}^{-1}\mathfrak{L})^{[\perp]} \tag{2.4}$$

$$(\mathcal{A}^\star \circ \mathcal{B}^\star)\mathfrak{L} \subset (\mathcal{B} \circ \mathcal{A})^\star \mathfrak{L} \tag{2.5}$$

We also introduce, as usual, ker $\mathcal{A} = \mathcal{A}^{-1}(0)$.

If $\mathcal{A}(0) = \{0\}$, then we identify the relation $\mathcal{A}$ with the operator A, uniquely defined by the condition gr $\mathcal{A}$ = gr A (or the equivalent condition: $\mathcal{A}x$ equals $\{Ax\}$ for $x \in$ dom A and is the empty set for all other x).

Lemma 2.1 *Let $\mathcal{A}_1$, $\mathcal{A}_2$ be linear relations from $\mathfrak{K}_1$ to $\mathfrak{K}_2$. Assume that for all closed subspaces $\mathfrak{L} \subset \mathfrak{K}_1$, $\mathfrak{M} \subset \mathfrak{K}_2$ it holds $\mathcal{A}_1\mathfrak{L} = \mathcal{A}_2\mathfrak{L}$, $\mathcal{A}_1^{-1}\mathfrak{M} = \mathcal{A}_2^{-1}\mathfrak{M}$. Then there is a $c \in \mathbb{C}\backslash\{0\}$ such that $\mathcal{A}_1 = c\mathcal{A}_2$.*

We omit the quite routine proof of this lemma. Now let $\mathcal{A}$ be a linear relation from a Kreĭn space $\mathfrak{K}$ to itself. It is called **selfadjoint** if $\mathcal{A}^\star = \mathcal{A}$. For any integer i we define the linear relations $\mathcal{A}^i$ as follows: $\mathcal{A}^0 := I$, $\mathcal{A}^i := \mathcal{A} \circ \mathcal{A}^{i-1}$ for $i > 0$ by induction, and $\mathcal{A}^i := (A^{-i})^{-1}$ for $i < 0$.

Below we give definitions of a regular point, a regular set and the spectrum of a linear relation, which slightly differ from the traditional ones. However, they are equivalent to the traditional definitions for linear operators with a dense domain. The purpose of our definitions is to include the notion of the spectrum of the restriction of a unitary relation to its neutral invariant subspaces. We say, that $z \in \mathbb{C}$ is a **regular point** of $\mathcal{A}$ if dom $(\mathcal{A} - zI)^{-1}$ = dom $\mathcal{A} \vee$ ran $\mathcal{A}$ and $(\mathcal{A} - zI)^{-1}$ is a continuous operator. We say, that $z = \infty$ is a regular point if $\mathcal{A}$ is a continuous operator with dom $\mathcal{A}$ = dom $\mathcal{A} \vee$ ran $\mathcal{A}$. The set of all regular points of $\mathcal{A}$ will be called a **regular set** and will be denoted by $\rho(\mathcal{A})$. We define the **spectrum** of $\mathcal{A}$ as $\sigma(\mathcal{A}) = \overline{\mathbb{C}} \backslash \rho(\mathcal{A})$, where $\overline{\mathbb{C}}$ means $\mathbb{C} \cup \{\infty\}$.

3 Completely Closed Linear Relations

In this section we introduce the class of completely closed linear relations, which will be useful for further investigations. It contains unitary relations in a Pontryagin space and some important classes of operators, and it is closed under the operations of composition and taking inverses and adjoints. In our investigation we need the class of completely closed relations only in Pontryagin spaces, but in this section and in the first part of the next section we bring several propositions about completely closed relations in Kreĭn spaces.

We say that a closed subspace X in a Kreĭn space is a **completely closed subspace** if for all closed subspaces Y the subspace $X + Y$ is closed. It is not difficult to see that the closed subspace X is completely closed iff either $\dim X < \infty$ or $\dim X^{[\perp]} < \infty$. It follows, that the set of completely closed subspaces of the given space is closed under the operations of sum, intersection and taking orthogonal complements.

Let $\mathfrak{K}_1$ and $\mathfrak{K}_2$ be Kreĭn spaces. We say that a linear relation $\mathcal{A}$ from $\mathfrak{K}_1$ to $\mathfrak{K}_2$ is a **completely closed relation** if it has the closed graph and the subspaces dom $\mathcal{A}$, ran $\mathcal{A}$, dom $\mathcal{A}^\star$ and ran $\mathcal{A}^\star$ are completely closed. We denote by $\mathfrak{R}(\mathfrak{K}_1, \mathfrak{K}_2)$ the set of all completely closed linear relations from $\mathfrak{K}_1$ to $\mathfrak{K}_2$; we write $\mathfrak{R}(\mathfrak{K})$ instead of $\mathfrak{R}(\mathfrak{K}, \mathfrak{K})$. If a completely closed linear relation is an operator, then it is continuous. Continuously invertible linear operators from $\mathcal{L}(\mathfrak{K}_1, \mathfrak{K}_2)$ and projections on completely closed subspaces are completely closed relations.

Proposition 3.1 *Let $\mathcal{A} \in \mathfrak{R}(\mathfrak{K}_1, \mathfrak{K}_2)$ be a completely closed linear relation. Then*

$$(\lambda \mathcal{A}) \in \mathfrak{R}(\mathfrak{K}_1, \mathfrak{K}_2) \tag{3.1}$$

$$\mathcal{A}^\star, \mathcal{A}^{-1} \in \mathfrak{R}(\mathfrak{K}_2, \mathfrak{K}_1) \tag{3.2}$$

$$\mathcal{A}^{\star\star} = \mathcal{A} \tag{3.3}$$

$$(\ker \mathcal{A})^{[\perp]} = \operatorname{ran} \mathcal{A}^\star \tag{3.4}$$

for any subspace $\mathfrak{L} \subset \mathfrak{K}_2$

$$\mathcal{A}^{\star}\mathfrak{L}^{[\perp]} = (\mathcal{A}^{-1}\mathfrak{L})^{[\perp]} \tag{3.5}$$

for any completely closed linear relation $\mathcal{B} \in \mathfrak{R}(\mathfrak{K}_2, \mathfrak{K}_3)$

$$(\mathcal{B} \circ \mathcal{A})^{\star} = \mathcal{A}^{\star} \circ \mathcal{B}^{\star} \tag{3.6}$$

$$(\mathcal{B} \circ \mathcal{A}) \in \mathfrak{R}(\mathfrak{K}_1, \mathfrak{K}_3) \tag{3.7}$$

A completely closed linear relation maps closed subspaces to closed subspaces and maps completely closed subspaces to completely closed subspaces.

Proof: Assertions (3.1)–(3.4) are quite simple and we omit the proofs. By (2.4), to obtain (3.5) it remains to prove that $\mathcal{A}^{\star}\mathfrak{L}^{[\perp]} \supset (\mathcal{A}^{-1}\mathfrak{L})^{[\perp]}$. Let us fix $x_0 \in (\mathcal{A}^{-1}\mathfrak{L})^{[\perp]}$. Since

$$x_0 \in (\mathcal{A}^{-1}\mathfrak{L})^{[\perp]} \subset (\ker \mathcal{A})^{[\perp]} = \operatorname{ran} \mathcal{A}^{\star},$$

we can find y_0 such that $x_0 \in \mathcal{A}^{\star} y_0$. For any $y \in \mathfrak{L} \cap \operatorname{ran} \mathcal{A}$ and $x \in \mathcal{A}^{-1}y$ we have $[y_0, y] = [x_0, x]$. The right side of the equality equals zero since $x_0 \in (\mathcal{A}^{-1}\mathfrak{L})^{[\perp]}$, therefore $y_0 \in (\mathfrak{L} \cap \operatorname{ran} \mathcal{A})^{[\perp]} = \overline{\mathfrak{L}^{[\perp]} + (\operatorname{ran} \mathcal{A})^{[\perp]}}$. Since $\operatorname{ran} \mathcal{A}$ is a completely closed subspace and $(\operatorname{ran} \mathcal{A})^{[\perp]} = \ker \mathcal{A}^{\star}$, we obtain $y_0 \in \mathfrak{L}^{[\perp]} + \ker \mathcal{A}^{\star}$ and $x_0 \in \mathcal{A}^{\star}\mathfrak{L}^{[\perp]}$.

It is easy to see that (3.5) implies that the completely closed linear relation maps closed subspaces to closed subspaces. Let $\dim \mathfrak{L} < \infty$ and let $\{l_i\}_1^N$ be a basis of $\mathfrak{L} \cap \operatorname{dom} \mathcal{A}$. One can choose $m_i \in \mathcal{A}l_i$. Then $\mathcal{A}\mathfrak{L} = \mathfrak{M} + \mathcal{A}(0)$, where $\mathfrak{M}$ is a linear span of the set $\{m_i\}_1^N$. The subspace $\mathcal{A}\mathfrak{L}$ is fully closed as a sum of two completely closed subspaces. It follows, that if $\dim \mathfrak{L}^{[\perp]} < \infty$, then $\mathcal{A}^{\star -1}\mathfrak{L}^{[\perp]}$ is completely closed, so $\mathcal{A}\mathfrak{L} = (\mathcal{A}^{\star -1}\mathfrak{L}^{[\perp]})^{[\perp]}$ is completely closed too. The last statement of the proposition is proved.

As a first step of the proof of (3.6) we shall verify that for any closed subspace $\mathfrak{L}$ the inclusion $(\mathcal{A}^{\star} \circ \mathcal{B}^{\star})\mathfrak{L} \supset (\mathcal{B} \circ \mathcal{A})^{\star}\mathfrak{L}$ holds:

$$\begin{aligned} \mathcal{A}^{\star}(\mathcal{B}^{\star}\mathfrak{L}) &= (\mathcal{A}^{-1}(\mathcal{B}^{\star}\mathfrak{L})^{[\perp]})^{[\perp]} = (\mathcal{A}^{-1}(\mathcal{B}^{-1}\mathfrak{L}^{[\perp]}))^{[\perp]} \\ &= ((\mathcal{B} \circ \mathcal{A})^{-1}\mathfrak{L}^{[\perp]})^{[\perp]} \supset (\mathcal{B} \circ \mathcal{A})^{\star}\mathfrak{L} \end{aligned}$$

With (2.5) it gives $(\mathcal{A}^{\star} \circ \mathcal{B}^{\star})\mathfrak{L} = (\mathcal{B} \circ \mathcal{A})^{\star}\mathfrak{L}$ for any closed subspace $\mathfrak{L}$. Applying this formula to the completely closed relations $\mathcal{B}^{-1}$, $\mathcal{A}^{-1}$ and a closed subspace $\mathfrak{M} \subset \mathfrak{K}_2$, we obtain

$$(\mathcal{A}^{\star} \circ \mathcal{B}^{\star})^{-1}\mathfrak{M} = ((\mathcal{B}^{-1})^{\star} \circ (\mathcal{A}^{-1})^{\star})\mathfrak{M} = (\mathcal{A}^{-1} \circ \mathcal{B}^{-1})^{\star}\mathfrak{M} = ((\mathcal{B} \circ \mathcal{A})^{\star})^{-1}\mathfrak{M}$$

So, using Lemma 2.1, we obtain $(\mathcal{A}^{\star} \circ \mathcal{B}^{\star}) = c(\mathcal{B} \circ \mathcal{A})^{\star}$, and using (2.5), we conclude that $c = 1$ and formula (3.6) holds. Now (3.7) follows from the last statement of the proposition and from the formula $\operatorname{gr}(\mathcal{B} \circ \mathcal{A}) = \operatorname{gr}(\mathcal{B}^{\star\star} \circ \mathcal{A}^{\star\star}) = \operatorname{gr}(\mathcal{A}^{\star} \circ \mathcal{B}^{\star})^{\star}$. □

Let $\mathfrak{K}$ be a Kreĭn space, $\mathcal{A} \in \mathfrak{R}(\mathfrak{K})$ and let $\mathfrak{L} \subset \mathfrak{K}$ be a closed subspace. Let $\mathcal{I}_{\mathfrak{L}}$ be an operator in $\mathfrak{K}$ with the graph $\operatorname{gr} \mathcal{I}_{\mathfrak{L}} = \left\{ \begin{bmatrix} x \\ x \end{bmatrix} : x \in \mathfrak{L} \right\}$. We define the **double restriction** of the linear relation $\mathcal{A}$ to the subspace $\mathfrak{L}$ as $\mathcal{A}\|_{\mathfrak{L}} := \mathcal{I}_{\mathfrak{L}} \circ \mathcal{A} \circ \mathcal{I}_{\mathfrak{L}}$. It is easy to see that $\operatorname{gr} \mathcal{A}\|_{\mathfrak{L}} = \operatorname{gr} \mathcal{A} \cap (\mathfrak{L} \boxplus \mathfrak{L})$. If the subspace $\mathfrak{L}$ is completely closed, then $\mathcal{I}_{\mathfrak{L}}$ and, therefore, $\mathcal{A}\|_{\mathfrak{L}}$ are completely closed relations in $\mathfrak{K}$.

Let $\mathfrak{L}$ be a regular subspace of the Kreĭn space $\mathfrak{K}$, then we can consider it as a Kreĭn space in the scalar product of $\mathfrak{K}$. We define the **embedding operator** $E_{\mathfrak{L}} := \mathcal{I}_{\mathfrak{L}}|_{\mathfrak{L}}$, then $E_{\mathfrak{L}}^{\star} = E_{\mathfrak{L}}^{-1} P_{\mathfrak{L}}$. Let the subspace $\mathfrak{L}$ be completely closed, then $E_{\mathfrak{L}} \in \mathfrak{R}(\mathfrak{L}, \mathfrak{K})$. In this case for all $\mathcal{A} \in \mathfrak{R}(\mathfrak{K})$

$$\operatorname{gr} (E_{\mathfrak{L}}^{-1} \mathcal{A} E_{\mathfrak{L}}) = (\operatorname{gr} \mathcal{A}) \cap (\mathfrak{L} \boxplus \mathfrak{L}) = \operatorname{gr} (\mathcal{A}\|_{\mathfrak{L}}) \tag{3.8}$$

$$\operatorname{gr} (E_{\mathfrak{L}}^{\star} \mathcal{A} E_{\mathfrak{L}}^{\star -1}) = P_{\mathfrak{L} \boxplus \mathfrak{L}} (\operatorname{gr} \mathcal{A}) \tag{3.9}$$

$$E_{\mathfrak{L}}^{-1} \mathcal{A} E_{\mathfrak{L}} \subset E_{\mathfrak{L}}^{\star} \mathcal{A} E_{\mathfrak{L}}^{\star -1} \tag{3.10}$$

$$\sigma(E_{\mathfrak{L}}^{-1} \mathcal{A} E_{\mathfrak{L}}) = \sigma(\mathcal{A}\|_{\mathfrak{L}}) \tag{3.11}$$

4 Isometric and Unitary Linear Relations

We first consider completely closed isometric and unitary relations in Kreĭn spaces.

A linear relation $\mathcal{V}$ is called **isometric** if $\operatorname{gr} \mathcal{V}^{-1} \subset \operatorname{gr} \mathcal{V}^{\star}$, or equivalently, $\operatorname{gr} \mathcal{V}$ is a neutral subspace: $\operatorname{gr} \mathcal{V} \subset (\operatorname{gr} \mathcal{V})^{[\perp]}$. A linear relation $\mathcal{U}$ is called **unitary** if $\mathcal{U}^{-1} = \mathcal{U}^{\star}$, or equivalently, $\operatorname{gr} \mathcal{U}$ is a hypermaximal neutral subspace: $\operatorname{gr} \mathcal{U} = (\operatorname{gr} \mathcal{U})^{[\perp]}$. It is easy to see that an isometric relation maps neutral subspaces to neutral subspaces, positive subspaces to positive subspaces and negative subspaces to negative subspaces.

Proposition 4.1 *Let $\mathcal{U}_1 \in \mathfrak{R}(\mathfrak{K}_1, \mathfrak{K}_2)$ and $\mathcal{U}_2 \in \mathfrak{R}(\mathfrak{K}_2, \mathfrak{K}_3)$ be unitary relations in Kreĭn spaces. Then the composition $\mathcal{U}_2 \circ \mathcal{U}_1 \in \mathfrak{R}(\mathfrak{K}_1, \mathfrak{K}_3)$ is also unitary.*

Proof: Using (3.6), we obtain: $(\mathcal{U}_2 \circ \mathcal{U}_1)^{-1} = \mathcal{U}_1^{-1} \circ \mathcal{U}_2^{-1} = \mathcal{U}_1^{\star} \circ \mathcal{U}_2^{\star} = (\mathcal{U}_2 \circ \mathcal{U}_1)^{\star}$, so $\mathcal{U}_2 \circ \mathcal{U}_1$ is unitary by definition. □

Let $\mathcal{U} \in \mathfrak{R}(\mathfrak{K})$ be a unitary relation. We say, that $\mathfrak{L}$ **reduces** $\mathcal{U}$ if

(1) $\mathfrak{L}$ is a regular subspace,

(2) $\operatorname{gr} \mathcal{U} = (\operatorname{gr} \mathcal{U}\|_{\mathfrak{L}}) [\dot{+}] (\operatorname{gr} \mathcal{U}\|_{\mathfrak{L}^{[\perp]}})$, and

(3) the relations $E_{\mathfrak{L}}^{-1} \mathcal{U} E_{\mathfrak{L}} \in \mathfrak{R}(\mathfrak{L})$ and $E_{\mathfrak{L}^{[\perp]}}^{-1} \mathcal{U} E_{\mathfrak{L}^{[\perp]}} \in \mathfrak{R}(\mathfrak{L}^{[\perp]})$ are unitary.

Proposition 4.2 *Let $\mathcal{U} \in \mathfrak{R}(\mathfrak{K})$ be a unitary relation in a Kreĭn space $\mathfrak{K}$ and let $\mathfrak{L}$ be a regular completely closed subspace of $\mathfrak{K}$.*

(1) *If $E_{\mathfrak{L}}^{-1} \mathcal{U} E_{\mathfrak{L}} \in \mathfrak{R}(\mathfrak{L})$ is unitary, then $\mathfrak{L}$ reduces $\mathcal{U}$.*

(2) *If $\mathcal{U}\mathfrak{L} \subset \mathfrak{L}$ and $\mathcal{U}^{\star}\mathfrak{L} \subset \mathfrak{L}$, then $\mathfrak{L}$ reduces $\mathcal{U}$.*

Proof: (1) Since $E_{\mathfrak{L}}^{-1}\mathcal{U}E_{\mathfrak{L}}$ is unitary, it is equal to $E_{\mathfrak{L}}^{\star}\mathcal{U}E_{\mathfrak{L}}^{\star\,-1}$. Using (3.8)–(3.10), we conclude that the neutral subspace gr $(E_{\mathfrak{L}}^{-1}\mathcal{U}E_{\mathfrak{L}})$ coincides with gr $(\mathcal{U}\|_{\mathfrak{L}})$ and with the orthogonal projection of the neutral subspace gr $\mathcal{U}$ onto $\mathfrak{L}\boxplus\mathfrak{L}$. Therefore, the projection of gr $\mathcal{U}$ onto $(\mathfrak{L}\boxplus\mathfrak{L})^{[\perp]}=\mathfrak{L}^{[\perp]}\boxplus\mathfrak{L}^{[\perp]}$ is also neutral and

$$\begin{aligned}(E_{\mathfrak{L}^{[\perp]}}^{\star}\mathcal{U}E_{\mathfrak{L}^{[\perp]}}^{\star\,-1})^{-1}\subset(E_{\mathfrak{L}^{[\perp]}}^{\star}\mathcal{U}E_{\mathfrak{L}^{[\perp]}}^{\star\,-1})^{\star} &= E_{\mathfrak{L}^{[\perp]}}^{-1}\mathcal{U}^{\star}E_{\mathfrak{L}^{[\perp]}}\subset E_{\mathfrak{L}^{[\perp]}}^{\star}\mathcal{U}^{\star}E_{\mathfrak{L}^{[\perp]}}^{\star\,-1}\\ = E_{\mathfrak{L}^{[\perp]}}^{\star}\mathcal{U}^{-1}E_{\mathfrak{L}^{[\perp]}}^{\star\,-1} &= (E_{\mathfrak{L}^{[\perp]}}^{\star}\mathcal{U}E_{\mathfrak{L}^{[\perp]}}^{\star\,-1})^{-1}\end{aligned}$$

i.e. the linear relation $E_{\mathfrak{L}^{[\perp]}}^{\star}\mathcal{U}E_{\mathfrak{L}^{[\perp]}}^{\star\,-1}$ is unitary and equals $E_{\mathfrak{L}^{[\perp]}}^{-1}\mathcal{U}^{\star}E_{\mathfrak{L}^{[\perp]}}$. Since its graph coincides with gr $(\mathcal{U}\|_{\mathfrak{L}^{[\perp]}})$, assertion (1) is proved.

(2) Conditions $\mathcal{U}\mathfrak{L}\subset\mathfrak{L}$ and $\mathcal{U}^{\star}\mathfrak{L}\subset\mathfrak{L}$ yield $E_{\mathfrak{L}}^{\star}\mathcal{U}E_{\mathfrak{L}}=E_{\mathfrak{L}}^{-1}\mathcal{U}E_{\mathfrak{L}}$ and $E_{\mathfrak{L}}^{\star}\mathcal{U}^{\star}E_{\mathfrak{L}}=E_{\mathfrak{L}}^{-1}\mathcal{U}^{\star}E_{\mathfrak{L}}$. So we can calculate:

$$\begin{aligned}(E_{\mathfrak{L}}^{-1}\mathcal{U}E_{\mathfrak{L}})^{-1} &= E_{\mathfrak{L}}^{-1}\mathcal{U}^{-1}E_{\mathfrak{L}}=E_{\mathfrak{L}}^{-1}\mathcal{U}^{\star}E_{\mathfrak{L}}=E_{\mathfrak{L}}^{\star}\mathcal{U}^{\star}E_{\mathfrak{L}}\\ &= (E_{\mathfrak{L}}^{\star}\mathcal{U}E_{\mathfrak{L}})^{\star}=(E_{\mathfrak{L}}^{-1}\mathcal{U}E_{\mathfrak{L}})^{\star}\end{aligned}$$

Thus, $E_{\mathfrak{L}}^{-1}\mathcal{U}E_{\mathfrak{L}}$ is unitary and we can apply assertion (1). □

In the rest of this section we speak about isometric and unitary relations only in Pontryagin spaces. The following result is well-known (see [AI]): If an isometric operator V in a Pontryagin space has a closed graph, then dom V and ran V are closed subspaces. If $\mathcal{V}$ is an isometric relation in a Pontryagin space, then $\mathcal{V}(0)$ and ker $\mathcal{V}$ are finite-dimensional.

Proposition 4.3 *If an isometric relation $\mathcal{V}$ in a Pontryagin space has a closed graph, then* dom $\mathcal{V}$, ran $\mathcal{V}$, dom $\mathcal{V}^{\star}$ *and* ran $\mathcal{V}^{\star}$ *are closed subspaces. All unitary relations in Pontryagin spaces are completely closed.*

Proof: Let $\mathcal{V}$ be an isometric relation from the Pontryagin space $\mathfrak{K}_1$ to the Pontryagin space $\mathfrak{K}_2$, and assume that its graph is closed. Then ker $\mathcal{V}\subset$ dom $\mathcal{V}$ and $\mathcal{V}(0)\subset$ ran $\mathcal{V}$, so we can find subspaces $\mathfrak{D}$ and $\mathfrak{R}$ such that dom $\mathcal{V}=\ker\mathcal{V}\dotplus\mathfrak{D}$ and ran $\mathcal{V}=\mathcal{V}(0)\dotplus\mathfrak{R}$. Since

$$\ker\mathcal{V}\boxplus\mathcal{V}(0)\subset\operatorname{gr}\mathcal{V}\subset\operatorname{dom}\mathcal{V}\boxplus\operatorname{ran}\mathcal{V}=(\ker\mathcal{V}\boxplus\mathcal{V}(0))\dotplus(\mathfrak{D}\boxplus\mathfrak{R})$$

the parallel projection of gr $\mathcal{V}$ onto $\mathfrak{D}\boxplus\mathfrak{R}$ coincides with (gr $\mathcal{V})\cap(\mathfrak{D}\boxplus\mathfrak{R})$ and it is a graph of some linear operator V_1, so gr $\mathcal{V}=(\ker\mathcal{V}\boxplus\mathcal{V}(0))\dotplus\operatorname{gr}V_1$. Since gr $\mathcal{V}$ is closed and ker $\mathcal{V}\boxplus\mathcal{V}(0)$ is finite-dimensional, gr V_1 is closed. Thus, V_1 is a closed isometric operator in a Pontryagin space, and consequently dom V_1 and ran V_1 are closed subspaces. Therefore, dom $\mathcal{V}=\ker\mathcal{V}\dotplus\operatorname{dom}V_1$ and ran $\mathcal{V}=\mathcal{V}(0)\dotplus\operatorname{ran}V_1$ are also closed subspaces. Using Lemma 1.1, we can find finite-dimensional subspaces $Z_1\subset\mathfrak{K}_1$, $Z_2\subset\mathfrak{K}_2$ and regular subspaces $X_1\subset\mathfrak{K}_1$, $X_2\subset\mathfrak{K}_2$, such that $(\operatorname{ran}\mathcal{V}^{-1}\dotplus Z_1)\,[\dotplus]\,X_1=\mathfrak{K}_1$ and $(\operatorname{dom}\mathcal{V}^{-1}\dotplus Z_2)\,[\dotplus]\,X_2=\mathfrak{K}_2$. Then $X_2\boxplus X_1\,[\perp]\operatorname{gr}\mathcal{V}^{-1}$, so

$$\operatorname{gr}\mathcal{V}^{-1}\,[\dotplus]\,(X_2\boxplus X_1)\subset(\operatorname{gr}\mathcal{V}^{-1})^{[\perp]}=\operatorname{gr}\mathcal{V}^{\star}\subset\mathfrak{K}_2\boxplus\mathfrak{K}_1,$$

and therefore

$$\operatorname{dom}\mathcal{V}^{-1}[\dot{+}]X_2 \subset \operatorname{dom}\mathcal{V}^\star \subset \mathfrak{K}_2 = (\operatorname{dom}\mathcal{V}^{-1}[\dot{+}]X_2) \dot{+} Z_2$$

$$\operatorname{ran}\mathcal{V}^{-1}[\dot{+}]X_1 \subset \operatorname{ran}\mathcal{V}^\star \subset \mathfrak{K}_1 = (\operatorname{ran}\mathcal{V}^{-1}[\dot{+}]X_1) \dot{+} Z_1$$

The left sides and the right sides are closed subspaces and Z_1, Z_2 are finite-dimensional. So dom $\mathcal{V}^\star$ and ran $\mathcal{V}^\star$ are closed too.

Let $\mathcal{U}$ be a unitary relation in a Pontryagin space, then $\ker\mathcal{U} = \mathcal{U}^\star(0)$ and $\mathcal{U}(0) = \ker\mathcal{U}^\star$ are finite-dimensional subspaces. Therefore, dom $\mathcal{U}$, ran $\mathcal{U}$, dom $\mathcal{U}^\star$ and ran $\mathcal{U}^\star$ are completely closed subspaces. The graph of $\mathcal{U}$ is closed since it is a hypermaximal neutral subspace: gr $\mathcal{U}$ = $(\text{gr }\mathcal{U})^{[\perp]}$. Thus, $\mathcal{U}$ is a completely closed relation. □

Let $\mathcal{U} \in \mathfrak{R}(\mathfrak{K})$ be a unitary relation in a Pontryagin space $\mathfrak{K}$ and $\kappa = \kappa^-[\mathfrak{K}]$. Let us consider an increasing sequence of neutral subspaces $\mathcal{U}(0) \subset \mathcal{U}^2(0) \subset \ldots \subset \mathcal{U}^i(0) \subset \ldots$. Since $\mathcal{U}^i(0)$ are neutral and their dimensions are bounded by κ, $\mathcal{U}^\kappa(0) = \mathcal{U}^{(\kappa+1)}(0) = \ldots$. Similarly, $\mathcal{U}^{\star\kappa}(0) = \mathcal{U}^{\star(\kappa+1)}(0) = \ldots$. In particular, $\mathcal{U}^\kappa(0) = (\mathcal{U}^\kappa)^2(0)$ and $\ker\mathcal{U}^\kappa = \ker(\mathcal{U}^\kappa)^2$.

Lemma 4.4 *Let $\mathcal{U} \in \mathfrak{R}(\mathfrak{K})$ be a unitary relation in a Pontryagin space $\mathfrak{K}$. Then $\mathfrak{K}$ admits a decomposition $\mathfrak{K} = \mathfrak{K}_s [\dot{+}] \mathfrak{K}_r$, where $\mathfrak{K}_s$ and $\mathfrak{K}_r$ reduce $\mathcal{U}$,* $\dim\mathfrak{K}_s \le 2\kappa$ *and*

$$\ker(\mathcal{U}\|_{\mathfrak{K}_s})^\kappa = (\mathcal{U}\|_{\mathfrak{K}_s})^\kappa(0) \text{ is a hypermaximal neutral subspace in } \mathfrak{K}_s$$

$$\ker(\mathcal{U}\|_{\mathfrak{K}_r})^\kappa \cap (\mathcal{U}\|_{\mathfrak{K}_r})^\kappa(0) = \{0\}$$

Proof: If $\ker\mathcal{U}^\kappa \cap \mathcal{U}^\kappa(0) = \{0\}$, then $\mathfrak{K}_r := \mathfrak{K}$ and $\mathfrak{K}_s := \{0\}$. Otherwise let n be the minimal number such that $\ker\mathcal{U} \cap \mathcal{U}^n(0) \ne \{0\}$. Then $\ker\mathcal{U} \cap \mathcal{U}^{n-1}(0) = \{0\}$ and, taking the orthogonal complement, we obtain that $\operatorname{dom}\mathcal{U} \vee \operatorname{ran}\mathcal{U}^n \ne \mathfrak{K} = \operatorname{dom}\mathcal{U} \vee \operatorname{ran}\mathcal{U}^{n-1}$. Let us fix some $y_n \in \operatorname{ran}\mathcal{U}^{n-1} \backslash (\operatorname{dom}\mathcal{U} \vee \operatorname{ran}\mathcal{U}^n)$. Then we can find $\{y_i\}_1^n$ such that $y_{i+1} \in \mathcal{U}y_i$, $1 \le i \le n-1$. We also can find $x_n \in \ker\mathcal{U} \cap \mathcal{U}^n(0)$ such that $[x_n, y_n] = 1$, then there is $\{x_i\}_1^n$ such that $x_{i+1} \in \mathcal{U}x_i$, $1 \le i \le n-1$, $0 \in \mathcal{U}x_n$, and $x_1 \in \mathcal{U}(0)$. It is easy to see that $\{y_i\}_1^n$ and $\{x_i\}_1^n$ are biorthogonal sets of neutral elements, so they are linearly independent and their linear spans are skewly linked. Let us denote by $\mathfrak{K}_{s1}$ their direct sum. Let us consider a linear span $\mathfrak{X}$ of the subset $\left\{\begin{bmatrix} y_i \\ y_{i+1}\end{bmatrix}\right\}_1^{n-1} \cup \left\{\begin{bmatrix} x_i \\ x_{i+1}\end{bmatrix}\right\}_1^{n-1} \cup \left\{\begin{bmatrix} 0 \\ x_1\end{bmatrix}, \begin{bmatrix} x_n \\ 0\end{bmatrix}\right\}$ in $\mathfrak{K}_{s1} \boxplus \mathfrak{K}_{s1}$. $\mathfrak{X}$ is neutral, $\dim\mathfrak{X} = 2n$ and $\dim\mathfrak{K}_{s1} \boxplus \mathfrak{K}_{s1} = 4n$, then (see [AI]) $\mathfrak{X}$ is a hypermaximal neutral subspace. Since $\mathfrak{X} \subset \text{gr }\mathcal{U}$, it follows that $E_{\mathfrak{K}_{s1}}^{-1}\mathcal{U}E_{\mathfrak{K}_{s1}}$ is a unitary relation and, by Proposition 4.2 (1), $\mathfrak{K}_{s1}$ reduces $\mathcal{U}$. Then we denote $\mathfrak{K}_{r1} := \mathfrak{K}_{s1}^{[\perp]}$, $\mathcal{U}_1 := E_{\mathfrak{K}_{r1}}^{-1}\mathcal{U}E_{\mathfrak{K}_{r1}}$. We repeat the above reasoning for unitary relation $\mathcal{U}_1$ in $\mathfrak{K}_{r1}$ to obtain $\mathcal{U}_2$ and $\mathfrak{K}_{r2}$, and so on, until at some N-th step the condition $\ker\mathcal{U}_N^\kappa \cap \mathcal{U}_N^\kappa(0) = \{0\}$ takes place. Then we set $\mathfrak{K}_r := \mathfrak{K}_{rN}$, $\mathfrak{K}_s := \mathfrak{K}_r^{[\perp]}$ and finish the proof. □

In the following lemma we investigate the structure of $\mathcal{U}\|_{\mathfrak{K}_r}$ more carefully.

Lemma 4.5 *Let $\mathcal{U} \in \mathfrak{R}(\mathfrak{K})$ be a unitary relation in a Pontryagin space $\mathfrak{K}$ with the property*

$$\ker \mathcal{U}^\kappa \cap \mathcal{U}^\kappa(0) = \{0\} \tag{4.1}$$

Let $\mathfrak{K}_p := \ker \mathcal{U}^\kappa \dotplus \mathcal{U}^\kappa(0)$, $\mathfrak{K}_{op} := \mathfrak{K}_p^{[\perp]}$. Then $\mathfrak{K}_p$ reduces $\mathcal{U}$, $\dim \mathfrak{K}_p \leq 2\kappa$, $\mathfrak{K} = \mathfrak{K}_p [\dotplus] \mathfrak{K}_{op}$ and

$$\ker(\mathcal{U}\|_{\mathfrak{K}_p})^\kappa \dotplus (\mathcal{U}\|_{\mathfrak{K}_p})^\kappa(0) = \mathfrak{K}_p, \qquad \ker \mathcal{U}\|_{\mathfrak{K}_{op}} = \mathcal{U}\|_{\mathfrak{K}_{op}}(0) = \{0\}$$

In other words, $E_{\mathfrak{K}_{op}}^{-1} \mathcal{U} E_{\mathfrak{K}_{op}}$ is a unitary operator in $\mathfrak{K}_{op}$.

Proof: At first we prove that the subspace $\mathfrak{K}_p$ is regular. Let us assume the contrary, then $\exists x \in \mathcal{U}^\kappa(0) : x\, [\perp] \ker \mathcal{U}^\kappa$. Therefore $x \in \operatorname{dom} \mathcal{U}^\kappa$ and $\mathcal{U}^\kappa x \subset (\mathcal{U}^\kappa)^2(0) = \mathcal{U}^\kappa(0)$. Hence $x \in \ker \mathcal{U}^\kappa$ and (4.1) implies $x = 0$. Thus, $\mathfrak{K}_p$ is regular, $\mathcal{U}\mathfrak{K}_p = \mathcal{U}(\ker \mathcal{U}^\kappa \dotplus \mathcal{U}^\kappa(0)) = \ker \mathcal{U}^{\kappa-1} + \mathcal{U}^{\kappa+1}(0) \subset \mathfrak{K}_p$. Similarly, $\mathcal{U}^*\mathfrak{K}_p \subset \mathfrak{K}_p$, so, by Proposition 4.2 (2), $\mathfrak{K}_p$ reduces $\mathcal{U}$. The other statements are obvious. □

Theorem 4.6 *Let $\mathcal{U} \in \mathfrak{R}(\mathfrak{K})$ be a unitary relation in a Pontryagin space $\mathfrak{K}$. Then $\mathfrak{K}$ admits a decomposition $\mathfrak{K} = \mathfrak{K}_s [\dotplus] \mathfrak{K}_p [\dotplus] \mathfrak{K}_{op}$, where $\mathfrak{K}_s$, $\mathfrak{K}_p$ and $\mathfrak{K}_{op}$ reduce $\mathcal{U}$, $\dim \mathfrak{K}_s + \dim \mathfrak{K}_p \leq 2\kappa$ and*

$$\begin{aligned} &\ker(\mathcal{U}\|_{\mathfrak{K}_s})^\kappa = (\mathcal{U}\|_{\mathfrak{K}_s})^\kappa(0) \ \textit{is a hypermaximal neutral subspace in } \mathfrak{K}_s \\ &\ker(\mathcal{U}\|_{\mathfrak{K}_p})^\kappa \dotplus (\mathcal{U}\|_{\mathfrak{K}_p})^\kappa(0) = \mathfrak{K}_p \\ &\ker \mathcal{U}\|_{\mathfrak{K}_{op}} = \mathcal{U}\|_{\mathfrak{K}_{op}}(0) = \{0\} \end{aligned}$$

Moreover, $\sigma(\mathcal{U}\|_{\mathfrak{K}_s}) = \overline{\mathbb{C}}$ if $\mathfrak{K}_s \neq \{0\}$ and $\sigma(\mathcal{U}\|_{\mathfrak{K}_p}) = \{0, \infty\}$ if $\mathfrak{K}_p \neq \{0\}$. The spectrum of the operator $\mathcal{U}\|_{\mathfrak{K}_{op}}$ has the property $\overline{\sigma(\mathcal{U}\|_{\mathfrak{K}_{op}})}^{-1} = \sigma(\mathcal{U}\|_{\mathfrak{K}_{op}})$ and $\sigma(\mathcal{U}\|_{\mathfrak{K}_{op}}) \cap \{|z| < 1\}$ consists of at most $\kappa^-[\mathfrak{K}_{op}]$ eigenvalues counting multiplicity.

Proof: Applying Lemma 4.4 to the unitary relation $\mathcal{U} \in \mathfrak{R}(\mathfrak{K})$, we obtain the subspaces $\mathfrak{K}_s$ and $\mathfrak{K}_r$, which reduce $\mathcal{U}$. Then applying Lemma 4.5 to the unitary relation $\mathcal{U}_r := E_{\mathfrak{K}_r}^{-1} \mathcal{U} E_{\mathfrak{K}_r} \in \mathfrak{R}(\mathfrak{K}_r)$, we obtain the subspaces $\mathfrak{K}_p$ and $\mathfrak{K}_{op}$. It remains to prove the statements about the spectrum. Let $\mathfrak{K}_s \neq \{0\}$, then there exists a nonzero sequence $\{x_i\}_1^N \subset \mathfrak{K}_s$, such that $x_{i+1} \in \mathcal{U}\|_{\mathfrak{K}_s} x_i$, $i = 0, \ldots, N$, where we assume $x_0 = x_{N+1} = 0$. Let us take the shortest of such sequences, then it is linearly independent. We omit the routine proof of this fact. Then for any $\lambda \in \mathbb{C} \backslash \{0\}$

$$\begin{aligned} \lambda \cdot (\lambda^{-1}x_1 + \lambda^{-2}x_2 + \cdots + \lambda^{-N}x_N) &\in \mathcal{U}\|_{\mathfrak{K}_s} \\ (0 + \lambda^{-1}x_1 + \lambda^{-2}x_2 + \cdots + \lambda^{-N}x_N) \end{aligned}$$

therefore $\sum_1^N \lambda^{-i} x_i \in \ker(\mathcal{U}\|_{\mathfrak{K}_s} - \lambda I)$ and $\lambda \in \sigma(\mathcal{U}\|_{\mathfrak{K}_s})$. We also have $\{0, \infty\} \subset \sigma(\mathcal{U}\|_{\mathfrak{K}_s})$ since $\ker \mathcal{U}\|_{\mathfrak{K}_s}$ and $\mathcal{U}\|_{\mathfrak{K}_s}(0)$ are nontrivial.

Let $\mathfrak{K}_p \neq \{0\}$, then $\{0, \infty\} \subset \sigma(\mathcal{U}\|_{\mathfrak{K}_p})$ for the same reason. If $\lambda \in \mathbb{C}\backslash\{0\}$ and $\lambda x \in \mathcal{U}\|_{\mathfrak{K}_p} x$, then $x \in \operatorname{dom}(\mathcal{U}\|_{\mathfrak{K}_p})^\kappa \cap \operatorname{ran}(\mathcal{U}\|_{\mathfrak{K}_p})^\kappa = \{0\}$, so $x = 0$. We proved that $(\mathcal{U}\|_{\mathfrak{K}_p} - \lambda I)^{-1}$ is an operator. Since $\dim \mathfrak{K}_p < \infty$, it is continuous, so λ is a regular point of $\mathcal{U}\|_{\mathfrak{K}_p}$ and $\sigma(\mathcal{U}\|_{\mathfrak{K}_p}) = \{0, \infty\}$.

By formula (3.11) the spectrum of $\mathcal{U}\|_{\mathfrak{K}_{op}}$ coincides with the spectrum of the unitary operator $E_{\mathfrak{K}_{op}}^{-1}\mathcal{U}E_{\mathfrak{K}_{op}} \in \mathcal{L}(\mathfrak{K}_{op})$. One can find the description of the spectrum of unitary operators in Pontryagin spaces in [IKL]. □

We note that for any unitary relation $\mathcal{U}$ in a Pontryagin space $\overline{\sigma(\mathcal{U})}^{-1} = \sigma(\mathcal{U})$, hence $(I - z\mathcal{U})^{-1}$ is a continuous operator for all $z \in \rho(\mathcal{U}^\star)\backslash\{\infty\}$.

Corollary 4.7 *Let $\mathcal{U}$ be a unitary relation in a Pontryagin space $\mathfrak{K}$. Then the following conditions are equivalent:*

(1) *$\mathcal{U}$ has at least one regular point $z \in \overline{\mathbb{C}}$;*

(2) *operator-functions $(\mathcal{U} - zI)^{-1}$ and $(I - z\mathcal{U})^{-1}$ are meromorphic inside the unit disk;*

(3) $\mathcal{U}^\kappa(0) \cap \ker \mathcal{U}^\kappa = \{0\}$

(4) $\operatorname{dom} \mathcal{U}^\kappa \vee \operatorname{ran} \mathcal{U}^\kappa = \mathfrak{K}$

(5) $\operatorname{dom} \mathcal{U}^\kappa \dot{+} \mathcal{U}^\kappa(0) = \mathfrak{K}$

Proposition 4.8 *Let $\mathcal{U}$ be a unitary relation in a Pontryagin space $\mathfrak{K}$, which has at least one regular point. Then $\mathfrak{K}$ admits a decomposition $\mathfrak{K} = \mathfrak{K}' [\dot{+}] (\Lambda^+ \dot{+} \Lambda^-)$, where $\Lambda^\pm$ are skewly linked finite-dimensional neutral subspaces, $\mathfrak{K}'$ is regular, and*

$$\begin{aligned} \mathcal{U}\mathfrak{K}' &= \mathfrak{K}', \quad |\sigma(\mathcal{U}\|_{\mathfrak{K}'})| = 1 \\ \mathcal{U}\Lambda^+ &\subset \Lambda^+, \quad |\sigma(\mathcal{U}\|_{\Lambda^+})| < 1 \\ \mathcal{U}^{-1}\Lambda^- &\subset \Lambda^-, \quad |\sigma(\mathcal{U}^{-1}\|_{\Lambda^-})| < 1 \end{aligned}$$

Proof: Applying Theorem 4.6, we can decompose $\mathfrak{K} = \mathfrak{K}_{op} [\dot{+}] (\ker(\mathcal{U}\|_{\mathfrak{K}_p})^\kappa \dot{+} (\mathcal{U}\|_{\mathfrak{K}_p})^\kappa(0))$, and $U := E_{\mathfrak{K}_{op}}^{-1}\mathcal{U}E_{\mathfrak{K}_{op}}$ is a unitary operator. Applying the classic results about the invariant subspaces and the spectrum of unitary operators in Pontryagin spaces, contained in [IKL], we can obtain a decomposition $\mathfrak{K}_{op} = \mathfrak{K}'_{op} [\dot{+}] (\Lambda^+_{op} \dot{+} \Lambda^-_{op})$, where $\Lambda^\pm_{op}$ are a skewly linked finite-dimensional neutral subspaces, $\mathfrak{K}'_{op}$ is regular, and the following properties hold

$$\begin{aligned} U\mathfrak{K}'_{op} &= \mathfrak{K}'_{op}, \quad |\sigma(U\|_{\mathfrak{K}'_{op}})| = 1 \\ U\Lambda^+_{op} &= \Lambda^+_{op}, \quad |\sigma(U\|_{\Lambda^+_{op}})| < 1 \\ U^{-1}\Lambda^-_{op} &= \Lambda^-_{op}, \quad |\sigma(U^{-1}\|_{\Lambda^-_{op}})| < 1 \end{aligned}$$

Thus, $\mathfrak{K} = \mathfrak{K}'_{op} [\dot{+}] ((\Lambda^+_{op} [\dot{+}] \ker(\mathcal{U}\|_{\mathfrak{K}_p})^\kappa) \dot{+} (\Lambda^-_{op} [\dot{+}] (\mathcal{U}\|_{\mathfrak{K}_p})^\kappa(0)))$. Now the other statements are obvious. □

5 Open and Closed Systems in Pontryagin Spaces

Below we give a generalization of the notion of a linear stationary conservative dynamical scattering system with discrete time (see [Ar]).

Let $\mathfrak{H}$ be a Pontryagin space, $\mathfrak{M}$ and $\mathfrak{N}$ be Hilbert spaces and let $\mathcal{W} \in \mathfrak{R}(\mathfrak{H} \oplus \mathfrak{M}, \mathfrak{H} \oplus \mathfrak{N})$ be a unitary relation. Then the quadruple $\alpha = (\mathfrak{H}, \mathfrak{M}, \mathfrak{N}, \mathcal{W})$ is called a **linear stationary conservative dynamical scattering system with discrete time and Pontryagin state space**. Here $\mathfrak{H}$, $\mathfrak{M}$ and $\mathfrak{N}$ are called state space, incoming space and outgoing space. The relation $\mathcal{W}$ is called a system relation. It describes the evolution of inner states of the system with some sequence of outer data:

$$\begin{pmatrix} h_{i+1} \\ n_{i+1} \end{pmatrix} \in \mathcal{W} \begin{pmatrix} h_i \\ m_i \end{pmatrix} \tag{5.1}$$

for all integer numbers i from some interval. Here h_i, m_i and n_i are interpreted as inner state, incoming and outgoing outer data at time i, respectively. We say that $\{h_i\}$ is a trajectory of states of the system with the sequence $\{(m_i; n_{i+1})\}$ of outer data. We shall consider the **future trajectory** $\{h_i\}_0^\infty$ with the sequence of outer data $\{(m_i; n_{i+1})\}_0^\infty$ and the **past trajectory** $\{h_i\}_{-\infty}^0$ with the sequence of outer data $\{(m_i; n_{i+1})\}_{-\infty}^{-1}$. The state of the system at time $i = 0$ will be called the **present state** of the trajectory. Since $\mathcal{W}$ is unitary, for any trajectory

$$\|n_{i+1}\|^2 - \|m_i\|^2 = [h_{i+1}, h_{i+1}] - [h_i, h_i]$$

Here $\|m_i\|^2$ is viewed as the incoming energy, $\|n_i\|^2$ as the outgoing energy and $[h_i, h_i]$ as the inner energy of the system at time i. We say that the system α is **regular** if for every finite trajectory $\{h_i\}_{i=0}^N$ with zero sequence of outer data the conditions $h_0 = h_N = 0$ imply $h_i = 0$ for $0 < i < N$. If the system relation of the system α is an operator, then α is regular.

Since in this paper we deal only with systems of the described type, below we use the term "open system" instead of "linear stationary conservative dynamical scattering system with discrete time and Pontryagin state space". We say that the open system $\alpha = (\mathfrak{H}, \mathfrak{M}, \mathfrak{N}, \mathcal{W})$ is a **closed system** if $\mathfrak{M} = \mathfrak{N} = \{0\}$. We shall write a closed system as a pair $\beta = (\mathfrak{K}, \mathcal{U})$ consisting of the state space $\mathfrak{K}$ and the system relation $\mathcal{U}$ (we usually use letters $\mathfrak{K}$ and $\mathcal{U}$ for closed systems, while $\mathfrak{H}$ and $\mathcal{W}$ are used for open systems, which may not be closed systems). The closed system itself differs from the unitary relation in a Pontryagin space only by terminology. In particular, the set of the present states of all future trajectories equals dom $\mathcal{U}^\kappa$, and the set of the present states of all past trajectories with a finite number of nonzero elements equals $\mathcal{U}^\kappa(0)$, where $\kappa := \kappa^-[\mathfrak{K}]$.

If the system relation of the closed system $\beta = (\mathfrak{K}, \mathcal{U})$ has at least one regular point, then by Corollary 4.7 β is regular and $(I - z\mathcal{U})^{-1}$ is meromorphic inside the unit disk. Each of the following conditions is equivalent to the regularity of the closed system:

(1) If the present state x_0^+ of the future trajectory $\{x_i^+\}_0^\infty$ and the present state x_0^- of the past trajectory $\{x_i^-\}_{-\infty}^0$ are equal to zero, then all states $x_i^\pm$ are equal to zero. This means that there is at most one infinite trajectory with a given present state, while finite trajectories, in general, are not uniquely defined by the present state.

(2) Every state x can be decomposed into a sum of the present state x_0^+ of some future trajectory $\{x_i^+\}_0^\infty$ and the present state x_0^- of some past trajectory $\{x_i^-\}_{-\infty}^0$.

(3) Every state x can be decomposed in a unique way into a sum of the present state x_0^+ of some future trajectory $\{x_i^+\}_0^\infty$ and the present state x_0^- of some past trajectory $\{x_i^-\}_{-\infty}^0$ having a finite number of nonzero elements.

Let $\beta = (\mathfrak{K}, \mathcal{U})$ be a regular closed system and let $\mathfrak{K} = \mathfrak{K}' \,[\dot{+}]\, (\Lambda^+ \dotplus \Lambda^-)$ be a decomposition as in Proposition 4.8. Then any state $x_0 \in \mathfrak{K}$ can be decomposed into the sum of a state $x_0^+ \in \mathfrak{K}' \,[\dot{+}]\, \Lambda^+$ and a state $x_0^- \in \Lambda^-$. Moreover, $U_+ := \mathcal{U}\|_{\mathfrak{K}'[\dot{+}]\Lambda^+}$ and $U_- := \mathcal{U}^{-1}\|_{\Lambda^-}$ are continuous operators with $|\sigma(U_+)| \leq 1$ and $|\sigma(U_-)| < 1$. Then it is easy to see that $\{U_+^i x_0^+\}_0^\infty$ is a future trajectory and $\{U_-^{-i} x_0^-\}_{-\infty}^0$ is a past trajectory. Let us fix some fundamental symmetry J in the Pontryagin space $\mathfrak{K}$. We denote by $\|\cdot\|_J$ the norm in the corresponding Hilbert space. Then $\|U_+^i x_0^+\|_J \leq \|x_0^+\|_J$ for $i \geq 0$ and, denoting $r := \max|\sigma(U_-)| < 1$ we obtain $\|U_-^{-i} x_0^-\|_J \leq r^{-i}\|x_0^-\|$ for $i \leq 0$. Therefore, the Laurent series

$$\sum_{i\geq 0} U_+^i x_0^+ z^i - \sum_{i<0} U_-^{-i} x_0^- z^i$$

converges in the ring $\{r < |z| < 1\}$. It is easy to verify, that the sum of this series equals $(I - zU_+)^{-1}x_0^+ + (I - zU_-^{-1})^{-1}x_0^- = (I - z\mathcal{U})^{-1}x_0$. Thus, we have proved the following

Proposition 5.1 *Let $\beta = (\mathfrak{K}, \mathcal{U})$ be a regular closed system in a Pontryagin space $\mathfrak{K}$. Let $x_0 \in \mathfrak{K}$ and let*

$$(I - z\mathcal{U})^{-1}x_0 = \sum_{i\geq 0} x_i^+ z^i - \sum_{i<0} x_i^- z^i \tag{5.2}$$

be a decomposition of $(I - z\mathcal{U})^{-1}x_0$ into a Laurent series in some ring $\{r < |z| < 1\}$ where it is holomorphic. Then $\{x_i^+\}_0^\infty$ is a future trajectory and $\{x_i^-\}_{-\infty}^0$ is a past trajectory of the system β, where $x_0^- := x_0 - x_0^+$. Furthermore, $x_i^+ \in \mathfrak{K}' \,[\dot{+}]\, \Lambda^+$ and $x_i^- \in \Lambda^-$, where $\mathfrak{K} = \mathfrak{K}' \,[\dot{+}]\, (\Lambda^+ \dotplus \Lambda^-)$ is a decomposition according to Proposition 4.8.

Remark 5.2 In the same notation one can obtain the following statements:

$$(5.3)\qquad (\mathcal{U}^\star - zI)^{-1}x_0 \in \mathcal{U}((I - z\mathcal{U})^{-1}x_0)$$

$$(5.4)\quad (\mathcal{U}^\star - zI)^{-1}x_0 = z^{-1}((I - z\mathcal{U})^{-1} - I)x_0 = \sum_{i\geq 0} x_{i+1}^{+} z^i - \sum_{i<0} x_{i+1}^{-} z^i$$

Let $\alpha = (\mathfrak{H}, \mathfrak{M}, \mathfrak{N}, \mathcal{W})$ be a regular open system. A **resolvent matrix** of the open system α is the linear relation $\Theta^\alpha(z)$ from $\mathfrak{H} \oplus \mathfrak{M}$ to $\mathfrak{H} \oplus \mathfrak{N}$ defined by

$$(5.5)\quad \forall z \in \mathbb{C} \quad \begin{pmatrix} h \\ n \end{pmatrix} \in \Theta^\alpha(z) \begin{pmatrix} h_0 \\ m \end{pmatrix} :\Longleftrightarrow \begin{pmatrix} h \\ n \end{pmatrix} \in \mathcal{W} \begin{pmatrix} h_0 + zh \\ m \end{pmatrix}$$

We also define the **transfer function** $\theta^\alpha(z) := P_{\mathfrak{N}} \Theta^\alpha(z)|\mathfrak{M}$ (in general, it is a linear relation). It is not difficult to see, that if $\mathcal{W}$ is an operator and $\mathcal{W} = \begin{pmatrix} T & F \\ G & H \end{pmatrix}$, then the resolvent matrix equals

$$\Theta^\alpha(z) = \begin{pmatrix} T(I - zT)^{-1} & (I - zT)^{-1}F \\ G(I - zT)^{-1} & H + zG(I - zT)^{-1}F \end{pmatrix}$$

and transfer function, as usual, equals $\theta^\alpha(z) = H + zG(I - zT)^{-1}F$.

Definition 5.3 The **embedding** τ of an open system $\alpha = (\mathfrak{H}, \mathfrak{M}, \mathfrak{N}, \mathcal{W})$ into a closed system $\beta = (\mathfrak{K}, \mathcal{U})$ is a pair of linear operators $(E_{\tau 1}, E_{\tau 2})$, where $E_{\tau 1} \in \mathcal{L}(\mathfrak{H}, \mathfrak{K})$, $E_{\tau 2} \in \mathcal{L}(\mathfrak{M} \boxplus \mathfrak{N}, \mathfrak{K} \boxplus \mathfrak{K})$, such that if $\begin{pmatrix} h_1 \\ n_1 \end{pmatrix} \in \mathcal{W} \begin{pmatrix} h_0 \\ m_0 \end{pmatrix}$, then $\begin{bmatrix} E_{\tau 1} h_0 \\ E_{\tau 1} h_1 \end{bmatrix} + E_{\tau 2} \begin{bmatrix} m_0 \\ n_1 \end{bmatrix} \in \operatorname{gr} \mathcal{U}$.

We introduce a **total embedding operator**

$$(5.6)\qquad E_\tau \begin{bmatrix} \begin{pmatrix} h_0 \\ m_0 \end{pmatrix} \\ \begin{pmatrix} h_1 \\ n_1 \end{pmatrix} \end{bmatrix} := \begin{bmatrix} E_{\tau 1} h_0 \\ E_{\tau 1} h_1 \end{bmatrix} + E_{\tau 2} \begin{bmatrix} m_0 \\ n_1 \end{bmatrix}$$

Then the condition of Definition 5.3 may be written as $E_\tau(\operatorname{gr} \mathcal{W}) \subset \operatorname{gr} \mathcal{U}$ or, equivalently, $\operatorname{gr} \mathcal{W} \subset E_\tau^{-1}(\operatorname{gr} \mathcal{U})$. Applying formula (2.4), we obtain

$$(5.7)\quad E_\tau^\star(\operatorname{gr} \mathcal{U}) = E_\tau^\star(\operatorname{gr} \mathcal{U})^{[\perp]} \subset (E_\tau^{-1}(\operatorname{gr} \mathcal{U}))^{[\perp]} \subset (\operatorname{gr} \mathcal{W})^{[\perp]} = \operatorname{gr} \mathcal{W}$$

Proposition 5.4 *Let τ be an embedding of the open system $\alpha = (\mathfrak{H}, \mathfrak{M}, \mathfrak{N}, \mathcal{W})$ into the regular closed system $\beta = (\mathfrak{K}, \mathcal{U})$. For any state $x_0 \in \mathfrak{K}$ of the closed system consider the state of the open system $h_0 := E_{\tau 1}^\star x_0$ and the following meromorphic operator-functions on the open unit disk:*

$$(5.8)\quad h(z) := E_{\tau 1}^\star (\mathcal{U}^\star - zI)^{-1} x_0 \qquad \begin{bmatrix} m(z) \\ n(z) \end{bmatrix} := E_{\tau 2}^\star \begin{bmatrix} (I - z\mathcal{U})^{-1} x_0 \\ (\mathcal{U}^\star - zI)^{-1} x_0 \end{bmatrix}$$

Then for all z in some ring $\{r < |z| < 1\} \subset \rho(\mathcal{U}^\star)$ it holds

$$\begin{pmatrix} h(z) \\ n(z) \end{pmatrix} \in \Theta^\alpha(z) \begin{pmatrix} h_0 \\ m(z) \end{pmatrix} \tag{5.9}$$

If, in addition, $h_0 = 0$, then $n(z) \in \theta^\alpha(z)m(z)$.

Proof: Applying Proposition 5.1, we can find a future trajectory $\{x_i^+\}_0^\infty$ and a past trajectory $\{x_i^-\}_{-\infty}^0$, which are the coefficients of the decomposition of $(I - z\mathcal{U})^{-1}x_0$ into a Laurent series. Then we introduce

$$h_i^\pm := E_{\tau 1}^\star x_i^\pm \begin{bmatrix} m_i^\pm \\ n_{i+1}^\pm \end{bmatrix} := E_{\tau 2}^\star \begin{bmatrix} x_i^\pm \\ x_{i+1}^\pm \end{bmatrix} \tag{5.10}$$

It follows from (5.7) that

$$\begin{pmatrix} h_{i+1}^+ \\ n_{i+1}^+ \end{pmatrix} \in \mathcal{W} \begin{pmatrix} h_i^+ \\ m_i^+ \end{pmatrix} \text{ for } i \geq 0 \quad \text{and} \quad \begin{pmatrix} h_{i+1}^- \\ n_{i+1}^- \end{pmatrix} \in \mathcal{W} \begin{pmatrix} h_i^- \\ m_i^- \end{pmatrix} \text{ for } i < 0$$

(5.11)

Substituting (5.2) and (5.4) into (5.8) and using (5.10), we conclude that

$$\begin{aligned} h(z) &= \sum_{i\geq 0} h_{i+1}^+ z^i - \sum_{i<0} h_{i+1}^- z^i, \, m(z) = \sum_{i\geq 0} m_i^+ z^i - \sum_{i<0} m_i^- z^i, \\ n(z) &= \sum_{i\geq 0} n_{i+1}^+ z^i - \sum_{i<0} n_{i+1}^- z^i \end{aligned}$$

Multiplying formulae (5.11) by z^i, summing over i and then taking the difference, we obtain:

$$\begin{pmatrix} h(z) \\ n(z) \end{pmatrix} \in \mathcal{W} \begin{pmatrix} h_0 + zh(z) \\ m(z) \end{pmatrix}$$

Now (5.9) follows from the definitions. □

Let $\alpha = (\mathfrak{H}, \mathfrak{M}, \mathfrak{N}, \mathcal{W})$ be an open system. We are going to build a special closed system $\mathring{\alpha} = (\mathring{\mathfrak{K}}, \mathring{\mathcal{U}})$, which is called a **canonical closure** of α, and an embedding of α into $\mathring{\alpha}$. We define the state space of $\mathring{\alpha}$ by $\mathring{\mathfrak{K}} := H^2(\mathfrak{M}) \oplus \mathfrak{H} \oplus H^2(\mathfrak{N})$ and the system relation by

$$\mathring{\mathcal{U}} \begin{pmatrix} \mu(\zeta) \\ h \\ \nu(\zeta) \end{pmatrix} := \left\{ \begin{pmatrix} \zeta^{-1}(\mu(\zeta) - \mu(0)) \\ h_1 \\ n_1 + \zeta\nu(\zeta) \end{pmatrix} : \begin{pmatrix} h_1 \\ n_1 \end{pmatrix} \in \mathcal{W} \begin{pmatrix} h \\ \mu(0) \end{pmatrix} \right\}$$

where H^2 is the Hardy space. In order to define the embedding, we introduce the isometric operators $E_\mathfrak{H} \in \mathcal{L}(\mathfrak{H}, \mathring{\mathfrak{K}})$, $E_\mathfrak{M} \in \mathcal{L}(\mathfrak{M}, \mathring{\mathfrak{K}})$ and $E_\mathfrak{N} \in \mathcal{L}(\mathfrak{N}, \mathring{\mathfrak{K}})$ by

$$E_\mathfrak{H} h := \begin{pmatrix} 0 \\ h \\ 0 \end{pmatrix} \quad E_\mathfrak{M} m := \begin{pmatrix} m \\ 0 \\ 0 \end{pmatrix} \quad E_\mathfrak{N} n := \begin{pmatrix} 0 \\ 0 \\ n \end{pmatrix}$$

Let $E_{\tau 1} := E_{\mathfrak{H}}$ and $E_{\tau 2} := \begin{bmatrix} E_{\mathfrak{M}} & 0 \\ 0 & E_{\mathfrak{N}} \end{bmatrix}$. Then $\tau = (E_{\tau 1}, E_{\tau 2})$ satisfies all the conditions of Definition 5.3.

Proposition 5.5 *Let* $\alpha = (\mathfrak{H}, \mathfrak{M}, \mathfrak{N}, \mathcal{W})$ *be an open system and let* $\overset{\circ}{\alpha} = (\overset{\circ}{\mathfrak{K}}, \overset{\circ}{\mathcal{U}})$ *be a canonical closure of* α. *Then* α *is regular if and only if* $\overset{\circ}{\alpha}$ *is regular. In this case the resolvent matrix and the transfer function of the open system* α *are operator-functions which are holomorphic in some ring* $\{r < |z| < 1\} \subset \rho(\overset{\circ}{\mathcal{U}}{}^{\star})$ *and are given by*

$$\Theta^{\alpha}(z) = \begin{pmatrix} E_{\mathfrak{H}}^{\star} \\ E_{\mathfrak{N}}^{\star} \end{pmatrix} (\overset{\circ}{\mathcal{U}}{}^{\star} - zI)^{-1} (E_{\mathfrak{H}} \ \ E_{\mathfrak{M}}) \tag{5.12}$$

$$\theta^{\alpha}(z) = E_{\mathfrak{N}}^{\star} (\overset{\circ}{\mathcal{U}}{}^{\star} - zI)^{-1} E_{\mathfrak{M}} \tag{5.13}$$

Proof: The statement about the regularity can be verified directly from the definitions. Let $\overset{\circ}{\mathfrak{K}}$ be a regular closed system. It is not difficult to verify that for all $z \in \rho(\overset{\circ}{\mathcal{U}}{}^{\star})$ it holds $(\Theta^{\alpha}(z))(0) \subset (I - z\overset{\circ}{\mathcal{U}})^{-1}\mathfrak{N} \cap \mathfrak{H} = \{0\}$, so $\Theta^{\alpha}(z)$ is an operator. Applying Proposition 5.4 to the embedding of α into $\overset{\circ}{\alpha}$, we obtain for z in some ring $\{r < |z| < 1\} \subset \rho(\overset{\circ}{\mathcal{U}}{}^{\star})$

$$\forall x_0 \in \overset{\circ}{\mathfrak{K}} \quad \begin{pmatrix} E_{\mathfrak{H}}^{\star} (\overset{\circ}{\mathcal{U}}{}^{\star} - zI)^{-1} x_0 \\ E_{\mathfrak{N}}^{\star} (\overset{\circ}{\mathcal{U}}{}^{\star} - zI)^{-1} x_0 \end{pmatrix} = \Theta^{\alpha}(z) \begin{pmatrix} E_{\mathfrak{H}}^{\star} x_0 \\ E_{\mathfrak{M}}^{\star} (I - z\overset{\circ}{\mathcal{U}})^{-1} x_0 \end{pmatrix}$$

Now let us take an arbitrary $h \in \mathfrak{H}$, $m \in \mathfrak{M}$ and assume $x_0 = E_{\mathfrak{H}} h + E_{\mathfrak{M}} m$. It is easy to verify that then $E_{\mathfrak{M}}^{\star} (I - z\overset{\circ}{\mathcal{U}})^{-1} x_0 = m$. Then (5.12) and (5.13) immediately follow. □

It follows that the equality in formula (5.9) holds in some ring $\{r < |z| < 1\} \subset \rho(\mathcal{U}^{\star}) \cap \rho(\overset{\circ}{\mathcal{U}}{}^{\star})$.

Acknowledgment

I would like to express deep gratitude to my supervisor Professor Damir Zyamovich Arov whose guidance and support were very important for me and very helpful for this paper. I thank my referee Professor Aad Dijksma for valuable remarks. The research described in this publication was made possible in part by Grant No. UCZ000 from International Science Foundation and by Grant No. UM1-298 from American CRDF and Ukrainian Government.

References

[AAK] V.M. Adamyan, D.Z. Arov and M.G. Krein, *Infinite Hankel Matrices and Generalized Problem of Caratheodory-Fejer and I.Schur*, Funk. Anal. i Prilozhen. **2:4** (1968), 1–17. English Transl.: Funct. Anal. Appl. **2** (1968), 269–281.

[AAK2] V.M. Adamyan, D.Z. Arov and M.G. Krein, *Analytic properties of the Schmidt pairs of a Hankel operator and the generalized Schur-Takagi problem*, Math. Sb. (N.S.) **86** (128) (1971), 34–75; English Transl.: Math. USSR-Sb. **15** (1971), no. 1, 31–73.

[A] V.M. Adamyan, *Non-degenerate Unitary Coupling of Semiunitary Operators*, Funk. Anal. i Prilozhen. **7** (1973), 1–16.

[Ar] D.Z. Arov, *Passive linear stationary dynamical systems*, Sibirsk. Mat. Zhurn. **20** (2) (1979), 211–228 (Russian).

[AG] D.Z. Arov and L.Z. Grossman, *Scattering matrices in the theory of unitary extensions of isometric operators*, Math. Nachr. **157** (1992), 105–123.

[AI] T.Ya. Azizov and I.S. Iohvidov, *Foundations of the Theory of Linear Operators in Spaces with an Indefinite Metric*, "Nauka", Moscow, 1986; English Transl.. Wiley, New York, 1989.

[B] P. Bruinsma, *Interpolation problems for Schur and Nevanlinna pairs*, Doctoral dissertation, University of Groningen, 1991.

[CG] T. Constantinescu and A. Geondea, *Minimal signature in lifting of operators, II*, J. Functional Analysis **103** (1992), 317–351.

[D] V. Derkach, *On Generalized Resolvents of Hermitian Relations in Kreĭn spaces*, to appear.

[DLS] A. Dijksma, H. Langer and H.S.V. de Snoo, *Generalized Coresolvents of Standard Isometric Operators and Generalized Resolvents of Standard Symmetric Relations in Kreĭn Spaces*, Operator Theory: Adv. Appl. **48** (1990), 261–274.

[IKL] I.S. Iohvidov, M.G. Krein and H. Langer, *Introduction to the Spectral Theory of Operators in Spaces with an Indefinite Metric*, volume 9 of *Mathematical Research*. Akademie Verlag, Berlin, 1982.

[KW] M. Kaltenbäck and H.Woracek, *The Kreĭn Formula for Generalized Resolvents in Degenerated Inner Product Spaces*, to appear.

[KL] M.G. Kreĭn and H. Langer, *On Defect Subspaces and Generalized Resolvents of a Hermitian Operator in Space Π_κ*, Funk. Anal. i Prilozhen. **5**(2) (1971), 59–71 (Russian); **5**(3) (1971), 54–69. English transl.: Funct. Anal. Appl. **5** (1971/1972), 136–146, 217–228.

Oleg Nitz
South-Ukrainian State Pedagogical University
Odessa
Ukraine
on@ibis.odessa.ua

Operator Theory:
Advances and Applications, Vol. 118

A Termwise Differentiation in the Inductive Scales of the Locally Convex Spaces

I.V. Orlov

There are obtained the sufficient conditions of a termwise differentiation for the series of differentiable mappings from a segment into the linear inductive scale of the locally convex spaces and the analogous ones for the series of homogeneously differentiable mappings from a segment into the arbitrary inductive scale of the locally convex spaces. Some examples are considered.

1 Inductive Scales of LCS. Examples of Inductive Scales

Introduction. As the classical conditions of a termwise differentiation for the mappings of a segment into a locally convex space (LCS) are, as it is well known, the uniform convergence of derivative series and the convergence of initial series at a point. In addition, the continuous differentiability of the series members implies the continuous differentiability of its sum.

However, already for the mappings of a segment into the linear inductive scales of LCS there is broken the standard topological result on continuity of a uniform limit (see example 2.1 below). It implies, obviously, the breakdown in the inductive scales of LCS of the abovementioned conditions of a termwise differentiation.

Our aim is to receive the additional conditions, ensuring such differentiation both in the linear and nonlinear inductive scales of LCS. Throughout the paper a convergence in any scale above is not connected with the passage to the inductive limit of the scale (see def. 1.2 below).

In item 1 of the work there is introduced the general notion of the inductive scale of LCS and there are discussed some nontrivial examples. In item 2 there are given the sufficient conditions of the continuity of a uniform limit. In item 3 and 4 there are considered the sufficient conditions of the termwise differentiation in the linear and nonlinear scales of LCS, respectively. Throughout we distinguish the important cases in which the classical conditions of termwise differentiation remain valid. At last, in the final item 5 the results above are applied to the termwise differentiation in parameter of measurable functions and linear functionals.

Definition 1.1 Call by *inductive scale of LCS* (or, simple, *LCS-scale*) any family of LCS $\{E_i\}_{i\in I}$ inductively ordered by continuous embedding: $(i_1 \preceq i_2) \Rightarrow (E_{i_1} \hookrightarrow E_{i_2})$. Call two scales, $\{E_i\}_{i\in I}$ and $\{E'_j\}_{j\in J}$, by *equivalent* ones if every space from the first scale is continuously embedded into some space from the second one, and vice versa. Call a scale $\{E_i\}_{i\in I}$ by *linear* one if, up to equivalence, the order in I is linear.

Definition 1.2 Let $E = \cup_{i\in I} E_i$ and $F = \cup_{j\in J} E_j$ be LCS-scales, $f : E \to F$. Say that $\lim_{x\to x_0} f(x) = y_0$ if ($x \to x_0$ in some E_i) $\Rightarrow$ ($f(x) \to y_0$ in some F_j). In analogous way one can define the continuity and the differentiability of f.

Note that our definition of the convergence is not in accordance with the topologies of the inductive limits of the scales (in addition, the limit above can be trivial one, see 2.9 below). Therefore we consider here some nonstandard (with the exception of 1.3) examples of LCS-scales.

Example 1.3 Any scale of Banach (or Hilbert) spaces ([?], [?]) is linear inductive scale. Up to the convergence for the mappings in such scales see ([?], [?]).

Example 1.4 Any countable LCS-scale is linear [?].

Example 1.5 *Normal decomposition of the dual space* ([?]). Let $\{\|\cdot\|_i\}_{i\in I}$ be a (inductively ordered) defining system of the prenorms in some LCS E, E^* be topological dual to E, $f \in E^*$. Set $\|f\|^i = \sup_{\|x\|_i \leq 1} |f(x)|$ $(i \in I)$ and

$$E_i^* = \{f \in E^* | \, \|f\|^i < \infty\}.$$

Then $\{(E_i^*, \|\cdot\|^i)\}_{i\in I}$ is an inductive scale of Banach spaces and $E^* = \cup_{i\in I} E_i^*$. Call this scale by *normal decomposition of* E^*. Note that this decomposition is minimal, up to equivalence, under the continuity of pairing $\langle f, x\rangle = f(x)$ in $E^* \times E$ ([?], §3). The scale above is linear iff I is linearly ordered.

Example 1.6 Normal decomposition of the operator space ([?]). Consider one more general construction. Let $\{\|\cdot\|_i\}_{i\in I}$ and $\{\|\cdot\|^j\}_{j\in J}$ be (inductively ordered) defining systems of the prenorms in some LCS E and F, respectively, $A \in \mathcal{L}(E, F)$. There exists such (increasing) mapping $q : J \to I$ that $\|A\|_j^q = \sup_{\|x\|_{q(j)}\leq 1} \|Ax\|^j < \infty$ $(j \in J)$. Let $Q = \{q\}$ be a set of such mappings (*normal indices*) and, for every $q \in Q$, set $\mathcal{L}_q(E, F) = \{A \in \mathcal{L}(E, F) | \, \|Ax\|_j^q < \infty \ (j \in J)\}$. Then $\{(\mathcal{L}_q(E, F); \{\|\cdot\|_j^q\}_{j\in J})\}_{q\in Q}$ is an inductive scale of (complete if F is complete) LCS and $\mathcal{L}(E, F) = \cup_{q\in Q} \mathcal{L}_q(E, F)$. Call this scale by *normal decomposition of* $\mathcal{L}(E, F)$. Note that this decomposition is minimal, up to equivalence, under the continuity of pairing $\langle A, x\rangle = Ax$ in $\mathcal{L}(E, F) \times E$. In addition, the composition map $\mathcal{L}(F, G) \times \mathcal{L}(E, F) \to \mathcal{L}(E, G)$ is also continuous with respect to the normal decompositions of the operator spaces ([?]).

Example 1.7 Space of measurable functions with almost everywhere convergence ([?]). Let S be a locally compact space with the finite measure μ; $M(S, \mu)$ be a space of measurable on S functions; $\{K_n\}_{n=1}^{\infty}$ be such an increasing sequence of the compact sets from S that $\mu(S \setminus \cup_{n=1}^{\infty} K_n) = 0$. If t is the equivalence class of such sequences (by the mutual embeddings of the compactae), $T = \{t\}$, then set $M_t(S, \mathrm{mod}\ \mu) = \{f \in M(S, \mu) | \text{ all restrictions } f|_{K_n} \text{ are continuous}$

$(n = 1, 2, \ldots)\}$. Then $M_t(S, \bmod \mu)$, $t \in T$, are complete LCS with respect to uniform convergence on K_n $(n = 1, 2, \ldots)$ forming a LCS-scale and $M(S, \mu)$ is a union of the scale. In addition, the convergence in the scale coincides with the sequential μ-almost everywhere convergence. Denote the considered scale by $M(S, \bmod \mu)$ and call it by *almost everywhere convergence space*. So,

$$M(S, \bmod \mu) = \bigcup_{t \in T} M_t(S, \bmod \mu).$$

This scale is essentially nonlinear, but it possesses a remarkable property of σ-*inductivity* (see 2.5 below). Note that the locally convex inductive limit of the scale $M(S, \bmod \mu)$ is trivial, at the same time the topological inductive limit of this scale is exactly the convergence in measure space $M(S, \mu)$.

Example 1.8 Pseudotopological LCS. ([?], [?]). Pseudotopologies in operator spaces provide in non-Banach case, on the contrary of topologies, a continuity of the pairing and composition maps and therefore are widely exploited in non-Banach differential calculus. Instead of a unique neighborhoods system, in pseudotopological space there is introduced some family of convergent filters (see [?]). It is shown by author [?] that any pseudotopological LCS one can replace by some LCS-scale preserving convergence in the space.

2 Uniform Convergence and Continuity in the Inductive Scales of LCS

Now we'll show that classical topological condition of uniform convergence no longer guarantees the continuity for the sum of the series of the continuous mappings into LCS-scale. There are deduced below sufficient conditions for the continuity of such sum. Restrict oneself, for simplicity, to the case of scalar argument (up to general case, see [?]).

Example 2.1 *A uniformly convergent series of the continuous mappings into LCS-scale, whose sum is discontinuous.* Let assign in the vector space E a decreasing sequence of the pairwise nonequivalent norms $\{\|\cdot\|_n\}_{n=0}^{\infty}$, here $(E, \|\cdot\|_0)$ is complete and in the ball $\|x\|_0 \leq 1$ for all $n \geq 1$ it is find such a sequence $\{x_k^{(n)}\}_{k=1}^{\infty}$, that $\|x_k^{(n)}\|_n \to 0$ for $k \to \infty$, but $\{\|x_k^{(n)}\|_{n-1}\}_{k=1}^{\infty}$ doesn't converge to zero.

In other words, for any $n \geq 1$ there exists such $\delta > 0$ that

$$\inf_{\|x\|_0 \leq 1, \|x\|_{n-1} \geq \delta_n} \frac{\|x\|_n}{\|x\|_{n-1}} = 0$$

In this case it is possible to assume that $\|x_k^{(n)}\|_{n-1} = 1$, $\|x_k^{(n)}\|_0 \leq 1/\delta_n$. Such a scale can be easily obtained by the projection of a nontrivial countable scale of Banach spaces onto the initial space of the scale; see also [?].

Let us choose in $[0, 1]$ an increasing sequence $t_k \nearrow 1$, $t_1 = 0$, and let us define mappings $f_n : [0, 1] \to E$ $(n = 1, 2, 3, \ldots)$ so that $f_n(t_k) = x_k^{(n)}$, f_n are linear on the segments $[t_k, t_{k+1}]$ $(k = 1, 2, \ldots)$, $f_n(1) = 0$. Then each f_n is continuous on $[0, 1]$ relative to $\|\cdot\|_n$, $(n = 1, 2, \ldots)$, but it is discontinuous for $t = 1$ relative to the previous norms.

It is easy to prove by induction the existence of such sequence of the positive numbers $\{\varepsilon_n\}_{n=0}^{\infty}$ that $\varepsilon_n > \sum_{m=n+1}^{\infty} \varepsilon_m/\delta_m$, $n = 0, 1, \ldots$. Let us consider the series $\sum_{n=1}^{\infty} \varepsilon_n f_n(t)$. Since $\sum_{n=1}^{\infty} \|\varepsilon_n f_n(t)\|_0 \le \sum_{n=1}^{\infty} \varepsilon_n/\delta_n < \varepsilon_0$, then the series is uniformly convergent even in $(E, \|\cdot\|_0)$. Its partial sums $F_n(t)$, by analogy with $f_n(t)$, are continuous on $[0, 1]$ relative to $\|\cdot\|_n$, but they are discontinuous for $t = 1$ relative to the previous norms. Let $R_n(t) = \sum_{m=n}^{\infty} \varepsilon_m f_m(t)$. Then:

$$\|R_n(t_k)\|_{n-1} = \left\| \sum_{m=n}^{\infty} \varepsilon_m f_m(t_k) \right\|_{n-1} = \left\| \sum_{m=n}^{\infty} \varepsilon_m x_k^{(n)} \right\|_{n-1}$$
$$\ge \varepsilon_n \|x_k^{(n)}\|_{n-1} - \sum_{m=n+1}^{\infty} \varepsilon_m \|x_k^{(m)}\|_{n-1} \ge \varepsilon_n - \sum_{m=n+1}^{\infty} \frac{\varepsilon_m}{\delta_m} > 0,$$

whence $R_n(t)$ is discontinuous at $t = 1$ concerning $\|\cdot\|_{n-1}$. Since $F_{n-1}(t)$ is continuous on $[0, 1]$ relative to $\|\cdot\|_{n-1}$, then the series sum $F(t) = F_{n-1}(t) + R_n(t)$ is also discontinuous at $t = 1$ relative to $\|\cdot\|_{n-1}$, $n = 1, 2, \ldots$.

So, the constructed above series uniformly converges, its members are continuous mappings from $[0, 1]$ into the scale $\{(E, \|\cdot\|_n)\}_{n=0}^{\infty}$, but the sum of the series is discontinuous at $t = 1$.

Definition 2.2 A family $\mathcal{F}$ of the mappings from $[a, b]$ into LCS-scale $\{E_i\}_{i\in I}$ is said to be homogeneously equicontinuous (*EH-continuous*) at $t_0 \in [a, b]$ if there exists such $i_0 \in I$ that for every $f \in \mathcal{F}$: $\Delta f(t_0, \Delta t) \to 0$ in E_{i_0} as $\Delta t \to 0$.

Evidently, EH-continuity of the family $\mathcal{F}$ at the point t_0 implies the continuity of each element $f \in \mathcal{F}$ at this point. However the inverse, in general, is false (see Example 2.1 above).

Theorem 2.3 *Let $E = \cup_{i\in I} E_i$ be LCS-scale, $f_n : [a, b] \to E$ $(n = 1, 2, \ldots)$. If f_n uniformly converges to f on $[a, b]$ and $\{f_n\}_{n=1}^{\infty}$ is EH-continuous at $t_0 \in [a, b]$ then f is continuous at t_0.*

As $\{f_n\}$ is uniformly fundamental, i.e. $f_n - f_m$ uniformly converges to 0 in some E_{i_1}, then $f_n - f_N : [a, b] \to E_{i_1}$ for some N and every $n \ge N$. By EH-continuity, there exists such $i_2 \in I$ that f_n (hence, $f_n - f_N$) are continuous at t_0 as mappings into E_{i_2} (locally). Let i_3 be any upper bound in I of the pair (i_1, i_2), then one can apply to the continuous mappings $(f_n - f_N)$ into E_{i_3} (locally) the well known result from the general topology, whence $(f - f_N)$ is continuous at t_0. Hence, f is continuous at t_0.

Corollary 2.4 *If, under the notation of Theorem, a series $\sum_{n=1}^{\infty} f_n$ uniformly converges on $[a, b]$ and the sequence $\{f_n\}$ is EH-continuous at t_0, then the sum of the series is continuous at t_0.*

Note two important classes of LCS-scales, for which one can except from the conditions of Theorem the requirement of EH-continuity.

Definition 2.5 LCS-scale $E = \cup_{i \in I} E_i$ is said to be *σ-inductive* if every countable subset of I possesses an upper bound in I.

Theorem 2.6 *Let $E = \cup_{i \in I} E_i$ be σ-inductive LCS-scale, $f_n : [a, b] \to E$ $(n = 1, 2. \dots)$. If f_n uniformly converges to f on $[a, b]$ and all f_n are continuous at t_0, then f is also continuous at t_0.*

If $\Delta f_n(t_0, \Delta t) \to 0$ in E_{i_n} as $\Delta t \to 0$ $(n = 1, 2, \dots)$ and i_0 is an upper bound of the set $\{i_1, i_2, \dots, i_n, \dots\}$, then $\Delta f_n(t_0, \Delta t) \to 0$ in E_{i_0} for all $n = 1, 2, \dots$, i.e. $\{f_n\}$ is EH-continuous at t_0.

Definition 2.7 LCS-scale is said to be *strong* if all the embeddings in the scale are strong.

Note that the passage to the inductive limit for the countable strong scales of Frèchet spaces (i.e. LF-spaces) conserves convergence in the scale; so, the situation is reduced to the topological one [?]. In the general case one can also except EH-continuity if all f_n are globally continuous.

Theorem 2.8 *Let $E = \cup_{i \in I} E_i$ be strong LCS-scale, $f_n : [a, b] \to E$ $(n = 1, 2, \dots)$. If f_n uniformly converges to f on $[a, b]$ and f_n $(n = 1, 2, \dots)$ are continuous on $[a, b]$, then f is also continuous on $[a, b]$.*

Fix n and choose for every $t \in [a, b]$ a neighborhood V_t such that $f_n(V_t)$ is contained in some $E_{i(n,t)}$. Choosing from the covering $\{V_t\}_{t \in [a,b]}$ a finite one $\cup_{j=1}^{k} V_{t_j} = [a, b]$ and denoting by $i(n)$ some upper bound in I of the set $\{i(n, t_j)\}_{j=1}^{k}$, we get $f_n([a, b]) \subset E_{i(n)}$.

Next, just as in the proof of Theorem ??, find such a number N and $i^1 \in I$ that $(f_n - f_N)([a, b]) \subset E_{i^1}$ for $n \geq N$. If i_0 is an upper bound of the pair $(i(N), i^1)$ then $f_n([a, b]) \subset (f_n - f_N)([a, b]) + f_N([a, b]) \subset E_{i_0}$ for $n \geq N$.

Here the strong embeddings in the scale allows to consider f_n as continuous mappings into E_{i_0} and the situation is also reduced to the topological one. #

Example 2.9 of σ-inductive LCS-scale. Let us show that the scale $M(S, \bmod \mu)$ (see ex. 1.6 above) is σ-inductive.

Let $\{M_{t_n}(S, \bmod \mu)\}_{n=1}^{\infty}$ be any sequence from the scale, $\{K_{nm}\}_{m=1}^{\infty} \in t_n$, $\varepsilon > 0$. Choose numbers $m_n^k(\varepsilon)$ that $\mu(S \backslash K_{n,m_n^k}) < \varepsilon / 2^{n+k}$ $(k = 1, 2, \dots)$,

$\{m_n^k\}_{k=1}^\infty \nearrow$. Then $\mu(S, \cap_{n=1}^\infty K_{n,m_n^k}) < \varepsilon/2^{k-1}$. Set $K_k^0 = \cap_{n=1}^\infty K_{n,m_n^k}$, then $\{K_k^0\}_{k=1}^\infty$ is increasing sequence of compact sets and $\mu(S\setminus \cup_{k=1}^\infty K_k^0) = 0$. Let $\{K_k^0\}_{k=1}^\infty \in t_0 \in T$, $f \in M_{t_n}(S, \bmod \mu)$. As, by construction, $K_k^0 \subset K_{n,m_n^k}$ ($k = 1, 2, \ldots$) then all restrictions $f|_{K_k^0}$ are continuous, whence $f \in M_{t^0}(S, \bmod \mu)$. Hence $M_{t_n}(S, \bmod \mu) \hookrightarrow M_{t^0}(S, \bmod \mu)$ ($n = 1, 2, \ldots$). #

Next, any inductive scale of the subspaces of an arbitrary LCS is a trivial example of the strong LCS-scale. Let in mention here one more essential example.

Definition 2.10 LCS E with (inductively ordered) defining system of prenorms $\{\|\cdot\|_i\}_{i\in I}$ is said to be *smooth* if

$$(i_1 \preceq i_2) \Longrightarrow \left(\liminf_{\|h\|_{i_1}\to 0} (\|x+h\|_{i_2} - \|x+h\|_{i_1}) = 0 \quad \text{for every} \quad x \in E \right).$$

A majority of current LCS in functional analysis are smooth.

Example 2.11 *of a strong LCS-scale.* Let us show that in the case of E is smooth LCS the normal decomposition $E^* = \cup_{i\in I} E_i^*$ (see ex.1.5 above) is strong LCS-scale.

For $i_1 \preceq i_2$, $l \in E_{i_1}^*$, $\varepsilon > 0$, choose $x \in E$ that $|\frac{|l(x)|}{\|x\|_{i_1}} - \|l\|^{i_1}| < \varepsilon$ and, according to def. 2.10, choose $h \in E$ that

$$\left| \frac{|l(x+h)|}{\|x+h\|_{i_2}} - \frac{|l(x+h)|}{\|x+h\|_{i_1}} \right| < \varepsilon \text{ and } \left| \frac{|l(x+h)|}{\|x+h\|_{i_1}} - \frac{|l(x)|}{\|x\|_{i_1}} \right| < \varepsilon.$$

Then $|\frac{|l(x)|}{\|x\|_{i_2}} - \|l\|^{i_1}| < 3\varepsilon$, whence $\|l\|^{i_2} \geq \|l\|^{i_1}$. As the inverse inequality is also true then $\|l\|^{i_2} = \|l\|^{i_1}$. #

Note that all the embeddings in the scale above are isometric.

3 A Termwise Differentiation in the Linear Scales of LCS

At first, let in mention an appropriate form of Hahn-Banach theorem for linear LCS-scales ([?], [?]). We need here an auxiliary definition.

Definition 3.1 Let $E = \cup_{i\in I} E_i$ be an inductive scale of topological vector spaces (TVS). Call a family of subsets $\{U_i | U_i \subset E_i\}_{i\in I}$ by *scale of sets* in E if $U_{i_2} \cap E_{i_1} = U_{i_1}$ for $i_2 \succeq i_1$. Analogously, call a family of functionals $\{f_i | f_i \in E_i^*\}_{i\in I}$ by *functional scale* in E if $f_{i_2}|_{E_{i_1}} = f_{i_1}$ for $i_2 \succeq i_1$.

Theorem 3.2 (Hahn-Banach theorem for linear scales of TVS). *Let* $\{E_i\}_{i\in I}$ *be a linear scale of TVS. Then:*

(i) *If* $\{F_i\}_{i\in I}$ *is a scale of affine subspaces of* E_i, $\{A_i\}_{i\in I}$ *is a scale of convex open subsets of* E_i, $A_i \cap F_i = \emptyset$ *for* $i \in I$, *then there exists such a scale* $\{H_i\}_{i\in I}$ *of closed hyperplanes from* E_i *that* $H_i \supset F_i$ *and* $H_i \cap A_i = \emptyset$ *for* $i \in I$.

(ii) *If* $\{F_i\}_{i\in I}$ *is a scale of vector subspaces of real* E_i, $\{A_i\}_{i\in I}$ *is a scale of convex open subsets of* E_i, $A_i \cap F_i = \emptyset$ *for* $i \in I$, *then there exists such a functional scale* $\{f_i\}_{i\in I}$ *that* $f_i(F_i) = 0 < f_i(A_i)$ *for* $i \in I$.

Corollary 3.3 *Let* $\{E_i\}_{i\in I}$ *be a linear scale of real LCS,* $\{A_i\}_{i\in I}$ *be a scale of convex closed subsets of* E_i *and* $x_0 \notin A_i$ *for* $i \in I$. *Then there exists such a functional scale* $\{f_i\}_{i\in I}$ *that* $f_i(x_0) > \sup f_i(A_i)$ *for* $i \in I$.

On the base of the last result ordinarily is deduced mean-value theorem for the mappings into the linear LCS-scales.

Theorem 3.4 (Mean-value theorem for linear LCS-scales). *Let* $E = \bigcup_{i\in I} E_i$ *be a linear scale of real LCS,* $f : [a,b] \to E$. *If* f *is continuous on* $[a,b]$ *and differentiable on* $[a,b]\setminus D$, *here* D *is finite or countable, then*

$$f(b) - f(a) \in (b-a) \cdot \overline{conv} f'([a,b]\setminus D). \tag{1}$$

Let us introduce now an additional condition, ensuring a differentiability of the uniform limit in the case of linear LCS-scales.

Definition 3.5 A family $\mathcal{F}$ of differentiable at $t_0 \in [a,b]$ mappings from $[a,b]$ into a real inductive LCS-scale $E = \bigcup_{i\in I} E_i$, is called to be *homogeneously equidifferentiable (EH-differentiable)* at t_0, if there exists such $i_0 \in I$ that

$$\alpha(f, t_0, \Delta t) = \frac{\Delta f}{\Delta t}(t_0) - f'(t_0) \longrightarrow 0 \quad \text{as} \quad \Delta \text{t} \longrightarrow 0 \quad \text{in} \quad \text{E}_{\text{i}_0} \tag{2}$$

for all $f \in \mathcal{F}$.

So, EH-differentiability of $\mathcal{F}$ at t_0 is equivalent to EH-continuity (see def.??) of the family $\{\alpha(f, t_0, \cdot)\}_{f\in\mathcal{F}}$ at zero. Let us formulate the main result of the work.

Theorem 3.6 (On termwise differentiation for linear LCS-scales). *Let* $E = \bigcup_{i\in I} E_i$ *be a linear scale of the real complete LCS and* $f_n : [a,b] \to E$ $(n = 1, 2, \ldots)$. *If* $\{f_n\}_{n=1}^{\infty}$ *is EH-differentiable at each point of* $[a,b]$ *and converges at some* $t_0 \in [a,b]$, $\{f_n'\}_{n=1}^{\infty}$ *uniformly converges on* $[a,b]$, *then* $\{f_n\}_{n=1}^{\infty}$ *uniformly converges to some* f, f *is differentiable on* $[a,b]$ *and* $f'(t) = \lim_{n\to\infty} f_n'(t)$.

1. Let us prove at first the uniform convergence of $\{f_n\}_{n=1}^{\infty}$ on $[a,b]$. Applying (1) to the segment $[t_0, t]$, we have for any $m, n \in \mathbb{N}$:

$$\begin{aligned} f_m(t) - f_n(t) &= [(f_m(t) - f_n(t)) - (f_m(t_0) - f_n(t_0))] + [f_m(t_0) - f_n(t_0)] \\ &\in (t - t_0) \cdot \overline{conv}(f_m' - f_n')([t_0, t]) + [f_m(t_0) - f_n(t_0)]. \end{aligned} \tag{3}$$

As $\{f_n'\}_{n=1}^{\infty}$ is uniformly fundamental on $[a, b]$ then $\overline{conv}\,(f_m' - f_n')([a, b]) \to 0$ as $m, n \to \infty$ in some E_{i_1}. Analogously, it follows from fundamentality of $\{f_n(t_0)\}_{n=1}^{\infty}$ that $f_m(t_0) - f_n(t_0) \to 0$ as $m, n \to \infty$ in some E_{i_2}. It follows from here and (3) that

$$(f_m - f_n)([a, b]) \subset ([a, b] - t_0) \cdot \overline{conv}(f_m' - f_n')([a, b]) + (f_m(t_0) - f_n(t_0)) \longrightarrow 0$$

in some E_{i_3} as $m, n \to \infty$, here i_3 is any upper bound of the pair (i_1, i_2). As E_{i_3} is complete then $\{f_n\}_{n=1}^{\infty}$ uniformly converges to some f in E_{i_3}, whence the same is true in E.

2. Let us check now the differentiability of f on $[a, b]$. Let us fix a point $t \in [a, b]$ and put

$$\alpha_n(t, \Delta t) = (\Delta f_n(t)/\Delta t) - f_n'(t). \tag{4}$$

Let us show that the sequence $\{\alpha_n(t, \cdot)\}_{n=1}^{\infty}$ is uniformly convergent at zero neighborhood. Applying (1), we find :

$$\alpha_m(t, \Delta t) - \alpha_n(t, \Delta t) = \left[\frac{f_m(t + \Delta t) - f_m(t)}{\Delta t} - f_m'(t)\right] - \left[\frac{f_n(t + \Delta t) - f_n(t)}{\Delta t} - f_n'(t)\right] = \frac{\Delta(f_m - f_n)}{\Delta t}(t) - (f_m - f_n)'(t) \in \overline{\mathrm{conv}}(f_m - f_n)'([t, t + \Delta t]) - (f_m - f_n)'(t).$$

From here, fixing zero neighborhood $V \ni \Delta t$ it follows

$$[\alpha_m(t, \cdot) - \alpha_n(t, \cdot)](V) \subset \overline{conv}\,(f_m' - f_n')(t + V) - (f_m' - f_n')(t) \longrightarrow 0$$

as $m, n \to \infty$ in E_{i_1} in accordance with item 1 of the proof. Hence, the sequence $\{\alpha_n(t, \cdot)\}_{n=1}^{\infty}$ is uniformly fundamental on V and, as E_{i_1} is complete, $\alpha_n(t, \cdot)$ uniformly converges to some $\alpha(t, \cdot)$ on V as $n \to \infty$. As $\{\alpha_n(t, \cdot)\}_{n=1}^{\infty}$ is EH-continuous at zero, then by Theorem **??** $\alpha(t, \cdot)$ is also continuous at zero, i.e. $\alpha(t, \Delta t) \to 0$ as $\Delta t \to 0$. Passing to the limit in (4) we get

$$\frac{\Delta f}{\Delta t}(t) - \left(\lim_{n\to\infty} f_n'(t)\right) = \alpha(t, \Delta t)$$

and Theorem is proved. #

Corollary 3.7 *If, under the notation of Theorem,* $\{f_n\}_{n=1}^{\infty}$ *is EH-differentiable at each point of* $[a, b]$, *the series* $\sum_{n=1}^{\infty} f_n$ *converges at some* $t_0 \in [a, b]$ *and the series* $\sum_{n=1}^{\infty} f_n'$ *uniformly converges on* $[a, b]$, *then the series* $\sum_{n=1}^{\infty} f_n$ *also uniformly converges, its sum is differentiable on* $[a, b]$ *and*

$$\frac{d}{dt}\sum_{n=1}^{\infty} f_n(t) = \sum_{n=1}^{\infty} \frac{df_n}{dt}(t). \tag{5}$$

In similar way one can obtain theorems on continuous differentiability of the uniform limit ([?], th.1.11–1.12).

Theorem 3.8 *If, under the assumptions of Theorem ??, $\{f_n'\}_{n=1}^{\infty}$ is EH-continuous at every point of $[a, b]$, then f is continuously differentiable on $[a, b]$.*

Corollary 3.9 *If, under the assumptions of Corollary ??, $\{f_n'\}_{n=1}^{\infty}$ is EH-continuous at every point of $[a, b]$, then sum of the series $\sum_{n=1}^{\infty} f_n$ is continuously differentiable on $[a, b]$.*

At the conclusion of the item, let us remark that in the case of σ-inductive scale, using Theorem ??, one can replace in the hypotheses of the statements 3.6–3.9 the EH-differentiability and the EH-continuity to the usual ones. In the case of the strong scales the same is true for the statements 3.8–3.9 by virtue of Theorem ??. Let us formulate these results for the series only.

Theorem 3.10 *Let $E = \bigcup_{i \in I} E_i$ be a linear σ-inductive LCS-scale and $f_n : [a, b] \to E$ $(n = 1, 2, \ldots)$. If all the f_n are differentiable on $[a, b]$, the series $\sum_{n=1}^{\infty} f_n$ converges at some $t_0 \in [a, b]$ and the series $\sum_{n=1}^{\infty} f_n'$ uniformly converges on $[a, b]$, then $\sum_{n=1}^{\infty} f_n$ uniformly converges and the equality* (5) *holds.*

If E is a linear σ-inductive or strong scale and all the f_n are continuously differentiable on $[a, b]$, then the sum of the series $\sum_{n=1}^{\infty} f_n$ is also continuously differentiable on $[a, b]$.

4 A Termwise Differentiation in Nonlinear Scales of LCS

To the nonlinear scales, as it was shown in [?], [?], Hahn-Banach theorem is not extended, that doesn't allow to transfer Theorem ?? to the differentiable mappings of a segment into such scales. Thus we restrict oneself to more narrow class of mappings, for which author have proved in ([?], Theorem 1, Corollary 4) the mean-value theorem.

Definition 4.1 Let $E = \bigcup_{i \in I} E_i$ be a LCS-scale and $f : [a, b] \to E$ be differentiable on $[a, b]$. Call f by *homogeneously differentiable (H-differentiable)* on $[a, b]$ if there exists such $i_0 \in I$ that

$$\alpha(f, t, \Delta t) \longrightarrow 0 \quad \text{in} \quad E_{i_0} \quad \text{as} \quad \Delta t \longrightarrow 0$$

for every $t \in [a, b]$.

Note that the H-differentiability of f on $[a, b]$ is equivalent to the EH-continuity at zero of the family $\{\alpha(f, t, \cdot)\}_{t \in [a,b]}$. In [?], it.2, there were considered general examples of H-differentiable and locally H-differentiable mappings. Let us formulate mean-value theorem for the H-differentiable mappings [?].

Theorem 4.2 (Mean-value theorem for nonlinear LCS-scales). *Let* $E = \cup_{i\in I} E_i$ *be a (linear or not) real LCS-scale. If* $f : [a, b] \to E$ *is continuous on* $[a, b]$ *and locally H-differentiable on* (a, b)*, then*

$$f(b) - f(a) \in (b-a) \cdot \overline{conv}\, f'((a, b)).$$

Definition 4.3 A family $\mathcal{F} = \{f : [a, b] \to E\}$ is said to be *homogeneously equidifferentiable on* $[a, b]$ *(EH-differentiable on* $[a, b]$*)* if there exists such $i_0 \in I$ that $\alpha(f, t, \Delta t) \to 0$ in E_{i_0} as $\Delta t \to 0$ for all $f \in \mathcal{F}$ and all $t \in [a, b]$.

Let us formulate the main result of the item.

Theorem 4.4 (On termwise differentiation for nonlinear scales of LCS). *Let* $E = \cup_{i\in I} E_i$ *be an inductive scale of the real complete LCS and* $f_n : [a, b] \to E$ $(n = 1, 2, \ldots)$*. If* $\{f_n\}_{n=1}^{\infty}$ *is EH-differentiable on* $[a, b]$ *and converges at some* $t_0 \in [a, b]$*,* $\{f'_n\}_{n=1}^{\infty}$ *uniformly converges on* $[a, b]$*, then* $\{f_n\}_{n=1}^{\infty}$ *uniformly converges to some* f*,* f *is H-differentiable on* $[a, b]$ *and* $f'(t) = \lim_{n\to\infty} f'_n(t)$.

The proof of the uniform convergence f_n to f and the differentiability of f repeats the proof of Theorem **??** with replace mean-value theorem for the linear scales 3.5 on its H-analog for the nonlinear scales 4.2.

Next, $\alpha_n(t, \cdot)$ uniformly converges to $\alpha(t, \cdot)$ and $\{\alpha_n(t, \cdot)\}_{n=1}^{\infty}$ is EH-continuous at zero locally on $[a, b]$. Hence, from the identity $\alpha(t, \cdot) = [\alpha(t, \cdot) - \alpha_n(t, \cdot)] + \alpha_n(t, \cdot)$ it follows, by analogy with the proof of Theorem **??**, that $\alpha(t, \cdot)$ is continuous mapping into some $E_{i(t_0)}$ in some neighborhood of every point $t_0 \in [a, b]$. Then $\alpha(t, \cdot)$ is continuous mapping of the whole $[a, b]$ into some space from the scale by virtue of the compactness of $[a, b]$, i.e. f is H-differentiable. #

Corollary 4.5 *If, under the notation of Theorem,* $\{f_n\}_{n=1}^{\infty}$ *is EH-differentiable on* $[a, b]$*, the series* $\sum_{n=1}^{\infty} f_n$ *converges at some* $t_0 \in [a, b]$ *and the series* $\sum_{n=1}^{\infty} f'_n$ *uniformly converges on* $[a, b]$*, then the series* $\sum_{n=1}^{\infty} f_n$ *uniformly converges, its sum is H-differentiable on* $[a, b]$ *and* (5) *holds.*

To obtain the conditions of continuous differentiability in nonlinear scales we need a global analog of EH-continuity.

Definition 4.6 A family $\mathcal{F} = \{f : [a, b] \to E\}$ is said to be *EH-continuous on* $[a, b]$ if there exists such $i_0 \in I$ that $\Delta f(t, \Delta t) \to 0$ in some E_{i_0} as $\Delta t \to 0$ for all $t \in [a, b]$ and all $f \in \mathcal{F}$.

Theorem 4.7 *If, under the hypotheses of Theorem* **??***,* $\{f'_n\}_{n=1}^{\infty}$ *is EH-continuous on* $[a, b]$*, then* f *is continuously differentiable on* $[a, b]$*.*

Corollary 4.8 *If, under the hypotheses of Corollary* **??***,* $\{f'_n\}_{n=1}^{\infty}$ *is EH-continuous on* $[a, b]$*, then sum of the series* $\sum_{n=1}^{\infty} f_n$ *is continuously differentiable.*

Note that the hypotheses of the statements 4.7–4.8 guarantee one more strict property: *H-continuity*, i.e. there exists such $i_0 \in I$ that $\Delta f(t, \Delta t) \to 0$ in E_{i_0} as $\Delta t \to 0$ for every $t \in [a, b]$.

In the case of σ-inductive or strong scales, similarly to the previous item, the conditions of termwise differentiation are also simplified. Formulate them only for the series.

Theorem 4.9 *Let $E = \cup_{i \in I} E_i$ be a real σ-inductive LCS-scale and $f_n : [a, b] \to E$ $(n = 1, 2, \dots)$. If all the f_n are H-differentiable on $[a, b]$, the series $\sum_{n=1}^{\infty} f_n$ converges at some $t_0 \in [a, b]$ and the series $\sum_{n=1}^{\infty} f_n'$ uniformly converges on $[a, b]$, then the series $\sum_{n=1}^{\infty} f_n$ uniformly converges on $[a, b]$, its sum is H-differentiable and equality* (5) *holds.*

If E is σ-inductive or strong scale and all the f_n are continuously differentiable on $[a, b]$, then sum of the series $\sum_{n=1}^{\infty} f_n$ is also continuously differentiable on $[a, b]$.

Finally, if E is strong scale and $f : [a, b] \to E$ is differentiable on $[a, b]$, then f is H-differentiable on $[a, b]$.

5 Some Examples

5.1 A Termwise Differentiation a.e. of the Measurable Functions in Parameter

It is shown in it.2 (2.9) that almost everywhere convergence space $M(S, \mathrm{mod}\, \mu)$ is σ-inductive scale. Applying to this case Theorem **??**, we get the following result (here only the conditions of continuous differentiability are written).

Theorem A. *Let (S, μ) be a locally compact space with finite measure μ and functions $f_n : S \times [a, b] \to \mathbb{R}$ satisfy conditions:*

(i) *every $f_n(x, \lambda)$ is differentiable in λ for μ-a.e. $x \in S$ and continuously in $\lambda \in [a, b]$ (i.e. $\frac{\partial f_n}{\partial \lambda}(x, \bar{\lambda}) \to \frac{\partial f_n}{\partial \lambda}(x, \lambda)$ μ-a.e. as $\bar{\lambda} \to \lambda$);*

(ii) *the series $\sum_{n=1}^{\infty} f_n(x, \lambda_0)$ converges for μ-a.e. $x \in S$ for some $\lambda_0 \in [a, b]$;*

(iii) *the series $\sum_{n=1}^{\infty} \frac{\partial f_n}{\partial \lambda}(x, \lambda)$ converges for μ-a.e. $x \in S$ uniformly for $\lambda \in [a, b]$ (i.e. all restrictions $\frac{\partial f_n}{\partial \lambda}|_{K_m}$ are continuous in x and all series $\sum_{n=1}^{\infty} \frac{\partial f_n}{\partial \lambda}|_{K_m}$ converge uniformly on $K_m \times [a, b]$ for some increasing sequence of compact sets $\{K_m\}_{m=1}^{\infty}$ such that $\mu(S \setminus \cup_{m=1}^{\infty} K_m) = 0$).*

Then the series $\sum_{n=1}^{\infty} f_n(x, \lambda)$ converges for μ-a.e. $x \in S$ uniformly for $\lambda \in [a, b]$, its sum is differentiable in λ for μ-a.e. $x \in S$ continuously in λ and

$$\frac{\partial}{\partial \lambda} \sum_{n=1}^{\infty} f_n(x, \lambda) = \sum_{n=1}^{\infty} \frac{\partial f_n}{\partial \lambda}(x, \lambda) \quad \text{for } \mu - a.e.\ x \in S.$$

5.2 A Termwise Differentiation of the Linear Functionals in Parameter

It is shown in it.2 (2.11) that the normal decomposition of dual space $E^* = \bigcup_{i\in I} E_i^*$ is, in the case of a smooth LCS E, a strong scale. Applying again Theorem **??**, we receive

Theorem B. *Let E be a smooth real LCS, $E^* = \bigcup_{i\in I} E_i^*$ be the normal decomposition of E^* and the mappings $l_n : E \times [a,b] \to \mathbb{R}$ be linear and continuous in the first variable. Suppose that:*

(i) *every $l_n(x,\lambda)$ is differentiable in λ uniformly in some ball $\|x\|_{i_1(n)} \le 1$ and continuously in $\lambda \in [a,b]$ (i.e. $\frac{\partial l_n}{\partial\lambda}(x,\bar\lambda) - \frac{\partial l_n}{\partial\lambda}(x,\lambda) \to 0$ as $\bar\lambda \to \lambda$ uniformly in some ball $\|x\|_{i_2(n,\lambda)\le 1}$);*

(ii) *the series $\sum_{n=1}^{\infty} l_n(x,\lambda_0)$ converges uniformly in some ball $\|x\|_{i_3(\lambda_0)} \le 1$ for some $\lambda_0 \in [a,b]$;*

(iii) *the series $\sum_{n=1}^{\infty} \frac{\partial l_n}{\partial\lambda}(x,\lambda)$ converges uniformly in $\|x\|_{i_4} \le 1$ and $\lambda \in [a,b]$ for some $i_4 \in I$.*

Then the series $\sum_{n=1}^{\infty} l_n(x,\lambda)$ converges uniformly in $\|x\|_{i_5} \le 1$ and $\lambda \in [a,b]$ for some $i_5 \in I$, its sum is differentiable in λ uniformly in some $\|x\|_{i_6} \le 1$ and

$$\frac{\partial}{\partial\lambda}\sum_{n=1}^{\infty} l_n(x,\lambda) = \sum_{n=1}^{\infty}\frac{\partial l_n}{\partial\lambda}(x,\lambda)$$

uniformly in some $\|x\|_{i_7} \le 1$ for all $\lambda \in [a,b]$.

Note in conclusion that the size of this paper doesn't allow us to consider the questions above for the mappings from one LCS-scale to another. Generalization of the results of ii.3–4 to this case see in [**?**].

References

[1] J.L. Lions and E. Magenes, *Problemes aux limites non homogenes et applications.* v. 1, Dunod, Paris, 1968.

[2] S.G. Krein, Ju.I. Petunin and E.M. Semenov, *Interpolation of linear operators*, Nauka, Moscow, 1978. (Russian)

[3] L.V. Ovsjannikov, *Analytic Groups*, Novosib. Univ. Publ., Novosibirsk, 1972. (Russian)

[4] A.M. Gomilko, *Continuous mappings in scales of Banach spaces*, Ukrainian Math. J. **41** (1989), N8, 1130–1134. (Russian)

[5] I.V. Orlov, *Hahn-Banach theorem for the inductive scales of spaces*, Reports of National Akademy of Science of Ukraine (Russian). (to appear)

[6] I.V. Orlov, *Hahn-Banach Theorem in linear and nonlinear scales of the topological vector spaces.* // Spectral and Evolutionary problems: Proceedings of the Seventh Crimean Autumn Mathematical School-Symposium (1997), no. 7, 27–31.

[7] I.V. Orlov, *Differentiability and homogeneous differentiability in pseudotopological vector spaces.* // Spectral and Evolutionary problems: Proceedings of the Sixth Crimean Autumn Mathematical School-Symposium (1996), no. 6, 292–303.

[8] I.V. Orlov, *The space of measurable functions with almost everywhere convergence is a nonlinear scale of the locally convex spaces.* // Spectral and Evolutionary problems: Proceedings of the Seventh Crimean Autumn Mathematical School-Symposium (1998), no. 8, 45–51.

[9] A. Frölicher and V. Bucher, *Differential calculus in vector spaces without norm*, Springer, Berlin-Heidelberg-New York, 1966.

[10] H.H. Keller, *Differential calculus in locally convex spaces*, Lect. Notes in Math., 1974.

[11] I.V. Orlov, *A termwise differentiation in linear and nonlinear scales of locally convex spaces. II. General case.* // Spectral and Evolutionary problems: Proceedings of the Sixth Crimean Autumn Mathematical School-Symposium (1996), no. 6, 303–314.

[12] I.V. Orlov, *An uniform limit continuity in a pseudotopological vector space.* // Spectral and Evolutionary problems: Proceedings of the Fifth Crimean Autumn Mathematical School-Symposium (1996), no. 5, 58–65.

[13] H. Schaefer, *Topological Vector Spaces*, Academic Press, New York, 1964.

[14] I.V. Orlov, *A termwise differentiation in linear and nonlinear scales of locally convex spaces. I. The case of scalar argument.* // Spectral and Evolutionary problems: Proceedings of the Fifth Crimean Autumn Mathematical School-Symposium (1996), no. 5, 116–124.

[15] I.V. Orlov, *Mean value theorem for the homogeneously differentiable mappings.* // Spectral and evolutionary problems : Proceedings of the Fourth Crimean Autumn Math. School-Symposium (1995), no. 4, 185–190.

V. Orlov Igor
Math. department of Simferopol State University
Yaltinskaja str. 4,
Simferopol, 333007
Ukraine
Phones: (0652) 230280 (office), (0652) 255306 (home)
FAX: (0652) 232310
knd@kopa.crimea.com (to Orlov I.V.)

1991 Mathematics Subject Classification:

Keywords and phrases: locally convex space, scale of spaces, Hahn-Banach theorem, termwise differentiation, mean-value theorem.

Operator Theory:
Advances and Applications, Vol. 118

Operator Relations, Dynamical Systems, and Representations of a Class of Wick Algebras

Vasyl Ostrovskyi and Daniil Proskurin*

We consider families of operators satisfying a general class of relations, whose solutions can be described in terms of orbits of some dynamical system acting on the spectrum of a commuting sub-family. In the first section we introduce a class of relations and show, how the representations of such relations are related to orbits of the corresponding dynamical system. Also, we discuss the problem of accurate sense of the relation for unbounded operators. In Section 2, we study the class of $*$-algebras allowing Wick ordering whose representations can be studied by using methods of Section 1. We classify such Wick $*$-algebras, and discuss their representations.

Introduction

Families of operators subjected to different kinds of relations appear in many problems of mathematics and its application, mostly in representation theory, and related problems of theoretical physics. In general, such problem may be very complicated (wild); for other classes of relations it may happen that complete description of all solutions is possible up to a unitary equivalence. If the relations are polynomial, the description of solutions of the relations is equivalent (at least, for bounded operators) to the description of representations of the corresponding $*$-algebra, generated by a family of generators satisfying these algebraic relations. In the case of non-polynomial relations, the corresponding algebraic objects are more complicated, and we will not discuss them here.

In this paper, we consider families of operators satisfying a general class of relations, whose solutions can be described in terms of orbits of some dynamical system acting on the spectrum of a commuting sub-family. The method for solving operator problems by using dynamical systems ascends to the works of J. von Neumann, G.W. Mackey (see, e.g., [4], and the bibliography therein). We apply ideas of this method to non-group relations, and non-unitary operators.

The simplest situation, related to one-dimensional dynamical system, arise from the study of relations of the form $XX^* = F(X^*X)$, where $F(\cdot)$ is a real measurable mapping; representations of such relations were studied in [3]; in [1], they were treated as a generalization of the canonical commutation relations.

The paper is organized as follows. In first section we introduce multi-dimensional version of the relations mentioned in previous paragraph. We show, how the representations of such relations are related to orbits of the corresponding dynamical

*Supported by CRDF grant no. UM1-311.

system. This gives the description of all irreducible representations of the relation. Also, we discuss the problem of accurate sense of the relation for unbounded operators.

In Section 2, we study the class of $*$-algebras allowing Wick ordering (see [5]) whose representations can be studied by using methods of Section 1. We classify such Wick $*$-algebras, and discuss their representations.

1 Operator Relations and Multi-dimensional Dynamical Systems

We consider representations of a family of operator relations between operators a_j, $j = 1, \ldots, n$, of the following form

$$
\begin{aligned}
a_j^* a_j &= F_j(a_1 a_1^*, \ldots, a_n a_n^*), \\
a_j^* a_k &= \mu_{jk}\, a_k a_j^*, \\
a_j a_k &= \lambda_{jk}\, a_k a_j,
\end{aligned} \tag{1}
$$

where $F_1(\cdot), \ldots, F_n(\cdot)\colon \mathbf{R}^n \to \mathbf{R}$ are measurable mappings, λ_{jk}, $\mu_{jk} > 0$, $1 \le j, k \le n$. Notice that the last two relations in (1) imply that the operators $a_1 a_1^*, \ldots, a_n a_n^*$ commute; this makes clear the sense of taking functions of them in the first relation.

Now we assume that $a_1, \ldots, a_n$ are, in general, unbounded, closed densely defined operators which satisfy relations (1). We need to take some care when studying relations between unbounded operators; our approach is to rewrite them in formally equivalent form, which involves only bounded operators (this is similar to the case of unbounded representations of a real Lie algebra, which can be described in terms of unitary representations of the corresponding Lie group). Write polar decompositions, $a_j = C_j U_j$, $j = 1, \ldots, n$, where C_j are non-negative, U_j are partial isometries, and each C_j is zero on vectors orthogonal to the range of U_j.

Lemma 1 *Let the operators a_j, $j = 1, \ldots, n$ be bounded. Then relations* (1) *are equivalent to the following*

$$
\begin{aligned}
C_j^2 U_k &= q_{jk} U_k C_j^2, \qquad j \neq k, \\
C_j^2 U_j &= U_j F_j(C_1^2, \ldots, C_n^2), \qquad j = 1, \ldots n, \\
U_j U_k &= U_k U_j, \quad U_j U_k^* = U_k^* U_j, \qquad j < k,
\end{aligned} \tag{2}
$$

where

$$
q_{jk} = \begin{cases} \mu_{jk}\lambda_{jk}, & j < k, \\ \mu_{jk}\lambda_{jk}^{-1}, & j > k. \end{cases}
$$

Moreover, operators $(U_j^*)^k U_j^k$, $U_l^m (U_l^*)^m$, $j, l = 1, \ldots, n$; $k, m = 1, 2, \ldots,$ form a commuting family (in particular, all U_j are centered partial isometries).

Proof: The proof is a straightforward calculation based on the equality $U_l U_l^* =$ sign C_l. □

According to [2] relations (2) can be rewritten in the following form, involving only bounded operators. Introduce mappings of $\mathbf{R}^n$ into itself

$$\begin{aligned} \mathbf{F}_l(x_1, \dots, x_n) &= (q_{1l}x_1, \dots, q_{l-1\,l}x_{l-1}, F_l(x_1, \dots, x_n), \\ &\qquad q_{l+1\,l}x_{l+1}, \dots, q_{nl}x_n), \quad l = 1, \dots, n. \end{aligned} \tag{3}$$

Then the relations are equivalent to the following

$$E(\Delta)U_l = U_l E(\mathbf{F}_l^{-1}(\Delta)),$$

where $E(\cdot)$ is a joint resolution of identity of the commuting family $C_1^2, \dots, C_n^2$, Δ ranges over all measurable subsets of $\mathbf{R}^n$, $l = 1, \dots, n$. The latter relations include only bounded operators, and will be used as a precise version of the relations in the unbounded case.

According to [2], it makes sense to consider such relations, for which $\mathbf{F}_j(\mathbf{F}_k(\cdot)) = \mathbf{F}_k(\mathbf{F}_j(\cdot))$, $j \neq k$, which is equivalent to the following equalities

$$F_j(\mathbf{F}_k(x_1, \dots, x_n)) = q_{jk}F_j(x_1, \dots, x_n).$$

In what follows, we are mostly interested in the case of the second order relations, i.e., linear functions $F_j(\cdot)$, $j = 1, \dots, n$. If

$$F_j(x_1, \dots, x_n) = \sum_{l=1}^{n} \phi_{jl}x_l + \alpha_j I, \quad j = 1, \dots, n,$$

the conditions are

$$\begin{aligned} &\phi_{jl}(q_{lk} - q_{jk}) + \phi_{jk}\phi_{kl} = 0, \qquad l \neq j,\ l \neq k, \\ &\phi_{jk}\phi_{kj} = 0, \quad \phi_{jk}(\phi_{kk} - q_{jk}) = 0, \quad \alpha_j(1 - q_{jk}) + \alpha_k\phi_{jk} = 0, \end{aligned} \tag{4}$$

for all $j, k = 1, \dots, n$, $j \neq k$.

In what follows, we will assume that the n-dimensional dynamical system generated by the mappings $\mathbf{F}_1(\cdot), \dots, \mathbf{F}_n(\cdot)$, possesses a measurable section, a measurable set which meets each orbit at a single point. In this case, for any irreducible representation of the relations, the spectral measure of the commuting family $C_1^2, \dots, C_n^2$ is concentrated on (a subset of) a single orbit, and we can classify all irreducible representations up to unitary equivalence. In the case of more complicated dynamical systems without a measurable section, non-trivial ergodic measures may occur, which gives rise to much more complicated structure of representations, including factor representations not of type I, etc. Notice that the linear dynamical system of the form (3) always possesses a measurable section.

We proceed with a more detailed study of irreducible collections a_j, $j = 1, \dots, n$, satisfying (1), which correspond to an orbit Ω. Denote by Δ the support of the spectral measure of the commuting family C_j^2, $j = 1, \dots, n$. It is a general fact that in the basis of eigenvectors of the commuting family the operators a_j act as weighted shift operators [2], but we need to take into account that $C_j \geq 0$, and that $U_j U_j^*$ is a projection on the co-kernel $(\ker C_j^2)^{\perp}$, $j = 1, \dots, n$.

Lemma 2 *For any* $\mathbf{x} = (x_1, \dots, x_n) \in \Delta$ *we have*

i) $x_j \geq 0$, $j = 1, \dots, n$;

ii) *either* $\mathbf{F}_j(\mathbf{x}) \in \Delta$, *or* $(\mathbf{F}_j(\mathbf{x}))_j = 0$;

iii) *similarly, either* $\mathbf{F}_j^{-1}(\mathbf{x}) \in \Delta$, *or* $\mathbf{x}_j = 0$.

Proof: i) Indeed, since $C_j^2 \geq 0$, we have $x_j \geq 0$, $j = 1, \dots, n$.

ii) If $\mathbf{F}_j(\mathbf{x}) \notin \Delta$, then $U_j e_{\mathbf{x}} = 0$, where $e_{\mathbf{x}}$ is the basis eigenvector of the commuting family corresponding to the joint eigenvalue $\mathbf{x}$. Then we also have $U_j U_j^* e_{\mathbf{F}_j(\mathbf{x})} = 0$, and

$$C_j^2 e_{\mathbf{F}_j(\mathbf{x})} = (\mathbf{F}_j(\mathbf{x}))_j \, e_{\mathbf{F}_j(\mathbf{x})} = 0,$$

which implies $(\mathbf{F}_j(\mathbf{x}))_j = 0$.

iii) Similarly, if $\mathbf{F}_j^{-1}(\mathbf{x}) \notin \Delta$, then $U_j^* e_{\mathbf{x}} = 0$. Then $U_j U_j^* e_{\mathbf{x}} = 0$, and $C_j^2 e_{\mathbf{x}} = x_j e_{\mathbf{x}} = 0$, which gives $x_j = 0$. □

Corollary 1 *If for some* $\mathbf{x} = (x_1, \dots, x_n) \in \Omega$ $x_j > 0$, *and* $(\mathbf{F}_j(x))_j < 0$, *then* $\mathbf{x} \notin \Delta$. *This condition implies that irreducible representations correspond only to orbits, for which* $x_j > 0$ *implies* $(\mathbf{F}_j(\mathbf{x}))_j \geq 0$, $(\mathbf{F}_j^{-1}(\mathbf{x}))_j \geq 0$. *Notice also that* (3) *implies that from* $x_j > 0$ *follows* $(\mathbf{F}_k(\mathbf{x}))_j > 0$ *for* $k \neq j$.

Consider possible types of orbits and describe corresponding irreducible representations of (1).

Theorem 1 *Any irreducible representation can be realized in the space* $l_2(\Delta)$. *For any* $l = 1, \dots, n$ *there can be one of the following:*

a). *Mapping* $\mathbf{F}_l(\cdot)$ *possesses a stationary point* $\mathbf{x} \in \Delta$ *(in this case all other points are also stationary). If* $x_l = 0$, *then* $a_l = 0$; *otherwise, the operator* a_l *has the form*

$$a_l e_{\mathbf{x}} = \beta_l \, x_l \, e_{\mathbf{x}},$$

where β_l *is a parameter equal to one by absolute value;*

b). *Mapping* $F_l(\cdot)$ *does not have stationary points. In this case the operator* a_l *has the form*

$$a_l e_{\mathbf{x}} = F_l(\mathbf{x}) \, e_{\mathbf{F}_l(\mathbf{x})}. \tag{5}$$

The kernel of the operator a_l *is generated by vectors* $e_{\mathbf{x}}$ *such that* $F_l(\mathbf{x}) = 0$; *the kernel of* b_l^* *is generated by vectors* $e_{\mathbf{x}}$ *for which* $x_l = 0$.

Proof: The proof is essentially based on the following statements from [2].

Theorem 2 *Let the dynamical system on $\mathbf{R}^n$ generated by the mappings $\mathbf{F}_l$, $l = 1, \ldots, n$, possesses a measurable section. Then for each irreducible collection of operators C_j, U_j, $j = 1, \ldots, n$ satisfying* (2), *the following holds.*

i. *There exists a unique orbit Ω of the dynamical system of full spectral measure of the commuting collection C_j, $j = 1, \ldots, n$, $E(\Omega) = 1$;*

ii. *If $\ker U_l = \{0\}$, then the spectral measure is quasi-invariant with respect to the mapping $\mathbf{F}_l(\cdot)$; in the case of unitary U_l, the measure is quasi-invariant with respect to $\mathbf{F}_l^{-1}(\cdot)$, too;*

iii. *The joint spectrum of the commuting family C_j, $j = 1, \ldots, n$, is simple.*

Theorem 3 *An irreducible collection C_j, U_j, $j = 1, \ldots, n$, satisfying* (2) *acts in the space $l_2(\Delta)$, $\Delta \subset \Omega$ is a subset of some orbit Ω (for unitary U_l, $l = 1, \ldots, n$, $\Delta = \Omega$) by the following formulas*

$$C_l e_{\mathbf{x}} = x_k e_{\mathbf{x}}, \quad U_l e_{\mathbf{x}} = u_l(\mathbf{x})\, e_{\mathbf{F}_l(\mathbf{x})},$$

where $u_l(\mathbf{x})$ are constants which determine the action of U_l. The subset Δ has the following "border conditions"

$$\begin{aligned} u_l(\mathbf{x}) &= 0, \quad \forall \mathbf{x} \in \Delta\colon \mathbf{F}_l(\mathbf{x}) \notin \Delta, \\ u_l(\mathbf{F}_l^{-1}(\mathbf{x})) &= 0, \quad \forall \mathbf{x} \in \Delta\colon \mathbf{F}_l^{-1}(\mathbf{x}) \notin \Delta, \end{aligned} \tag{6}$$

and is "connected" in the following sense: $u_l(\mathbf{x}) \neq 0\ \forall \mathbf{x} \in \Delta\colon \mathbf{F}_l(\mathbf{x}) \in \Delta$, $l = 1, \ldots, n$.

Let $\mathbf{x}$ be a stationary point of the mapping $\mathbf{F}_l(\cdot)$. If $x_l = 0$, then for all points $\mathbf{y} \in \Delta$, the commutation of $\mathbf{F}_l(\cdot)$ and $\mathbf{F}_k(\cdot)$ implies $y_l = 0$ as well. Then $a_l = 0$. If $x_l \neq 0$, then also $y_l \neq 0$ for all $\mathbf{y} \in \Delta$. In this case, the operator U_l commutes with all operators a_j, a_j^*, and therefore, is a multiple of the identity.

In the case when the mapping $\mathbf{F}_l(\cdot)$ does not have stationary points, the operator U_l is unitary equivalent to the shift operator; taking into account that $a_l a_l^* = C_l^2$, we get the needed formula for a_l. □

2 Wick Algebras

In this section we consider some Wick algebras connected with dynamical systems.

It was noted above that relations (4) are needed in some consistency condition. They arised in the previous section as a commutation of vector functions obtained from the basic relations.

In this section we look at this consistency from the point of view of Wick algebras.

Recall that Wick algebras are defined by elements a_j, $j = 1, \dots, d$, and by relations between a_i^*, a_j,

$$a_i^* a_j = \delta_{ij} 1 + \sum_{k,l=1}^{d} t_{ij}^{kl} a_k a_l^*, \qquad i, j = 1, \dots, d; \tag{7}$$

and possible relations between a_i, a_j are described by so-called Wick ideals (see [5] and [6] for more details).

Let us consider the following class of Wick algebras:

$$\begin{aligned} a_i^* a_i &= 1 + \sum_{j=1}^{d} \alpha_{ij} a_j a_j^*, \\ a_i^* a_j &= \lambda_{ij} q_{ij} a_j a_i^*, \qquad i \neq j, \end{aligned}$$

$0 < \alpha_{ii} < 1$, $q_{ij} = q_{ji} \in \mathbf{R}_+$, $\bar{\lambda}_{ij} = \lambda_{ji}$, $|\lambda_{ij}| = 1$, denoted by $\mathcal{U}(A, \Lambda, Q)$, where $A = (\alpha_{ij})$, $\Lambda = (\lambda_{ij})$, $Q = (q_{ij})$.

The purpose of this section is to describe algebras from this class which have the quadratic ideal of the maximal possible rank and to classify $*$-representations of these algebras by bounded operators.

2.1 Quadratic Ideals

Let $\mathcal{U} = \mathcal{U}(A, \Lambda, Q)$, then the operator $T = (t_{ij}^{kl}) \colon \mathcal{H} \to \mathcal{H}$ defined by coefficients from (7) (see [5]) has the form:

$$\begin{aligned} T e_i \otimes e_i &= \alpha_{ii} e_i \otimes e_i, \\ T e_i \otimes e_j &= \alpha_{ij} e_i \otimes e_j + \lambda_{ji} q_{ji} e_j \otimes e_i. \end{aligned}$$

Then

$$\begin{aligned} \mathcal{H} \otimes \mathcal{H} &= \bigoplus_{i=1}^{d} \mathcal{H}_i \oplus \bigoplus_{i,j=1}^{d} \mathcal{H}_{ij}, \\ \mathcal{H}_i &= \langle e_i \otimes e_i \rangle, \quad \mathcal{H}_{ij} = \langle e_i \otimes e_j, e_j \otimes e_i \rangle. \end{aligned}$$

According to [5], quadratic Wick ideals are determined by the projections P on them; these projections are completely defined by the "linear" and "quadratic" conditions of general form

$$(\mathbf{1} + T)P = 0, \quad (\mathbf{1} \otimes (\mathbf{1} - P))(T \otimes \mathbf{1})(\mathbf{1} \otimes T)(P \otimes \mathbf{1}) = 0,$$

see [5]. We write these conditions explicitly in our cases. The "linear" condition means in our case, that P must be the projection on the subspace, generated by eigenvectors of T with eigenvalue -1. Since $\alpha_{ii} \neq -1$ then the rank of P is the maximal possible one, if and only if equalities

$$(\alpha_{ij}+1)(\alpha_{ji}+1) = q_{ij}q_{ji} \tag{8}$$

hold for all $i \neq j$, and

$$P\mathcal{H}\otimes\mathcal{H} = \left\langle e_j \otimes e_i - \frac{\lambda_{ij}q_{ij}}{\alpha_{ij}+1} e_i \otimes e_j,\ i < j \right\rangle.$$

Denote the algebra $\mathcal{U}(A, \Lambda, Q)$ for which the equations (8) hold, by $\mathcal{U}(A, \Lambda)$. The "quadratic condition" from [5] takes the form:

$$\begin{aligned} &\alpha_{ij}\alpha_{ji} = 0, \quad \alpha_{ij}(\alpha_{ij}+1-\alpha_{jj}) = 0, \qquad i \neq j, \\ &\alpha_{ik}(\alpha_{kj}-\alpha_{ij}) + \alpha_{ij}\alpha_{jk} = 0, \qquad i \neq j,\ i \neq k,\ j \neq k. \end{aligned} \tag{9}$$

It is convenient to consider $\{\alpha_{ij}\}$ as a function

$$\alpha: \mathcal{I} \times \mathcal{I} \longmapsto \mathbf{R}, \qquad \mathcal{I} = \{1, \ldots, d\},$$

and to denote $\alpha_{ij} = \alpha(i, j)$.

Remark 1 If α is a solution of the system (9) then for all permutations $\pi \in S_d$, $\alpha^{\pi}(i, j) = \alpha(\pi(i), \pi(j))$ is also a solution, and if $\hat{a}_i = a_{\pi_i}$, then the "structural constants" for $\hat{a}_i$ are $\hat{\lambda}_{ij} = \alpha^{\pi}(i, j)$, $\hat{\lambda}_{ij} = \lambda^{\pi}(i, j)$.

Consequently it is suffice to describe solutions of (9) up to the action of the permatation group S_d.

Definition 1 Solution α is called canonical if $\alpha(i, j) = 0$ for all $i < j$.

Proposition 1 *Let α be an arbitrary solution of* (9) *then exists $\pi \in S_d$ such that α^{π} is a canonical solution.*

We may suppose now that $\alpha_{ij} = 0\ \forall i < j$. Then (9) is reduced to the following:

$$\begin{aligned} \alpha_{ij}(\alpha_{jk}-\alpha_{ik}) &= 0, \quad 1 \le k < j < i \le d, \\ \alpha_{ij}(1+\alpha_{ij}-\alpha_j) &= 0, \quad 1 \le j < i \le d, \\ \alpha_j &= \alpha_{jj}, \end{aligned}$$

where the second equation means only the fact that all non-zero α_{ij} are equal to the same parameter $\alpha_j - 1$ for fixed j and $i > j$.

Definition 2 Canonical solution is called decomposable if

$$\mathcal{I} = \mathcal{I}_1 \cup \mathcal{I}_2, \quad \mathcal{I}_1 \cap \mathcal{I}_2 = \emptyset,$$

and for all $i \in \mathcal{I}_1$, $j \in \mathcal{I}_2$, $\alpha_{ij} = \alpha_{ji} = 0$.

Remark 2 If a canonical solution is decomposable, then $\exists \pi \in S_d$ such that α^π is decomposable, canonical and

$$\mathcal{I}_1 = \{1, \dots, m\}, \quad \mathcal{I}_2 = \{m+1, \dots, d\}$$

It is clear that if $\alpha_{21} = \cdots = \alpha_{d1} = \alpha_1 - 1$, then α is indecomposable.

Proposition 2 *Let α be a canonical solution, then it is indecomposable if and only if $\alpha_{21} = \cdots = \alpha_{d1} = \alpha_1 - 1$.*

Let α be a canonical solution, $A = (\alpha_{ij})$. It follows from Proposition 2 and Remark 2, that we may suppose that for any fixed j all non-zero α_{ij}, $i > j$ are placed before all zeroes. Consider $\vec{k} = (k_1, \dots, k_{d-1})$, where $i \le k_i \le d$ are natural numbers, which is constructed by the following rule: if for a fixed j and all $i > j, \alpha_{ij} = 0$ then $k_j = 0$, else k_j is the greatest number l for which $\alpha_{lj} = \alpha_j - 1$. The characteristic property of $\vec{k}$ is the following.

Proposition 3 *If $i > j$ and $i \le k_j$, then $k_i \le k_j$.*

Conversely, let $\vec{k}$ be a vector with the characteristic property, and $A = (\alpha_{ij})$ be a matrix, such that $\alpha_{ii} = \alpha_i, \alpha_{ij} = 0, i < j$. Then

$$k_j = j \Longrightarrow \alpha_{ij} = 0, \qquad \forall i > j;$$

else

$$\alpha_{lj} = \alpha_j - 1, \quad j < l \le k_j; \quad \alpha_{lj} = 0, \quad l > k_j.$$

Then it is easy to verify, that A is a matrix of the canonical solution. We will denote such a matrix by $A(\vec{k})$. We have proved the following statement.

Theorem 4 *Let α be a solution of the system (9), then $\exists \pi \in S_d$, and $\vec{k}$ with the characteristic property, such that $(\alpha^\pi_{ij}) = A(\vec{k})$. Conversely, for any $\vec{k}$ with the characteristic property $A = A(\vec{k})$ gives a solution.*

Note, that for fixed $\vec{k}$ with the characteristic property we have the following relations:

$$\begin{aligned}
a_i^* a_i &= 1 + \alpha_i \, a_i a_i^* + \sum_{j<i, k_j \ge i} (\alpha_j - 1) a_j a_j^*, \\
a_i^* a_j &= \lambda_{ij} q_{ij} a_j a_i^*, \qquad i < j, \quad q_{ij} = (\alpha_{ji} + 1)^{1/2}, \\
\alpha_{ji} &= \begin{cases} \alpha_i - 1, & i < j \le k_i, \\ 0, & j > k_i. \end{cases}
\end{aligned}$$

2.2 Representations

Let $A = A(\vec{k}), \mathcal{U} = \mathcal{U}(A, \Lambda)$. Then $\mathcal{U}$ has the largest quadratic ideal, generated by

$$A_{ij} = a_j a_i - \lambda_{ij} q_{ij} a_i a_j, \quad i < j,$$

Theorem 5 *Let π be any bounded representation of $\mathcal{U}(A, \Lambda)$, then $\pi(A_{ij}) = 0$.*

Remark 3 The basic steps of the proof coincide with the proof of the analogous fact for the twisted commutation relations (see [5]), which are determined by $\Lambda = 1$, $A = A(d, \dots, d)$ and $\alpha_j = \mu^2$.

Consequently, to describe irreducible representations of $\mathcal{U}$, we must describe the families of operators $\{A_i, \ i = 1, \dots, d\}$, for which the following relations hold:

$$\begin{aligned} A_i^* A_i &= 1 + \alpha_i A_i A_i^* + \sum_{j<i, k_j \geq i} (\alpha_j - 1) A_j A_j^*, \\ A_i^* A_j &= \lambda_{ij} q_{ij} A_j A_i^*, \qquad i < j, \\ A_j A_i &= \lambda_{ij} q_{ij} A_i A_j, \qquad i < j, \quad q_{ij}^2 = \alpha_{ji} + 1, \quad (\alpha_{ij}) = A(\vec{k}). \end{aligned} \tag{10}$$

Let $A_i^* = U_i C_i$ be the polar decomposition. Then the system (10) can be rewritten in the equivalent form:

$$\begin{aligned} \vec{C} U_i^* &= U_i^* \vec{\mathcal{F}}_i(\vec{C}), \qquad \vec{C} = (C_1^2, \dots, C_d^2), \\ [C_i, C_j] &= 0, \quad U_i U_j = \bar{\lambda}_{ij} U_j U_i, \quad U_i U_j^* = \lambda_{ij} U_j^* U_i, \quad i < j, \\ \vec{\mathcal{F}}_i(x_1, \dots, x_i, \dots, x_d) &= \Bigg(x_1, \dots, x_{i-1}, 1 + \alpha_i x_i + \sum_{j<i, k_j \geq i} \\ &\qquad (\alpha_j - 1) x_j, q_{ii+1}^2 x_{i+1}, \dots, q_{id}^2 x_d \Bigg). \end{aligned}$$

Using the dynamical systems technique, we can reduce the problem of the describing of irreducible representations of $\mathcal{U}(A, \Lambda)$ to the analogous problem for the finite families of the unitary operators $\{U_i\}$, which satisfy relations $U_i U_j = \lambda_{ij} U_j U_i$, $i < j$.

First, we introduce some notations: $D(\mu)$ denotes the operator in $l_2(N)$:

$$\begin{aligned} D(\mu) e_n &= \mu^{n-1} e_n, \qquad n \in \mathbf{N}. \\ D(j, k_i) &= \begin{cases} 1, & j > k_i, \\ D(\alpha_j), & j \leq k_i, \end{cases} \end{aligned}$$

S stands for the unilateral shift,

$$D(f_j)\, e_n = f_j^{n-1}(0)\, e_n, \quad f_j(x) = 1 + \alpha_j x,$$

and f^n denotes the n-th iteration of f.

Let $1 \leq i_1 \leq \cdots \leq i_l \leq d$ be natural numbers, such that $k_{i_j} + 1 \leq i_{j+1}$, $j = 1, \ldots, l-1$.

Fix one of the such families. Write $\Phi = \cup_{j=1}^{l} \{i_j + 1, \ldots, k_{i_j}\}$ and construct the following irreducible representation for a fixed family $\{i_1, \ldots, i_l\}$:

$$\begin{aligned}
C_j &= U_j = 0, \qquad \forall j \in \Phi, \\
C_j^2 &= \bigotimes_{i=1, i \notin \Phi}^{j-1} D(j, k_i) \otimes D(f_j) \otimes 1 \cdots \otimes 1, \qquad j \neq i_k, \\
U_j^* &= \bigotimes_{i=1, i \notin \Phi}^{j-1} D(\lambda_{ij}) \otimes S \otimes 1 \otimes \cdots \otimes 1, \qquad j \neq i_k, \\
U_{i_k}^* &= \bigotimes_{i < i_k, i \notin \Phi} D(\lambda_{i i_k}) \otimes \bigotimes_{i > i_k, i \notin \Phi} D(\overline{\lambda}_{i i_k}) \otimes \hat{U}_{i_k}^*, \qquad k = 1, \ldots, l, \\
C_{i_k}^2 &= \frac{1}{1 - \alpha_{i_k}} \bigotimes_{i < i_k, i \notin \Phi} D(i, k_{i_k}) \otimes 1 \cdots \otimes 1, \qquad k = 1, \ldots, l,
\end{aligned}$$

where $\{\hat{U}_{i_k}\}$ is irreducible family of the unitary operators, which satisfy relations:

$$\hat{U}_i \hat{U}_j = \lambda_{ij} \hat{U}_j \hat{U}_j.$$

Theorem 6 *All irreducible representations can be obtained in the way described above; moreover, two representations are unitary equivalent if and only if they correspond to the same family $\{i_1, \ldots, i_l\}$, and the corresponding unitary families are unitary equivalent.*

Remark 4 1. If at least one of λ_{ij} is not a root of 1, then there exists a not of type I representation.

2. If all λ_{ij} are roots from 1, then the problem of the classification of families $\{U_i\}$ can be reduced to the case $\lambda_{ij}^q = 1$, where $q = p^m$ for some prime p. In this case families $\{U_i\}$ can be described by simple reduction algorithm.

The authors express their gratitude to prof. Yuriĭ Samoĭlenko for his kind attention and helpful discussions of the topics of the paper.

References

[1] C. Daskaloyanis, *Generalized deformed oscillator and nonlinear algebras*, J. Phys. A **24** (1991), L789–L794.

[2] V.L. Ostrovskyĭ and L.B. Turovskaya, *Representations of ∗-algebras and multidimensional dynamical systems*, Ukr. Mat. Zhurn. **47** no. 4 (1995), 488–497.

[3] E.Ye. Vaisleb, *Representations of relations which connect a family of commuting operators with non-sefadjoint one*, Ukrain. Math. Zh. **42** (1990), 1258–1262, (Russian).

[4] G.W. Mackey, *Induced representations of locally compact groups*, Ann. Math. **55** no. 1 (1952), 101–139.
[5] P.E.T. Jorgensen, L.M. Schmitt and R.F. Werner, *Positive representations of general commutation relations allowing Wick ordering*, J. Funct. Anal. **134** (1995), 33–99.
[6] D.P. Proskurin, *Homogeneous ideals in Wick $*$-algebras*, Proc. AMS. **126**, no. 11 (1998), 3371–3376.

Vasyl Ostrovskyi
Institute of Mathematics
Ukrainian National Acad. Sci.
Kyiv
Ukraine
vo@imath.kiev.ua

Daniil Proskurin
Kyiv National T. Shevchenko University
Kyiv
Ukraine
prosk@imath.kiev.ua

AMS Subject Classification: Primary 46K10, 46L55

Operator Theory:
Advances and Applications, Vol. 118

On Representations of $\mathcal{F}_n$-Algebras and Invertibility Symbols

Slavik Rabanovich and Yuriĭ Samoĭlenko

We study algebras generated by n flips all but two of which pairwise commute or anticommute, as well as other $\mathcal{F}_n$-algebras. Irreducible representations or $*$-representations of such algebras are found. By using representation theory, we show that they are $\mathcal{F}_n$-algebras (for some n). We also study the Invertibility Symbol for these algebras.

Banach algebras and $*$-algebras generated by idempotents or flips are directly related to numerous problems of functional analysis (see, for example, [1] and the bibliography therein). If such an algebra is $\mathcal{F}_n$ (see Section 1), then it has an Invertibility Symbol [2]. Banach $\mathcal{F}_4$-algebras generated by two flips u, v (or two idempotents $p = \frac{u+1}{2}, q = \frac{v+1}{2}$) are considered in many analytic examples. In [3, 4] and other articles, the authors studied algebras generated by three flips, $\mathbb{C}\langle u, v, s \mid u^2 = v^2 = s^2 = e, us = h_u su, vs = h_v sv, h_u = \pm 1, h_v = \pm 1\rangle$. In this article we will be dealing with a more general class of algebras, namely the algebras generated by n flips all but two of which pairwise commute or anticommute, and other examples. All examples are obtained as a generalization of the group algebras of the affine Coxeter groups. By using representations and $*$-representations of these algebras, we show that they are $\mathcal{F}_n$-algebras (for some n), and describe their irreducible representations or $*$-representations. We also study the Invertibility Symbol (Section 3) for these algebras.

1 Representations of $\mathcal{F}_n$-algebras

In what follows, all linear spaces are considered over the complex field $\mathbb{C}$. In the sequel, we will be interested in representations and $*$-representations of algebras generated by idempotents. We use standard definitions from the representation theory: algebras and $*$-algebras, representations and $*$-representations, irreducible, indecomposable, equivalent, unitarily equivalent representations, etc. In particular, a set $\mathcal{L}$ of representations of an algebra A forms a residual family of representations (we abbreviate it as res. fam.) if for each $x \in A$, $x \neq 0$, there exists $\pi \in \mathcal{L}$ such that $\pi(x) \neq 0$ ($\pi(x)$ is an operator on a Hilbert space H_π). For example, all irreducible representations of a C^*-algebra form a residual family.

Let $\mathcal{F}_n$ denote the standard polynomial of degree n in n non-commuting variables:

$$\mathcal{F}_n(x_1, x_2, \dots, x_n) = \sum_{\sigma \in S_n} (-1)^{p(\sigma)} x_{\sigma(1)} \dots x_{\sigma(n)},$$

where $p(\sigma)$ is the parity of the permutation σ, S_n is the symmetric group. In the sequel, we say that A is an $\mathcal{F}_n$-algebra, if $\forall x_1, \dots, x_n \in A$, we have that $\mathcal{F}_n(x_1, \dots, x_n) = 0$.

The following Amitsur-Levitsky's theorem takes place [5]: *$M_n(\mathbb{C})$ is an $\mathcal{F}_{2n}$-algebra, but not an $\mathcal{F}_{2n-1}$-algebra.*

Remark 1.1 If a Banach algebra $\mathcal{A}$ contains a dense $\mathcal{F}_n$-subalgebra A, i.e. $\mathcal{A} = \overline{A}$, then $\mathcal{A}$ is also an $\mathcal{F}_n$-algebra.

$\mathcal{F}_n$-algebras form one of the most simple class of algebras, if considered from the standpoint of the structure of irreducible representations.

Theorem 1.1 *Consider the following statements:*

(i) *there exists a residual family $\mathcal{L}$ of irreducible representations* $\mathrm{IrRep}A \supset \mathcal{L} \ni \pi$ *such that* $\dim H_\pi \leq n$ *for all* $\pi \in \mathcal{L}$;

(ii) *there exists a residual family $\mathcal{L}$ of representations* $\mathrm{Rep}A \supset \mathcal{L} \ni \pi$ *such that* $\dim H_\pi \leq n$ *for all* $\pi \in \mathcal{L}$;

(iii) *A is an $\mathcal{F}_{2n}$-algebra;*

(iv) *for any* $\pi \in \mathrm{IrRep}A$, $\dim H_\pi \leq n$.

We have the following implications: (i) $\Rightarrow$ (ii) $\Rightarrow$ (iii) $\Rightarrow$ (iv).

Here we will only prove that (ii) $\Rightarrow$ (iii), since it is this statement that will be later used in examples to prove that the corresponding algebra is an $\mathcal{F}_{2n}$-algebra.

Assume that (ii) holds, but A is not an $\mathcal{F}_{2n}$-algebra. Then there exist $x_1, \dots, x_{2n} \in A$ such that $\mathcal{F}_{2n}(x_1, \dots, x_{2n}) = x \neq 0$. Choose $\pi \in \mathcal{L}$ such that $\pi(x) \neq 0$. Then we get

$$\pi(\mathcal{F}_{2n}(x_1, \dots, x_{2n})) = \mathcal{F}_{2n}(\pi(x_1), \dots, \pi(x_{2n})) = \pi(x) \neq 0.$$

But since $\dim H_\pi \leq n$, this contradicts the Amitsur-Levitsky's theorem.

Remark 1.2 Neither of the inverse implications of Theorem 1 holds. Indeed, to see that (ii) does not imply (i), and (iv) does not imply (iii), consider the nilpotent algebra of complex $n \times n$ matrices of the form

$$\begin{pmatrix} 0 & & * \\ & \ddots & \\ 0 & & 0 \end{pmatrix},$$

which has only the trivial irreducible representation for any $n \in \mathbb{N}$.

Condition (iii) does not imply (ii). For example, the algebra of matrices $X \in M_n(\mathbb{C})$ of the form

$$X = \begin{pmatrix} a_{11} & * & \dots & * \\ 0 & \vdots & & \vdots \\ \vdots & * & \dots & * \\ 0 & \dots & 0 & a_{11} \end{pmatrix}$$

is an $\mathcal{F}_{2n-2}$-algebra. But for any representation π, $\dim H_\pi \leq n-1$, and the nilpotent element

$$S = \begin{pmatrix} 0 & 1 & & 0 \\ & \ddots & \ddots & \\ & & 0 & 1 \\ 0 & & & 0 \end{pmatrix},$$

we have that $\pi(S^{n-1}) = (\pi(S))^{n-1} = 0$, since $\pi(S)$ is a nilpotent element in $M_{n-1}(\mathbb{C})$. Hence such representations do not separate S^{n-1} and the zero element of the algebra.

Remark 1.3 If $\mathfrak{A}$ is a $*$-algebra, and one only considers its $*$-representations, then, evidently, (i) $\Leftrightarrow$ (ii).

(iii) does not imply (ii), since, for example, the algebra $M_n(\mathbb{C})$ with a nonstandard involution does not have nonzero $*$-representations.

Condition (iv) does not imply (iii). For example, the Weyl $*$-algebra $\mathbb{C}\langle P = P^*, Q = Q^* \mid [P, Q] = iI\rangle$ of differential operators with the coefficients being polynomials in one variable does not have $*$-representations in bounded operators, but it is not an $\mathcal{F}_n$-algebra for any $n \in \mathbb{N}$.

Remark 1.4 If $\mathfrak{A}$ is a Banach semisimple algebra (in particular, a C^*-algebra), then all conditions of the lemma are equivalent, since for a Banach semisimple algebra $\mathfrak{A}$, the set $\mathrm{IrRep}\mathfrak{A}$ of its irreducible representations is a residual family and, therefore, (iv) $\Rightarrow$ (i) (see [2]).

2 Examples of $\mathcal{F}_n$-algebras Generated by Idempotents and their Representations

Here, we give a number of examples of algebras and $*$-algebras generated by idempotents, and for each of them construct a residual family of representations or $*$-representations π such that $\dim H_\pi \leq n$, and, therefore, show that these algebras are $\mathcal{F}_{2n}$-algebras.

$*$-algebras

Example 1.1 Representations of the $\mathcal{F}_4$-algebra with unit generated by two idempotents q_1, q_2,

$$\begin{aligned} Q_2 &= \mathbb{C}\langle q_1, q_2 \mid q_1{}^2 = q_1, q_2{}^2 = q_2\rangle \\ &= \mathbb{C}\langle u = 2q_1 - e,\ v = 2q_2 - e \mid u^2 = e, v^2 = e\rangle \end{aligned}$$

are well known, nevertheless we present a description of irreducible representation of the algebra to show the scheme of investigations we will follow in more complicated examples.

All finite dimensional irreducible representations of Q_2, up to equivalence, are:

a) four one-dimensional representations: $\pi_{0,0}(q_1) = 0$, $\pi_{0,0}(q_2) = 0$; $\pi_{1,0}(q_1) = 1$, $\pi_{1,0}(q_2) = 0$; $\pi_{0,1}(q_1) = 0$, $\pi_{0,1}(q_2) = 1$; $\pi_{1,1}(q_1) = 1$, $\pi_{1,1}(q_2) = 1$;

b) the family, parameterized by $z \in \mathbb{C}\backslash\{0, 1\}$, of two-dimensional representations:

$$\pi_z(q_1) = \begin{pmatrix} 1 & 0 \\ 0 & 0 \end{pmatrix}, \quad \pi_z(q_2) = \begin{pmatrix} z & 1 \\ z - z^2 & 1 - z \end{pmatrix}.$$

Every $\pi \in \operatorname{IrRep} Q_2$ is one- or two-dimensional. Indeed, the space $H = \mathbb{C}\langle e_\lambda, \pi(v)e_\lambda\rangle$ (e_λ is an eigen vector of $\pi(u) \cdot \pi(v)$, $\|e_\lambda\| = 1$, $\lambda \neq 0$) is invariant for the representation π.

Remark 2.1 In the algebra Q_2, one can introduce two natural structures of an algebra with involution: 1)

$$\begin{aligned} \mathcal{P}_2 &= \mathbb{C}\langle p_1^{*_1} = p_1, p_2^{*_1} = p_2 \mid p_1^2 = p_1, p_2^2 = p_2\rangle \\ &= \mathbb{C}\langle u^{*_1} = u, v^{*_1} = v \mid u^2 = v^2 = e\rangle \\ &= \mathbb{C}[\mathbb{Z}_2 * \mathbb{Z}_2] \end{aligned}$$

is a group $*_1$-algebra generated by two unitary selfadjoint generators. Irreducible 2-dimensional $*$-representations of $\mathcal{P}_2$(up to a unitary equivalence) are:

$$\pi_\phi(p_1) = \begin{pmatrix} 1 & 0 \\ 0 & 0 \end{pmatrix}, \quad \pi_\phi(p_2) = \begin{pmatrix} \cos^2\phi & \cos\phi\sin\phi \\ \cos\phi\sin\phi & \sin^2\phi \end{pmatrix}, \phi \in (0, \pi/2).$$

They are equivalent to the representations π_z, $z \in (0, 1) \subset \mathbb{R}$.

2)

$$Q_2 = \mathbb{C}\langle q_1^{*_2} = q_2 \mid q_1^2 = q_1\rangle = \mathbb{C}\langle u^{*_2} = v \mid u^2 = e\rangle$$

is a $*_2$-algebra generated by an idempotent and its adjoint. Irreducible two-dimensional $*$-representations of Q_2 are:

$$\pi_\alpha = \begin{pmatrix} 1 & \alpha \\ 0 & 0 \end{pmatrix}, \quad (\alpha > 0).$$

They are equivalent to the representations π_z, $z \in (1, \infty) \subset \mathbb{R}$.

Remark 2.2 A complete list of indecomposable representations of Q_2 can be obtained, for example, by using the results from [6].

Since the 2-dimensional representations π_z, $z \in \mathbb{C} \setminus \{0, 1\}$, (as well as π_ϕ, $\phi \in (0, \frac{\pi}{2})$, and π_α, $\alpha > 0$) form a residual family, Q_2 is an $\mathcal{F}_4$-algebra.

Example 1.2 The algebra Q_2 is the group algebra of the Coxeter group $\bullet \overset{\infty}{-} \bullet$, generated by two flips without any relations.

Consider the Coxeter group G_M with a matrix $M = (m_{ij})_{i,j=1}^m$, ($m_{ij} \in \mathbb{N} \cup \{\infty\}$, $m_{ii} = 1$; $m_{ij} = m_{ji} > 1$, $i \neq j$; $i, j = 1, \dots, m$), which is defined in terms of generators $(w_i)_{i=1}^m$ and the relations $(w_i w_j)^{m_{ij}} = e$, $i, j = 1, \dots, m$; if $m_{ij} = \infty$, then there is no relation between the generators w_i and w_j. If the Cartan matrix $K = (-\cos \pi / m_{ij})_{i,j=1}^m$, which corresponds to M, is positive definite (all its principal minors are positive), then the group G_M is finite; if $\det K = 0$, but the other principal minors are positive, then the group G_M is infinite, but G_M is a semidirect product of the lattice $\mathbb{Z}^{p(M)}$ by a finite group $G_f(M)$, $G_M = \mathbb{Z}^{p(M)} \ltimes G_f(M)$.

Since the Coxeter group G_M is generated by flips $w_i^2 = e, i = 1, \dots, m$, $\mathbb{C}[G_M]$ gives also an example of an algebra generated by m projections (see, for example [7]). There is a natural involution in $\mathbb{C}[G_M]$ with which all of the group elements are unitary, $g^* = g^{-1}$. (Generally speaking, this is not the unique involution that can be defined on $\mathbb{C}[G_M]$).

The dimensions of the irreducible $*$-representations π_α of the group $*$-algebra of the Coxeter group $G_M = \mathbb{Z}^{p(M)} \ltimes G_f$ are majorized by the number $|G_f|$. These representations form a residual family, because irreducible $*$-representations of $\mathbb{C}[G_M]$, with the involution $g^* = g^{-1}$, make a residual family. Hence $\mathbb{C}[G_M]$ is an $\mathcal{F}_{2|G_f|}$-algebra which is generated by flips.

Remark 2.3 It is a every difficult problem to describe indecomposable representations of $\mathbb{C}[G_M]$ (except for the case where the Coxeter group G_M is a finite group or is $\mathbb{Z} \ltimes \mathbb{Z}_2$) [8].

Central extensions of the group $\mathcal{F}_n$-algebras generated by idempotents significantly extend the class of $\mathcal{F}_n$-algebras. For example the "non-commutative circle",

$$\mathbb{C}\langle x, y \mid x^2 + y^2 = 1 \rangle,$$

the "hyperbola",

$$\mathbb{C}\langle x, y \mid x^2 - y^2 = 1 \rangle,$$

the "pair of intersecting lines",

$$\mathbb{C}\langle x, y \mid x^2 = y^2 \rangle,$$

are $\mathcal{F}_4$-algebras.

For a description of irreducible representations of these algebras, see for example [9, 10].

Let now $\mathfrak{A}_k = \mathbb{C}\langle u_1^{(k)}, \dots, u_{n_k}^{(k)} \mid ()_k\rangle$ be $\mathcal{F}_{2m_k}$-algebras generated by flips $u_1^{(k)}, \dots, u_{n_k}^{(k)}$ and relations $()_k$, having a residual family of $\pi^{(k)}$ with $\dim H_{\pi^{(k)}} \leq m_k$ $(k = 1, \dots, n)$. Of course, the algebra

$$\mathbb{C}\langle u_1^{(1)}, \dots, u_{n_n}^{(n)} \mid ()_1, \dots, ()_n, [u_i^{(k)}, u_j^{(l)}] = 0 (k \neq l)\rangle$$

is an $\mathcal{F}_{2m_1\cdot\ldots\cdot m_n}$-algebra that has the residual family of $\pi^{(1)}\otimes\cdots\otimes\pi^{(n)}$. Examples of algebras that we will further consider are also defined by generators $u_1^{(1)}, \dots, u_{n_n}^{(n)}$, but if the upper indices are not equal, the generators pairwise commute or anticommute. In examples 2.1 and 2.2, these relations are as follows: $u_i^{(k)}u_j^{(l)} = \epsilon_{kl}u_j^{(l)}u_i^{(k)}, k \neq l$ $(\epsilon_{kl} = +1 \text{ or } -1, \epsilon_{kl} = \epsilon_{lk}), (k, l = 1, \dots, n)$ and do not depend on $i = 1, \dots, n_k$ and $j = 1, \dots, n_l$.

Example 2.1 Let $\mathfrak{A}_{n,\epsilon}$ be an algebra generated by $s_1, \dots, s_n$:

$$\mathfrak{A}_{n,\epsilon} = \mathbb{C}\langle s_1, \dots, s_n \mid s_i^2 = 1, s_is_j = \epsilon_{ij}s_js_i, (i, j = 1, \dots n)\rangle,$$
$$(\epsilon = (\epsilon_{ij}), \quad \epsilon_{ii} = 1).$$

The algebra $\mathfrak{A}_{n,\epsilon}$ is finite dimensional and semisimple, it has a finite residual family of irreducible $*$-representations π_p $(s_i^* = s_i (i = 1, \dots, n))$ and is an $\mathcal{F}_m$-algebra, where $m \geq 2^n$.

Example 2.2 Let $\mathfrak{B}_{(\mathfrak{A}_k),\epsilon} = \mathbb{C}\langle u_1^{(1)}, \dots, u_{n_n}^{(n)} \mid ()_1, \dots, ()_n \; u_i^{(k)}u_j^{(l)} = \epsilon_{kl}u_j^{(l)}u_i^{(k)} \; (k \neq l), \; k, l = 1 \dots, n; \; i = 1, \dots, n_k, \; j = 1, \dots, n_l\rangle$.

This is an $\mathcal{F}_{2^{n+1}m_1\cdot\ldots\cdot m_n}$-algebra that has a residual family of $*$-representations $\pi^{(1)} \otimes \cdots \otimes \pi^{(n)} \otimes \pi_p$ with $\dim H_{\pi^{(1)}\otimes\cdots\otimes\pi^{(n)}\otimes\pi_p} \leq 2^n \cdot m_1 \cdot \ldots \cdot m_n$ $(\pi^{(1)} \otimes \cdots \otimes \pi^{(n)} \otimes \pi_p(u_i^{(k)}) = 1 \otimes \cdots \otimes \pi^{(k)}(u_i^{(k)}) \otimes \cdots \otimes 1 \otimes \pi_p(s_k))$.

In Example 2.2, the generators $u_i^{(k)}$ and $u_j^{(l)}$ commute or anticommute independently of i and j. In Example 2.3, whether the generators $u_i^{(k)}$ and $u_j^{(l)}$ commute or not depends on i, j.

Example 2.3 Let $A_k = \mathbb{C}\langle u, v, s_1, \dots, s_k \mid u^2 = v^2 = s_i^2 = e \; (i = 1, \dots, k)$, $us_i = \alpha_i s_i u, \; vs_i = \beta_i s_i v, \; s_is_j = \epsilon_{ij}s_js_i \; (i, j = 1, \dots, k)\rangle$, where $\alpha_i = \pm 1$, $\beta_i = \pm 1; \epsilon_{ij} = \epsilon_{ji} = \pm 1 \; (i \neq j), \epsilon_{ii} = 1$.

Of course, this algebra should have been denoted by $A_{k,\alpha,\beta,\epsilon}$, but we leave out the symbols α, β, and ϵ for brevity.

Remark 2.4 The algebras $\mathbb{C}\langle u, v, s_1 \mid us_1 = -s_1u, \; vs_1 = -s_1v\rangle$ and $\mathbb{C}\langle u, v, s_1 \mid us_1 = -s_1u, \; vs_1 = s_1v\rangle$ were considered in [3, 4].

Lemma 2.1 *The algebra A_k is an $\mathcal{F}_{2^{k+2}}$-algebra, and it has a residual family $\mathcal{L}_k$ subject to the condition: $\forall \pi \in \mathcal{L}_k$, $\dim H_\pi \leq 2^{k+1}$.*

Proof: For $k = 0$, A_0 is an $\mathcal{F}_4$-algebra, and has a residual family $\mathcal{L}_0$, since the algebra $A_0 = Q_2$. Let $k = n$, and assume that all A_n are $\mathcal{F}_{2^{n+2}}$-algebras and there exists $\mathcal{L}_n$ with $\dim H_\pi \leq 2^{n+1}$. By induction, consider the algebra A_{n+1}. It contains the subalgebra $B = C\langle u, v, s_1, \ldots, s_n \mid \alpha_i, \beta_i, \epsilon_{ij}\rangle$. Clearly, B is isomorphic to some A_n. Therefore, the claim is true for B. Let $\mathcal{L}_n$ be a residual family for B with $\dim H_\pi \leq 2^{n+1}$. We construct a residual family for A_{n+1} by applying the following procedure: $\forall \pi \in \mathcal{L}_n, \pi : H_\pi \to H_\pi$, introduce $\hat{\pi} \in \mathcal{L}_{n+1}$, $\hat{\pi} : H_\pi \oplus H_\pi \to H_\pi \oplus H_\pi$, by

$$\hat{\pi}(u) = \begin{pmatrix} \pi(u) & 0 \\ 0 & \alpha_{n+1}\pi(u) \end{pmatrix}, \quad \hat{\pi}(v) = \begin{pmatrix} \pi(v) & 0 \\ 0 & \beta_{n+1}\pi(v) \end{pmatrix},$$

$$\hat{\pi}(s_i) = \begin{pmatrix} \pi(s_i) & 0 \\ 0 & \epsilon_{n+1i}\pi(s_i) \end{pmatrix}, \quad i = 1, \ldots, n,$$

$\hat{\pi}(s_{n+1}) = \begin{pmatrix} 0 & I \\ I & 0 \end{pmatrix}$, $I : H_\pi \to H_\pi$ is the identity operator. Let us show that $\mathcal{L}_{n+1}$ is indeed a residual family. For A_{n+1}, $\forall x \in A_{n+1}$ there exists an expansion $x = b_1 + s_{n+1}b_2$ $(b_1, b_2 \in B)$. If $b_2 \neq 0$, then $\exists \pi \in \mathcal{L}_n$ such that $\pi(b_2) \neq 0 \Rightarrow$

$$\hat{\pi}(x) = \begin{pmatrix} \pi(b_1) & * \\ \pi(b_2) & * \end{pmatrix} \neq 0 \qquad (\hat{\pi} \in \mathcal{L}_{n+1}).$$

If $b_2 = 0$, $b_1 \neq 0$, then $\exists \pi \in \mathcal{L}_n$: $\pi(b_1) \neq 0 \Rightarrow \hat{\pi}(b_1) \neq 0$. Note that $\forall \hat{\pi} \in \mathcal{L}_{n+1}$, $\dim \hat{\pi} \leq 2^{n+2}$. □

Remark 2.5 There is a natural involution in A_k given by $u^* = u$, $v^* = v$, $s_i^* = s_i$. Since there exists a residual family for Q_2 such that $\forall \pi \in \mathcal{L}_0$ the operators $\pi(u)$, $\pi(v)$ are selfadjoint, there exists a residual family for A_k satisfying the condition: $\forall \pi \in \mathcal{L}_k$ the operators $\pi(u), \pi(v)\, \pi(s_i)$ are selfadjoint. There is a residual family for A_k consisting only of irreducible $*$-representations (see Remark 1.3).

For a description of irreducible $*$-representations of A_k, also see [11].

Remark 2.6 $\forall\, k, \alpha_i, \beta_i, \epsilon_{ij}$, the algebra A_k is semisimple.

Moreover we have the next theorem which we give without a proof (for a proof, see [12, 13]).

Theorem 2.1 *Let $Q_{2,m} = A_m$ with $\alpha_i = 1$, $\beta_i = 1$, $\epsilon_{ij} = 1$, for all i, j. Then every algebra A_k is isomorphic to $M_{2^n}(Q_{2,m})$ or to $M_{2^n}(Z(A_k))$, where $Z(A_k)$ is the center of A_k.*

By using Theorem 2.1, it is possible to obtain a description for all irreducible representations of the algebras A_k (for such a description of the algebra $Q_{2,m}$, see Section 3).

Corollary 2.1 *If a Banach algebra $\mathcal{A}$ contains a dense subalgebra isomorphic to A_k, then $\mathcal{A} \cong M_{2^n}(\overline{Q_{2,m}})$ or else it is isomorphic to $M_{2^n}(\overline{Z(A_k)})$.*

Proof: A proof follows from the fact that if a Banach algebra $\mathcal{A}$ contains a dense subalgebra isomorphic to $M_m(\mathfrak{A})$ ($\mathfrak{A}$ is a subalgebra in $\mathcal{A}$), then $\mathcal{A} \cong M_m(\overline{\mathfrak{A}})$. □

Remark 2.7 A study of examples of algebras generated by flips $s_1, \dots, s_k$, which commute or anticommute, with the flips w_{ij}, $i, j = 1, \dots, n$, generating $\mathbb{C}[G_M]$, and more complicated examples of algebras generated by flips will be carried out elsewhere.

3 Invertibility Symbol for Banach Algebras $\overline{A}_k$

The methods developed in the preceding sections will now be used to solve problems of invertibility in Banach algebras $\overline{A}_k$.

Definition 3.1 (see, for example, [2]). Let $\mathcal{A}$ be a Banach algebra with a unit. The set of homomorphisms $\Phi_\alpha : \mathcal{A} \to L(H_\alpha)$, $\alpha \in \Lambda$, is said to be an Invertibility Symbol of order n, $IS(n)$, if

a) for all $\alpha \in \Lambda$, $\dim H_\alpha \leq n$;

b) for all $x \in \mathcal{A}$, x is invertible if and only if $\Phi_\alpha(x)$ is invertible for all $\alpha \in \Lambda$.

For properties of the invertibility symbol and its applications, see for example [2]. We will only recall a few facts related to the invertibility symbol. If $\mathcal{A}$ is a Banach $\mathcal{F}_{2n}$-algebra, then the set of all its irreducible representations forms an $IS(n)$ [2]. It is also known (see [14]) that if a Banach algebra $\mathcal{B}$ is isomorphic to $M_k(\mathcal{A})$ and the Banach algebra $\mathcal{A}$ has an invertibility symbol $IS(n)$ of order n, then the algebra $\mathcal{B}$ has $IS(k \cdot n)$. A construction of $IS(kn)$ for $\mathcal{B}$ can be carried out as follows. Let Φ_α be an $IS(n)$ for $\mathcal{A}$ and $\chi = (\chi_{ij}) : \mathcal{B} \to M_k(\mathcal{A})$ be an isomorphism. Then $IS(nk)$, for $\mathcal{B}$, will be the following function: $\hat{\Phi}_\alpha(x) = (\Phi_\alpha(\chi_{ij}(x)))$, $x \in \mathcal{B}$, where $(\Phi_\alpha(\chi_{ij}(x)))$ is a matrix with the element $\Phi_\alpha(\chi_{ij}(x))$ at the (i, j)th place.

Let A_k be the algebra defined in Example 2.3, and let A_k be everywhere dense in the Banach algebra $\mathcal{A}$. The remaining part of the article is dealing with a construction of an invertibility symbol for the algebra $\mathcal{A}$. By Lemma 2.1 and Remark 1.1, we see that $\mathcal{A}$ is an $\mathcal{F}_{2^{k+2}}$-algebra. Thus, on the one hand, $\mathcal{A}$ has $IS(2^{k+1})$, and on the other hand, $\mathcal{A} \cong M_{2^n}(\mathcal{A}_1)$, where $\mathcal{A}_1$ is a Banach subalgebra of $\mathcal{A}$ (see Corollary 2.1). There are many articles dealing with construction of $IS(2)$ for the algebra $\mathcal{A}_1 = \overline{Q}_{2,0}$. In [2, 3] the authors have constructed an invertibility symbol for the algebras $M_2(\overline{Z(A_1)})$ and $M_2(\overline{Q}_{2,0})$. In order to construct an invertibility symbol for $\mathcal{A}$, it is sufficient to construct an invertibility symbol for $\mathcal{A}_1$.

Now, $\mathcal{A}_1$ is isomorphic to either $\overline{Q}_{2,m}$ or $\overline{Z(A_k)}$. The second case is simple, since $\overline{Z(A_k)}$ is a commutative algebra, and an invertibility symbol for it can be found from the Gelfand homomorphisms. Thus we will only be considering the algebras $\overline{Q}_{2,m}$. To construct an invertibility symbol for $\mathcal{A}_1 = \overline{Q}_{2,m}$ $(m > 0)$, we use ideas and constructions given in [14]. In the sequel, $\sigma(\cdot)$ denotes the spectrum of the element in $\mathcal{A}_1$.

Let write down all irreducible representations of the algebra $Q_{2,m}$. All irreducible representations of $Q_{2,m}$, up to equivalence, are as follows: one-dimensional f_n and two-dimensional $\pi_{z,n}$. They are given by: $f_n(u) = (-1)^{1+n_{m+1}}$, $f_n(v) = (-1)^{1+n_{m+2}}$, $f_n(s_i) = (-1)^{1+n_i}$, $n = n_{m+2}n_{m+1}\ldots n_1$, is the binary notation for n $(n = 0, \ldots, 2^{m+2} - 1)$,

$$\pi_{z,n}(u) = \begin{pmatrix} 1 & 0 \\ 0 & -1 \end{pmatrix}, \quad \pi_{z,n}(v) = \begin{pmatrix} z & 1 \\ 1 - z^2 & -z \end{pmatrix},$$

$\pi_{z,n}(s_i) = (-1)^{1+n_i} \begin{pmatrix} 1 & 0 \\ 0 & 1 \end{pmatrix}$, $z \neq \pm 1$, $n = n_m \ldots n_1$, is written in the binary notation $(n = 0, \ldots, 2^m - 1)$.

Consider those irreducible representations $Q_{2,m}$ which can be lifted to IrRep$\mathcal{A}_1$ =IrRep$\overline{Q}_{2,m}$. If there exists an element x in $Q_{2,m}$ which is not invertible in $\mathcal{A}_1$ and such that there exists a unique, up to equivalence, representation $\psi \in$ IrRep$Q_{2,m}$ not invertible on x, i.e. the operator $\psi(x)$ is not invertible, then ψ is a restriction of a certain representation of the algebra $\mathcal{A}_1$. In what follows we will find elements a_1 and a_2 from $Q_{2,m}$ and a correspondence between $\lambda \in \sigma(a_1)$ and one-dimensional representations $\psi \in$ IrRep $Q_{2,m}$ which would not be invertible on $a_1 - \lambda e$, and find a correspondence between $\lambda \in \sigma(a_2)$ and two-dimensional representations $\psi \in$ IrRep $Q_{2,m}$ which is not invertible on x.

Define the elements $c = \frac{1}{2}(uv + vu)$, $a_1 = \frac{1}{2}((\frac{c}{2\|c\|} + u2^m + s_1 + \cdots + s_m 2^{m-1} + 2^{m+1} - 1)$, $a_2 = \frac{1}{2}((\frac{c}{2\|c\|} + s_1 + \cdots + s_m 2^{m-1} + 2^m - 1)$. In order to find representations that correspond to points of spectrum of a_1 and a_2, introduce the following sets: $M_+ = \{t \in \sigma(a_1) \subset \mathbb{C} \mid 4\|c\|(t - [|t| + 1/2]) = 1\}$, $M_- = \{t \in \sigma(a_1) \subset \mathbb{C} \mid 4\|c\|(t - [|t| + 1/2]) = -1\}$ and

$$\mathcal{R}_1 = \left\{ n \in \mathbb{Z} \mid n = \left\{ \begin{array}{ll} 2^{m+1} + [|t| + 1/2] & t \in M_+, \ |t| > 2^m - 1/2 \\ [|t| + 1/2] & t \in M_-, \ |t| > 2^m - 1/2 \\ [|t| + 1/2] & t \in M_+, \ |t| < 2^m - 1/2 \\ 2^{m+1} + [|t| + 1/2] & t \in M_-, \ |t| < 2^m - 1/2 \end{array} \right\} \right.$$

($[\cdot]$ is the integer part of a real number). Let also $M = \{t \in \sigma(a_2) \subset \mathbb{C} \mid 4\|c\| (t - [|t| + 1/2]) \neq \pm 1\}$ and

$$\mathcal{R}_2 = \{(\mu, n) \in \mathbb{C} \times \mathbb{Z} \mid \mu = 4\|c\|(t - [|t| + 1/2]), n = [|t| + 1/2], t \in M\}$$

There is a representation f_n for each $n \in \mathcal{R}_1$ ($n = n_{m+2}n_{m+1}\dots n_1$ is the binary notation), and for a pair $(\mu, n) \in \mathcal{R}_2$ there is a representation $\pi_{\mu,n}$. Hence we can formulate the following theorem.

Theorem 3.1 *Let $\mathcal{A}_1$ be a Banach algebra generated by flips $u, v, s_1, \dots, s_m$ that contained the dense subalgebra $Q_{2,m}$. Then the representations f_n, $n \in \mathcal{L}_1$, and $\pi_{\mu,n}, (\mu, n) \in \mathcal{L}_2$ are all continuous and define a $IS(2)$ for algebra $\mathcal{A}_1$.*

Proof: For $\mathcal{A}_1$, let $\mathcal{L}$ be $IS(2)$ that consists only of the representations f_n and $\pi_{z,n}$. In part 1) of the proof, we will be looking for f_n that lie in $\mathcal{L}$; in part 2) we do the same for $\pi_{z,n}$.

1) Let $t_0 \in \sigma(a_1)$. It is easy to show that $t_0 \in \cup_{j=0}^{2^{m+1}-1} B(j, 1/2)$ ($B(j, 1/2) = \{z \in \mathbb{C} \mid |z - j| < 1/2\}$). Suppose also that t_0 lies, say, in M_-, and $|t_0| > 2^m - 1/2$. Then there exists $n \in \mathcal{R}_1$ such that $n = [|t_0| + 1/2]$. The element $a_1 - t_0 e$ is not invertible in $\mathcal{A}_1$ by definition. On the other hand, $\forall k\ \forall z \neq \pm 1, |z| \leq \|c\|$, the element $\pi_{z,k}(a_1 - t_0 e) = \begin{pmatrix} b_1 & 0 \\ 0 & b_2 \end{pmatrix}$ is invertible, where $b_1 = \frac{1}{2}(\frac{c}{2\|c\|} + 2^m + \epsilon_1 + \dots + \epsilon_m \cdot 2^{m-1} + 2^{m+1} - 1) - t_0$, $b_2 = \frac{1}{2}(\frac{c}{2\|c\|} - 2^m + \epsilon_1 + \dots + \epsilon_m \cdot 2^{m-1} + 2^{m+1} - 1) - t_0$, $\epsilon_i = (-1)^{1+k_i}$. Indeed, suppose that this is not the case. Then $b_1 = 0$ or $b_2 = 0$. Whence it follows that $(\frac{z}{4\|c\|} - t_0)$ lie in $\mathbb{Z}$. But $\frac{|z|}{4\|c\|} \leq 1/4$, so $(\frac{z}{4\|c\|} = \pm(t_0 - [|t_0| + 1/2])$ and $z = \pm 1$. A contradiction.

By a direct calculation one can check that $f_n(a_1 - t_0 e) = 0$, and for any other $n_1 \in \mathcal{R}_1$, $n_1 \neq n$, the number $f_{n_1}(a_1 - t_0 e)$ is invertible. So f_n is necessarily in $IS(2)$. Therefore $\forall n \in \mathcal{R}_1$, $f_n \in IS(2)$ and f_n is continuous on A. It could happen that $n^* \notin \mathcal{R}_1$, but f_{n^*} is continuous on $\mathcal{A}_1$. We assert that this is impossible. Indeed, if this were the case, then we would add f_{n^*} to $IS(2)$, and this set would also be $IS(2)$ for $\mathcal{A}_1$. Then, for $b = (a_1 - f_{n^*}(a_1) \cdot e)$ we have that $f_{n^*}(b) = 0$, and this would imply that $t_0 = f_{n^*}(a_1) \in \sigma(a_1)$. Hence, $(\frac{f_{n^*}(c)}{4\|c\|} - t_0)$ lies in $\mathbb{Z}$ with $\frac{|f_{n^*}(c)|}{4\|c\|} \leq 1/4$ (since f_{n^*} is continuous). As a result we obtain that $\frac{f_{n^*}(c)}{4\|c\|} = \pm(t_0 - [\mid t_0 \mid +1/2])$. A direct calculation shows that $f_{n^*}(c) = \pm 1$ and $\pm(t_0 - [\mid t_0 \mid +1/2]) \cdot 4 \parallel c \parallel = \pm 1$. Therefore $\exists n \in \mathcal{R}_1$ such that $f_n(b) = 0$ and $f_n(a_1) = f_{n^*}(a_1)$. This is impossible.

2) Let $t_0 \in \sigma(a_2)$ and $t_0 \in M$. Then $\exists(\mu, n) \in \mathcal{R}_2$ such that $(\frac{\mu}{4\|c\|} + n) = t_0$. We assert that $\forall n \in \mathcal{R}_1$, $f_n(a_2 - t_0 \cdot e)$ is invertible. The proof is similar to that in 1), and we omit it. It is obvious that $\pi_{\mu,n}(a_2 - t_0 \cdot e)$ is not invertible and, as in 1), one can show that if $(z, k) \neq (\mu, n)$, $|z| \leq \|c\|$, then $\pi_{z,k}(a_2 - t_0 \cdot e)$ is invertible. So for all $(\mu, n) \in \mathcal{R}_2$, $\pi_{\mu,n}$ is continuous (they are the only continuous representations) on $\mathcal{A}_1$, and with f_n, $n \in \mathcal{R}_1$, they make $IS(2)$. □

We thus have constructed $IS(2)$ for the Banach algebra $\mathcal{A}_1$. For the Banach algebra $M_{2^n}(\mathcal{A}_1)$, the construction of the Invertibility Symbol is given above, and it follows the construction in [14].

References

[1] A. Böttcher, I. Gohberg, Yu. Karlovich, N. Krupnik, S. Roch, B. Silbermann and I. Spitkovsky, *Banach algebras generated by N idempotents and applications*. Operator Theory: Advances and Applications, vol. **90**, Birkhäuser Verlag, Basel 1996, 19–54.

[2] N. Krupnik, *Banach algebras with symbol and singular integral operators*. Operator Theory: Advances and Applications, vol. **90**, Birkhäuser Verlag, Basel, 1987.

[3] S. Roch and B. Silbermann, *Algebras generated by idempotents and the symbol calculus for singular integral operators*. IEOT **11** (1988), 385–419.

[4] N. Krupnik and E. Spigel, *Invertibility symbols for a Banach algebra generated by two idempotents and a shift*. IEOT **17** (1993), 597–578.

[5] S.A. Amitzur and J. Levitski, *Minimal identities for algebras*. Proc. Amer. Math. Soc. **1** (1950), 449–463.

[6] I.M. Gel'fand and V.A. Ponomarev, *Quadruples of subspaces of a finite dimensional vector space*. Doklady Academii Nauk SSSR **197** (1971), N. 4, 762–765 (In Russian).

[7] N. Bourbaki, *Groupes et Algebres de Lie*. Chapitres IV–VI. Hemann, Paris VI, 1968.

[8] P. Donovan and M.R. Freislich, *Some evidence for an extension of the Brauer–Thrall conjecture*. Sonderforschungsbereich Theor. Math., Bonn **40** (1972), 24–26.

[9] V.L. Ostrovskiĭ and Yu.S. Samoĭlenko, *Structure theorems for a pair of unbounded self-adjoint operators satisfying a quadratic relation*. Adv. Sov. Math. **9** (1992), 131–149.

[10] V.L. Ostrovskiĭ and Yu.S. Samoĭlenko, *Representations of quadratic $*$-algebras by bounded and unbounded operators*. Reports Math. Phys. **35** (1995), 283–301.

[11] Samoĭlenko Yuriĭ, Turowska Lyudmila and Popovych Stanislav, *Representations of a Cubic Deformation of $su(2)$ and Parasupersymmetric Commutation Relations*. Symmetry in Nonlinear Mathematical Physics vol. **2** (1997), 272–383.

[12] Slavik Rabanovich and Yuriĭ Samoĭlenko, *Represetations of $\mathcal{F}_n$-algebras and applications*. MFAT, vol. **4** (1998), N. 4, 86–96.

[13] V.L. Ostrovskiĭ and Yu. Samoĭlenko, *Introduction to the theory of representations of finitely presented $*$-algebras*, Harwood Academic Publisher, 1999.

[14] I. Gohberg and N. Krupnik, *Extension theorem for invertibility symbols in Banach algebras*. IEOT **15** (1992), 991–1010.

Institute of Mathematics
Ukrainian National Academy of Sciences
Tereshchinkivs'ka
3, Kiev, 252601
Ukraine
Yurii_sam@imath.kiev.ua

AMS Classification Numbers: Primary 46H15, 46K10, Secodary 46K99

Operator Theory:
Advances and Applications, Vol. 118

Ergodic Theorems with Rates for r-times Integrated Solution Families

Sen-Yen Shaw

We deduce strong and uniform ergodic theorems for r-times integrated solution families for the linear Volterra equation $u(t) = f(t) + \int_0^t a(t-s)Au(s)ds$. Rates of optimal convergence and non-optimal convergence, and sharpness of non-optimal rate of ergodic limits and approximate solutions of the linear functional equation $Ax = y$ are discussed. Specialization of the result to r-times integrated semigroups is observed.

1 Introduction

Let X be a Banach space and $B(X)$ the space of all bounded linear operators on X. Let A be a (not necessarily densely defined) closed linear operator in X and $a \in L^1_{loc}(\mathbb{R}^+)$ be a positive kernel such that $\int_0^\infty a(t)dt = \infty$. Consider the Volterra equation:

$$(VE, A, a, f) \qquad u(t) = f(t) + \int_0^t a(t-s)Au(s)ds, \; t \geq 0,$$

for $f \in C([0, \infty); X)$.

Let $r \in [0, \infty)$. A family $\{S(t); t \geq 0\}$ in $B(X)$ is called a *r-times integrated solution family* for (VE, A, a, f) (see [2], [19] for the case $r = n \in \mathbb{N}$) if

(S1) $S(\cdot)$ is strongly continuous on $[0, \infty)$, and $S(0) = I$ if $r = 0$ and 0 if $r > 0$;

(S2) $S(t)x \in D(A)$ and $AS(t)x = S(t)Ax$ for all $x \in D(A)$ and $t \geq 0$;

(S3) $a * S(t)x \in D(A)$ and $S(t)x = \frac{t^r}{\Gamma(r+1)}x + A\int_0^t a(t-s)S(s)xds$ for all $x \in X$ and $t \geq 0$.

A 0-times integrated solution family is also called a *solution family* or *resolvent family* ([3], [10], [14], [20]). Since A is closed, (S2) and (S3) imply

(S4) $S(t) = \frac{t^r}{\Gamma(r+1)}x + \int_0^t a(t-s)S(s)Axds$ for all $x \in D(A)$ and $t \geq 0$.

The notion of a r-times integrated solution family is an extension of the concepts of *r-times integrated semigroups* (see [1], [12], [17]) and *n-times integrated cosine functions* [30] (corresponding to the cases $a \equiv 1$ and $a(t) = t$, respectively). The existence of an n-times integrated solution family enables one to find the solution for the equation (VE, A, a, f) (see [19]).

Ergodic properties of resolvent families have been discussed in papers [3], [10], [14], [26], and [29]. In this paper, we shall generalize the results to r-times integrated solution families. What we are concerned with now is the convergence of some Cesáro type means $Q_m(t)$, $m \geq 1$ and Abel means of $S(\cdot)$ as $t \to \infty$. For the existence of the limits, the fulfilment of the condition $S(t) = O(t^r)$ $(t \to \infty)$ is required. Thus, throughout the paper we assume that

$$\|S(t)\| \leq Mt^r \text{ for all } t \geq 0. \tag{1.1}$$

We shall characterize the strong convergence, the uniform convergence, the rates of optimal and non-optimal convergence, and verify the sharpness of non-optimal convergence rates. We shall deduce the results from our results for general ergodic nets which have been established in a series of papers [22], [23], [24], [28]. We first quote these general results in Section 2, then in Sections 3 and 4 we deduce Cesáro ergodic theorems and Abelian ergodic theorems for r-times integrated solution families. Finally, in Section 5 we discuss specialization to r-times integrated semigroups. Similar results for integrated cosine functions can also be deduced.

2 Preliminaries

In [22–24] and [28] we considered the following framework for discussing general strong ergodic theorem, uniform ergodic theorem, and ergodic theorems with rates.

Let $A : D(A) \subset X \to X$ be a closed linear operator, and let $\{A_\alpha\}$ and $\{B_\alpha\}$ be two nets in $B(X)$ satisfying:

(C1) $\|A_\alpha\| \leq M$ for all α;

(C2) $R(B_\alpha) \subset D(A)$ and $B_\alpha A \subset AB_\alpha = I - A_\alpha$ for all α;

(C3) $R(A_\alpha) \subset D(A)$ for all α, and $\|AA_\alpha\| = O(e(\alpha))$;

(C4) $B_\alpha^* x^* = \varphi(\alpha) x^*$ for all $x^* \in R(A)^\perp$, and $|\varphi(\alpha)| \to \infty$;

(C5) $\|A_\alpha x\| = O(f(\alpha))$ (*resp.* $o(f(\alpha))$) implies $\|B_\alpha y\| = O(\frac{f(\alpha)}{e(\alpha)})$ (resp. $o(\frac{f(\alpha)}{e(\alpha)})$),

where e and f are positive functions satisfying $0 < e(\alpha) \leq f(\alpha) \to 0$. We call $\{A_\alpha\}$ an *A-ergodic net* and $\{B_\alpha\}$ its *companion net.*

The functions e and f are to act as estimators of the convergence rates of $\{A_\alpha x\}$ and $\{B_\alpha y\}$, which, in practical applications, approximate the ergodic limit and the solution x of $Ax = y$, respectively.

Let P and B_1 be the operators defined respectively by

$$\begin{cases} D(P) := \{x \in X; \lim_\alpha A_\alpha x \text{ exists}\}; \\ Px := \lim_\alpha A_\alpha x \text{ for } x \in D(P), \end{cases} \qquad \begin{cases} D(B_1) := \{y \in X; \lim_\alpha B_\alpha y \text{ exists}\}; \\ B_1 x := \lim_\alpha B_\alpha y \text{ for } y \in D(B_1). \end{cases}$$

$\{A_\alpha\}$ is said to be strongly (resp. uniformly) ergodic if $D(P) = X$ and $A_\alpha x \to Px$ for all $x \in X$ (resp. $\|A_\alpha - P\| \to 0$).

In [22, Theorem 1.1, Corollary 1.4 and Remark 1.7] we proved the following theorem.

Theorem 2.1 (Strong Ergodic Theorem). *Under conditions* (C1)–(C4) *the following are true.*

(i) *P is a bounded linear projection with range $R(P) = N(A)$, null space $N(P) = \overline{R(A)}$, and domain $D(P) = N(A) \oplus \overline{R(A)} = \{x \in X; \{A_\alpha x\}$ has a weak cluster point$\}$.*

(ii) *B_1 is the inverse operator A_1^{-1} of the restriction $A_1 := A|\overline{R(A)}$ of A to $\overline{R(A)}$; it has range $R(B_1) = D(A_1) = D(A) \cap \overline{R(A)}$ and domain $D(B_1) = R(A_1) = A(D(A) \cap \overline{R(A)})$. Moreover, for each $y \in D(B_1)$, $B_1 y$ is the unique solution of the functional equation $Ax = y$ in $\overline{R(A)}$.*

(iii) *$\{A_\alpha\}$ is strongly ergodic if and only if $N(A)$ separates $R(A)^\perp$, if and only if $\{A_\alpha x\}$ has a weak cluster point for each $x \in X$. In this case, we have $R(A) = R(A_1)$. These conditions are satisfied in particular when X is reflexive.*

Note that when A is densely defined, the condition $R(A) = R(A_1)$ is also equivalent to the strong ergodicity. The next theorem is proved in [23] under the assumption that A is densely defined. It can be shown that the conclusion still holds without this assumption.

Theorem 2.2 (Uniform Ergodic Theorem). *Under conditions* (C1)–(C3), *we have: $D(P) = X$ and $\|A_\alpha - P\| \to 0$ if and only if $\|B_\alpha|_{R(A)}\| = O(1)$, if and only if B_1 is bounded and $\|B_\alpha|_{R(A)} - B_1\| \to 0$, if and only if $R(A)$ (or $R(A_1)$) is closed, if and only if $R(A^2)$ (or $R(A_1^2)$) is closed, if and only if $X = N(A) \oplus R(A)$. Moreover, the convergence of these limits has order $O(e(\alpha))$.*

A Banach space X is called a *Grothendieck space* if every weakly* convergent sequence in X^* is weakly convergent (see e.g. [27] for equivalent definitions), and is said to have the *Dunford-Pettis property* if every weakly compact operator from X to any Banach space maps weakly compact sets into norm compact sets, or equivalently, if $\langle x_n, x_n^* \rangle \to 0$ whenever $x_n \to 0$ weakly in X and $x_n^* \to 0$ weakly in X^*. It is known that a Banach space X is a Grothendieck space with the Dunford-Pettis property if and only if every weak* convergent sequence in X^* converges weakly and uniformly on weakly compact subsets of X, if and only if every bounded linear operator from X into c_0 is weakly compact and maps weakly compact sets into norm compact sets. The spaces L^∞, H^∞, and $B(S, \Sigma)$ are particular examples of Grothendieck spaces with the Dunford-Pettis property. An interesting phenomenon in such spaces is that strong operator convergence often implies uniform operator convergence (see e.g., [16], [21], [25]). The following theorem slightly generalizes Theorem 2 in [23], which deals with the case that $\overline{D(A)} = X$.

Theorem 2.3 *Let* $\{\{A_\alpha\}, \{B_\alpha\}, A\}$ *satisfy conditions* (C1)–(C3), *and suppose* $\{x \in D(A); Ax \in Y\}$ *is dense in* $Y := \overline{D(A)}$. *When* Y *is a Grothendieck space with the Dunford-Pettis property,* $\{A_\alpha\}$ *is uniformly ergodic on* Y *if and only if it is strongly convergent on* Y.

Proof: We first see that $A_\alpha Ax = AA_\alpha x$ for all $x \in D(A)$. Indeed, if $x \in D(A)$, then by (C2) we have $A_\alpha x = x - B_\alpha Ax \in D(A)$ and $AA_\alpha x = Ax - AB_\alpha Ax = Ax - (I - A_\alpha)Ax = A_\alpha Ax$. Hence $A_\alpha A \subset AA_\alpha$. Note that Y is an invariant subspace for A_α and B_α. Let A_α° and B_α° denote their restrictions to Y, and let A° denote the part of A in Y. Being the intersection of the closed graph of A and $Y \times Y$, the graph of A° is closed. Using this and the fact that $R(A_\alpha) \subset D(A)$ and $A_\alpha A \subset AA_\alpha$, we easily see that $R(A_\alpha^\circ) \subset A^\circ$, $A_\alpha^\circ A^\circ \subset A^\circ A_\alpha^\circ$ and $\|A^\circ A_\alpha^\circ\|_Y = O(e(\alpha))$, i.e., (C3) holds with A_α and A replaced by A_α° and A°, respectively. Similarly, (C2) with the closedness of A° implies that it holds with A_α, B_α, and A replaced by A_α° B_α°, and A°, respectively. Hence $\{A_\alpha^\circ\}$ is an A°-ergodic net on Y and $\{B_\alpha^\circ\}$ is its companion net. Since $D(A^\circ) = \{x \in D(A); Ax \in Y\}$ is assumed to be dense in Y, and Y is assumed to be a Grothendieck space with the Dunford-Pettis property, it follows from Theorem 2 of [23] that the strong convergence of $\{A_\alpha^\circ\}$ on Y implies its operator-norm convergence. □

The rates of covergence of ergodic limits are characterized by means of *K-functional* and *relative completion*, which we recall as below.

Let X be a Banach space with norm $\|\cdot\|_X$, and Y a submanifold with seminorm $\|\cdot\|_Y$. The *K-functional* is defined by

$$K(t, x) := K(t, x, X, Y, \|\cdot\|_Y) = \inf_{y \in Y}\{\|x - y\|_X + t\|y\|_Y\}.$$

If Y is a Banach space with norm $\|\cdot\|_Y$, the *completion of* Y *relative to* X is defined as

$$\tilde{Y}^X := \Big\{x \in X : \exists\{x_m\} \subset Y \text{ such that } \lim_{m\to\infty}\|x_m - x\|_X = 0 \text{ and } \sup\|x_m\|_Y < \infty\Big\}.$$

It is known [4] that $K(t, x)$ is a bounded, continuous, monotone increasing and subadditive function of t for each $x \in X$, and $K(t, x, X, Y, \|\cdot\|_Y) = O(t)$ $(t \to 0^+)$ if and only if $x \in \tilde{Y}_X$.

We next specify the required notations. Let $X_1 := \overline{R(A)}$ and $X_0 := D(P) = N(A) \oplus X_1$. Since the operator $B_1 : D(B_1) \subset X_1 \to X_1$ is closed, its domain $D(B_1)\,(= R(A_1))$ is a Banach space with respect to the norm $\|x\|_{B_1} := \|x\| + \|B_1 x\|$. Let $B_0 : D(B_0) \subset X_0 \to X_0$ be the operator $B_0 := 0 \oplus B_1$. Then its domain

$$D(B_0)\,(= N(A) \oplus D(B_1) = N(A) \oplus A(D(A) \cap \overline{R(A)}))$$

is a Banach space with norm $\|x\|_{B_0} := \|x\| + \|B_0 x\|$, and $[D(B_0)]\tilde{}_{X_0} = N(A) \oplus [D(B_1)]\tilde{}_{X_1}$.

Now we can state the following theorem [28, Theorem 1], which is concerned with optimal convergence and non-optimal convergence rates of ergodic limits and approximate solutions.

Theorem 2.4 *Under conditions* (C1)–(C5) *the following statements hold.*

(i) *For* $x \in X_0 = N(A) \oplus \overline{R(A)}$, *one has*

$$\begin{aligned}\|A_\alpha x - Px\| = O(f(\alpha)) &\iff K(e(\alpha), x, X_0, D(B_0), \|\cdot\|_{B_0}) \\ = O(f(\alpha)) &\iff x \in [D(B_0)]\tilde{}_{X_0} incase f = e).\end{aligned}$$

(ii) *For* $x \in \overline{R(A)}$, *one has*

$$\begin{aligned}\|A_\alpha x\| = O(f(\alpha)) &\iff K(e(\alpha), x, X_1, D(B_1), \|\cdot\|_{B_1}) \\ = O(f(\alpha)) &\iff x \in [D(B_1)]\tilde{}_{X_1} \text{ (in case } f = e).\end{aligned}$$

(iii) *For* $y \in D(B_1) = R(A_1)$ *one has*

$$\begin{aligned}&\|B_\alpha y - B_1 y\| = O(f(\alpha)) \iff K(e(\alpha), B_1 y, X_1, D(B_1), \|\cdot\|_{B_1}) \\ &= O(f(\alpha)) \iff y \in A(D(A) \cap [D(B_1)]\tilde{}_{X_1}) \text{ (in case } f = e).\end{aligned}$$

The saturation case ($f = e$) was proved in [24]. It was also shown there that for $x \in X_0$, $\|A_\alpha x - Px\| = o(e(\alpha))$ if and only if $x \in N(A)$; for $x \in X$ $\|B_\alpha x\| = o(1)$ if and only if $x = 0$; for $y \in D(B_1) = R(A_1)$, $\|B_\alpha y - B_1 y\| = o(e(\alpha))$ if and only if $y = 0$. Thus, when $A \neq 0$, the rate of optimal convergence of $\|A_\alpha y\| = O(e(\alpha))$ is sharp everywhere on $[D(B_1)]\tilde{}_{X_1} \backslash \{0\}$. The following theorem [28, Theorem 2] shows that the non-optimal convergence rate: $\|A_\alpha y\| = O(f(\alpha))$ with f satisfying $f(\alpha)/e(\alpha) \to \infty$ is sharp.

Theorem 2.5 *Suppose that A,* $\{A_\alpha\}$, *and* $\{B_\alpha\}$ *satisfy conditions* (C1)–(C5), *with* $f(\alpha)/e(\alpha) \to \infty$. *Then* $R(A)$ *is not closed if and only if there exists an element* $y_f \in X_1$ *such that*

$$\|A_\alpha y_f\| \begin{cases} = O(f(\alpha)); \\ \neq o(f(\alpha)). \end{cases}$$

Moreover,

$$\|A_\alpha(x + y_f) - P(x + y_f)\| \begin{cases} = O(f(\alpha)); \\ \neq o(f(\alpha)) \end{cases}$$

for all $x \in N(A)$.

3 Cesáro Ergodic Theorems

Let $S(\cdot)$ be a r-times integrated solution family as defined in Section 1. We shall use the following notations from [15].

Put $j_r(t) = \frac{t^r}{\Gamma(r+1)}$ for $t \geq 0$ and $r \geq 0$ and denote by a_0 the *Dirac measure* δ_0 at 0. For each $m \geq 0$, let $a_{m+1}(t) = a * a_m(t)$ for $t \geq 0$, let $k_m(t) = \frac{a_{m+1} * j_r(t)}{a_m * j_r(t)}$ for $t > 0$, and define

$$Q_m(t)x = \frac{a_m * S(t)x}{a_m * j_r(t)} \quad \text{for } x \in X \text{ and } t > 0.$$

By the assumption (1.1) we have

$$\begin{aligned} \|Q_m(t)x\| &\leq \frac{1}{a_m * j_r(t)} \int_0^t a_m(t-s)\|S(s)x\|ds \\ &\leq \frac{M\|x\|}{a_m * j_r(t)} \int_0^t a_m(t-s)s^r ds = M\Gamma(r+1)\|x\| \end{aligned}$$

for all $x \in X$, so that

(3.1) $$\|Q_m(t)\| \leq M\Gamma(r+1) \qquad (m \geq 0,\ 0 < t \leq 1).$$

The next lemma relates A, $Q_m(t)$, and $Q_{m+1}(t)$.

Lemma 3.1 [15, Lemma 2.1 and Prop. 2.2]. *Suppose that $\|S(t)\| \leq Mt^r$ for all $t \geq 0$. Then*

(a) *$Q_m(t)x \in D(A)$ and $AQ_m(t)x = Q_m(t)Ax$ for all $x \in D(A)$, $m \geq 0$ and $t > 0$;*

(b) *$R(Q_{m+1}(t)) \subset D(A)$, and $Q_{m+1}(t)A \subset AQ_{m+1}(t) = \frac{1}{k_m(t)}[Q_m(t) - I]$ for all $m \geq 0$ and $t > 0$;*

(c) *The part A° of A in $Y := \overline{D(A)}$ is densely defined in Y.*

Proof: (S2) implies that $Q_0(t)(D(A)) \subset D(A)$ and $AQ_0(t)x = Q_0(t)Ax$ for $x \in D(A)$ and $t > 0$, which and (S3) imply that

$$Q_1(t)X \subset D(A) \text{ and } Q_1(t)A \subset AQ_1(t) = \frac{1}{k_0(t)}(Q_0(t) - I).$$

Then by induction and the closedness of A we have $Q_{m+1}(t)X \subset D(A)$ and $Q_{m+1}(t)A \subset AQ_{m+1}(t)$ for $m \geq 0$. Moreover, we have, by (S3),

$$\begin{aligned} AQ_{m+1}(t)x &= \frac{1}{a_{m+1} * j_r(t)}[a_m * A(a * S(t)x)] \\ &= \frac{1}{a_{m+1} * j_r(t)}[(a_m * S(t)x) - a_m * j_r(t)x] \\ &= \frac{1}{k_m(t)}[Q_m(t)x - x] \end{aligned}$$

for all $x \in X$. This proves (a) and (b).

(c) From (b) we can derive that

$$(3.2)\qquad \begin{aligned} &Q_{m+1}(t)Y \subset D(A^\circ) \text{ and } Q_{m+1}(t)A^\circ \subset A^\circ Q_{m+1}(t)|_Y \\ &= \frac{1}{k_m(t)}(Q_m(t) - I)|_Y \end{aligned}$$

for all $m \geq 0$ and $t > 0$. Since $k_m(t) \to 0$ as $t \to 0^+$, it follows from (b) that for $m \geq 0$, $Q_{m+1}(t)x \to x$ if and only if $x \in Y$. This fact together with (3.2) shows in particular that $D(A^\circ)$ is dense in Y.

For $m \geq 0$ and $t > 0$ let A_t and B_t be operators defined respectively by

$$A_t := Q_{m+1}(t) \quad \text{and} \quad B_t x := -k_{m+1}(t)Q_{m+2}(t) = -\frac{a_{m+2} * S(t)}{a_{m+1} * j_r(t)}.$$

Then (b) of Lemma 3.1 becomes

$$R(B_t) \subset D(A) \text{ and } B_t A \subset AB_t = I - A_t \text{ for } t > 0,$$

that is, condition (C2) is satisfied. Moreover, $A_t A \subset AA_t = (k_m(t))^{-1}(Q_m(t) - I)$ for $t > 0$. Hence conditions (C1) and (C3) hold with $e(t) = 1/k_m(t)$.

We suppose $k_m(t) \to \infty$ and $k_{m+1}(t) \to \infty$ as $t \to \infty$. To check (C4) let $x^* \in R(A)^\perp$. Then, by (S3) we have

$$\langle x, S^*(t)x^*\rangle = j_r(t)\langle x, x^*\rangle + \left\langle A\int_0^t a(t-s)S(s)xds, x^*\right\rangle = j_r(t)\langle x, x^*\rangle$$

for all $x \in X$, so that $S^*(t)x^* = j_r(t)x^*$ and hence

$$(B_t)^*(t)x^* = -\frac{a_{m+2} * S^*(t)x^*}{a_{m+1} * j_r(t)} = -\frac{a_{m+2} * j_r(t)}{a_{m+1} * j_r(t)}x^* = -k_{m+1}(t)x^*$$

for all $t \geq 0$. Thus condition (C4) is satisfied.

Finally, to see (C5) with $f(t) = e(t)^\beta = (k_m(t))^{-\beta}$, $0 < \beta \leq 1$, let x be such that $\|A_t x\| \leq M_x(k_m(t))^{-\beta}$. Then

$$\begin{aligned} \|B_t x\| &= \left\|\frac{a_{m+2} * S(t)}{a_{m+1} * j_r(t)}\right\| = \left\|\frac{a * [a_{m+1} * j_r(t)A_t x]}{a_{m+1} * j_r(t)}\right\| \\ &\leq \frac{a * a_m * j_r(t)k_m(t)(k_m(t))^{-\beta}M_x}{a_{m+1} * j_r(t)} = M_x(k_m(t))^{1-\beta} \end{aligned}$$

for all $t \geq 0$. Similarly, $\|A_t x\| = o((k_m(t))^{-\beta})$ implies $\|B_t x\| = o((k_m(t))^{1-\beta})$.

Now, applying Theorems 2.1, 2.4, and 2.5, we can formulate the following strong ergodic theorem with rates. □

Theorem 3.2 *Let $a \in L^1_{loc}(\mathbb{R}^1)$ be a positive function such that $k_m(t) \to \infty$ and $k_{m+1}(t) \to \infty$ as $t \to \infty$, and let $S(\cdot)$ be a r-times integrated solution family for* *the equation* $u = f + k * Au$ *satisfying* $\|S(t)\| \le Mt^r$ forall$t \ge 0$. *For* $m \ge 0$ *we have:*

(i) *The mapping* $P : x \to \lim_{t\to\infty} Q_{m+1}(t)x$ *is a bounded linear projection with* $R(P) = N(A)$, $N(P) = \overline{R(A)}$, *and*

$$D(P) = N(A) \oplus \overline{R(A)} = \{x \in X; \{Q_{m+1}(t)x\} \text{ has a weak cluster point}\}.$$

For $0 < \beta \le 1$ *and* $x \in X_0 := D(P) = N(A) \oplus \overline{R(A)}$, *one has:*

$$\begin{aligned} &\|Q_{m+1}(t)x - Px\| = O((k_m(t))^{-\beta}) \\ &\iff K((k_m(t))^{-1}, x, X_0, D(B_0), \|\cdot\|_{B_0}) = O((k_m(t))^{-\beta}) \\ &\iff x \in [D(B_0)]\tilde{}_{X_0} \ (\text{in case } \beta = 1). \end{aligned}$$

(ii) *The mapping* $B_1 : y \to -\lim_{t\to\infty} k_{m+1}(t)Q_{m+2}(t)y$ *is the inverse operator* A_1^{-1} *of the restriction* $A_1 := A|\overline{R(A)}$ *of* A *to* $\overline{R(A)}$*; it has range* $R(B_1) = D(A) \cap \overline{R(A)}$, *and domain* $D(B_1) = A(D(A) \cap \overline{R(A)})$. *For each* $y \in A(D(A) \cap \overline{R(A)})$, $B_1 y$ *is the unique solution of the functional equation* $Ax = y$ *in* $\overline{R(A)}$, *and we have for* $0 < \beta \le 1$

$$\begin{aligned} &\|k_{m+1}(t)Q_{m+2}(t)y + A_1^{-1}y\| = O((k_m(t))^{-\beta})\ (t \longrightarrow \infty) \\ &\iff K((k_m(t))^{-1}, B_1y, X_1, D(B_1), \|\cdot\|_{B_1}) = O((k_m(t))^{-\beta})\ (t \longrightarrow \infty) \\ &\iff y \in [D(B_1)]\tilde{}_{X_1} \ (\text{in case } \beta = 1). \end{aligned}$$

(iii) $R(A)$ *is not closed if and only if for every (some)* $0 < \beta < 1$ *there is a* $y_\beta \in \overline{R(A)}$ *such that*

$$\|Q_{m+1}(t)y_\beta ds\| \begin{cases} = O((k_m(t))^{-\beta}) \\ \ne o((k_m(t))^{-\beta}) \end{cases} (t \to \infty).$$

In view of (c) of Lemma 3.1 we can apply Theorems 2.2 and 2.3 to deduce the following uniform ergodic theorem.

Theorem 3.3 *Under the hypothesis in Theorem* 3.2 *we have:*

(i) $\|Q_{m+1}(t) - P\| \to 0$ *if and only if* $\|k_{m+1}(t)Q_{m+2}(t) + A_1^{-1}\|_{R(A)} \to 0$, *if and only if* $R(A)$ *is closed, if and only if* $R(A^2)$ *is closed, if and only if* $X = N(A) \oplus R(A)$. *In this case, we have* $\|Q_{m+1}(t) - P\| = O((k_m(t))^{-1})$ *and* $\|k_{m+1}(t)Q_{m+2}(t) + A_1^{-1}\|_{R(A)} = O((k_m(t))^{-1})$.

(ii) *If* $Y := \overline{D(A)}$ *is a Grothendieck space with the Dunford-Pettis property, and if* $Y \subset D(P)$, *then* $\|Q_{m+1}(t)|_Y - P|_Y\| = O((k_m(t))^{-1})$.

4 Abelian Ergodic Theorems

Next, we consider the case that the kernel $a \in L^1_{loc}(\mathbb{R}^+)$ has the Laplace transform $\hat{a}(\lambda) = \int_0^\infty e^{-\lambda t}a(t)dt < \infty$ for all $\lambda > 0$. Under the assumption (1.1), we can take Laplace transform of the equation in (S3) to obtain

$$\hat{S}(\lambda)x = \begin{cases} \frac{1}{\lambda^{n+1}}x + \hat{a}(\lambda)\hat{S}(\lambda)Ax, & x \in D(A), \\ \frac{1}{\lambda^{n+1}}x + A\hat{a}(\lambda)\hat{S}(\lambda)x, & x \in X \end{cases}$$

for $\lambda > 0$. Thus

$$\lambda^{n+1}\hat{a}(\lambda)\hat{S}(\lambda)((\hat{a}(\lambda))^{-1} - A) \subset ((\hat{a}(\lambda))^{-1} - A)\lambda^{n+1}\hat{a}(\lambda)\hat{S}(\lambda) = I,$$

that is, $(\hat{a}(\lambda))^{-1} \in \rho(A)$ and $((\hat{a}(\lambda))^{-1} - A)^{-1} = \lambda^{n+1}\hat{a}(\lambda)\hat{S}(\lambda)$ for all $\lambda > 0$. Moreover, (1.1) implies

$$\begin{aligned} \|(\hat{a}(\lambda))^{-1}((\hat{a}(\lambda))^{-1} - A)^{-1}\| &= \|\lambda^{n+1}\hat{S}(\lambda)\| \\ &= \left\|\lambda^{n+1}\int_0^\infty e^{-\lambda t}S(t)dt\right\| \le M\Gamma(r+1) \end{aligned}$$

for all $\lambda > 0$.

By the assumption $\int_0^\infty a(t)dt = \infty$ we have $(\hat{a}(\lambda))^{-1} \to 0^+$ as $\lambda \to 0^+$. It is easy to show that $A_\lambda := (\hat{a}(\lambda))^{-1}((\hat{a}(\lambda))^{-1} - A)^{-1}$ is an A-ergodic net and $B_\lambda := -((\hat{a}(\lambda))^{-1} - A)^{-1}$ is a companion net, with $e(\lambda) = (\hat{a}(\lambda))^{-1} \to 0^+$ and $\phi(\lambda) = \hat{a}(\lambda) \to \infty$ as $\lambda \to 0^+$.

Hence we can apply Theorems 2.1, 2.4, and 2.5 to formulate the following strong ergodic theorem with rates.

Theorem 4.1 *Suppose that* $\int_0^\infty a(t)dt = \infty$ *and* $\hat{a}(\lambda) = \int_0^\infty e^{-\lambda t}a(t)dt < \infty$ *for all* $\lambda > 0$, *and suppose* $\|S(t)\| \le Mt^r$ *for all* $t \ge 0$. *Then the following are true for* $0 < \beta \le 1$:

(i) *For* $x \in X_0$, *one has*

$$\begin{aligned} &\|(\hat{a}(\lambda))^{-1}((\hat{a}(\lambda))^{-1} - A)^{-1}x - Px\| = O((\hat{a}(\lambda))^{-\beta})(\lambda \longrightarrow 0^+) \\ &\iff K(\lambda, x, X_0, D(B_0), \|\cdot\|_{B_0}) = O(\lambda^\beta)\ (\lambda \longrightarrow 0^+) \\ &\iff x \in [D(B_0)]^\sim_{X_0}\ (\textit{in case } \beta = 1). \end{aligned}$$

(ii) *For* $y \in D(B_1) = R(A_1)$, *one has*

$$\begin{aligned} &\|((\hat{a}(\lambda))^{-1} - A)^{-1}y + B_1y\| = O((\hat{a}(\lambda))^{-\beta})\ (\lambda \longrightarrow 0^+) \\ &\iff K(\lambda, B_1y, X_1, D(B_1), \|\cdot\|_{B_1}) = O(\lambda^\beta)\ (\lambda \longrightarrow 0^+) \\ &\iff x \in [D(B_1)]^\sim_{X_1}\ (\textit{in case } \beta = 1). \end{aligned}$$

(iii) $R(A)$ *is not closed if and only if for each (some)* $0 < \beta < 1$ *there exists an element* $y_\beta \in \overline{R(A)}$ *such that*

$$\|(\hat{a}(\lambda))^{-1}((\hat{a}(\lambda))^{-1} - A)^{-1}y_\beta\| \begin{cases} = O((\hat{a}(\lambda))^{-\beta}) \\ \ne o((\hat{a}(\lambda))^{-\beta}) \end{cases} (\lambda \to 0^+).$$

¿From Theorems 2.2 and 2.3 one can deduce the following uniform ergodic theorem.

Theorem 4.2 *Under the hypothesis in Theorem* 4.1 *we have:*

(i) $\|(\hat{a}(\lambda))^{-1}((\hat{a}(\lambda))^{-1}-A)^{-1}-P\| \to 0$ *if and only if* $\|((\hat{a}(\lambda))^{-1}-A)^{-1}+A_1^{-1}\|_{R(A)} \to 0$, *if and only if* $R(A)$ *is closed, if and only if* $R(A^2)$ *is closed, if and only if* $X = N(A) \oplus R(A)$. *In this case, we have* $\|(\hat{a}(\lambda))^{-1}((\hat{a}(\lambda))^{-1}-A)^{-1}-P\| = O((\hat{a}(\lambda))^{-\beta})\,(\lambda \to 0^+)$ *and* $\|((\hat{a}(\lambda))^{-1}-A)^{-1}+A_1^{-1}\|_{R(A)} = O((\hat{a}(\lambda))^{-\beta})\,(\lambda \to 0^+)$.

(ii) *If* $Y := \overline{D(A)}$ *is a Grothendieck space with the Dunford-Pettis property, and if* $Y \subset D(P)$, *then* $\|(\hat{a}(\lambda))^{-1}((\hat{a}(\lambda))^{-1}-A)^{-1}|_Y - P|_Y\| = O((\hat{a}(\lambda))^{-1})\,(\lambda \to 0^+)$.

5 Specialization to r-times Integrated Semigroups

For a $r > 0$ a family $\{T(t); t \geq 0\}$ in $B(X)$ is called a *r-times integrated semigroup* if $T(\cdot)$ is strongly continuous on $[0,\infty)$ and $T(0) = 0$ and

$$T(t)T(s) = \frac{1}{\Gamma(r)}\left(\int_0^{t+s} - \int_0^t - \int_0^s\right)(t+s-r)^{r-1}T(r)dr$$

for $s, t \geq 0$. A C_0-semigroup is called a 0-times integrated semigroup.

N-times integrated semigroups were introduced in [1], and fractionally integrated semigroups have been studied in [12] and [17].

If $T(\cdot)$ is *nondegenerate* in the sense that $x = 0$ whenever $T(t)x = 0$ for all $t > 0$, then there exists uniquely a closed (but not necessarily densely defined) operator A such that $x \in D(A)$ and $Ax = y$ if and only if $T(t)x = \int_0^t T(s)yds + \frac{t^r}{\Gamma(r+1)}x$ for all $t \geq 0$. This operator A is called the *generator* of $T(\cdot)$. It satisfies $T(t)D(A) \subset D(A)$ and $T(t)A \subset AT(t)$, and

$$\int_0^t T(s)xds \in D(A) \text{ and } A\int_0^t T(s)xds = T(t)x - \frac{t^r}{\Gamma(r+1)}x,\ x \in X,\ t \geq 0.$$

Thus $T(\cdot)$ ia a r-times integrated solution family for (VE, A, a, f), with $a \equiv 1$. In this case, we have $k_0(t) = t/(r+1)$, $Q_1(t) = \Gamma(r+2)t^{-r-1}\int_0^t T(s)ds$ and $Q_2(t) = \Gamma(r+3)t^{-r-2}\int_0^t\int_0^s T(u)duds$, and $\hat{a}(\lambda) = 1/\lambda$. Hence Theorems 3.2 and 4.1 reduce to the following theorem.

Theorem 5.1 *Let* $\{T(t); t \geq 0\}$ *be a nondegenerate r-times integrated semigroup with generator A, and suppose* $\|T(t)\| \leq Mt^r$ *for all* $t \geq 0$.

(i) *The mapping* $P : x \to \lim_{t\to\infty}\Gamma(r+2)t^{-r-1}\int_0^t T(s)xds$ *is a bounded linear projection with* $R(P) = N(A)$, $N(P) = \overline{R(A)}$, *and* $D(P) = N(A)\oplus\overline{R(A)}$. *For* $0 < \beta \leq 1$ *and* $x \in D(P)$, *we have* $\|\Gamma(r+2)t^{-r-1}\int_0^t T(s)xds - Px\| = O(t^{-\beta})\,(t \to \infty)$ *if and only if* $\|\lambda(\lambda-A)^{-1}x - Px\| = O(\lambda^\beta)\,(\lambda \to 0^+)$, *if and only if* $K(\lambda, x, X_0, D(B_0), \|\cdot\|_{B_0}) = O(\lambda^\beta)$ $(\lambda \to 0^+)$. *Moreover, in case* $\beta = 1$, *these conditions are equivalent to that* $x \in [D(B_0)]^{\sim}_{X_0}$.

(ii) *The mapping* $B_1 : y \to -\lim_{t\to\infty} \Gamma(r+2)t^{-r-1} \int_0^t \int_0^s T(u)xduds$ *is the inverse operator* A_1^{-1} *of the restriction* $A_1 := A|\overline{R(A)}$ *of* A *to* $\overline{R(A)}$; *it has range* $R(B_1) = D(A) \cap \overline{R(A)}$, *and domain* $D(B_1) = A(D(A) \cap \overline{R(A)})$. *For each* $y \in A(D(A)\cap\overline{R(A)})$ $B_1 y$ *is the unique solution of the functional equation* $Ax = y$ *in* $\overline{R(A)}$. *For* $0 < \beta \le 1$ *we have* $\|\Gamma(r+2)t^{-r-1} \int_0^t \int_0^s T(u)yduds + A_1^{-1}y\| = O(t^{-\beta})$ $(t \to \infty)$ *if and only if* $\|(\lambda - A)^{-1}y + A_1^{-1}y\| = O(\lambda^\beta)$ $(\lambda \to 0^+)$, *if and only if* $K(\lambda, B_1 y, X_1, D(B_1), \|\cdot\|_{B_1}) = O(\lambda^\beta)$ $(\lambda \to 0^+)$. *Moreover, in case* $\beta = 1$, *these conditions are equivalent to that* $y \in [D(B_1)]^{\sim}_{X_1}$.

(iii) $R(A)$ *is not closed if and only if for every (some)* $0 < \beta < 1$ *there are* $y_\beta, y'_\beta \in \overline{R(A)}$ *such that*

$$\left\|\Gamma(r+2)t^{-r-1}\int_0^t T(s)y_\beta ds\right\| \begin{cases} = O(t^{-\beta}) \\ \neq o(t^{-\beta}) \end{cases} (t \longrightarrow \infty),$$

$$\|\lambda(\lambda - A)^{-1}y'_\beta\| \begin{cases} = O(\lambda^\beta) \\ \neq o(\lambda^\beta) \end{cases} (\lambda \longrightarrow 0^+).$$

Remarks: When $r = 0$, (i) reduces to Butzer and Dickmeis' result [5] on C_0-semigroups, and (iii) becomes the result in [18]. When $r = n > 0$ is a natural number, the theorem is proved in [28] (see also [24, Theorem 5] for (i) and (ii) for the case $\beta = 1$).

Theorems 3.3 and 4.2 reduce to the following theorem.

Theorem 5.2 *Under the hypothesis in Theorem* 5.1, *we have:*

(i) $\|\Gamma(r+2)t^{-r-1}\int_0^t T(s)ds - P\| \to 0$ *if and only if* $\|\lambda(\lambda - A)^{-1} - P\| \to 0$, *if and only if* $\|\Gamma(r+2)t^{-r-1}\int_0^t\int_0^s T(u)duds + A_1^{-1}\|_{R(A)} \to 0$, *if and only if* $\|(\lambda - A)^{-1} + A_1^{-1}\|_{R(A)} \to 0$, *if and only if* $R(A)$ *is closed, if and only if* $R(A^2)$ *is closed, if and only if* $X = N(A) \oplus R(A)$. *In this case, we have* $\|\Gamma(r+2)t^{-r-1}\int_0^t T(s)ds - P\| = O(t^{-1})(t \to \infty)$ *and* $\|\lambda(\lambda - A)^{-1} - P\| = O(\lambda)(\lambda \to 0^+)$.

(ii) *If* $Y := \overline{D(A)}$ *is a Grothendieck space with the Dunford-Pettis property, and if* $Y \subset D(P)$, *then* $\|\Gamma(r+2)t^{-r-1}\int_0^t T(s)|_Y ds - P|_Y\|_Y = O(t^{-1})(t \to \infty)$ *and* $\|\lambda(\lambda - A)^{-1}|_Y - P|_Y\|_Y = O(\lambda)(\lambda \to 0^+)$.

References

[1] W. Arendt, *Vector-valued Laplace transforms and Cauchy problems*, Israel J. Math. **59** (1987), 327–352.

[2] W. Arendt and H. Kellerman, *Integrated solutions of Volterra integrodifferential equations and applications*, in: Volterra Integrodifferential Equations in Banach Spaces, Pitman Res. Notes in Math., vol. **190** (1989), pp. 21–51.

[3] W. Arendt and J. Prüss, *Vector-valued Tauberian theorems and asymptotic behavior of linear Volterra equations*, SIAM J. Math. Anal. **23** (1992), 412–448.

[4] H. Berens, *Interpolationsmethoden zur Behandlung von Approximationsprozessen auf Banachräumen*, Lecture Notes in Mathematics **64**, Springer-Verlag, Berlin, Heidelberg, New York, 1968.

[5] P.L. Butzer and W. Dickmeis, *Direct and inverse mean ergodic theorems with rates for semigroups of operators*, in: Approximation and Function Spaces (Proc. Conf. Gdansk (Poland), August, 27–31, 1979) North-Holland, Amsterdam 1981, xiv + 897, 191–206.

[6] P.L. Butzer and A. Gessinger, *The mean ergodic theorem for cosine operator functions with optimal and non-optimal rates*, Acta Math. Sci. Hungarica (4) **68** (1995).

[7] P.L. Butzer and A. Gessinger, *Ergodic theorems for semigroups and cosine operator functions at zero and infinity with rates*; applications to partial differential equations. A survey, *Contemporary Mathematics* **190** (1995), 67–93.

[8] P.L. Butzer and U. Westphal, *The mean ergodic theorem and saturation*, Indiana Univ. Math. J. **20** (1971), 1163–1174.

[9] P.L. Butzer and U. Westphal, *Ein Operatorenkalkül für das approximations theoretische Verhalten des Ergodensatzes in Mittel*, in Linear Operators and Approximation I (Proc. Conf. Oberwolfach 1971; P.L. Butzer, J.P. Kahane and B. Sz.-Nagy, Eds.) ISNM 20, Birkhäuser Verlag, Basel, 1972, 102–113.

[10] J.-C. Chang and S.-Y. Shaw, *Rates of approximation and ergodic limits of resolvent families*, Arch. Math. **66** (1996), 320–330.

[11] O.V. Davydov, *Some condensation theorems for semigroup operators*, Manuscripta Math. **79** (1993), 435–446.

[12] M. Hieber, *Laplace transforms and α-times integrated semigroups*, Forum Math. **3** (1991), 595–612.

[13] Y.-C. Li and S.-Y. Shaw, *An abstract ergodic theorem and some inequalities for operators on Banach spaces*, Proc. Amer. Math. Soc. **125** (1997), 111–119.

[14] C. Lizama, *A mean ergodic theorem for resolvent operators*, Semigroup Forum **47** (1993), 227–230.

[15] H. Liu and S.-Y. Shaw, *Rates of local ergodic limits of r-times integrated solution families*, preprint.

[16] H.P. Lotz, *Tauberian theorems for operators on L^∞ and similar spaces*, in Functional Analysis: Surveys and Recent Results III, 1984, pp. 117–133.

[17] I. Miyadera, M. Okubo and N. Tanaka, *α-times integrated semigroups and abstract Cauchy problems*, Memoirs of the School of Science & Engineering **57** (1993), 256–289.

[18] D. Nasri-Roudsari, R.J. Nessel and R. Zeler, *Resonance principles with applications to mean ergodic theorems and projection operators*, Acta Math, Sci. Hungarica (3) **68** (1995).

[19] H. Oka, *Linear Volterra equations and integrated solution families*, Semigroup Forum **53** (1996), 278–297.

[20] J. Prüss, *Evolutionary Integral Equations and Applications*, Birkhäuser Verlag, 1993.

[21] S.-Y. Shaw, *Asymptotic behavior of pseudo-resolvents on some Grothendieck spaces*, Publ. RIMS, Kyoto Univ. **24** (1988), 277–282.

[22] S.-Y. Shaw, *Mean ergodic theorems and linear functional equations*, J. Funct. Anal. **87** (1989), 428–441.

[23] S.-Y. Shaw, *Uniform convergence of ergodic limits and approximate solutions*, Proc. Amer. Math. Soc. **114** (1992), 405–411.

[24] S.-Y. Shaw, *Convergence rates of ergodic limits and approximate solutions*, J. Approximation Theory **75** (1993), 157–166.

[25] S.-Y. Shaw, *On uniform ergodic theorems in Grothendieck spaces with the Dunford-Pettis property*, Houston J. Math. **20** (1994), 313–320.

[26] S.-Y. Shaw, *Ergodic properties of integrated semigroups and resolvent families*, Proceedings IMC '94, World Scientific, 1996, 171–178.

[27] S.-Y. Shaw, *Grothendieck space, in*: Encyclopaedia of Mathematics, supplement I, Kluwer Academic Publishers, 1997, p. 278.

[28] S.-Y. Shaw, *Non-optimal rates of ergodic limits and approximate solutions*, J. Approximation Theory **94** (1998), 285–299.

[29] S.-Y. Shaw, *Abstract ergodic theorems*, Proceedings of the 3rd International Conference on Functional Analysis and Approximation Theory, Acquafredda di Maratea, Italy, September 23–28, 1996; Rendiconti del Circolo Matematico di Palermo, Serie II, Suppl. **52** (1998), 141–155.

[30] S.-Y. Shaw and Y.-C. Li, *On n-times integrated C-cosine functions*, in: Evolution Equations, Marcel Dekkar, 1995, pp. 393–406.

Department of Mathematics
National Central University
Chung-Li
Taiwan
shaw@math.ncu.edu.tw

AMS Classification: 45D05, 41A25, 47A35, 47D06, 47D09.

Operator Theory:
Advances and Applications, Vol. 118

Stochastic Equations in the Space of Formal Series: Convergence of Solution Series

I. Spectorsky

We consider stochastic equations in the space of formal series (an analogue of power series in a Hilbert space with no requirements of convergence). Such an equation is treated as a countable system of linear equations in a Hilbert space. The solution is defined as formal series, whose components solve the given system. The existence and the uniqueness of solution is proved. The main result of this paper is a local convergence of the solution (the convergence of the solution series over a random time interval) in some particular cases. We apply this result to study classical stochastic equations in a Hilbert space with analytical coefficients, whose solutions exist only within a random time interval.

Introduction

In this paper we consider stochastic equations in the space of formal series. Formal series and formal mappings are non-convergent analogues of power series and analytical functions respectively (exact definitions will be given in Preliminaries). An equation in the space of formal series is naturally derived, when we apply power series method to the 'classical' equation in a Hilbert space with analytical coefficients.

Deterministic differential equations in the space of formal operator series are studied, for example, in [1]. Sufficient conditions for solution to converge in deterministic case are given by Cauchy-Kovalevskaya theorem (see [1] and, for finite-dimensional case, [9]).

The main purpose of this paper is to give conditions, sufficent for local convergence of solution of stochastic equation. The result allows to apply power series method to construct 'classical' stochastic equations in a Hilbert space with analytical coefficients, without additional restrictions on drifth growth. The main difficulty, which may face traditional approaches to study equations of such a type, is that the solution has living time, which is random and cannot be separated from 0. The technique of formal series is quite easy way to give sense and to study stochastic equations with non-Lipschitzian analytical coefficients, when the living time of solution is not separated from 0. In the essence, we represent the solution pathwise as a power series by initial condition.

Stochastic equations with non-Lipschitzian coefficients in finite-dimensional (Euclidian) spaces are considered, for example, in [7, 10]. The case of locally existing solutions is studied in [7]. The case of strong global solutions with restrictions on coefficients growth was considered in [7, 10]. (In [10] the diffusion and drift coefficients are continuous functions of the whole path).

Stochastic equations in a Hilbert spaces with restrictions on coefficients growth are studied, e.g., in [3]. In [3] one can find the theorems about the existence and the uniqueness of solutions, as well as consideration of some properties of solution.

Analyticity property of stochastic flows in finite-dimensional case for equations with bounded coefficients is studied in [8].

This paper consists of 5 sections. In the first section we give the necessary definitions and some results on formal series and formal mappings. In the second section we prove the existence and the uniqueness of stochastic equation in the space of formal series. In the third section we prove theorem of local convergence of solution for these equations in the case of 'linear diffusion' and 'analytical drift'. In the fourth section we give some extension of the convergence result onto the case of 'non-linear' diffusion (in scalar case). In the fifth section we apply the convergence result to solve 'classical' stochastis equations in a Hilbert space for the case of linear diffusion and analytical drift.

1 Preliminaries

In this section we give the necessary definitions and describe some properties of formal series and formal mappings.

Let Y, Z be Hilbert spaces (all Hilbert spaces are supposed to be real and separable). Further $\mathcal{L}(Y, Z)$ denotes the space of linear continuous operators, acting from Y to Z.

Definition 1 The sequence $y = (y_k)_{k \geq 1}$, for all $k \geq 1$ $y_k \in Y$, is called formal series in Y.

Definition 2 Let a_k $(k \geq 1)$ be linear continuous operators, acting from $Y^{\otimes k}$ to Z $(a_k \in \mathcal{L}(Y^{\otimes k}, Z))$. Then we call the sequence $a = (a_k)_{k \geq 1}$ formal mapping from Y to Z. For $k \in \mathbf{N}$ element a_k is called k-th component of a.

For formal series and formal mapping there are defined component-wise operations of summing and multiplying by scalar constant. Let Y and Z be Hilbert spaces, y^1 and y^2 be formal series in Y, a and b be formal mappings from Y to Z, $c \in \mathbf{R}$. Then $y^1 + y^2$ and cy^1 are defined as formal series in Y, $a + b$ and ca are defined as formal mappings from Y to Z:

$$\begin{aligned} (y^1 + y^2)_k &= y^1_k + y^2_k, \ (cy^1)_k = cy^1_k, \\ (a + b)_k &= a_k + b_k, \ (ca)_k = cb_k \quad (k \geq 1). \end{aligned}$$

So, the set of formal series in Y and set of formal mappings from Y to Z are equipped with a structure of linear spaces.

The space of formal series in Y we denote by Y^∞, the space of formal mappings from Y to Z is denoted by $\mathcal{L}_\infty(Y, Z)$.

Definition 3 Let Y_1, Y_2, Y_3 be Hilbert spaces, $a \in \mathcal{L}_\infty(Y_1, Y_2)$ and $b \in \mathcal{L}_\infty(Y_2, Y_3)$. Composition $b \circ a$ of formal mappings b and a is a formal mapping $b \circ a \in \mathcal{L}_\infty(Y_1, Y_3)$, defined by:

$$(b \circ a)_n := \sum_{k=1}^{n} \sum_{\substack{(j_1,\dots,j_k)\in \mathbf{N}^k \\ j_1+\cdots+j_k=n}} b_k(a_{j_1} \otimes a_{j_2} \otimes \cdots \otimes a_{j_k}).$$

Here $a_{j_1} \otimes \cdots \otimes a_{j_k} \in \mathcal{L}(Y_1^{\otimes n}, Y_2^{\otimes k})$ denotes a tensor product of linear continuous operators $a_{j_1}, \dots, a_{j_k}$ ($a_l \in \mathcal{L}(Y_l^{\otimes l}, Y_2)$), defined, for example, in [2]. We remind that it is defined for $y^1 \otimes y^2 \otimes \cdots \otimes y^n \in Y_1^{\otimes n}$ by

$$\begin{aligned}(a_{j_1} \otimes \cdots \otimes a_{j_k})(y^1 \otimes \cdots \otimes y^n) \\ = (a_{j_1}(y^1 \otimes \cdots \otimes y^{j_1}) \otimes \cdots \otimes a_{j_k}(y^{j_1+\cdots+j_{k-1}+1} \otimes \cdots \otimes y^n))\end{aligned}$$

and then expanded for an arbitrary $\zeta \in Y_1^{\otimes n}$ by continuity, see [2] for details.

Because a_j have no sense for $j \notin \mathbf{N}$, further we omit restriction $(j_1, \dots, j_k) \in \mathbf{N}^k$. This means that further we write $\sum_{j_1+\cdots+j_k=n} b_k(a_{j_1} \otimes \cdots \otimes a_{j_k})$ instead of $\sum_{k=1}^n \sum_{\substack{(j_1,\dots,j_k)\in \mathbf{N}^k \\ j_1+\cdots+j_k=n}} b_k(a_{j_1} \otimes \cdots \otimes a_{j_k})$.

One can prove an associativity of composition: $(a \circ b) \circ c = a \circ (b \circ c)$, where $a \in \mathcal{L}_\infty(Y_3, Y_4)$, $b \in \mathcal{L}_\infty(Y_2, Y_3)$, $c \in \mathcal{L}_\infty(Y_1, Y_2)$ (the proof is completely analogous to one in [1], where formal mappings of topological spaces are considered).

Definition 4 Let Y be a Hilbert space, $a \in \mathcal{L}_\infty(Y, Y)$, $a_1 = \mathrm{id}_Y$, $a_k = 0$ for $k \geq 2$, where id_Y denotes an identical operator in Y. Then a is called an identical formal mapping in Y. We denote the identical formal mapping in Y by Id_Y.

Note that $b \circ \mathrm{Id}_Y = b$, $\mathrm{Id}_Y \circ c = c$ for any $b \in \mathcal{L}_\infty(Y, Z)$, $c \in \mathcal{L}_\infty(Z, Y)$.

Definition 5 Let Y, Z be Hilbert spaces, $a \in \mathcal{L}_\infty(Y, Z)$, $b \in \mathcal{L}_\infty(Z, Y)$, $a \circ b = \mathrm{Id}_Z$, $b \circ a = \mathrm{Id}_Y$. Then the formal mapping a is called inverse to formal mapping b (as well as b is called inverse to a), and formal mapping a is called invertible (respectively, b is also called invertible).

It is easy to prove that the formal mapping a is invertible if and only if the operator a_1 is invertible, and in this case the inverse formal mapping a^{-1} is uniquely defined (the proof for the case of formal mappings of topological spaces can be found in [1]).

Definition 6 Let Y, Z be Hilbert spaces, $a \in \mathcal{L}_\infty(Y, Z)$, $y \in Y^\infty$. We say that the formal mapping a acts on formal series y and its action $a(y)$ is given by:

$$(1) \quad a(y) \in Z^\infty, \quad (a(y))_n := \sum_{k=1}^{n} \sum_{j_1+\cdots+j_k=n} a_k(y_{j_1} \otimes y_{j_2} \otimes \cdots \otimes y_{j_k}).$$

Using the procedure of proving associativity formula $(a \circ b) \circ c = a \circ (b \circ c)$ (by analogy as in [1]), one can prove that $(a \circ b)(y) = a(b(y))$, where $a \in \mathcal{L}_\infty(Y_1, Y_2)$, $b \in \mathcal{L}_\infty(Y_2, Y_3)$ $y \in Y_1^\infty$. This property means the consistency of two operation: the action of a formal mapping on formal series, and composition of two formal mappings.

Definition 7 Let Y, Z be Hilbert spaces. A formal mapping $a \in \mathcal{L}_\infty(Y, Z)$ is called analytical with analyticity radius r, if $\limsup_{n\to\infty} \sqrt[n]{\|a_n\|_{\mathcal{L}(Y^{\otimes n}, Z)}} = r < \infty$.

It is easy to see that the composition of analytical formal mappings is analytical, and an inverse formal mapping to the analytical one (if exists) is analytical.

Definition 8 Formal series $y \in Y^\infty$ is locally convergent with convergence radius r, if $\limsup_{n\to\infty} \sqrt[n]{\|y_n(t)\|_Y} = r < \infty$.

It is easy to see that for $a \in \mathcal{L}_\infty(Y, Z)$ being analytical and $y \in Y^\infty$ being locally convergent, the formal series $a(y)$ is locally convergent.

The following example shows the naturality of notions of formal series and formal mappings.

Example: Let Y_1, Y_2, Y_3 be Hilbert spaces, $A : Y_2 \to Y_1$ be an analytical function in a neighbourhood of 0 and $A(0) = 0$. it means that there exists an expansion of $A(\xi)$ in power series by ξ in a neighbourhood of 0: $A(\xi) = \sum_{k\geq 1} a_k(\xi^{\otimes k})$, $a_k \in \mathcal{L}(Y_2^{\otimes k}, Y_1)$ $(k \geq 1)$, and the power series converges in Y_1 uniformly in ξ in some neighbourhood of 0. In such a case we say that analytical function A corresponds to analytical formal mapping $a \in \mathcal{L}_\infty(Y_2, Y_1)$, i.e. components $(a_k)_{k\geq 1}$ of formal mapping a are coefficients of the Tailor series in the neighbourhood of 0 for the function A.

Let argument ξ of A be represented in the form of a sum of locally convergent power series $\xi = \sum_{n\geq 1} y_n$, where $y_n = b_n(\theta^{\otimes n})$, $b_n \in \mathcal{L}(Y_3^{\otimes n}, Y_2)$, $\theta \in Y_3$, convergent in Y_2 (note that any $\xi \in Y_2$ can be represented in the form of sum of convergent power series, and not in a unique way). Let us consider

$$A(\xi) = A\left(\sum_{n\geq 1} b_n(\theta^{\otimes n})\right) = \sum_{n\geq 1} A_{b,n}(\theta^{\otimes n}), \quad A_{b,n} \in \mathcal{L}(Y_3^{\otimes n}, Y_2), \quad n \geq 1.$$

We calculate expansion coefficients $A_{b,n}$ $(n \geq 1)$ directly:

$$\begin{aligned} A_{b,1}(\theta) &= a_1 b_1(\theta) = a_1 y_1 = (a(y))_1; \\ A_{b,2}(\theta^{\otimes 2}) &= a_1 b_2(\theta^{\otimes 2}) + a_2(b_1(\theta) \otimes b_1(\theta)) \\ &= a_1 y_2 + a_2(y_1 \otimes y_1) = (a(y))_2; \\ &\vdots \end{aligned}$$

$$A_{b,n}(\theta^{\otimes n}) = \sum_{k=1}^{n} \sum_{j_1+\cdots+j_k=n} a_k(b_{j_1}(\theta^{\otimes j_1}) \otimes \cdots \otimes b_{j_k}(\theta^{\otimes j_k}))$$

$$= \sum_{k=1}^{n} \sum_{j_1+\cdots+j_k=n} a_k(y_{j_1} \otimes \cdots \otimes y_{j_k}) = (a(y))_n;$$

$$\vdots$$

We obtain that the expansion of $A(\xi)$ in powers of θ has the following form: $A(\xi) = \sum_{n\geq 1}(a(y))_n$, where $(a(y))_n = A_{b,n}(\theta^{\otimes n})$ is given by (1). So, the action of the analytical function A on the sum of convergent power series $\xi = \sum_{n\geq 1} y_n$ coincides with the action of the respective formal mapping a on the respective formal series y.

Note that there exists the well-developed apparatus to deal with formal mappings (see [1]).

2 Stochastic Equations in the Space of Formal Series

In this section we introduce a stochastic equation in the space of formal series. For this equation we prove the existence and the uniqueness of a non-anticipative solution.

Let $H_+ \subset H_0 \subset H_-$ be Gelfand triple (see [4] for details) and $\mathcal{H}_+ \subset \mathcal{H}_0 \subset \mathcal{H}_-$ be another Gelfand triple, given by

$$\mathcal{H}_0 = H_0 \otimes L_2([0,T]), \quad \mathcal{H}_+ = H_+ \otimes W_1^2, \quad \mathcal{H}_- = \mathcal{H}_+^*,$$

where $W_1^2 = \{\int_0^{\cdot} h(t)dt|\ h(\cdot) \in L_2([0,T])\}$ is the Sobolev space. One can construct (see, e.g., [6]) the canonical Gaussian measure μ on the measurable space $(\mathcal{H}_-, \mathbf{B}(\mathcal{H}_-))$ ($\mathbf{B}(\mathcal{H}_-)$ denotes Borel σ-algebra).

We deal with a standard Wiener process $w(t)$, associated with $H_+ \subset H_0 \subset H_-$ ($w(t) \in H_-$) on the probability space $(\mathcal{H}_-, \mathbf{B}(\mathcal{H}_-), \mu)$. The standard Wiener process $w(t)$ is defined (see [3]) as a Gaussian process with the first moment $\mathrm{E}w(t) = 0$ and correlation operator $\mathcal{L}(H_0, H_0) \ni K_w(t,\tau) = \min(t,\tau)\mathrm{id}_{H_0}$.

Denote by $\mathbf{F}_t = \sigma\{w(\tau),\ 0 \leq \tau \leq t\}$ $(0 \leq t \leq T)$ σ-algebra, generated by Wiener process. The process $\xi(t)$ is called adaptive (non-anticipative), if $\xi(t)$ is $\mathbf{F}_t$-measurable for any $t \in [0,T]$.

We further deal with stochastic integrals of Ito type treating them in the sense of L_2-theory. This means that the Ito integral $\int_s^t Q(\tau)dw(\tau)$ is well-defined for any adaptive process $Q(t)$ in $\mathcal{L}_2(H_0, Y)$, such that $\int_s^t \mathrm{E}\|Q(\tau)\|^2_{\mathcal{L}_2(H_0,Y)}d\tau < \infty$. By $\mathcal{L}_2(H_0, Y)$ we denote the space of Hilbert-Schmidt type operator from H_0 to Y. There is the well-developed apparatus to deal with Ito integrals and stochastic equations in Hilbert spaces (see, e.g., [3]).

Definition 9 A sequence $y(t) = (y_n(t))_{n\geq 1}$ is called a random process in Y^∞ if each y_n $(n \geq 1)$ is random process in Y, i.e. random process in the space of formal series is treated component-wise.

Let us consider stochastic equation in the space of formal series:

$$\begin{aligned} y(t) &= y_s + \int_s^t a(\tau)(y(\tau))d\tau + \int_s^t b(\tau)(y(\tau))dw(\tau), \\ &\qquad 0 \leq s \leq t \leq T, \end{aligned} \tag{2}$$

where each a_k and b_k $(k \geq 1)$ is a measurable bounded mapping from $[0, T]$ to $\mathcal{L}(Y^{\otimes k}, Y)$ and $\mathcal{L}(Y^{\otimes k}, \mathcal{L}_2(H_0, Y))$ respectively, y_s is nonrandom initial condition. Both integrals in (2) are random processes in Y^∞, whose components are defined by:

$$\left(\int_s^t a(\tau)(y(\tau))d\tau\right)_n := \int_s^t (a(\tau)(y(\tau)))_n \, d\tau;$$

$$\left(\int_s^t b(\tau)(y(\tau))dw(\tau)\right)_n := \int_s^t (b(\tau)(y(\tau)))_n \, dw(\tau).$$

We assume that for all $k \geq 2$ $y_{s,k} = 0$, and we denote $y_{s,1} = y_1^s$.

Equation (2) is treated component-wise. i.e. it is equivalent (by definition) to the countable system of linear stochastic equations:

$$\begin{aligned} y_1(t) &= y_1^s + \int_s^t a_1(\tau)y_1(\tau)d\tau + \int_s^t b_1(\tau)(y_1(\tau), dw(\tau)), \\ y_n(t) &= \int_s^t \sum_{k=1}^n \sum_{j_1+\cdots+j_k=n} a_k(\tau)(y_{j_1}(\tau) \otimes y_{j_2}(\tau) \otimes \cdots \otimes y_{j_k}(\tau))d\tau \\ &\quad + \int_s^t \sum_{k=1}^n \sum_{j_1+\cdots+j_k=n} b_k(\tau)(y_{j_1}(\tau) \otimes y_{j_2}(\tau) \otimes \cdots \otimes y_{j_k}(\tau))dw(\tau), \\ &\quad n \geq 2, \end{aligned} \tag{3}$$

where $a_k(t) \in \mathcal{L}(Y^{\otimes k}, Y)$, $b_k(t) \in \mathcal{L}(Y^{\otimes k}, \mathcal{L}_2(H_0, Y))$

Definition 10 We call the process $y(t) = (y_n(t))_{n\geq 1} \in Y^\infty$ the solution of equation (2), if $y_n(t)$ $(n \geq 1)$ solve (3).

Theorem 1 *Equation* (2) *has a unique adaptive solution* $y(t) = (y_k(t))_{k\geq 1} \in Y^\infty$.

Proof: Rewrite system (3) in a more compact form:

$$
\begin{aligned}
y_1(t) &= y_1^s + \int_s^t a_1(\tau)y_1(\tau)d\tau + \int_0^t b_1(\tau)(y_1(\tau), dw(\tau)), \\
y_n(t) &= \int_s^t a_1(\tau)y_n(\tau)d\tau + \int_s^t b_1(\tau)(y_n(\tau), dw(\tau)) \\
&\quad + \int_s^t f_n(\tau)d\tau + \int_s^t g_n(\tau)dw(\tau), \quad n \geq 2,
\end{aligned} \tag{4}
$$

$$
\begin{aligned}
\text{where } f_n(t) &= \sum_{k=2}^{n} \sum_{j_1+\cdots+j_k=n} a_k(t)(y_{j_1}(t) \otimes y_{j_2}(t) \otimes \cdots \otimes y_{j_k}(t)), \\
g_n(t) &= \sum_{k=2}^{n} \sum_{j_1+j_2+\cdots+j_k=n} b_k(t)(y_{j_1}(t) \otimes y_{j_2}(t) \otimes \cdots \otimes y_{j_k}(t)).
\end{aligned}
$$

□

As far as f_n and g_n depend only on y_k with $k \leq n-1$, each n-th equation of system (4) is a linear equation with respect to y_n. Therefore, each n-th equation of system (4) has a unique non-anticipative (adaptive) solution y_n with y_k ($k \leq n-1$) being known (see. e.g. [3]). So, the solution of (4) can be found recursively, by calculating $y_1, y_2, \ldots, y_n$, etc.

Theorem 1 states the existence and the uniqueness of solution y for (2), but say nothing about convergence of the series $\sum_{n\geq 1} y_n$. Convergence of this series in some particular case is proved in the next section. An analogue of (1) for the deterministic case is proved in [1].

3 Convergence of Solution Series in the Case of 'Linear Diffusion' and 'Analytical Drift'

In this section we obtain some sufficient conditions for convergence of series $\sum_{n\geq 1} y_k(t)$, where $y(t) = (y_k(t))_{k\geq 1}$ is the solution of (2). We prove convergence of this series in the particular case of 'linear diffusion' b ($b_k = 0$ for $k \geq 2$).

Consider equation (2) under the following assumption: $b_k \equiv 0$ for all $k \geq 2$ (the case of 'linear diffusion'), $a_1(t) \in \mathcal{L}_2(Y, Y)$, $b_1(t) \in \mathcal{L}_2(Y \otimes H_0, Y)$.

In such a case the solution of system (4) can be written in the form:

$$
y_1(t) = S_1(t, s)y_1^s, \; y_n(t) = \int_s^t S_1(t, \tau)(f_n(\tau))d\tau, \quad n \geq 2, \tag{5}
$$

where $S_1(t, \tau)$ is the evolution operator for the first linear homogeneous equation of (3).

Due to Hilbert-Schmidt type of $a_1(t)$ and $b_1(t)$, $S_1(t, s) \in \mathcal{L}(Y, Y)$ a.s. (see [3]). Here and further we treat all pathwise relations almost surely with respect to the canonical Gaussian measure μ.

Note that $S_1(t,s)$ is an a.s. continuous function with respect to t and s, valued $\mathcal{L}(Y,Y)$ (see, e.g. [3]). So, we can introduce $c(t)$ by:

$$c(t) = \max_{s \le \tau_1 \le \tau_2 \le t} \|S_1(\tau_2, \tau_1)\|_{\mathcal{L}(Y,Y)}.$$

Since $c(t)$ depends on $w(\tau)$ only with $\tau \le t$, $c(t)$ is measurable with respect to $\mathbf{F}_t$ $(s \le t \le T)$.

Due to boundness of $a_k(\cdot)$, we can introduce $\alpha_k = \sup_{t\in[0,T]} \|a_k(t)\|_{\mathcal{L}(Y^{\otimes k},Y)}$ $(k \ge 1)$.

From (5) we obtain:

$$\begin{aligned} \|y_1\|_Y &\le c(t)\|y_1^s\|_Y; \\ \forall n \ge 2 : \|y_n(t)\|_Y &\le c(t)\int_s^t \sum_{k=2}^n \alpha_k \sum_{j_1+j_2+\dots+j_k=n} \\ &\qquad \|y_{j_1}(\tau)\|_Y \dots \|y_{j_k}(\tau)\|_Y d\tau. \end{aligned} \tag{6}$$

Let $(S_n(t,s,(y_1^s)^{\otimes n}))_{n\ge1}$ solves (5):

$$S_n(t,s,(y_1^s)^{\otimes n}) = y_n(t).$$

It follows from (5) and (6) that the mapping $(y_1^s)^{\otimes n} \mapsto S_n(t,s,(y_1^s)^{\otimes n})$ belongs to $\mathcal{L}(Y^{\otimes n}, Y)$.

To estimate $\|y_n(t)\|_Y$ in a proper way, and, as a consequence, to obtain sufficient conditions for convergence of $\sum_{n\ge1} S_n(t,s,(y_1^s)^{\otimes n})$, we need the following simple result.

Lemma 1 *There exists some $\beta > 0$ such that*

$$\forall k \ge 1 \forall n \ge 1 : \varphi_{n,k} = \sum_{j_1+j_2+\dots+j_k=n} \frac{1}{j_1^2}\frac{1}{j_2^2}\cdots\frac{1}{j_k^2} \le \frac{\beta^{k-1}}{n^2}.$$

Proof: The lemma is proved by induction. Denote by P_k $(k \ge 1)$ the following sequence of statements:

$$P_k = \left(\forall n \ge k : \varphi_{n,k} \le \frac{\beta^{k-1}}{n^2}\right).$$

The base of induction is obvious: $\varphi_{n,1} = \frac{1}{n^2}$.

Give the induction assumption: let P_l holds for all $l \le k$.

To make the step of induction, note that for $k = 2$:

$$\varphi_{n,2} = \sum_{j=1}^{n-1} \frac{1}{j^2}\frac{1}{(n-j)^2} \le \frac{4}{n^2}\sum_{j=1}^{\infty}\frac{1}{j^2} = \frac{\beta}{n^2},$$

where $\beta = 4\sum_{j=1}^{\infty}\frac{1}{j^2} = \frac{2}{3}\pi^2$.

Now we prove statement P_l for $l = k + 1$.

$$\begin{aligned}
\varphi_{n,k+1} &= \sum_{j_1+j_2+\cdots+j_{k+1}=n} \frac{1}{j_1^2}\frac{1}{j_2^2}\cdots\frac{1}{j_{k+1}^2} \\
&= \sum_{j_{k+1}=1}^{n-k} \frac{1}{j_{k+1}^2} \sum_{j_1+j_2+\cdots+j_k=n-j_{k+1}} \frac{1}{j_1^2}\frac{1}{j_2^2}\cdots\frac{1}{j_k^2} \\
&\leq \sum_{j=1}^{n-k} \frac{1}{j^2}\frac{\beta^{k-1}}{(n-j)^2} = \beta^{k-1}\varphi_{n,2} \leq \frac{\beta^k}{n^2}.
\end{aligned}$$

The Lemma is proved. □

Now we can obtain the sufficient conditions to ensure the estimate:

$$\|y_n(t)\|_Y \leq \frac{r}{n^2}\left(\frac{c(t)\|y_1^s\|_Y}{r}\right)^n, \tag{7}$$

where $r > 0$ is a nonrandom constant.

Lemma 2 *Let the following estimate hold:*

$$(t-s)c(t)\sum_{k=2}^{\infty}\alpha_k r^{k-1}\beta^{k-1} \leq 1.$$

Then the components of the solution $y(t) = (y_n(t))_{n\geq 1}$ *of system* (4) *satisfy* (7), *i.e.:*

$$\|y_n(t)\|_Y \leq \frac{r}{n^2}\left(\frac{c(t)\|y_1^s\|_Y}{r}\right)^n, \quad n \geq 1.$$

Proof: We check (7), using induction in n. The base of induction is obvious:

$$\|y_1(t)\|_Y \leq c(t)\|y_1^s\|_Y.$$

Assume that (7) holds for any $n \leq m$ and prove (7) for $n = m+1$. For $n = m+1$ (6) has the form

$$\|y_{m+1}(t)\|_Y \leq c(t)\int_s^t \sum_{k=2}^{m+1}\alpha_k \sum_{j_1+j_2+\cdots+j_k=m+1} \|y_{j_1}(\tau)\|_Y \ldots \|y_{j_k}(\tau)\|_Y d\tau.$$

□

Taking into account the induction assumption and using Lemma 1, we obtain the estimate for $\|y_{m+1}(t)\|_Y$:

$$\|y_{m+1}(t)\|_Y \leq \left(\frac{c(t)\|y_1^s\|_Y}{r}\right)^{m+1}\frac{c(t)(t-s)}{(m+1)^2}\sum_{k=2}^{m+1}\alpha_k r^k\beta^{k-1}.$$

So, induction statement holds due to the assumption:

$$(t-s)c(t)\sum_{k=2}^{\infty}\alpha_k r^{k-1}\beta^{k-1} \leq 1.$$

Lemma 2 immediately implies the main theorem, which gives sufficient conditions for the convergence of power series $\sum_{n\geq 1} y_n(t) = \sum_{n\geq 1} S_n(t,s,(y_1^s)^{\otimes n})$.

Theorem 2 *Let $a(t)$ be an analytical formal mapping with analyticity radius r independent of t, $c(t) = \max_{s\leq\tau_1\leq\tau_2\leq t}\|S_1(\tau_2,\tau_1)\|_{\mathcal{L}(Y,Y)}$ be a.s. continuous.*

Then the power series $\sum_{n\geq 1} y_n(t) = \sum_{n\geq 1} S_n(t,s,(y_1^s)^{\otimes n})$ is locally convergent within random a.s. positive time interval, with random a.s. positive convergence radius.

Theorem 2 allows to apply the power series method to solve stochastic equations in a Hilbert space. Note that an analogue of 2 for the deterministic case is nothing but the variant of Cauchy-Kovalevskaya theorem (see, e.g, in [1] and, for scalar case, [9]).

4 The Case of Nonlinear Diffusion

In this section we extend the result of Theorem 1 to the case of a nonlinear diffusion, when $Y = H_0 = \mathbf{R}$ (scalar case).

Rewrite system (3) for this case.

$$\begin{aligned}
y_1(t) &= y_1^s + \int_s^t a_1(\tau)y_1(\tau)d\tau + \int_s^t b_1(\tau)y_1(\tau)dw(\tau),\\
y_n(t) &= \int_s^t \sum_{k=1}^{n}\sum_{j_1+\cdots+j_k=n} a_k(\tau)y_{j_1}(\tau)y_{j_2}(\tau)\ldots y_{j_k}(\tau)d\tau\\
&\quad + \int_s^t \sum_{k=1}^{n}\sum_{j_1+\cdots+j_k=n} b_k(\tau)y_{j_1}(\tau)y_{j_2}(\tau)\ldots y_{j_k}(\tau)dw(\tau),\\
&\quad n\geq 2.
\end{aligned} \tag{8}$$

Denote $S_n(t,s,(y_1^s)) = y_n(t)$ $(n\geq 1)$.

Theorem 3 *Let $a(t)$ and $b(t)$ be analytical formal mappings and $|b_1(t)| > \varepsilon > 0$ for any $t\in[0,T]$. Let each $a_k(t)$ and $b_k(t)$ $(k\geq 1)$ be differentiable for any $t\in[0,T]$, and formal mappings $a'(t) = (a_n'(t))_{n\geq 1}$ and $b'(t) = (b_n'(t))_{n\geq 1}$ be analytical.*

Then the series $\sum_{n\geq 1} y_n(t) = \sum_{n\geq 1} S_n(t,s,(y_1^s)^n)$ is locally convergent within positive time interval.

Proof: To prove the theorem, for any $t \in [s, T]$ we find an analytical invertible formal mapping $F(t) \in \mathcal{L}_\infty(\mathbf{R}, \mathbf{R})$, such that the process $z(t) = F(t)(y(t))$ obeys stochastic equation with 'linear diffusion'. Using Ito's formula, we find next conditions for $F(t)$:

$$b_1(t)F_n(t) = \sum_{k+j-1=n} kF_k(t)b_j(t).$$

It is easy to see that this system can be solved recursively:

$$F_1(t) = 1, \quad F_{n+1}(t) = \frac{1}{(1-n)b_1(t)} \sum_{k=1}^{n-1} kF_k(t)b_{n-k+1}(t).$$

Note that F_1 can be arbitrary nonzero.

One can easily check that the process $z(t) = F(t)(y(t))$ solves the following stochastic equation:

$$z(t) = z^s + \int_s^t \tilde{a}(\tau)(z(\tau))d\tau + \int_s^t \tilde{b}(\tau)(z(\tau))dw(\tau), \tag{9}$$

where

$$\begin{aligned} \tilde{a}(t) &= \hat{a}(t) \circ F^{-1}(t), \quad \hat{a}_n(t) = \sum_{k+j-1=n} F_k(t)a_j(t) \\ &+ \sum_{k+j_1+j_2-2=n} F_k(t)a_{j_1}(t)a_{j_2}(t) + F_n'(t); \\ &\tilde{b}_1(t) = b_1(t), \; b_n(t) = 0 \;\; \forall n \geq 2. \end{aligned}$$

Therefore, equation (9) obeys requirements of Theorem 2. So, formal series $z(t)$, and, consequently, formal series $y(t) = F^{-1}(t)(z(t))$ is locally convergent. □

5 Local Existence of Solution for Stochastic Equation in Hilbert Space with Analytical Coefficients

In this section we apply the convergence result, obtained in Sections 3 and 4, to solve stochastic equation in a Hilbert space, whose solution exists only within random time interval.

Consider a stochastic equation in Hilbert space:

$$
\xi(t) = \xi^s + \int_s^t A(\tau)(\xi(\tau))d\tau + \int_s^t B(\tau)(\xi(\tau), dw(\tau)),
\tag{10}
$$
$$
0 \le s \le t \le T.
$$

Here A is continuous function from $[0, T] \times Y$ to Y, B is continuous function from $[0, T]$ to $\mathcal{L}(Y, \mathcal{L}_2(H_0, Y))$, ξ^s is nonrandom initial condition. We suppose that $A(t)(\theta)$ is an analytical function of θ over the set $U_\varepsilon(0) = \{\theta \in Y | \|\theta\|_Y \le \varepsilon\}$, and $\|\xi^s\|_Y < \varepsilon$.

Let formal mapping $a(t) \in \mathcal{L}_\infty(Y, Y)$ correspond to the analytical function $A(t)$, i.e.:

$$
A(t)(\theta) = \sum_{n \ge 1} a_n(t)(\theta^{\otimes n})
$$

for $\theta \in U_\varepsilon(0)$. We also suppose that $a_1(t) \in \mathcal{L}_2(Y, Y)$.

Let formal mapping $b(t) \in \mathcal{L}_\infty(Y, \mathcal{L}_2(H_0, Y))$ correspond to $B(t)$, i.e.

$$
b(t) = (B(t), 0, 0, \dots, 0, \dots),
$$

ν be the stopping time, $A_\nu(t) = A(t)\chi_\nu(t)$, $B_\nu(t) = B(t)\chi_\nu(t)$, where $\chi_\nu(t) = 1$ for $t < \nu$ and $\chi_\nu(t) = 0$ for $t \ge \nu$.

Definition 11 Equation (10) has a solution $\xi_\nu(t)$ for $t < \nu$, if the process $\xi_\nu(t)$ is a strong solution to

$$
\xi_\nu(t) = \xi^s + \int_s^t A_\nu(\tau)(\xi_\nu(\tau))d\tau + \int_s^t B_\nu(\tau)(\xi_\nu(\tau), dw(\tau)),
$$
$$
0 \le s \le t \le T.
$$

Note that the similar definition for solution of equations like (10) (in scalar case) is used in [7].

Now we find solution of (10) in the form of power series with respect to $\xi^s \in Y^\infty$.

Let y be the solution of:

$$
y(t) = y_s + \int_s^t a(\tau)(y(\tau))d\tau + \int_s^t b(\tau)(y(\tau), dw(\tau)),
$$

with $y_s = (\xi^s, 0, 0, \dots, 0)$. Rewrite this equation component-wise:

$$
y_1(t) = \xi^s + \int_s^t a_1(\tau)y_1(\tau)d\tau + \int_s^t B(\tau)(y_1(\tau), dw(\tau)),
$$
$$
y_n(t) = \int_s^t \sum_{k=1}^n \sum_{j_1+j_2+\dots+j_k=n} a_k(\tau)(y_{j_1}(\tau) \otimes y_{j_2}(\tau) \otimes \dots \otimes y_{j_k}(\tau))d\tau
\tag{11}
$$
$$
+B(\tau)(y_1(\tau), dw(\tau)), \quad n \ge 2.
$$

Note that (11) is a system of type (3).

For y_n we use notations of the previous section: $y_n(t) = S_n(t, s, (y_1^s)^{\otimes n})$, $S_1(t, s)$ equals to the evolution operator for the first equation of (11) ($y_1(t) = S_1(t, s)\xi^s$), and $c(t) = \max_{s \leq \tau_1 \leq \tau_2 \leq t} \|S_1(\tau_2, \tau_1)\|_{\mathcal{L}(Y,Y)}$.

Denote $\alpha_k = \sup_{0 \leq t \leq T} \|a_k(t)\|_{\mathcal{L}(Y^{\otimes k}, Y)}$ $(k \geq 1)$, $\beta = 4 \sum_{j=1}^{\infty} \frac{1}{j^2} = \frac{2}{3}\pi^2$.

Theorem 4 *Let* $\sum_{k=2}^{\infty} \alpha_k r^{k-1} \beta^{k-1} = \gamma < \infty$ *for* $r > 0$, *and*

$$\nu = \inf\{t \in [s, T] | c(t)(t - s)\gamma > 1 \text{ or } \|\xi^s\|_Y c(t) > r\}.$$

Then the series $\sum_{n \geq 1} y_n(t)\chi_\nu(t)$ *converges in* Y *a.s., and the random process* $\xi_\nu(t) = \sum_{n \geq 1} y_n(t)\chi_\nu(t) + \sum_{n \geq 1} y_n(\nu)(1 - \chi_\nu(t))$ *solves* (10) *for* $t < \nu$.

Proof: Note that Theorem 2 immediately yields a.s. convergence of series $\sum_{n \geq 1} y_n(t)\chi_\nu(t)$ and $\sum_{n \geq 1} y_n(\nu)(1 - \chi_\nu(t))$. Convergence of series $A(\xi_\nu(t)) = \sum_{n \geq 1} (a(t)(y(t)))_n \chi_\nu(t)$ is an immediate consequence of Lemmas 2 and 1:

$$\sum_{n \geq 1} \|(a(t)(y(t)))_n\|_Y \chi_\nu(t) \leq \sum_{n \geq 1} \gamma r \frac{1}{n^2} < \infty$$

Now we have to prove that

$$\xi_\nu(t) = \xi^s + \int_s^t A_\nu(\tau)(\xi_\nu(\tau))d\tau + \int_s^t B_\nu(\tau)(\xi_\nu(\tau), dw(\tau)) \tag{12}$$
$$0 \leq s \leq t \leq T.$$

Let $\xi_\nu^n(t) = \sum_{k=1}^n y_k(t)\chi_\nu(t) + \sum_{k=1}^n y_k(\nu)(1 - \chi_\nu(t))$. Remind that formal series $y(t) = (y_k(t))_{k \geq 1}$ is the solution of system (11). □

Since $\|y_n(t)\|_Y \leq \frac{r}{n^2}(\frac{c(t)}{r}\|\xi^s\|_Y)^n$ for $c(t)(t-s)\gamma \leq 1$ by virtue of Theorem 2, we have:

$$\begin{aligned} &\xi_\nu^n(t) \longrightarrow \xi_\nu(t), \\ &\sum_{k=1}^n (a(t)(y(t)))_k \chi_\nu(t) \longrightarrow A_\nu(t)(\xi_\nu(t)), \\ &\sum_{k=1}^n (b(t)(y(t)))_k \chi_\nu(t) \longrightarrow B_\nu(t)(\xi_\nu(t)) \end{aligned} \tag{13}$$

in the mean-square sense.

As far as $\xi_\nu^n(t)$ is a.s. continuous with respect to $t \in [s, T]$,

$$\xi_\nu^n(t) \longrightarrow \xi_\nu(t),$$

and, therefore, we have mean–square convergence for the stochastic integral:

$$\int_s^t B_\nu(\tau)(\xi_\nu^n(\tau), dw(\tau)) \longrightarrow \int_s^t B_\nu(\tau)(\xi_\nu(\tau), dw(\tau)). \tag{14}$$

Summing up first n equations of (11), we obtain the following formula:

$$\xi_\nu^n(t) = \xi^s + \int_s^t \sum_{k=1}^n (a(\tau)(y(\tau)))_k \chi_\nu(\tau) d\tau + \int_s^t B_\nu(\tau)(\xi_\nu^n(\tau), dw(\tau)). \tag{15}$$

Using (13) and (14), we can pass to the mean–square limit in the both side of (15):

$$\xi_\nu(t) = \xi^s + \int_s^t A_\nu(\tau)(\xi_\nu) d\tau + \int_s^t B_\nu(\tau)(\xi_\nu(\tau), dw(\tau)).$$

We have obtained (12), so the theorem is proved

Example: Put in (10) the drift operator $A(t)(\xi(t)) = A(\xi(t)) = \xi(t)\ln(1 - \|\xi(t)\|_Y^2)$ and diffusion $B(t)(\xi(t), \cdot) = B(\xi(t), \cdot) = \xi(t)\langle h, \cdot\rangle_H$ ($h \in H_+$). So, we consider the following stochastic equation in Y:

$$\begin{aligned} \xi(t) &= \xi^s + \int_s^t \xi(\tau)\ln(1 - \|\xi(\tau)\|_Y^2) d\tau \\ &\quad + \int_s^t \xi(\tau)\langle h, dw(\tau)\rangle_H, \quad 0 \le s \le t \le T, \end{aligned} \tag{16}$$

where $h \in H_+$, $\xi^s \in Y$ is nonrandom. In this case, we have the following expansion for $A(\xi(t))$ within $\|\xi(t)\|_Y < 1$:

$$A(\xi(t)) = \xi(t)\ln(1 - \|\xi(t)\|_Y^2) = \sum_{n\ge1} \xi(t)\|\xi(t)\|_Y^{2n} \frac{(-1)^n}{n} = \sum_{n\ge1} a_n((\xi(t))^{\otimes n}),$$

where $a_1(\theta) = a_{2k}(\theta^{\otimes 2k}) = 0$, $a_{2k+1}(\theta^{\otimes(2k+1)}) = \frac{(-1)^k}{k}(\theta,\theta)_Y^k\theta$, $k \in \mathbf{N}$, $\theta \in Y$. So, $\alpha_1 = \alpha_{2k} = 0$, $\alpha_{2k+1} = \frac{1}{k}$, and the power series

$$\gamma = \sum_{k=2}^\infty \alpha_k r^{k-1}\beta^{k-1} = \sum_{k=1}^\infty \frac{1}{k} r^{2k}\beta^{2k},$$

defined in Theorem 4, converges for positive constant $r \le \beta^{-1} = \frac{3}{2\pi^2}$. For this example, evolution operator $S_1(t,\tau) \in \mathcal{L}(Y,Y)$ is defined by linear stochatsic equation

$$\xi(t) = \xi^s + \int_s^t \xi(\tau)\langle h, dw(\tau)\rangle_H \quad 0 \le s \le t \le T.$$

Therefore, if we fix the constant $0 < r \le \frac{2}{3}\pi^2$, then the solution of (16) exists for $\|\xi^s\|_Y \le r$ till stopping time $\nu = \inf\{t \in [s,T] \| c(t)(t-s)\gamma > 1$ or $\|\xi^s\|_Y c(t) > r\}$.

Note that Theorem 4 remains valid for the scalar case of non-linear diffusion. In such a case the convergence of a solution of the respective equation in formal series is implied by Theorem 3.

Moreover, Theorem 4 gives the possibility to use power series method to solve stochastic equations with the case of linear diffusion coefficient and analytical (in the neighbourhood of zero) drift coefficient in a Hilbert space. In a scalar case, it could be applied to equations with nonlinear analytic diffusion and drift coefficients.

References

[1] A.M. Baranovitch and Yu.L. Daletskiĭ, *Differential-geometric relations for a class of formal operator power series* // MFAT **2**, 1 (1996), pp. 1–16.

[2] Yu.M. Berezanskiĭ and Yu.G. Kondratyev, *Spectral Methods in Infinite Dimensional Analysis*, Kiev, Naukova Dumka, 1988 (in Russian).

[3] Yu.L. Daletsky and Ya.I. Belopolskaya, *Stochastic Equations and Differential Geometry*, Kyyiv, Vyshcha Shkola, 1989 (in Russian).

[4] I.M. Gelfand and N.Ya. Vilenkin, *Some applications of garmonic analysis. Equipped Hilbert spaces*, Moscow, Fizmatgiz, 1961 (in Russian).

[5] G.H. Hardy, *Divergent Series*, Oxford, 1949.

[6] T. Hida, *Brownian motion*, New York, Springer-Verlag, 1980.

[7] N. Ikeda and Sh. Watanabe, *Stochastic differential equations and diffusion processes*, North-Holland Publishing Company, Amsterdam-Oxford-New York, 1981.

[8] A. Ishiwata and S. Taniguchi, *On the analyticity of stochastic flows* // Osaka Journal of mathematics, 1999, to appear.

[9] I.G. Petrovsky, *Lections on partial derivative equitation*, Moscow-Leningrad, Gousdarstvennoye Izdatelstvo Tekhniko-Tvorcheskoy Literatury, 1950 (in Russian).

[10] S.I. Pisanets, *On the existence of the strong solution for diffusion type stochastic equations in* $\boldsymbol{R}^m$ // Teoriya veroyatnostey i eye primeneniya **42**, 1 (1997), pp. 189–194 (in Russian).

Igor Ya. Spectorsky
National Technical University of Ukraine
Apt.76, Zalomov St. 2
252069, Kiev
Ukraine

AMS Classification Numbers: 60H10, 58D25.

Operator Theory:
Advances and Applications, Vol. 118

A Trigonometric Moment Problem for $\mathbf{Z}^{2*}$

Ion Suciu

A rectangular $n \times m$-matrix of complex numbers can be extended to a full quadrant of the Fourier coefficients of a representing measure on the bitorus only in very special conditions. The aim of this paper is to describe the nature of the obstructions which appear when we try to make this extension line by line by using the method of *choice sequences* from the operator extrapolation theory. The obstructions are described by a sequence of Hankel operators which appear in the recursive construction of an Ando dilation of the pair of the null contractions on **C**, canonically associated to the representing measure.

1 Formulation of the problem

Consider the bitorus $\mathbf{T}^2 = \mathbf{T}\times\mathbf{T}$ as the dual group of the discrete group $\mathbf{Z}^2 = \mathbf{Z}\times\mathbf{Z}$. A regular Borel measure μ on $\mathbf{T}^2$ is well defined by the array $[c_{n,m}(\mu)]_{(n,m)\in\mathbf{Z}^2}$ of its Fourier coefficients:

$$c_{n,m}(\mu) = \int_{\mathbf{T}^2} e^{-int}e^{-ims}d\mu(t,s), \quad (n,m) \in \mathbf{Z}^2. \tag{1.1}$$

A *representing measure* on the bitorus is a probability (positive of total mass 1) measure μ on $\mathbf{T}^2$ verifying

$$c_{n,m}(\mu) = 0, \quad (n,m) \in Z^2, \; n,m \geq 0, \; n+m > 0. \tag{1.2}$$

Let us denote by $\mathcal{M}_0$ the set of all such measures. The elements of $\mathcal{M}_0$ represent the origin as a point in the maximal ideal space of the bidisc algebra $A(\mathbf{D}^2)$, i.e.

$$f(0,0) = \int_{\mathbf{T}^2} f(e^{it}, e^{is}d\mu(t,s), \quad f \in A(\mathbf{D}^2). \tag{1.3}$$

The normalized product of Lebesgue arc measure on $\mathbf{T}$ by itself, as the Haar measure of the compact group $\mathbf{T}^2$, is an element in $\mathcal{M}_0$. So $\mathcal{M}_0$ is a non void set of probability measures on the bitorus $\mathbf{T}^2$, which is convex and weak* - compact.

The moment problem considered here is the following:

Let

$$[a_{n,m}]_{\substack{1\leq n<p\leq\infty \\ \leq m<q\leq\infty}}$$

---

*Partially supported by Romanian Academy by GAR 109/1998.

be a $p-1 \times q-1$ matrix of complex numbers. Give necessary and sufficient conditions for the existence of a representing measure μ on the bitorus such that

$$a_{n,m} = c_{-n,m}(\mu), \qquad 1 \leq n < p,\ 1 \leq m < q. \tag{1.4}$$

In the case the set of all solutions is non void, find a system of free parameters which label it and give algorithms to produce recurrently, by successive choice of a free parameter, a desired solution.

The matrix $[a_{n,m}]_{\substack{1\leq n<p \\ 1\leq m<q}}$ defines *the data* of the moment problem. We say that the matrix $[a_{n,m}]_{\substack{1\leq n<p \\ 1\leq m<q}}$ is representable by the measure $\mu \in \mathcal{M}_0$ provided μ is a solution of our moment problem for data $[a_{n,m}]_{\substack{1\leq n<p \\ 1\leq m<q}}$.

The matrix of the Fourier coefficients of a solution μ looks like

$$\begin{array}{cccc|c|ccc}
 & \cdot & \cdot & & \cdot & \cdot & \cdot & \\
 & \cdot & \cdot & & \cdot & \cdot & \cdot & \\
 & \cdot & \cdot & & \cdot & \cdot & \cdot & \\
\cdots & \overline{a}_{22} & \overline{a}_{21} & & 0 & 0 & 0 & \cdots \\
\cdots & \overline{a}_{12} & \overline{a}_{11} & & 0 & 0 & 0 & \cdots \\
\hline
\cdots & 0 & 0 & & 1 & 0 & 0 & \cdots \\
\hline
\cdots & 0 & 0 & & 0 & a_{11} & a_{12} & \cdots \\
\cdots & 0 & 0 & & 0 & a_{21} & a_{22} & \cdots \\
 & \cdot & \cdot & & \cdot & \cdot & \cdot & \\
 & \cdot & \cdot & & \cdot & \cdot & \cdot & \\
 & \cdot & \cdot & & \cdot & \cdot & \cdot &
\end{array}$$

The problem, in this general formulation, seems to be very difficult, as any moment problem concerning the group $\mathbf{Z}^2$ is. In [11] we connected this problem with some topics from operator extrapolation theory (cf. [7]) and we constructed a class of particular solutions for special given data.

In this paper we shall describe the nature of the obstructions which appear in a recursive construction of a general solution.

Following an idea from [4], the obstructions are described by a sequence of Hankel operators which appear in the recursive construction of an Ando dilation of the pair of the null contractions on $\mathbf{C}$, canonically associated to the representing measure.

2 Preliminary Notations and Results

We shall denote by L^2 the Lebesgue space on the unit circle (one dimensional torus) **T**, and by H^2 the Hardy subspace of L^2, considered also as a space of analytic functions in the open unit disc **D**. By H^∞ we shall denote the space of all bounded analytic functions on **D**, considered also as a subspace of L^∞ - the space of all measurable and essentially bounded functions on **T**.

For a function f in H^∞ we shall denote by T_f the operator of multiplication by f on H^2 - the analytic Toeplitz operator of symbol f.

For a Hilbert space H we denote by $B(H)$ the C^*-algebra of all linear bounded operators on H. A map $g \mapsto T_g$ of a group G into $B(H)$ is called *positive definite* provided

$$\sum_{g,\gamma\in G} (T_{g-\gamma}h(g), h(\gamma))_H \geq 0, \tag{2.1}$$

for any finite supported function h from G into H.

Bochner Theorem asserts that the array $[c_{n,m}]_{(n,m)\in\mathbf{Z}^2}$ of complex numbers is the array of the Fourier coefficients of a positive measure μ on $\mathbf{T}^2$ iff the map $(n,m) \mapsto c_{n,m}$ from $\mathbf{Z}^2$ into **C** is positive definite (cf. [12]). Also the Naimark dilation Theorem asserts that a map $g \mapsto T_g$ of a group G into $B(H)$ is positive definite if and only if it is the compression to H of a unitary representation of G on a larger space K of H.

It follows that our moment problem can be formulated as a problem of extension to a positive definite function on $\mathbf{Z}^2$. We try to reduce it to a succession of contractive and **Z**-positive definite extension problems.

Let H be a Hilbert space and let $(T_n)_{n\geq 1}$ be a sequence of operators on H. We say that the sequence $(T_n)_{n\geq 1}$ is *positive definite* if the map $n \mapsto T_n$ of the group **Z** into $B(H)$, were $T_0 = I_H$ and $T_{-n} = T_n^*$ for $n \geq 1$, is positive definite.

From the Naimark dilation Theorem (cf. [7], [12]) we deduce the existence of the minimal isometric dilation $[K, V]$ of $(T_n)_{n\geq 1}$ which means that there exist a Hilbert space K, containing H as a closed subspace and an isometry V on K such that

$$K = \bigvee_{n\geq 0} V^n H, \quad T_n = P_H V^n | H, \quad n \geq 1. \tag{2.2}$$

It follows that for any $n \geq 1$, T_n is a contraction on H.

In case $T_n = T^n$ for a contraction T on H and any $n \geq 1$, the known B.Sz.-Nagy Theorem asserts that the sequence $(T_n)_{n\geq 1}$ is positive definite. In this case the minimal isometric dilation $[K, V]$ has a simple geometric structure, known from Sz.-Nagy-Foiaş model theory for contractions.

In the general case of an arbitrary positive definite sequence $(T_n)_{n\geq 1}$ we have the following model based on the notion of choice sequence.

For a contraction A on H, let us denote here and in all what follows, by $D_A = [I - A^*A]^{1/2}$ - the defect operator of A and by $\mathcal{D}_A = \overline{D_A H}$ - the defect space of A.

A *choice sequence* on H (cf. [3], [7]) is a sequence $(R_n)_{n\geq 1}$, where R_1 is a contraction on H and for any $n \geq 2$, R_n is a contraction from $\mathcal{D}_{R_{n-1}}$ into $\mathcal{D}_{R^*_{n-1}}$, but otherwise unrestricted.

To any choice sequence we can attach the pair $[K, V] = [K(R), V(R)]$ where $K = K(R)$ is the Hilbert space containing H as a closed subspace, defined by

$$K = K(R) = H \oplus \mathcal{D}_{R_1} \oplus \mathcal{D}_{R_2} \oplus \cdots, \tag{2.3}$$

and $V = V(R)$ is the isometry on $K = K(R)$ defined by the matrix

$$V = \begin{bmatrix} R_1 & D_{R_1^*}R_2 & D_{R_1^*}D_{R_2^*}R_3 & \cdots \\ D_{R_1} & -R_1^*R_2 & -R_1^*D_{R_2^*}R_3 & \cdots \\ 0 & D_{R_2} & -R_2^*R_3 & \cdots \\ \vdots & \vdots & \vdots & \end{bmatrix}. \tag{2.4}$$

One of the most important result of the operator extrapolation theory (cf. [3], [7]) establishes a bijective correspondence between the set of all choice sequences $R = (R_n)_{n\geq 1}$ on H and the set of all positive definite sequences $T = (T_n)_{n\geq 1}$ on H given by $R \to T(R) = (T_n)_{n\geq 1}$, where

$$T_n = P_H V^n | H, \qquad n \geq 1, \tag{2.5}$$

and $V = V(R)$ is the isometry on $K = K(R)$ associated to R as above.

Clearly $[K, V] = [K(R), V(R)]$ is (coincides with) the minimal isometric dilation of $T(R) = (T_n)_{n\geq 1}$.

In this correspondence, for any $p \geq 1$ the string $\{T_1, T_2, \ldots, T_p\}$ depends only on the string $\{R_1, R_2, \ldots, R_p\}$ and conversely. There are recurrent formulas which establish this correspondence (cf. [7]). These formulas imply that $T_n = T^n, n \geq 1$, for a contraction T on H, if and only if $R_1 = T$ and $R_n = 0$ for any $n \geq 2$.

It follows that the choice sequences are free parameters in producing positive definite sequences of contractions on H. They play the role which Schur numbers played in the classical Carathéodory interpolation problem.

3 Equivalent Formulations of the Moment Problem

The first result is the following:

Theorem 1 *The $p - 1 \times q - 1$ -matrix $[a_{nm}]$ is representable by a measure μ in $\mathcal{M}_0$ if and only if for any $n \geq 1$ there exists $f_n \in H^\infty$, $f_n(0) = 0$, $\|f_n\| \leq 1$, such that:*

1) *for any n, $1 \leq n < p$ and any m, $1 \leq m < q$, the m-th Taylor coefficient of f_n is a_{nm}:*

$$a_{nm} = \frac{1}{2\pi}\int_0^{2\pi} e^{-imt} f_n(e^{it})dt, \qquad 1 \leq n < p, 1 \leq m < q. \tag{3.1}$$

2) *the sequence $(T_{f_n})_{n\geq 1}$ of analytic Toeplitz operators is positive definite.*

Proof: Suppose the matrix $[a_{n,m}]$ is representable by a measure μ on the bitorus. Let U, V be the isometries defined on the Hilbert space $K = H^2(d\mu)$ - the closure in $L^2(d\mu)$ of the set of all analytic polynomials, by multiplication by the coordinate functions e^{it}, e^{is} respectively.

Denote

$$K_0 = \bigvee_{m \geq 0} U^m \mathbf{C}, \quad U_0 = U|K_0, \quad T_n = P^K_{K_0} V^n|K_0.$$

Using the special form of μ we can identify K_0 with H^2 such that U_0 becomes the shift S. Clearly H^2 reduces U to S and consequently $T_n = T_{f_n}$ with f_n as above.

Conversely, if for any $n \geq 1$, the function f_n is in H^∞, $f_n(0) = 0$, and the sequence $(T_{f_n})_{n \geq 1}$ of analytic Toeplitz operators is positive definite, then, taking V, acting on K, to be the minimal isometric (Naimark) dilation of this sequence, we can extend the shift operator S on H^2 to an isometry U on K such that $UV = VU$ and H^2, as a subspace of K, reduces U. We can then extend the commuting pair $[U, V]$ of the isometries on K to its minimal unitary extension -a unitary representation of the group $\mathbf{Z}^2$ on a supra space $\tilde{K}$ of K generated by a pair $[\tilde{U}, \tilde{V}]$ of commuting unitary operators on $\tilde{K}$, verifying $\tilde{U}|K = U$, $\tilde{V}|K = V$.

Let E be the spectral measure, supported on $\mathbf{T}^2$, associated to this representation. Setting for any Borel subset σ of $\mathbf{T}^2$, $\mu(\sigma) = (E(\sigma)1, 1)_{\tilde{K}}$, 1 being here the constant function 1 in H^2, as a vector in $\tilde{K}$, we obtain a positive measure μ on $\mathbf{T}^2$. For any $n \geq 0$ and $m \in \mathbf{Z}$ we have

$$\begin{aligned} c_{-n,m}(\mu) &= \int e^{int} e^{-ims} d\mu(t, s) \\ &= \int e^{int} e^{-ims} d(E(t, s)1, 1)_{\tilde{K}} = (\tilde{U}^{*m} V^n 1, 1)_{\tilde{K}}. \end{aligned}$$

Hence for $m \geq 0$ we have

$$c_{-n,-m}(\mu) = (V^n e^{imt}, 1)_K = (T_{f_n} e^{imt}, 1)_{H^2} = 1/2\pi \int e^{imt} f_n(e^{it}) dt,$$

with $f_0 = 1$, which clearly implies $\mu \in \mathcal{M}_0$, and

$$\begin{aligned} c_{-n,m}(\mu) &= (\tilde{U}^{*m} V^n 1, 1)_{\tilde{K}} = (V^n 1, e^{imt})_K = (T_{f_n} 1, e^{imt})_{H^2} \\ &= 1/2\pi \int e^{-imt} f_n(e^{it}) dt = a_{nm}, \end{aligned}$$

which means that μ represents the matrix $[a_{nm}]$.

The theorem is completely proved. □

The commuting pair of the isometries U, V, acting on K, associated to a solution of our moment problem as in the proof of the above theorem, satisfies the following properties: Considering $\mathbf{C}$, the subspace of the constant functions in H^2, as a subspace of K, then

$$\text{(i)} \qquad K = \bigvee_{n,m \geq 0} U^n V^m \, \mathbf{C},$$

$$\text{(ii)} \qquad P_{\mathbf{C}} U^n V^m | \, \mathbf{C} = 0,$$

which means that $(K, [U, V])$ is an *Ando dilation* of the pair of the null contractions on $\mathbf{C}$ (cf. [12]).

In fact we have:

Theorem 2 *There exists a bijective correspondence between the set of all representing measures for the matrix* $[a_{nm}]$ *and the set of all Ando dilations* $(K, [U, V])$ *of the pair* $(\mathbf{C}, [0, 0])$ *of the null contractions on* $\mathbf{C}$ *verifying*

$$(3.2) \qquad a_{nm} = (U^{*m} V^n 1, 1)_K, \quad 1 \leq n \leq p-1, 1 \leq m \leq q-1.$$

Proof: According to Theorem 1, in order to produce solutions for our moment problem we have to extend simultaneously the polynomials

$$\sum_{m=1}^{q-1} a_{nm} z^m, \quad n = 1, \ldots, p-1,$$

to the functions $f_n \in H^\infty$, $\|f_n\| \leq 1$, such that the string $(T_{f_n})_{1 \leq n \leq p-1}$ of analytic Toeplitz operators on H^2 can be extended to a positive definite sequence of analytic Toeplitz operators. As in the proof of Theorem 1, we can produce an Ando dilation $(K, [U, V])$ of the pair $(\mathbf{C}, [0, 0])$ of the null contractions on $\mathbf{C}$ verifying (3.2).

Conversely, from an Ando dilation $(K, [U, V])$ of the pair $(\mathbf{C}, [0, 0])$ of the null contractions on $\mathbf{C}$, verifying (3.2), we obtain f_n by

$$T_{f_n} = P_{K_0} V^n | K_0, \quad n \geq 1,$$

where $K_0 = \vee_{k \geq 0} U^k \, \mathbf{C}$, identified in a obvious way to H^2.

The theorem is completely proved. □

Theorem 1 reduces our moment problem to a succession of some scalar Carathéodory problems together with an extension problem to a positive definite sequence of analytic Toeplitz operators. Theorem 2 reduces the moment problem under consideration to the construction of an Ando dilation of $(\mathbf{C}, [0, 0])$ verifying (3.2).

We shall adapt to this special context, some techniques from the operator extrapolation theory (cf. [7]) and Ando dilations theory (cf. [4], [5]) to obtain solutions for our moment problem.

4 Producing Positive Definite Sequences of Analytic Toeplitz Operators and Ando Dilations

We are interested in producing, by free parameters, positive definite sequences of analytic Toeplitz operators. Following the way of producing positive definite sequences by choice sequences, we try to give conditions under which a choice sequence on H^2 produces positive definite sequence of analytic Toeplitz operators.

We begin with the following geometric characterization of the choice sequences which produce positive definite sequences of analytic Toeplitz operators.

Theorem 3 *Let $R = (R_n)_{n\geq 1}$ be a choice sequence on H^2, $(T_n)_{n\geq 1}$ be its corresponding positive definite sequence and $[K, V] = [K(R), V(R)]$ be the minimal isometric dilation of $(T_n)_{n\geq 1}$. The following assertions are equivalent:*

(i) *For any $n \geq 1$, $T_n = T_{f_n}$ for some $f_n \in H^\infty$, $f_n(0) = 0$.*

(ii) *$V^n \mathbf{C} \perp \mathbf{C}$, $n \geq 1$ and there exists an isometry $U = U(R)$ on K such that $UV = VU$ and, H^2, as a subspace of K, reduces U to the shift operator S on H^2.*

(iii) *There exists an isometry $U = U(R)$ on K such that $UV = VU$ and $(K, [U, V])$ is an Ando dilation of $(\mathbf{C}, [0, 0])$.*

Proof: Suppose that the assertion (ii) holds. Then for any $n \geq 1$ and any $h \in H^2$ we have

$$T_n Sh = P_{H^2} V^n Sh = P_{H^2} V^n Uh = P_{H^2} UV^n h = UP_{H^2} V^n h = ST_n h.$$

Hence T_n commutes with the shift S on H^2 which implies $T_n = T_{f_n}$ for some $f_n \in H^\infty$. The assertion (i) results.

Also we have

$$\begin{aligned}(V^n U^m c_1, c_2)_K &= (V^n e^{imt} c_1, c_2)_K = (f_n e^{imt} c_1, c_2)_{H^2} \\ &= \int_{\mathbf{T}} f_n e^{imt} c_1 \bar{c}_2 = 0,\end{aligned}$$

for any $n, m \geq 0$, $n + m > 0$, and $c_1, c_2 \in \mathbf{C}$.

It follows

$$P_{\mathbf{C}} V^n U^m | \mathbf{C} = 0, \quad n, m \geq 0, n + m > 0,$$

which means that $(K, [U, V])$ is an Ando dilation of $(\mathbf{C}, [0, 0])$, i.e. (iii) holds.

Suppose now that (i) holds. Setting for any $n \geq 0$ and any $h_0, h_1, \ldots, h_n$ in H^2

$$U \sum_{j=0}^{n} V^j h_j = \sum_{j=0}^{n} V^j S h_j, \tag{4.1}$$

we have

$$
\begin{aligned}
\left\| U \sum_{j=0}^{n} V^j h_j \right\|_K^2 &= \left(\sum_{j=0}^{n} V^j S h_j,\ \sum_{k=0}^{n} V^k S h_k \right)_K \\
&= \sum_{j \geq k} (V^{j-k} S h_j,\ S h_k)_K + \sum_{j<k} (S h_j,\ V^{k-j} S h_k)_K \\
&= \sum_{j \geq k} (T_{f_{j-k}} S h_j,\ S h_k)_{H^2} + \sum_{j<k} (S h_j,\ T_{f_{k-j}} S h_k)_{H^2} \\
&= \sum_{j \geq k} (S T_{f_{j-k}} h_j,\ S h_k)_{H^2} + \sum_{j<k} (S h_j,\ S T_{f_{k-j}} h_k)_{H^2} \\
&= \sum_{j \geq k} (T_{f_{j-k}} h_j,\ h_k)_{H^2} + \sum_{j<k} (h_j,\ T_{f_{k-j}} h_k)_{H^2} \\
&= \cdots = \left\| \sum_{j=1}^{n} V^j h_j \right\|_K^2 .
\end{aligned}
$$

It follows that (4.1) defines an isometry U on K. Clearly U commutes with V and coincides with S on H^2.

For $h \in H^2$ with $h(0) = 0$ we have $h = Sg = Ug$ for some $g \in H^2$. Hence

$$U^* h = U^* U g = g \in H^2.$$

In case $h = c$ is a constant function in H^2 then we have

$$
\begin{aligned}
(U^* h, V^n g)_K &= (h, U V^n g)_K = (h, V^n U g)_K \\
&= (h, P_{H^2} V^n S g)_{H^2} = (h, T_{f_n} S g)_{H^2} = 0.
\end{aligned}
$$

We showed that H^2 is invariant for U^* hence it reduces U. The assertion (ii) is proved. The implication (iii) $\rightarrow$ (i) it follows from the proof of Theorem 2.

The theorem is completely proved. □

According to Theorem 3 any Ando dilation of the pair of the null contractions on **C** can be obtained by a choice sequence initiated on H^2 verifying one of the equivalent conditions (i)–(iii). From Theorem 2 it follows that any solution of our moment problem can be obtained from a suitable chosen choice sequence.

Some recursive methods for producing representing measures on the bitorus, along this line, were presented in [10] and [11].

Our main goal here is to identify the obstructions in choices when we want to produce row by row the array of Fourier coefficients of a solution μ.

5 Hankel Operators of the Obstructions

Let μ be a measure in $\mathcal{M}_0$. Consider in $L^2(d\mu)$ the following subspaces:

$$\hat{K} = \bigvee_{n\geq 0} e^{ins} \bigvee_{m\in\mathbf{Z}} e^{imt}\,\mathbf{C};\quad K = \bigvee_{n,m\geq 0} e^{ins}e^{imt}\,\mathbf{C},$$

$$\hat{K}_0 = \bigvee_{m\in\mathbf{Z}} e^{imt}\,\mathbf{C};\quad K_0 = \bigvee_{m\geq 0} e^{imt}\,\mathbf{C}$$

were e^{it}, e^{is} are the coordinate functions on the bitorus, as elements in $L^2(d\mu)$.

Clearly we can identify $K_0 \subset \hat{K}_0$ to $H^2 \subset L^2$ - the usual (one variable) Hardy and Lebesgue spaces, respectively.

Denote $\hat{U}$, $\hat{V}$ the operators on $\hat{K}$ of multiplication by e^{it}, e^{is}, respectively, and let $U = \hat{U}|K$, $V = \hat{V}|K$. It is clear that U and V are two commuting isometries on K. Moreover, $\hat{U}$ is unitary on $\hat{K}$ and coincides with the minimal unitary extension of U and $\hat{V}$ is an isometry on $\hat{K}$ such that $\hat{U}\hat{V} = \hat{V}\hat{U}$, the unique extension of V to a commutant of $\hat{U}$ (cf. [6]).

Using the above identifications, it is clear that L^2 reduces $\hat{U}$ and H^2 reduces U to the multiplication by coordinate function on the corresponding space. Let us denote

$$\hat{T}_n = P^{\hat{K}}_{L^2}\hat{V}^n|L^2,\quad T_n = P^{K}_{H^2}V^n|H^2.$$

The sequences $(\hat{T}_n)_{n\geq 1}$ and $(T_n)_{n\geq 1}$ are positive definite sequences of operators on L^2 and H^2 respectively. As in the proof of Theorem 1, for any $n \geq 1$, we have $T_n = T_{f_n}$ for some $f_n \in H^\infty$, $f_n(0) = 0$, $\|f_n\| \leq 1$. It is easy to see that we also have $\hat{T}_n = M_{f_n}$ - the multiplication by f_n on L^2.

Denote also, for $n \geq 0$,

$$\hat{K}_n = \bigvee_{0\leq k\leq n} \hat{V}^k L^2;\quad \hat{U}_n = \hat{U}|\hat{K}_n,\ \hat{V}_n = P_{\hat{K}_n}\hat{V}|\hat{K}_n, \tag{5.1}$$

$$K_n = \bigvee_{0\leq k\leq n} V^k H^2;\quad U_n = U|K_n,\ V_n = P_{K_n}V|K_n. \tag{5.2}$$

We have $K_n \subset \hat{K}_n$ and

$$U_n = \hat{U}_n|K_n,\quad V_n = P_{K_n}\hat{V}_n|K_n, \tag{5.3}$$

$$\hat{U}_n\hat{V}_n = \hat{V}_n\hat{U}_n,\quad V_n = U_n^* V_n U_n. \tag{5.4}$$

Note also that for any $n \geq 1$ we have

$$\hat{V}|\hat{K}_{n-1} = \hat{V}_n|\hat{K}_{n-1},\quad V|K_{n-1} = V_n|K_{n-1}, \tag{5.5}$$

and

$$VK \perp L^2 \ominus e^{it}H^2. \tag{5.6}$$

Let $(\hat{R}_n)_{n\geq 1}$ be the choice sequence initiated on L^2 corresponding to $(\hat{T}_n)_{n\geq 1}$ and $(R_n)_{n\geq 1}$ be the choice sequence initiated on H^2 corresponding to $(T_n)_{n\geq 1}$. From the geometry of the space of the minimal isometric dilation we know that

$$\hat{R}_1 = \hat{V}_0 = M_{f_1}, \quad R_1 = V_0 = T_{f_1}$$

and for $n \geq 1$

$$\hat{K}_n = \hat{K}_{n-1} \oplus \mathcal{D}_{\hat{R}_n},$$

$$\hat{V}_n = \begin{pmatrix} \hat{V}_{n-1} & \hat{D}_{*n} \\ \hat{D}_n & -\hat{R}_n^* \hat{R}_{n+1} \end{pmatrix}$$

where $\hat{D}_1 = D_{\hat{R}_1}$, $\hat{D}_{*1} = D_{\hat{R}_1^*} \hat{R}_2$ and for $n \geq 2$, $\hat{D}_n$ and $\hat{D}_{*n}$ are the following row and column matrices, respectively:

$$\hat{D}_n = (0 \ \cdots \ 0 \ D_{\hat{R}_{n-1}}),$$

$$\hat{D}_{*n} = \begin{pmatrix} D_{\hat{R}_1^*} D_{\hat{R}_2^*} \cdots D_{\hat{R}_n^*} \hat{R}_{n+1} \\ -\hat{R}_1^* D_{\hat{R}_2^*} \cdots D_{\hat{R}_n^*} \hat{R}_{n+1} \\ \vdots \\ -\hat{R}_{n-1}^* D_{\hat{R}_n^*} \hat{R}_{n+1} \end{pmatrix}.$$

Similar formulas hold for K_n and V_n.

The embeddings Z_n of K_n into $\hat{K}_n$ are given by the following recurrent formulas:

$$Z_n = \begin{pmatrix} Z_{n-1} & a_n \\ 0 & b_n \end{pmatrix}, \quad n \geq 1 \tag{5.7}$$

where Z_0 is the usual inclusion of H^2 in L^2 and for $n \geq 1$,

$$b_n D_{R_n} \cdots D_{R_1} h = D_{\hat{R}_n} \cdots D_{\hat{R}_1} h, \quad h \in H^2, \tag{5.8}$$

$a_1 = 0$ and for $n \geq 2$,

$$a_n D_{R_n} \cdots D_{R_1} h = [\hat{V}_{n-1} Z_{n-1} - Z_{n-1} V_{n-1}] \begin{pmatrix} 0 \\ D_{R_{n-1}} \cdots D_{R_1} h \end{pmatrix}, h \in H^2. \tag{5.9}$$

The following relations are also easy to verify:

$$\hat{U}_n Z_n = Z_n \hat{U}_n, \quad V_n = Z_n^* \hat{V}_n Z_n. \tag{5.10}$$

In fact, for each $n \geq 0$, $\hat{U}_n$ is the minimal unitary extension of the isometry U_n and $\hat{V}_n$ is the unique dilation to a commutant of $\hat{U}_n$ (the symbol) of the U_n-Toeplitz operator V_n. Recall also the formula

$$\hat{V}_n k_n = \lim_{m \to \infty} \hat{U}_n^{*m} Z_n V_n U_n^m k_n, \ k_n \in K_n, \tag{5.11}$$

which uniquely determines $\hat{V}_n$.

It follows that the pairs $[\hat{U}_n, \hat{V}_n]$, $[U_n, V_n]$ are well determined one by another and each of them is well determined by any string $\{\hat{R}_1, \dots, \hat{R}_n\}$ or $\{R_1, \dots, R_n\}$, separately.

The following result describes the obstructions in choosing $\hat{R}_n$.

Theorem 4 *Let μ be a measure in $\mathcal{M}_0$ and $(\hat{R}_n)_{n\geq 1}$, $(R_n)_{n\geq 1}$, $(Z_n)_{n\geq 0}$ be as above. Then we have:*

(i) *$\hat{R}_1$ is the multiplication by a function $f_1 \in H^\infty$, $f_1(0) = 0$, $\|f_1\| \leq 1$.*

(ii) *For any $n \geq 2$, the formula*

$$(5.12)\qquad \begin{aligned}(H_n h, g) &= -\left(\hat{V}_{n-1}^2 Z_{n-1}\begin{pmatrix} 0 \\ D_{R_{n-1}} \cdots D_{R_1} h\end{pmatrix}, g\right)_{\hat{K}_{n-1}}, \\ &\qquad h \in H^2,\ g \in L^2 \ominus H^2,\end{aligned}$$

defines a contractive Hankel operator H_n.

(iii) *For any $n \geq 2$, the operator $D_{\hat{R}_1^*} \cdots D_{\hat{R}_{n-1}^*} \hat{R}_n D_{\hat{R}_{n-1}} \cdots D_{\hat{R}_1}$ is (the multiplication by) a contractive symbol of the Hankel operator H_{n-1}, i.e.*

$$(5.13)\qquad \begin{aligned}&(D_{\hat{R}_1^*} \cdots D_{\hat{R}_{n-1}^*} \hat{R}_n D_{\hat{R}_{n-1}} \cdots D_{\hat{R}_1} h, g)_{L^2} \\ &= (H_{n-1} h, g)_{L^2}, h \in H^2, g \in L^2 \ominus H^2,\end{aligned}$$

where H_1 is the null Hankel operator.

Proof: Since $\hat{U}\hat{V} = \hat{V}\hat{U}$, and L^2, as the subspace of $\hat{K}$, reduces $\hat{U}$ to the multiplication by e^{it}, it follows that $\hat{R}_1$ commutes with the multiplication by e^{it}. Hence $\hat{R}_1 = M_{r_1}, r_1 \in L^\infty$, $\|r_1\| \leq 1$.

Since H^2, as a subspace of K, reduces U to the shift operator, it follows $r_1 = f_1$, with $f_1 \in H^\infty$, $f_1(0) = 0$. It is also clear that $D_{\hat{R}_1}$ commutes with the bilateral shift on L^2. It follows that $\mathcal{D}_{\hat{R}_1} = \mathcal{D}_{\hat{R}_1^*}$ reduces the bilateral shift. By induction we can prove that for any $n \geq 2$, the operator $D_{\hat{R}_1^*} \cdots D_{\hat{R}_{n-1}^*} \hat{R}_n D_{\hat{R}_{n-1}} \cdots D_{\hat{R}_1}$ on L^2 commutes with the multiplication by e^{it}, hence it is given by the multiplication by a function in L^∞. Using (5.7) and (5.6), we deduce

$$\begin{aligned}(D_{\hat{R}_1^*}\hat{R}_2 D_{\hat{R}_1} h, g)_{L^2} &= \left(\hat{V}\begin{pmatrix} 0 \\ D_{\hat{R}_1} h\end{pmatrix}, g\right)_{\hat{K}_1} = \left(\hat{V} Z_1\begin{pmatrix} 0 \\ D_{R_1} h\end{pmatrix}, g\right)_{\hat{K}_1} \\ &= \left(V\begin{pmatrix} 0 \\ D_{R_1} h\end{pmatrix}, g\right) = 0,\end{aligned}$$

for any $h \in H^2$ and $g \in L^2 \ominus H^2$.

It results that $D_{\hat{R}_1^*}\hat{R}_2 D_{\hat{R}_1}$ is a contractive symbol of the null Hankel operator. Using again (5.7) and (5.6) we obtain:

$$(D_{\hat{R}_1^*}\cdots D_{\hat{R}_{n-1}^*}\hat{R}_n D_{\hat{R}_{n-1}}\cdots D_{\hat{R}_1}h, g)_{L^2} = \left(\hat{V}\begin{pmatrix} 0 \\ D_{\hat{R}_{n-1}}\cdots D_{\hat{R}_1}h \end{pmatrix}, g\right)_{\hat{K}}$$

$$= \left(\hat{V}\left[Z_{n-1}\begin{pmatrix} 0 \\ D_{R_{n-1}}\cdots D_{R_1}h \end{pmatrix} - a_{n-1}D_{R_{n-1}}\cdots D_{R_1}h\right], g\right)_{\hat{K}}$$

$$= \left(\hat{V}Z_{n-1}\begin{pmatrix} 0 \\ D_{R_{n-1}}\cdots D_{R_1}h \end{pmatrix}, g\right)_{\hat{K}} - (\hat{V}a_{n-1}D_{R_{n-1}}\cdots D_{R_1}h, g)_{\hat{K}}$$

$$= \left(V\begin{pmatrix} 0 \\ D_{R_{n-1}}\cdots D_{R_1}h \end{pmatrix}, g\right)_{K} - \Bigg(\hat{V}[\hat{V}_{n-2}Z_{n-2} - Z_{n-2}V_{n-2}]$$

$$\begin{pmatrix} 0 \\ D_{R_{n-2}}\cdots D_{R_1}h \end{pmatrix}, g\Bigg)_{\hat{K}}$$

$$= \left(\hat{V}_{n-2}^2 Z_{n-2}\begin{pmatrix} 0 \\ D_{R_{n-2}}\cdots D_{R_1}h \end{pmatrix}, g\right)_{\hat{K}} = (H_{n-1}h, g)_{L^2},$$

for any $n \geq 3$.

It follows that, for any $n \geq 2$, formula (5.12) defines a contractive Hankel operator H_n and $D_{\hat{R}_1^*}\cdots D_{\hat{R}_{n-1}^*}\hat{R}_n D_{\hat{R}_{n-1}}\cdots D_{\hat{R}_1}$ is a a contractive symbol of H_{n-1}.

The theorem is completely proved. □

Let now $(\hat{R}_n)_{n\geq 1}$ be a choice sequence initiated on L^2, $(\hat{T}_n)_{n\geq 1}$ be its corresponding positive definite sequence on L^2 and $[\hat{K}, \hat{V}]$ be the minimal isometric dilation of $(\hat{T}_n)_{n\geq 1}$. Let us denote

$$K = \bigvee_{n\geq 0} \hat{V}^n H^2, \quad V = \hat{V}|K. \tag{5.14}$$

Setting

$$T_n = P_{H^2}V^n|H^2, \; n \geq 1,$$

we obtain a positive definite sequence $(T_n)_{n\geq 1}$ on H^2. Let $(R_n)_{n\geq 1}$ be the choice sequence initiated on H^2 corresponding to $(T_n)_{n\geq 1}$.

Denote again, for any $n \geq 0$, $[\hat{K}_n, \hat{V}_n]$ as in (5.1) and $[K_n, V_n]$ as in (5.2). Then $K_n \subset \hat{K}_n$ and, in the recursive matricial forms of $\hat{K}_n$ and K_n, this embedding is given by the matrix Z_n defined recurrently as in (5.7)–(5.9).

We can state now the following converse of Theorem 4:

Theorem 5 *Let $(\hat{R}_n)_{n\geq 1}$ be a choice sequence initiated on L^2 verifying:*

(i) *$\hat{R}_1$ is the multiplication by a function $f_1 \in H^\infty$, $f_1(0) = 0$, $\|f_1\| \leq 1$.*

(ii) *For any* $n \geq 2$, *the formula*

$$(5.15)\qquad \begin{aligned}(H_n h, g) &= -\left(\hat{V}_{n-1}^2 Z_{n-1}\begin{pmatrix} 0 \\ D_{R_{n-1}}\cdots D_{R_1} h\end{pmatrix}, g\right)_{\hat{K}_{n-1}},\\ &\qquad h \in H^2,\ g \in L^2 \ominus H^2,\end{aligned}$$

defines a contractive Hankel operator H_n.

(iii) *For any* $n \geq 2$, *the operator* $D_{\hat{R}_1^*}\cdots D_{\hat{R}_{n-1}^*}\hat{R}_n D_{\hat{R}_{n-1}}\cdots D_{\hat{R}_1}$ *is (the multiplication by) a contractive symbol of the Hankel operator* H_{n-1}, *i.e.*

$$(5.16)\qquad \begin{aligned}&(D_{\hat{R}_1^*}\cdots D_{\hat{R}_{n-1}^*}\hat{R}_n D_{\hat{R}_{n-1}}\cdots D_{\hat{R}_1} h, g)_{L^2}\\ &= (H_{n-1}h, g)_{L^2},\ h \in H^2,\ g \in L^2 \ominus H^2,\end{aligned}$$

where H_1 *is the null Hankel operator.*

Then, for each $n \geq 1$, *there exists* f_n *in* H^∞, $f_n(0) = 0$, $\|f_n\| \leq 1$, *such that*

$$(5.17)\qquad \hat{T}_n = M_{f_n}, \quad T_n = T_{f_n}.$$

Proof: First, from (i) and (iii) it follows that for any $n \geq 1$ we have $\hat{T}_n e^{it} = e^{it}\hat{T}_n$. Hence $\hat{T}_n$ is given by the multiplication by a function f_n from L^∞. It remains to show that for any $n \geq 1$ we have $f_n \in H^\infty$, $f_n(0) = 0$.

This will results from the relation

$$(5.18)\qquad VK \perp L^2 \ominus e^{it}H^2.$$

Indeed, if (5.18) holds then for $p \geq 0$ we have:

$$\begin{aligned}1/2\pi \int_0^{2\pi} e^{ipt} f_n(e^{it})dt &= (f_n, e^{-ipt})_{L^2} = (\hat{V}^n 1, e^{-ipt})_{\hat{K}}\\ &= (V^n 1, e^{-ipt})_{\hat{K}} = 0.\end{aligned}$$

Hence $f_n \in H^\infty$, $f_n(0) = 0$.

In order to prove (5.18) let us firstly remark that it is equivalent to

$$\hat{V}K \perp L^2 \ominus e^{it}H^2,$$

which is clearly a consequence of

$$(5.19)\qquad \hat{V}_n K_n \perp L^2 \ominus e^{it}H^2, \quad n \geq 0.$$

We shall prove (5.19) by induction. For $n = 0$ we have $\hat{V}_0|H^2 = T_{f_1}$ because of (i). Since $f_1(0) = 0$, (5.19) holds for $n = 0$. Suppose it holds for any $n \leq p - 1$,

$p \geq 1$. Then for any $k_p \in K_p$ of the form $k_{p-1} \oplus D_{R_p} \cdots D_{R_1} h$ with $k_{p-1} \in K_{p-1}$ and $h \in H^2$ and for any $g \in L^2 \ominus e^{it} H^2$, we have:

$$(\hat{V}_p k_p, g)_{\hat{K}_p} = \left(\hat{V}_p Z_p \begin{pmatrix} k_{p-1} \\ D_{R_p} \cdots D_{R_1} h \end{pmatrix}, g \right)_{\hat{K}_p}$$

$$= \left(\begin{pmatrix} \hat{V}_{p-1} & \hat{D}_{*p} \\ \hat{D}_p & -\hat{R}_p^* \hat{R}_{p+1} \end{pmatrix} \begin{pmatrix} Z_{p-1} k_{p-1} + a_p D_{R_p} \cdots D_{R_1} h \\ b_p D_{R_p} \cdots D_{R_1} h \end{pmatrix}, g \right)_{\hat{K}_p}$$

$$= (\hat{V}_{p-1} Z_{p-1} k_{p-1}, g)_{\hat{K}_p} + (\hat{V}_{p-1} a_p D_{R_p} \cdots D_{R_1} h, g)_{\hat{K}_p}$$

$$+ (\hat{D}_{*p} b_p D_{R_p} \cdots D_{R_1} h, g)_{\hat{K}_p} = (V_{p-1} k_{p-1}, g)_{\hat{K}_p}$$

$$+ \left(\hat{V}_{p-1} [\hat{V}_{p-1} Z_{p-1} - Z_{p-1} V_{p-1}] \begin{pmatrix} 0 \\ D_{R_{p-1}} \cdots D_{R_1} h \end{pmatrix}, g \right)_{\hat{K}_p}$$

$$+ (D_{\hat{R}_1^*} \cdots D_{\hat{R}_p^*} \hat{R}_{p+1} D_{\hat{R}_p} \cdots D_{\hat{R}_1} h, g)_{L^2}$$

$$= (V_{p-1} k_{p-1}, g)_{\hat{K}_p} - \left(\hat{V}_{p-1} Z_{p-1} V_{p-1} \begin{pmatrix} 0 \\ D_{R_{p-1}} \cdots D_{R_1} h \end{pmatrix}, g \right)_{\hat{K}_p}$$

$$+ \left(\hat{V}_{p-1}^2 Z_{p-1} \begin{pmatrix} 0 \\ D_{R_{p-1}} \cdots D_{R_1} h \end{pmatrix}, g \right)_{\hat{K}_p} + (D_{\hat{R}_1^*} \cdots D_{\hat{R}_p^*} \hat{R}_{p+1}$$

$$D_{\hat{R}_p} \cdots D_{\hat{R}_1} h, g)_{L^2} = 0.$$

In the last equality we used induction hypothesis and (iii). Hence (5.19) is proved.
This completes the proof of Theorem 5. □

Using Theorem 1, we conclude that the hypotheses of Theorem 4 or Theorem 5 are sufficient conditions for a choice sequence $(\hat{R}_n)_{n \geq 1}$ to produce a representing measure μ in $\mathcal{M}_0$. These conditions also suggest how to choose $\hat{R}_n$ at each step $n \geq 1$: take $\hat{R}_1 = T_{f_1}$, $f_1 \in H^\infty$, $f_1(0) = 0$, $\|f_1\|| \leq 1$, and for $n \geq 2$, choose a contractive symbol for the Hankel operator H_{n-1} as the operator $D_{\hat{R}_1^*} \cdots D_{\hat{R}_{n-1}^*} \hat{R}_n D_{\hat{R}_{n-1}} \cdots D_{\hat{R}_1}$. In the Adamian-Arov-Krein theory there are algorithms to construct an arbitrary contractive symbol for H_{n-1}.

Let us mention the following situation in which the choice is blocked at the first row:

Corollary 1 *Let μ be a measure in $\mathcal{M}_0$ and let $(\hat{R}_n)_{n\geq 1}$, $(R_n)_{n\geq 1}$, $(f_n)_{n\geq 1}$ be as above. If f_1 is an extreme point in the unit ball of H^∞, i.e.*

$$\int_0^{2\pi} \log(1 - |f_1(e^{it})|)dt = -\infty, \tag{5.20}$$

then for any $n \geq 2$ we have $\hat{R}_n = 0$, $R_n = 0$ and $f_n = f_1^n$.

Proof: (cf. [4], [11]). If (5.20) holds then according to Beurling Theorem we have

$$\overline{D_{\hat{R}_1} H^2} = \mathcal{D}_{\hat{R}_1}, \quad \overline{D_{\hat{R}_1^*}[L^2 \ominus H^2]} = \mathcal{D}_{\hat{R}_1^*}.$$

¿From (5.13) it follows firstly that $\hat{R}_2 = 0$, and then, by induction, that $\hat{R}_n = 0$, for any $n \geq 2$. Clearly then $f_n = f_1^n$ for any $n \geq 2$ and consequently $R_n = 0$ for any $n \geq 2$.

In solving our moment problem the obstructions described above play an important role. In the case the matrix of the data has rows of infinite dimension - the band case - the procedure described above furnishes the set of all solutions. In the case the matrix of the data is finite in both dimensions we have the freedom to extend each row to the full line of the Taylor coefficients of a function in H^∞ such that to the obtained band we can apply the above procedure. □

References

[1] V.M. Adamian, D.Z. Arov and M.G. Krein, *Infinite Hankel block matrices and related extension problems*, Izv. Acad. Nauk Armjan SSR, Matematika **6** (1971), pp. 87–112.

[2] T. Ando, *On a pair of commutative contractions*, Acta Sci. Math. **24** (1963), pp. 88–90.

[3] Z. Ceauşescu and C. Foiaş, *On intertwining dilations V*, Acta Sci. Math. **40** (1978), pp. 9–32.

[4] Z. Ceauşescu and I. Suciu, *Isometric dilations of commuting contractions I*, J. Operator Theory **12** (1984), pp. 65-88.

[5] Z. Ceauşescu and I. Suciu, *Isometric dilations of commuting contractions IV*, Advances in invariant subspaces and other results of operator theory, Operator Theory; Advances and Applications **17** (1986), Birkhäuser Verlag, Basel, pp. 55–77.

[6] R.G. Douglas, *On the operator equation $S^*XT = T$ and related topics*, Acta Sci. Math. **30** (1969), pp. 19–32.

[7] C. Foiaş, and A.E. Frazho, *The Commutant Lifting Approach to Interpolation Problems*, Birkhäuser Verlag, Basel-Boston-Berlin, 1990.

[8] H. Helson, *Lectures on Invariant Subspaces,* New York: Academic Press, 1964.

[9] K. Hoffman, *Banach spaces of analytic functions*, Englewood cliffs, Prentice Hall, New Jersey, 1962.

[10] I. Suciu, *On the matrix of the Fourier coefficients of a representing measures on the bitorus*, Preprint IMAR, no. 22, 1994.

[11] I. Suciu, *Maximum entropy truncable positive definite sequences of analytic Toeplitz operators*, Preprint IMAR, no. 10, 1997.

[12] B.-Sz. Nagy and C. Foiaş, *Harmonic Analysis of Operators on Hilbert Space*, North-Holland Publishing Co., Amsterdam-Budapest, 1970.

Ion Suciu
Institute of Mathematics
Romanian Academy
P.O. Box 1–764
RO–70700
Bucharest
Romania
isuciu@stoilow.imar.ro

AMS Classification Numbers: Primary 47A20 Secondary 46E20

Operator Theory:
Advances and Applications, Vol. 118

Inner-outer Factorization of J-Contractive-Valued Functions

Alexey Tikhonov

A new approach to obtain the existence and uniqueness theorem of inner-outer factorization is proposed. This approach does not employ the Sz.-Nagy-Foiaş-Naboko functional model nor the Shmul'jan-Pekarev regularity factorization criterion.

0 Introduction

The inner-outer factorization, i.e. a representation of an analytic function $f(z)$ on the unit disk in the form $f(z) = f_i(z)f_e(z)$, where the boundary values for the inner function $f_i(z)$ are unitary on the unit circle and the outer function $f_e(z)$ has no inner divisors, is well known for scalar functions [1] and for contractive-valued analytic functions [2]. This factorization plays important role in function theory, operator theory, and related subjects.

It is interesting to extend these results to the case of J-contractive-valued functions. The positive answer was obtained in [3], [4].

In this paper a new proof of the existence and uniqueness of the inner-outer factorization theorem is given. This approach does not employ the Sz.-Nagy-Foiaş-Naboko function model [2], [5] nor the Shmul'jan-Pekarev regularity factorization criterion [6], which take a lot of pages in original proof of the author in [4].

The basic techniques used in the present paper are the following: (1) The theory of absolutely continuous and singular subspaces of linear operators [5], [7], [8], [9], [10]. The correspondence between these subspaces and inner-outer factorization was established for contractive (dissipative) operators by Sz.-Nagy, Foiaş, Sakhnovich, Pavlov. (2) The theory of operator nodes and its characteristic functions, which was developed by Livsic, Brodski and others. In this paper we use the notion of a J-node and its properties from [11]. (3) Some facts concerning the existence of boundary values for operator-valued functions that originated from the Birman-Entina result. Using these technical tools we obtain our main theorem:

Theorem: *Let $W(z) \in B(J, \mathfrak{S}_1)$. Then there exists a unique (up to a J-unitary factor) factorization $W(z) = W_i(z)W_e(z)$, where the boundary values of the J-inner function $W_i(z)$ are a.e. J-unitary on the unit circle and the J-outer function $W_e(z)$ has only trivial left J-inner divisors.*

Here $B(J, \mathfrak{S}_1)$ is the class of meromorphic on the unit disk J-contractive-valued functions such that $J - W(z)JW(z)^* \in \mathfrak{S}_1$, ($\mathfrak{S}_1$ is the trace class) and $W(0)$ is an invertible two-sided J-contractive operator.

1 Absolutely Continuous Subspace

Let H be a Hilbert space, $S \in [H]$ be invertible, and $\sigma_c(S) \subset \Gamma$, where $\sigma_c(S)$ is the continuous spectrum of the operator S and Γ is the unit circle. For an arbitrary Borel set $\omega \subset \Gamma$, we define a linear manifold [9]:

$$\widetilde{N}_\omega(S) = \Big\{ f \in H : \forall g \in H\ \exists e(f,g,w) \in L^1(\omega) \forall w_0 \in \rho(S)$$
$$((S - w_0)^{-1} f, g) = \int_\omega \frac{e(f,g,w)dw}{w - w_0} \Big\}$$

In this case, $e(f,g,w)$ can be calculated using boundary values of its Cauchy transform:

$$e(f,g,w) = e_S(f,g,w) = 1/2\pi[((S-w)^{-1}f,g)_+ - ((S-w)^{-1}f,g)_-],$$

where $((S-w)^{-1}f,g)_\pm$ are interior and exterior boundary angular limits of $((S-w)^{-1}f,g)$ on Γ. In the sequel, we use the existence of such angular limits a.e. on $\Gamma\ \forall f,g \in H$.

By definition, put: $\widetilde{M}_\omega(S) = \{f \in H : \forall g \in H\ e_S(f,g,w) = 0 \text{ a.e. } w \in \omega\}$.

Lemma 1 1) $S\widetilde{N}_\omega(S) = \widetilde{N}_\omega(S)$;

2) $S\widetilde{M}_\omega(S) = \widetilde{M}_\omega(S)$;

3) $SH_1 \subset H_1, S^{-1}H_1 \subset H_1 \Rightarrow \widetilde{N}_\omega(S|H_1) = \widetilde{N}_\omega(S) \cap H_1,\ \widetilde{M}_\omega(S|H_1) = \widetilde{M}_\omega(S) \cap H_1$.

Proof: 1) $((S-w_0)^{-1}Sf,g) = ((S-w_0)^{-1}f, S^*g) = \int_\omega \frac{e(f,S^*g,w)dw}{w-w_0}$.

2) $e_S(Sf,g,w) = we_S(f,g,w)$.

3) $e_S(f_1, g_1 + g_2, w) = e_S(f_1, g_1, w),\ f_1, g_1 \in H_1,\ g_2 \perp H_1$.

□

Definition: The space $N_\omega(S) = \operatorname{clos} \widetilde{N}_\omega(S)$ is called the absolutely continuous subspace of the operator S corresponding to the set ω. The space $M_\omega(S) = \operatorname{clos} \widetilde{M}_\omega(S)$ is called the complementary subspace to $N_\omega(S)$. We put $N(S) = N_\Gamma(S)$ and $M(S) = M_\Gamma(S)$.

Definition: We shall call ω a similarity set if $N_\omega(S) = \widetilde{N}_\omega(S)$.

A sufficient condition for ω to possess this property is the following

Condition A_ω: For all $f, g \in H$ we have $\|e_S(f, g, w)\|_{L^1(\omega)} \leq C(\omega)\|f\|\|g\|$. We write $S \in A_\omega$, if S satisfies the condition A_ω. Evidently, $S \in A_\omega \Leftrightarrow S^* \in A_{\bar{\omega}}$.

Theorem 2 [9]. *Let $S \in A_\omega$. Then there exists a bounded projection*

$$(E_S(\omega)f, g) = \int_\omega e_S(f, g, w)dw \quad \textit{onto subspace } N_\omega(S).$$

In this case, $M_\omega(S) = \widetilde{M}_\omega(S) = \operatorname{Ker} E_S(\omega)$, $N_\omega(S)^\perp = M_{\bar{\omega}}(S^*)$, $E_{S^*}(\bar{\omega}) = E_S(\omega)^*$, $SE_S(\omega) = E_S(\omega)S$.

Theorem 3 [9]. *Let S be an invertible operator, U be an unitary operator, $S-U \in \mathfrak{S}_1$. Then $\forall \varepsilon > 0\, \exists \omega \subset \Gamma$ such that* $\operatorname{mes}(\Gamma \setminus \omega) < \varepsilon$, $S \in A_\omega$.

Corollary: *There exists an increasing sequence of sets $\omega_k \subset \Gamma$ such that* $\omega_k \subset \omega_{k+1}$, $S \in A_{\omega_k}$, $\lim_{k\to\infty} \operatorname{mes}(\Gamma \setminus \omega_k) = 0$.

Notice that the unitary operator from the polar decomposition of S can be regarded as U in Theorem 3. If $I - S^*S \in \mathfrak{S}_1$, then $I - |S| = (I + |S|)^{-1}(I - |S|^2) = (I + |S|)^{-1}(I - S^*S) \in \mathfrak{S}_1$. From this it follows that $U - S \in \mathfrak{S}_1$.
Notice also that, $I - S_1^*S_1 \in \mathfrak{S}_1$ if $I - S^*S \in \mathfrak{S}_1$, $SH_1 \subset H_1$, $S_1 = S|H_1$.

Theorem 4 *Let S be an invertible operator, $I - S^*S \in \mathfrak{S}_1$. Then $\forall \omega \subset \Gamma$*

$$1)\ M_\omega(S) = \widetilde{M}_\omega(S) = \bigcap_{k=1}^{\infty} M_{\omega\cap\omega_k}(S),$$

$$2)\ N_\omega(S) = \operatorname{clos} \bigcup_{k=1}^{\infty} N_{\omega\cap\omega_k}(S) = M_{\bar{\omega}}(S^*)^\perp,$$

where $\omega_k \subset \omega_{k+1}$, $S \in A_{\omega_k}$, $\lim_{k\to\infty} \operatorname{mes}(\Gamma \setminus \omega_k) = 0$.

Proof: 1) Since $\lim_{k\to\infty} \operatorname{mes}(\omega \setminus \omega_k) = 0$ we have $\widetilde{M}_\omega(S) = \cap_{k=1}^{\infty} \widetilde{M}_{\omega\cap\omega_k}(S)$. By Theorem 2, $\widetilde{M}_{\omega\cap\omega_k}(S) = M_{\omega\cap\omega_k}(S)$. Then $\widetilde{M}_\omega(S)$ is closed as the intersection of closed sets.
2)We have

$$\begin{aligned} M_{\bar{\omega}}(S^*)^\perp &= \left(\bigcap_{k=1}^{\infty} M_{\bar{\omega}\cap\bar{\omega}_k}(S^*)\right)^\perp = \operatorname{clos} \bigcup_{k=1}^{\infty} (M_{\bar{\omega}\cap\bar{\omega}_k}(S^*))^\perp \\ &= \operatorname{clos} \bigcup_{k=1}^{\infty} N_{\omega\cap\omega_k}(S) = \operatorname{clos} \bigcup_{k=1}^{\infty} \widetilde{N}_{\omega\cap\omega_k}(S) \subset \operatorname{clos} \widetilde{N}_\omega(S) = N_\omega(S). \end{aligned}$$

Let $f \in \widetilde{N}_\omega(S)\,,\ g \in M_{\bar\omega}(S^*)$. Then

$$\begin{aligned}(f,g) &= -\lim_{w_0\to\infty} w_0((S-w_0)^{-1}f,g) = -\lim_{w_0\to\infty}\int_\omega \frac{w_0 e_S(f,g,w)dw}{w-w_0}\\ &= \int_\omega e_S(f,g,w)dw = \int_{\bar\omega}\overline{e_{S^*}(g,f,\bar w)}d\bar w = \int_{\bar\omega} 0\,d\bar w = 0\,.\end{aligned}$$

Therefore, $\widetilde{N}_\omega(S) \subset M_{\bar\omega}(S^*)^\perp$. □

Corollary: *Let* $SH_1 \subset H_1,\ S^{-1}H_1 \subset H_1$. *Then*

1) $M_\omega(S|H_1) = M_\omega(S) \cap H_1$;
2) $M_\omega(S|M_\omega(S)) = M_\omega(S)$;
3) $N_\omega(S|H_1) \subset N_\omega(S) \cap H_1$;
4) $N_\omega(S|N_\omega(S)) = N_\omega(S)$.

Proof: 1) By Theorem 4, $\widetilde{M}_\omega(S) = M_\omega(S)$, $\widetilde{M}_\omega(S|H_1) = M_\omega(S|H_1)$. Combining this with Lemma 1, we obtain the required identity.

2) $H_1 = M_\omega(S)$.

3) Obvious from Lemma 1.

4) By Lemma 1 and Theorem 2, we have

$$\begin{aligned}N_{\omega\cap\omega_k}(S|N_\omega(S)) &= \widetilde{N}_{\omega\cap\omega_k}(S|N_\omega(S)) = \widetilde{N}_{\omega\cap\omega_k}(S) \cap N_\omega(S)\\ &= \widetilde{N}_{\omega\cap\omega_k}(S) = N_{\omega\cap\omega_k}(S).\end{aligned}$$

Hence, according Theorem 4, we may conclude that the required identity is fulfilled. □

Lemma 5 *Let* $\operatorname{clos} D_1 = \operatorname{clos} D_2 = H$. *Then*

$$M_\omega(S) = M_{\bar\omega}(S^*) = H \iff \forall f \in D_1,\ g \in D_2 \quad e_S(f,g,w) = 0 \ \textit{a.e. on}\ \omega.$$

Proof: The necessity is obvious. To prove the sufficiency, let $\nu \subset \omega, S \in A_\nu$. Then

$$\forall f \in D_1,\ g \in D_2 \quad (E_S(\nu)f,g) = \int_\nu e_S(f,g,w)dw = 0\,.$$

Therefore, $E_S(\nu) = 0$, $N_\nu(S) = \{0\}$. Whence, by Theorem 4, $N_\omega(S) = \{0\}$. □

2 J-nodes and Characteristic Functions

In this section we survey the basic properties for J-nodes and their characteristic functions [11]. The aggregate of bounded operators and Hilbert spaces $\alpha = (S, H, G, E, J, W)$ is called a J-node if the operators $S : H \to H$, $W : E \to E$ are invertible and $I - S^*S = GJG^*$, $WJW^* = J - G^*G$, $J = J^* = J^{-1}$. The following $I \oplus J$-unitary operator

$$\alpha = \begin{pmatrix} S & Q \\ P & R \end{pmatrix} : H \oplus E \longrightarrow H \oplus E, \quad P = -G^*,$$

$$Q = SG(W^{-1})^*J, \quad R = W$$

can be assigned to a J-node. Conversely, if S and R are invertible, then the J-node is uniquely determined by the $I \oplus J$-unitary operator α. In the sequel, we shall use the notation in terms of $I \oplus J$-unitary operators.

Let α_1, α_2 be J-nodes. The J-node

$$a = \begin{pmatrix} S_1 & Q_1P_2 & Q_1R_2 \\ 0 & S_2 & Q_2 \\ P_1 & R_1P_2 & R_1R_2 \end{pmatrix}$$

is called the product of J-nodes α_1, α_2. Here $H = H_1 \oplus H_2$, $S : H \to H$ and H_1 is an invariant subspace for the operators S and S^{-1}. Conversely, if $SH_1 \subset H_1$, $S^{-1}H_1 \subset H_1$, $H_1 \subset H$, then α can be represented in the form of the product of J-nodes α_1 and α_2. In this case, α_1, α_2 can be determined up to a J-unitary operator U:

$$Q_1P_2 = Q_1UU^{-1}P_2, \quad Q_1R_2 = Q_1UU^{-1}R_2,$$
$$R_1P_1 = R_1UU^{-1}P_2, \quad R_1R_2 = R_1UU^{-1}R_2.$$

The J-nodes α_1 and α_2 are called projections of the J-node α.

The subspace

$$\widehat{H} = \left(\bigvee_{k\geq 0} S^kQ(E)\right) \bigvee \left(\bigvee_{k\geq 0} (S^*)^kP^*(E)\right)$$

is called the simple subspace of a J-node α. It is not difficult to verify that

$$\widehat{H} = \bigvee_{k\in\mathbf{Z}} S^kQ(E) = \bigvee_{k\in\mathbf{Z}} (S^*)^kP^*(E).$$

A J-node α is called simple if $H = \widehat{H}$. We put $H_0 = H \ominus \widehat{H}$.

Lemma 6 *Let* $\alpha = \alpha_1\alpha_2$. *Then* $H_1 = \widehat{H_1} \oplus (H_1 \cap H_0)$, $H_2 = \widehat{H_2} \oplus (H_2 \cap H_0)$.

Proof: We shall prove only the first assertion. Let $f \in H_1, g \in E$, then from the equality

$$\begin{aligned}(f, (S_1^*)^k P_1^* g) &= (S_1^k f, P_1^* g) = (S^k f, P_{H_1} P^* g) \\ &= (S^k f, P^* g) = (f, (S^*)^k P^* g), \ k \in \mathbf{Z},\end{aligned}$$

we have $f \bot \widehat{H}_1 \Leftrightarrow f \bot \widehat{H}$. This means that $H_1 \ominus \widehat{H}_1 = H_1 \cap H_0$. □

Corollary: *If the J-nodes α_1, α_2 are simple, then $H_0 \cap H_1 = \{0\}$, $H_0 \cap H_2 = \{0\}$.*

The product of the simple J-nodes α_1, α_2 is called regular if the J-node $\alpha = \alpha_1\alpha_2$ is simple.

The operator-valued function $W_\alpha(z) = R + zP(I - zS)^{-1}Q$ is called the characteristic function of a J-node α. The class of characteristic functions for J-nodes will be denoted by $B(J)$.

A simple J-node is uniquely (up to unitary equivalence) determined by its characteristic function.

We have the following Multiplication Theorem $W_{\alpha_1\alpha_2}(z) = W_{\alpha_1}(z)W_{\alpha_2}(z)$.

An invariant subspace $H_1 \subset H$ for the operators S and S^{-1} generates a factorization $W_\alpha(z) = W_{\alpha_1}(z)W_{\alpha_2}(z)$, where the projections α_1, α_2 are generated by H_1.

The product $W_1(z)W_2(z)$, $W_1(z), W_2(z) \in B(J)$ is called regular if the product of the corresponding simple J-nodes α_1, α_2 is regular, where $W_1(z) = W_{\alpha_1}(z)$, $W_2(z) = W_{\alpha_2}(z)$.

The next assertion follows directly from these facts.

Proposition 7 *Suppose α is a simple J-node, $W_\alpha(z) = W_1(z)W_2(z)$, $W_1(z)$, $W_2(z) \in B(J)$, and the product $W_1(z)W_2(z)$ is regular. Then there exists a subspace $H_1 \subset H$ such that $SH_1 \subset H_1$, $S^{-1}H_1 \subset H_1$ and $W_1(z) = W_{\alpha_1}(z)$, $W_2(z) = W_{\alpha_2}(z)$, where the projections α_1, α_2 of the J-node α are generated by H_1.*

A J-node is called weak if $P, Q \in \mathfrak{S}_2$. A J-node is weak iff $J - RJR^* \in \mathfrak{S}_1$. If α is weak, then $I - S^*S = P^*JP \in \mathfrak{S}_1$. Projections of a weak J-node are weak. The class of characteristic functions for weak J-nodes will be denoted by $B(J, \mathfrak{S}_1)$. This class can be described as the class of meromorphic on the unit disk J-contractive-valued functions such that $J - W(0)JW(0)^* \in \mathfrak{S}_1$ and $W(0)$ is an invertible two-sided J-contractive operator.

3 Correspondence between Subspaces and Characteristic Functions

We shall consider only operator-valued functions of the class $B(J, \mathfrak{S}_1)$. Since $P, Q \in \mathfrak{S}_2$ and $S - U \in \mathfrak{S}_1$, where U is unitary, the operator-valued function $P(I - zS)^{-1}Q$ has angular boundary values a.e. on Γ (see, for example [9]).

For this reason, $W(z) \in B(J, \mathfrak{S}_1)$ possesses boundary values a.e. on the unit circle Γ.

Definition: The operator-valued function $W(z) \in B(J, \mathfrak{S}_1)$ is called J-inner if its boundary values are J-unitary a.e. on Γ.

Lemma 8 *Let* $W(z) = W_1(z)W_2(z)$, $W_1(z), W_2(z) \in B(J)$, $W(z) \in B(J, \mathfrak{S}_1)$. *Then*

1) $W_1(z), W_2(z) \in B(J, \mathfrak{S}_1)$;
2) $W(0)JW(0)^* = J \Leftrightarrow W_1(0)JW_1(0)^* = J,\ W_2(0)JW_2(0)^* = J$;
3) $W(z)$ *is J-inner* $\Leftrightarrow W_1(z), W_2(z)$ *is J-inner.*

Proof: The proof follows from the equality

$$\begin{aligned} J - W(z)JW(z)^* &= (J - W_1(z)JW_1(z)^*) + W_1(z) \\ &\quad (J - W_2(z)JW_2(z)^*)W_1(z)^*, \ |z| < 1 \end{aligned}$$

and nonnegativity of each term. □

Theorem 9 *Let* $W(z) \in B(J, \mathfrak{S}_1)$. *Then* $N_\omega(S) = N_{\bar\omega}(S^*) = \{0\}$ *if and only if* $W(e^{it})$ *are J-unitary a.e. on* $\bar\omega$. *Here S is the main operator of the simple J-node* α *with characteristic function* $W_\alpha(z) = W(z)$.

Proof: It is easy to verify the following formulas:

$$\begin{aligned} P(S - 1/z)^{-1}S^kQ &= -1/z^k(W(z) - W(0) \\ &\quad -W'(0)z \cdots - 1/k!\, W^{(k)}(0)\, z^k), \\ P(S - 1/z)^{-1}S^{-l}Q &= -z^l(W(z) - W_\infty(0) \\ &\quad -W'_\infty(0)\, 1/z \cdots - 1/(l-1)!\, W_\infty^{(l-1)}(0)\, 1/z^{l-1}), \end{aligned}$$

where $k \geq 0, l \geq 1, W_\infty(z) = W(1/z) = JW^{-1}(\bar z)^*J$. Then $\forall f, g \in E\ \forall m, n \in \mathbf{Z}$

$$\begin{aligned} &2\pi i\, es(S^mQf, S^{*n}P^*g, e^{-it}) = \lim_{r\to 1-0}(P[(S - 1/z)^{-1} \\ &\quad - (S - 1/z^*)^{-1}]S^{n+m}Qf, g) \\ &= \lim_{r\to 1-0} e^{-i(m+n)t}((W(z^*) - W(z))f, g), \ z = re^{it}, z^* = 1/\bar z. \end{aligned}$$

Taking this into account, Lemma 5 and the identities

$$\begin{aligned} H &= \bigvee_{m\in\mathbf{Z}} S^mQ(E) = \bigvee_{n\in\mathbf{Z}} S^{*n}P^*(E), \quad J - W(z)^*JW(z) \\ &= W(z)^*J(W(z^*) - W(z)), \end{aligned}$$

we obtain the desired assertion. □

Corollary: *The function* $W(z) \in B(J, \mathfrak{S}_1)$ *is J-inner* $\Leftrightarrow$ $M(S) = H$ $\Leftrightarrow$ $M(S^*) = H$.

Lemma 10 *Suppose* $SH_1 \subset H_1$, $S^{-1}H_1 \subset H_1$, *and that the factorization* $W_\alpha(z) = W_1(z)W_2(z)$ *is generated by* H_1 *and* $H_1 \subset M(S)$. *Then* $W_1(z)$ *is J-inner.*

Proof: By Corollary of Theorem 4, $M(S|H_1) = M(S) \cap H_1 = H_1$. Hence, by Corollary of Theorem 9 the function $W_1(z)$ is J-inner. □

Theorem 11 *Let* α_1, α_2 *be simple weak J-nodes,* $M(S_1) = H_1, N(S_2^*) = H_2$. *Then the product* $\alpha = \alpha_1\alpha_2$ *is regular.*

Proof: Let $H_0 = H \ominus \widehat{H}$. Then, by Corollary of Theorem 4 and Theorem 4,

$$H_1 = M(S_1) = M(S|H_1) = M(S) \cap H_1 \subset M(S),$$

$$H_2 = N(S_2^*) = N(S^*|H_2) \subset N(S^*) \cap H_2 \subset N(S^*),$$

$$H_1 \subset M(S) = N(S^*)^\perp \subset H_2^\perp = H_1 .$$

Whence, $M(S) = H_1$, $N(S^*) = H_2$. Using Theorem 4, Corollary of Theorem 4 and Corollary of Lemma 6, we get

$$\begin{aligned} H_0 &= M(S|H_0) \oplus N((S|H_0)^*) = M(S|H_0) \oplus N(S^*|H_0) \\ &\subset (H_0 \cap M(S)) \oplus (H_0 \cap N(S^*)) \\ &= (H_0 \cap H_1) \oplus (H_0 \cap H_2) = \{0\} \oplus \{0\} = \{0\}. \end{aligned}$$

□

Definition: The operator-valued function $W(z) \in B(J, \mathfrak{S}_1)$ is called J-outer if it has only trivial (equal to a J-unitary constant) J-inner left divisors.

Lemma 12 *Let* $W(z) \in B(J)$. *The following assertions are equivalent.*

1) $W(z) \equiv U$, *U is a J-unitary operator;*
2) $W(0)JW(0)^* = J$;
3) $P = 0$;
4) $\widehat{H} = \{0\}$.

Proof: This follows from the indentities

$$\begin{aligned} J - W(0)JW(0)^* &= PP^*, \quad \widehat{H} = \bigvee_{k \in \mathbf{Z}} S^{*k}P^*(E), W_\alpha(z) \\ &= W(0) + zP(I - zS)^{-1}Q. \end{aligned}$$

□

Theorem 13 *The operator-valued function $W(z) \in B(J, \mathfrak{S}_1)$ is J-outer $\Leftrightarrow$ $N(S^*) = H$, where S is the main operator of the simple J-node α with characteristic function $W_\alpha(z) = W(z)$.*

Proof: *Necessity.* Let $W(z)$ be J-outer. Assume $N(S^*) \neq H$. Let $W(z) = W_1(z)W_2(z)$ be the factorization, generated by $H_1 = M(S) = N(S^*)^\perp \neq \{0\}$. Let α_1 be the projection of the J-node α corresponding to subspace H_1. By Lemma 6, α_1 is simple. Hence, we have $\widehat{H}_1 = H_1 \neq \{0\}$. Therefore, by Lemma 12, $W_1(z)$ is nontrivial. By Lemma 10, $W_1(z)$ is J-inner. Therefore $W(z)$ has nontrivial J-inner divisors. This contradiction proves the necessity.

Sufficiency. Let $N(S^*) = H$. Assume that $W(z) = W_1(z)W_2(z)$, where $W_1(z)$ is a nontrivial J-inner operator-valued function. The subspace $M(S_2) \subset H_2$ generates the factorization $W_2(z) = W_{2i}(z)W_{2e}(z)$. By Lemma 10, $W_{2i}(z)$ is J-inner. By Lemma 12, we have $W_1(0)JW_1(0)^* \neq J$, because $W_1(z)$ is nontrivial. Hence, by Lemma 8, $W_i(0)JW_i(0)^* \neq J$, where $W_i(z) = W_1(z)W_{2i}(z)$. Therefore $W_i(z)$ is nontrivial J-inner. By Corollary of Theorem 4, we have $N(S_2^*|N(S_2^*)) = N(S_2^*)$. Thus, the J-nodes corresponding to operator-valued functions $W_i(z)$, $W_{2e}(z)$ satisfy the conditions of Theorem 11. This implies that the factorization $W(z) = W_i(z)W_{2e}(z)$ is regular. Therefore, by Proposition 7, $H = H_i \oplus H_{2e}$. By Corollary of Theorem 4, $H_i = M(S|H_i) = M(S) \cap H_i \subset M(S)$. Hence, $M(S) \neq \{0\}$ contrary to the condition $N(S^*) = H$. This contradiction concludes the proof of sufficiency. □

Corollary: *The product of J-inner and J-outer operator-valued functions of the class $B(J, \mathfrak{S}_1)$ is regular.*

4 Inner-outer Factorization

Theorem 14 *Let $W(z) \in B(J, \mathfrak{S}_1)$. Then there exist a unique (up to a J-unitary constant factor) factorization*

$$W(z) = W_i(z)W_e(z),$$

where $W_i(z) \in B(J, \mathfrak{S}_1)$ is J-inner, $W_e(z) \in B(J, \mathfrak{S}_1)$ is J-outer.

Proof: The decomposition $H = M(S) \oplus N(S^*)$ generates the factorization $W(z) = W_i(z)W_e(z)$, which, by Corollary of Theorem 4, Corollary of Theorem 9, and Theorem 13, is inner-outer. Suppose $W(z) = W_i'(z)W_e'(z)$, where $W_i'(z)$ is J-inner, $W_e'(z)$ is J-outer. By Corollary of Theorem 13 this factorization is regular. By Proposition 7, it generates the decomposition $H = H_i' \oplus H_e'$. By Corollary of Theorem 4, $H_i' = M(S|H_i') \subset M(S)$, $H_e' = N(S^*|H_e') \subset N(S^*)$. Hence, $M(S) = H_i'$, $N(S) = H_e'$. Since the projections determine the factors up to a J-unitary constant, we have $W_i(z) = W_i'(z)U$, $W_e(z) = U^{-1}W_e'(z)$. □

Corollary: *Let* $W(z)$, $B_k(z) \in B(J, \mathfrak{S}_1)$, $W(z) = \prod_{k=1}^{\infty} B_k(z)$, $B_k(z)$ *be J-inner. Then* $W(z)$ *is J-inner. In particular, the Blaschke-Potapov product is J-inner* [*12*].

Proof: Let $W(z) = W_n(z)V_n(z)$, where $W_n(z) = \prod_{k=1}^{n} B_k(z)$. It is easy to see that $W_n(z)$, $V_n(z) \in B(J, \mathfrak{S}_1)$, $W_n(z)$ is J-inner. By Theorem 14, $W_i(z) = W_n(z)V_{ni}(z)$, where $W_i(z)$, $V_{ni}(z)$ are the J-inner factors from the corresponding inner-outer factorizations. Further, there exists

$$V_i(z) = \lim_{n\to\infty} V_{ni}(z) = \lim_{n\to\infty} W_n(z)^{-1} W_i(z) \in B(J).$$

Hence, $W_i(z) = W(z)V_i(z)$. Finally, by Lemma 8, $W(z)$ is J-inner. □

References

[1] J. Garnett, *Bounded analytic functions*, Academic Press, N.-Y.-London, 1981.

[2] B. Szökefalvi-Nagy and C. Foiaş, *Analyse harmonique des opérateurs de l'espace de Hilbert*, Masson, Paris; Akad. Kiadó, Budapest, 1967; English transl. *Harmonic analysis of operators on Hilbert space*, North-Holland, Amsterdam-London; American Elsevier Publishing Co., Inc., N.-Y.; Akad. Kiadó, Budapest, 1970.

[3] P.M. Yuditski, *Outer-inner factorization of J-stretching matrix-valued functions*, Teoria Funk., Funktional. analiz i Prilozh. **46** (1986), 132–135 (in Russian).

[4] A.S. Tikhonov, *Regular factorization of J-contractive-valued functions*, Simferopol, 1986, p. 29, Dep. in UkrNIINTI, no. 2580–Uk86 (in Russian).

[5] S.N. Naboko, *Function model for the perturbation theory and its applications to the scattering theory*, Trudy Mat. Inst. Steklov **147** (1980), 84–114; English transl. in Proc. Steklov Inst. Math., 1981, no. 2.

[6] Yu.L. Shmul'jan, *Some questions of the theory of operators with finite non-Hermitian rank*, Matem. Sbornik **57** (1962), no. 1, 105–136 (in Russian).

[7] N.G. Makarov and V.I. Vasjunin, *A model for noncontraction and stability of the continuous spectrum*, Lect. Notes in Math., 1981, v. **864**, 365–412.

[8] A.S. Tikhonov, *Absolutely continuous spectrum of linear operator*, Simferopol, 1988, p. 43, Dep. in UkrNIINTI, no. 2471–Uk88 (in Russian).

[9] A.S. Tikhonov, *Absolutely continuous spectrum and scattering theory for operators with spectrum on curve*, Algebra i analiz, **7** (1995), no. 1, 200–220; English transl. in St. Petersburg Math. J. **7** (1996), no. 1, 169–184.

[10] V.F. Veselov and S.N. Naboko, *Determinant of characteristic function and singular spectrum of non-selfajoint operator*, Matem. Sbornik **129** (1986), no. 1, 20–39 (in Russian).

[11] V.M. Brodski, I.C. Gohberg and M.G. Krein, *Definition and basic properties of the characteristic function of J-node, Funktional. Analiz i Prilozh.* **4** (1970), no. 1, 88–90 (in Russian); English transl. in Funct. Anal. Appl. **4** (1970), no. 1, 78–80.

[12] D.Z. Arov and L.A. Simakova, *Concerning boundary values of convergent sequence of J-contractive matrix-valued functions*, Matem. Zametki **19** (1976), no. 4, 490–500 (in Russian).

Alexey Tikhonov
Section of Mathematical Analysis
Department of Mathematics
Simferopol State University
4, Yaltinskaya str.
Simferopol, 334036
Crimea, Ukraine

AMS Classification Numbers: 47A56, 47A48, 47A68, 47A15

Table of Contents of Volume I

MIX
Papier aus verantwortungsvollen Quellen
Paper from responsible sources
FSC® C105338
FSC
www.fsc.org

If you have any concerns about our products,
you can contact us on
ProductSafety@springernature.com

In case Publisher is established outside the EU,
the EU authorized representative is:
Springer Nature Customer Service Center GmbH
Europaplatz 3, 69115 Heidelberg, Germany

Printed by Libri Plureos GmbH
in Hamburg, Germany